Concepts of Human Physiology

Concepts of Human Physiology

Richard L. Malvin
University of Michigan

Michael D. Johnson
West Virginia University

Gary M. Malvin
The Lovelace Institutes

An imprint of Addison Wesley Longman, Inc.

Menlo Park, California • Reading, Massachusetts • New York • Harlow, England
Don Mills, Ontario • Sydney • Mexico City • Madrid • Amsterdam

Executive Editor: Johanna Schmid
Developmental Editor: Maxine Effenson Chuck
Supplements Editor: Cyndy Taylor
Project Coordination and Text Design: Electronic Publishing Services Inc.
Cover Designer: Yvo Riezebos
Cover Photo: © David Madison 1996
Art Coordinator: Electronic Publishing Services Inc.
Art Manager: Nadine B. Sokol
Art Studio: Nadine B. Sokol
Photo Researcher: Mira Schachne
Electronic Production Manager: Valerie L. Zaborski
Manufacturing Manager: Helene G. Landers
Electronic Page Makeup: Electronic Publishing Services Inc.
Printer and Binder: R.R. Donnelley & Sons Company
Cover Printer: Phoenix Color Corp.

For permission to use copyrighted material, grateful acknowledgment is made to the copyright holders on pp. 433–434 , which are hereby made part of this copyright page.

Library of Congress Cataloging-in-Publication Data
Malvin, Richard L.
    Concepts of human physiology / Richard L. Malvin, Michael D. Johnson, Gary Malvin.
        p.   cm.
    Includes index.
    ISBN 0-673-98562-8
    1. Human physiology. I. Johnson, Michael D. II. Malvin, Gary. III. Title.
    QP34.5.M25   1997   96-32232
612—dc20   CIP

ISBN 0-673-98562-8

12345678910—DOW—99989796

About the Authors

Richard L. Malvin
Richard L. Malvin is a Professor of Physiology at the University of the Michigan School of Medicine. He received a B.S. from McGill University in 1950, a M.S. in Physiology from New York University, and a Ph.D. in Physiology from the University of Cincinnati. He is an internationally recognized expert in kidney physiology, fluid and electrolyte balance, and cardiovascular physiology. He has taught physiology for over forty years to undergraduate, graduate, and medical students. Dr. Malvin has been a visiting professor at a number of universities; the University of Cantebury, Christchurch in New Zealand and the University of Hawaii just to name a few. In addition, he has developed and taught for several years a course for undergraduate students in the history of the scientific method as it applies to biomedical research. He has been the recipient of many fellowships and has won several prestigious awards including a Gold Medal from the British Medical Association Film Competition, The Fogarty International Award, and the USPHS Research Career Development Award. Dr. Malvin has successfully authored an impressive list of published papers and textbook chapters. Not only is he a member of various societies including the American Physiological Society and the American Association for the Advancement of Science, but he is also one of the founders of the American Society for Nephrology.

Michael D. Johnson
Michael D. Johnson is a professor of Physiology at West Virginia University. He received his B.S. in Zoology from Washington State University in 1970 and his Ph.D. in Physiology from the University of Michigan. He was the recipient of the Distinguished Teacher Award of the School of Medicine in 1991 as well as a West Virginia University Foundation Outstanding Teacher Award in 1992. He is a member of the American Physiological Society and the American Association for the Advancement of Science. Dr. Johnson is also the author of an extensive list of published papers in the life sciences field.

Gary M. Malvin
Gary M. Malvin is a Research Scientist at the Lovelace Institutes, in Albuquerque, New Mexico. He received his B.A. from the University of Michigan in 1976 and his Ph.D. in Physiology from the University of New Mexico in 1983. He is an internationally recognized expert in comparative physiology, particularly in the areas of respiratory, cardiovascular and smooth muscle physiology. In addition, he has developed a series of visual aids and study materials for a course in human physiology. He has an established reputation as an outstanding teacher. Dr. Malvin is an active member of the American Physiological Society and the author of numerous papers and text book chapters.

This book is dedicated to all of our students—
past, present, and future

Contents In Brief

Contents In Detail

CHAPTER 3

CHAPTER 4

CHAPTER 5

Endocrine System 111

CHAPTER 6

Skeletal and Muscular Systems 143

CHAPTER 7

Sensory Systems   179

CHAPTER 8

Cardiovascular System: Heart 215

CHAPTER 9

Cardiovascular System: Blood Vessels 235

CHAPTER 10

Defense Systems 262

CHAPTER 11

Respiration   278

CHAPTER 12

Renal Excretory System   304

Chapter 15

Reproductive System   *384*

Preface

PHILOSOPHY AND GOALS

This book was written expressly for you, the beginning student of human physiology. Our goal was to write a book that is readable and interesting, yet accurate, current, and complete. We intended to write a text that could be completed in one term. At times you should be able to read passages like a novel; at other times you will treat it like a textbook. Perhaps you are interested in a career in a scientific field such as allied health; perhaps you are simply interested in how the human body works but do not yet have a strong background in science. No matter this book is for you.

Our philosophy is that we never stop learning, and thus the processes of teaching and learning are inseparable and intertwined. We would not choose to be either a teacher or a student if we did not enjoy the process or find the subject interesting. We use a conversational style whenever possible, liberally sprinkled with historical examples, just as we do when lecturing or explaining a particular topic to our own students. Complex or difficult subjects are developed slowly, starting with the essential facts. Our intention is to teach you the facts and how to use them so you will understand the underlying principles and concepts of how the body works. The development of subjects from a historical perspective allows you to share the thoughts of scientists of the past as they wrestled with the unanswered questions of their day. As a result you should gain an appreciation of the scientific method and of how scientific research is conducted, enabling you to become a better decision-maker and a better consumer of health care.

We hope you will share our enthusiasm for the subject of human physiology and the wonder and excitement that we feel when, even today, we learn something new about how the human body works.

PEDAGOGICAL FEATURES

Each chapter incorporates a number of pedagogical elements designed to make the text a more effective learning tool for instructors and students alike. The organization of the pedagogical elements follows a logical sequence of preparing the student for the subject to be covered, presenting the material first as a general overview and then in detail, summarizing the material, and then testing the student's understanding at both basic and conceptual levels.

- **Chapter opening pedagogy** The chapter opening pedagogy orients the student to the topics covered with the following elements:

  Chapter outline An outline provides the structure for the chapter and orients the student to the sequence of topics to be covered.

  Chapter objectives A list of learning objectives provides the student with information about what they are expected to know by the time they complete the chapter.

  Chapter introduction The introductory section of each chapter presents a general overview of the subject, helping the student to understand its importance in the context of the overall function of the human body.

- **Boxed features** Two types of boxed features accompany the text: *Highlights* Material in Highlights includes some common or especially interesting examples of the application of human physiology to daily life, presents examples of how scientific research is conducted in human physiology, or describes medical or health issues related to the subject. *Milestones* Subjects of special historical significance, whether they be single events or the development of our knowledge over time, are described in Milestones. Milestones reinforce for the student that human physiology is a dynamic, ever-changing field.

- **Summary** A short narrative summary at the end of each chapter reinforces the key facts and concepts that have been covered in the chapter.

- **End of chapter questions** To facilitate the student's self-assessment of whether they have mastered the material in the chapter; each chapter concludes with some questions that address the material covered. Questions are of two types:

Conceptual and Factual Review These questions test the student's recall of the key concepts and facts presented in the chapter.

Applying What You Know These questions are designed to test how well the student understands the concepts presented in the chapter and whether the student can apply those concepts in a broader context.

CONTENT AND ORGANIZATION

The human body is a complex system of mutually interdependent cells, tissues organs and organ systems. We have chosen to organize the book around the time-honored organ system approach, starting with cells and moving up to the more complex organ systems. In keeping with our goal that the text be designed to be completed in one term, we have limited the number of chapters to 15. A positive outcome of this approach is that some important topics that are truly integrated across several organ systems are also integrated in their presentation in the text. An example is the maintenance of fluid volume and composition. How cells maintain their cellular volume and composition is discussed in Chapter 3, as the student is learning about cell membrane ion transport pumps. Selected topics relating to the overall regulation of total body fluid volume and composition are discussed in the endocrine, cardiovascular, renal excretory, and digestive systems.

- **Chapter 1;** *Introduction* This short chapter lays the foundation on which subsequent chapters are based. The scientific method is described and its application to the advancement of knowledge in human physiology is emphasized. The concept of homeostasis is introduced and the principles of positive and negative feedback systems are described.

- **Chapters 2 and 3;** *The Cell* and *Cell Communication* Together, these two chapters cover the material necessary to understand how cells function. Because the student is not expected to have a strong background in chemistry, the necessary knowledge in chemistry is developed step by step in Chapter 2. The structure and functions of the cellular elements are then discussed. Chapter 3 provides the foundation for understanding membrane excitability. This is followed by a discussion of electrical and chemical forms of com-

munication between cells, leading directly into the subjects of the next two chapters.

- **Chapters 4 and 5;** *Neurophysiology* and *Endocrine System* The structure and functions of the specialized neural and endocrine communications systems follows logically after Chapter 3. These subjects are placed early in the text so that students will be better able to understand how each of the subsequent organ systems are controlled.

- **Chapter 6;** *Skeletal and Muscle Systems* The subject of the systems involved in mobility are integrated together in one chapter. The inclusion of the structure, function and dynamic nature of bone into this chapter is somewhat different from many other texts, which ignore the subject of bones altogether.

- **Chapter 7;** *Sensory Systems* The Sensory Systems chapter goes beyond a discussion of the traditional four "special senses" of taste, smell, hearing, and vision, and includes the sensory mechanisms located in the skin, muscles, tendons and joints.

- **Chapters 8 and 9;** *Cardiovascular System: Heart* and *Cardiovascular System: Blood Vessels* The two major but structurally and functionally different components of the cardiovascular system are treated in separate, closely-linked chapters, to keep this important subject from becoming too cumbersome a unit.

- **Chapter 10;** *Defense Systems* The general nature of defense systems is discussed, followed by a discussion of the immune system. This chapter follows logically after Chapters 6, 8, and 9 because many of the cells of the immune system originate from bone and circulate with the blood.

- **Chapters 11, 12, and 13;** *Respiration, Renal Excretory System,* and *Digestive System* These three major organ systems are grouped together because they are all, to some degree, involved in the exchange (either by intake or excretion, or both) of materials between the organism and the environment.

- **Chapter 14;** *Energy Exchange and Temperature Regulation* This chapter integrates the general subject of energy, including energy balance, storage, utilization, and management of energy waste (temperature regulation). This latter subject is not always included in texts of human physiology.

- **Chapter 15;** *Reproductive System* A general discussion of cellular reproduction is followed by a discussion of male and female reproductive physiology. Pregnancy, childbirth, lactation and fertility control are treated in this chapter as well.

SUPPLEMENTS

The following supplementary materials are available to accompany *Concepts of Human Physiology*. This well-rounded support system for the text includes printed materials for both instructors and students, as well as visual aids and software. For complete information on any of the following items, please contact your local sales representative.

Instructor's Manual

Prepared by Izak Paul of Mount Royal College, each chapter in the Instructor's Manual includes a chapter synopsis, student objectives, a lecture outline, teaching tips and activities, and lists of audiovisual materials and software.

Testbank

The testbank, written by William D. Blaker of Furman University, consists of multiple-choice, essay, short-answer, matching and true-false questions for each chapter of the book. The testbank will be available in a printed form as well as on *Testmaster* for use with IBM or Macintosh computers.

Transparencies

A set of 227 transparencies is available to adopters of the text. Care has been taken to create large labels and to assure that the transparencies are clear and usable for overhead projection, even in a large lecture hall.

Student Study Guide

Yvette Swendson of Mount Royal College is the author of this valuable aid for students. Features for each text chapter include the following: an outline and objectives, "Wordbytes" (a group of parts of common words and their meanings), checkpoints, answers to the checkpoints, and a self-test for students, including the answers to these questions.

ACKNOWLEDGMENTS

We wish to acknowledge our students past and present, for whom this book was written and without whom we never would have attempted to write this book. Your interest and enthusiasm in learning how the human body works have been an inspiration to us. We also thank our colleagues and friends who shared their ideas and knowledge with us as we wrote this book. Finally, we thank the many reviewers of the initial drafts of the book and the survey participants; your time and patience, your honest assessments, and your insights into science have contributed heavily to the final outcome. Though we did not know who you were at the time, the efforts of the following reviewers and survey participants are especially appreciated:

Reviewers and Survey Participants

Ronald Adkins, Washington State University
Matthew Berria, Weber State University
J. Gordon Betts, East Texas State University
William Blaker, Furman University
Mildred Brammer, Ithaca College
Elaine Brubacher, University of Redlands
Phyllis Callahan, Miami University
Thomas Collins, Moorhead State University
Jack Cummings, Waynesburg College
Dwayne Curtis, California State University
William DeGraw, University of Nebraska
P.A. George Fortes, University of California
Leon Goldstein, Brown University
Harriet Gray, Hollins College
Robert S. Greene, Niagra University
Lewis Greenwald, Ohio State University
Don Hay, Stephen F. Austin State University
Richard Heninger, Brigham Young University
Cindy Hoorn, Western Michigan University
Vincent Johnson, St. Cloud State University
William B. Keith, University of Mississippi
Loren W. Knapp, University of South Carolina
Helen Lambert, Northeastern University
John Lepri, University of North Carolina
Mary Katherine Lockwood, Amherst College
David Marsh, Angelo State University
Stu Matz, University of Oregon
Donald McEachron, Drexel University
John T. Morse, California State University
Richard Moss, University of Wisconsin
Allen Norton, University of Southern California
Richard Olsen, Kalamazoo Valley Community
     College
Stephan R. Overmann, Southeast Missouri State
     University
Karen Peterson, University of Washington
Onkar Phalora, Anderson University
Dell Redding, Evergreen Valley College
Gary Ritchison, Eastern Kentucky University
Jack Rose, Idaho State University
Richard Satterlie, Arizona State University
Wayne Savage, San Jose State University

David Saxon, Morehead State University
Edward I. Shaw, University of Kansas
Robert Sherman, Miami University
Allan Wade, Triton College
Edward Wickersham, Pennsylvania State
    University
Marcus Youngowl, California State University

We would also like to thank Benjamin Cummings for their help and support; especially Cyndy Taylor, Maxine Chuck, and Val Zaborski. Thanks also to Nadine Sokol who provided us with the beautiful artwork found in this text. Finally, we thank Electronic Publishing Services Inc., especially Patty Andrews, Jason Jones, and Sheri Hyman, for all of their time and effort.

Richard Malvin
Michael Johnson
Gary Malvin

Concepts of Human Physiology

Chapter 1
Introduction to Human Physiology

Objectives

By the time you complete this chapter you should be able to

◆ Understand why it is important to study human physiology.

◆ Describe the scientific method and explain how it can be used to investigate hypotheses.

◆ Explain the role of animal research in scientific studies and how it helps us better understand our systems.

◆ Explain the concept of homeostasis and how it applies to our daily lives.

◆ Describe the difference between positive feedback and negative feedback control systems, and give an example of each.

Why Study Physiology?

Every day we are bombarded with information about what is and isn't good for us. We are told that taking an aspirin a day for life may prevent blood clots and heart disease; that being thin is not only "in," but important for a longer life, especially for women; that eating leafy green vegetables will prevent cancer; that too much salt causes high blood pressure. Is all of this information really true? How do "they" know all of this? Why should we listen?

There are even more complex issues that we must all address together as a society. How do our bodies defend against toxins, including those that might cause cancer, in our drinking water and in the air? What might be the future benefits to human health of knowing the entire sequence of all of the human genes? Given limited financial resources, should we spend a billion tax dollars to map the entire set of human genes, or would it be better spent on AIDS research or a way to prevent hypertension? What is the basis for DNA fingerprinting, and when and how should it be used in the courts? Although questions such as these may seem unanswerable at the moment, how we frame the questions, how we seek information to answer the questions, and how we act on our knowledge will have broad social, political, and even biological consequences for ourselves and for future generations. Regardless of your education or career field, you may be asked to make an informed choice. In the privacy of the voting booth, your decision is as important as that of a Nobel prize-winning scientist.

What Is Physiology?

Physiology is the study of how living organisms function. The word is derived from the Greek word *physiologia,* meaning "to study nature." Human physiology, then, is the study of how the human body works. Along with its sister science, **anatomy,** which is the study of the *structure* of living organisms, it forms the fundamental backbone of all life sciences. It is an older science than the newer fields of molecular biology or biochemistry, but it is not dull. Physiology tells the story of how individual cells live and reproduce within our bodies, how billions of cells are joined together to form tissues and organs, how the heart can pump blood for 80 years without ever needing an overhaul, how our breathing is controlled, and how we fight infections. In this book we hope to share with you our own sense of wonder and amazement at the intricacies of the human body.

We hope that once you complete this text, you will be better able to make educated choices in the future, regardless of your chosen field. Throughout life, one of the few things to remain your own is your body. Because it is important to keep it running properly, you should know something about how it functions. Understanding normal function will also enable you to distinguish normal physiology from **pathophysiology,** the branch of science that deals with how the body functions under abnormal conditions, when disease is present.

There are several recurrent questions regarding physiology that must be addressed before we begin: What is the scientific method and how does it work? How does our knowledge change over time? and What is homeostasis? This last topic is the very cornerstone of any discussion of the function of the human body.

The Scientific Method

We mentioned that human physiology is a science. By that we mean that our current knowledge of how the body works has come about as a consequence of applying the **scientific method.** Indeed, one often hears about scientific discoveries as if they were different from any other sort of discovery. The implication is that there is something mysterious about the scientific method and that only scientists with special training are knowledgeable in its use. Nothing could be further from the truth.

The scientific method is nothing more than the rigorous and repetitive application of a series of simple steps to any problem, without bias or deviation. In the simplest terms, the scientific method uses the following steps in arriving at a decision:

1. List the current facts.
2. Make a prediction, or **hypothesis,** based on the current facts.
3. Design an experiment such that alternative outcomes are possible. At least one of the possible outcomes must be able to prove the original hypothesis false.
4. Carry out the experiment or the observation so as to get an unambiguous result.
5. If the initial hypothesis is shown to be false, repeat steps 1–4.

An important element along the path to scientific discovery is the willingness, indeed the necessity, of attempting to prove the current hypothesis wrong. This is essentially what distinguishes a hypothesis from a **theory,** science from pseudoscience, and the acquisition of the truth from unexamined belief. Theories need not be tested, whereas scientific hypotheses *must* be rigorously examined if they are to have any validity. In contrast, virtually nothing can persuade the pseudoscientist of the false nature of his or her hypothesis or belief, because he or she does not want the hypothesis to be falsified. That is why the Flat Earth Society still exists, despite what most of us would agree is considerable evidence that the earth is essentially spherical.

How Our Knowledge Changes Over Time

Knowledge advances as a series of small steps: proposing hypotheses, rigorously testing them, and modifying or replacing them as they are proved false. Consequently, what we accept as true about human physiology has undergone significant change over

---

HIGHLIGHT 1.1

An Example of the Scientific Method

The scientific method may be illustrated by the following example. Suppose a class is asked to solve a problem presented by the teacher. She tells the class that they are to determine a truth, or rule, of three numbers that only she knows. They are given a set of three numbers that fits the rule. From that one set they are to form a hypothesis regarding the rule and design experiments to test the hypothesis. The number set is 2,4,6.

Propose a Hypothesis

The first student proposes the hypothesis that the rule is, "Any three even numbers ascending by 2." That hypothesis certainly fits the facts. He then goes on to design an experiment and asks the teacher whether the number set 6,8,10 fits the rule. (Note that the teacher plays the role of nature by presenting data in the form of a "Yes" or "No" to the student's question-experiment). The teacher answers, "Yes."

"Does the set 12, 14, 16 fit the rule?"

"Yes."

Is the student justified in thinking that he now knows the rule? He has shown that when he presents a number set conforming to his hypothesis, the teacher agrees that the set fits her rule and is consistent with the hypothesis proposed by the student. Note, however, that the student did not test his hypothesis by trying any set of numbers that violated his own hypothesis; that is, he did not attempt to prove the hypothesis false.

Design an Experiment to Prove the Hypothesis False

A second student recognizes that the rule may not include the qualification that all the numbers be even; she asks whether the number set 3,5,7 fits the rule. The teacher states that it does. The class now knows that the original hypothesis has been proven to be false and so must be modified. Note that only by attempting to prove the hypothesis false have the students actually made progress.

Repeat the Process

They attempt a new experiment presenting the number sets 15, 17, 19 and 311, 313, 315, both of which are consistent with the modified hypothesis. Again, all that can be said at this juncture is that the data are consistent with the modified hypothesis. They do not prove the new hypothesis to be correct.

Another student suggests that separation by 2 is not necessary and asks whether the set 3,10,11 fits the rule. It does. Once again the students have moved closer to discerning the rule. They now know that the numbers can be even or odd and that the rule does not require separation by 2. Another modification of the hypothesis is needed.

One can go on and on with this example, in which one moves toward a solution of the problem step by step. Note that no matter how many confirmatory experiments the class might have made of the original hypothesis (6,8,10, 12,14,16, 102, 104, 106, etc.), they do not *prove* the original hypothesis to be true. The true test of a hypothesis occurs only when an experiment attempts to *disprove* the hypothesis. Following modification, the experiments go on, with each experiment coming closer to the truth. We leave this example in your hands. You can go on with it if one plays nature and another plays experimenter.

The important point to remember from this demonstration is that although the first student presents a few number sets that fit the rule, all he is justified in saying about his data is that they fit his hypothesis. He is not justified in claiming that his data prove his hypothesis. This is a most important distinction: Scientists do not try to prove their hypotheses; they attempt to test their hypotheses with appropriate experiments. (By the way, the rule is any set of positive integers in ascending order.)

time. The story of the death of president George Washington is a good example of how far we have come in the past several centuries. President Washington was the most illustrious man in the country at the time, and his physicians must have tried their best to do all they could for their patient. The physicians based their therapy on the then-current belief that illness was the result of "evil humors" accumulating in the body. Bleeding, purging, and blistering were seen as ways to remove those evil humors. Despite their good intentions and their medical training, it is very likely that their treatment contributed to his death.

As a responsible citizen and a consumer of health care, you need to be able to understand and evaluate the newest information as it becomes available. Textbooks can be one source of information because they generally include the very latest, most accurate information. However, the danger in presenting only the facts as we now know them is that it may lead you to lose sight of the fact that our knowledge continues to change, even today. In this book, we use historical examples to reinforce how our knowledge has changed over time, so that you will become comfortable with the notion that knowledge constantly evolves. Figures 1.1 and 1.2 illustrate that dramatic changes have occurred in the last 100 years alone.

Figure 1.1 shows a kindly physician sitting at the bedside of a seriously ill child. Although he is clearly concerned, there is nothing he can do but watch and wait. Figure 1.2 shows a modern medical facility with a patient undergoing state-of-the-art diagnosis with a computed axial tomography (CAT) scan. The availability of advanced diagnostic equipment is but one example of change. Polio, smallpox, diphtheria, measles, and whooping cough are all essentially under control. That was not the case 100 years ago. Corneal transplants are able to make the blind see, microsurgery can restore hearing, and artificial hip joints allow disabled people to walk. Open-heart surgery and even heart transplants are almost routine. This century has seen a greater change in medical treatment than all the previous centuries of human experience combined.

How is it that medical knowledge was essentially stationary for most of recorded history, but advanced so rapidly in the last century? Until almost the beginning of this century, medical theory and treatment rested largely on anecdotal evidence. If an ill person was treated with any medicine and recovered, it was taken as convincing evidence that that particular medicine was a cure for that illness. A single positive outcome was enough to convince physicians of the

---

MILESTONE 1.1

The Death of George Washington

Painting of President George Washington by Stuart.

It was a cold wintry day in 1799 when President George Washington mounted his horse to survey his plantation. In spite of rain followed by snow, he remained in the fields from 10 A.M to 3 P.M. Arriving home late, he complained of a sore throat. The following morning he had a fever, and had difficulty speaking and swallowing. The overseer of his plantation administered a folk remedy of the day, a mixture of molasses, vinegar, and water. Unfortunately, Washington almost choked to death trying to swallow the mixture. Then, following the best tradition of the times, the overseer bled Washington of one pint of blood and wrapped his neck in flannel. This exhausted the overseer's medical skills, and when Washington did not improve, he called a physician.

The physician immediately bled Washington again, applied blisters to his throat (blisters were formed by applying a caustic paste made from the shell of the blister beetle), and gave him a second mixture of molasses, vinegar, and water, with the same result. Two hours later, when Washington showed no sign of improvement, two other physicians came to attend him. They recommended more bleeding and with Washington's approval, took an additional quart of blood. The physicians noted, "It came thick and slow." Still Washington showed no improvement. They gave him calomel to increase his urine flow, and tartar to induce vomiting, both in an attempt to remove "evil humors" from his body. Now Washington, in great pain, was weakening rapidly. They then applied blisters to his feet. Realizing that he did not have long to live, Washington said, "I die hard but am not afraid to go. I feel myself going. Thank you for your attention, but I pray you to take no trouble for me. Let me go quietly. I cannot last long." He died shortly thereafter.

Figure 1.1 "The Doctor" by Sir Luke Fildes.

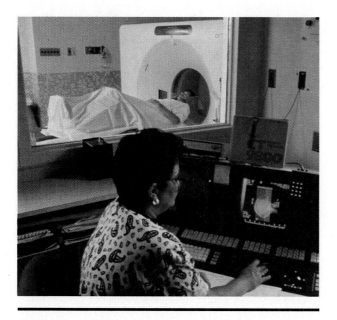

Figure 1.2 A patient undergoing a CAT scan in a modern medical facility.

efficacy of the cure. That is not sufficient evidence today. Modern medical practice requires a good knowledge of the inner workings of the body as well as good scientific evidence of efficacy of any treatment. That knowledge came from the use of the scientific method applied to physiologic research.

A cornerstone of good scientific information-gathering, then, is the need for a well-designed experiment that tests a specific hypothesis. In order to determine whether a new drug is effective in treating a specific disease, one might design the following experiment:

1. Divide a large set of individuals randomly into two groups. The random assignment ensures that both groups are identical at the start. The groups must be large enough that standard statistical tests can be applied effectively to the results.

2. Give the drug to one of the groups. This group is called the **experimental group** because it is receiving the experimental treatment.

3. Give a **placebo** to the other group; this group is called the **control group** because it allows the experimenter to account for, and thus control, any unknown factors in the initial large set of individuals that might affect the outcome of the experiment.

Only by comparing the incidence of the disease between these two groups could one tell whether the drug had any effect. For example, if the incidence of the disease were no different between the two groups after the treatment period, one would conclude that the drug did *not* have the desired effect, and the hypothesis would be rejected.

The very idea of conducting such experiments raises an ethical problem; what if the drug is dangerous in some unknown way? Certainly one would not want to begin investigating the action of a new drug by giving it to half of the individuals in a large group of people.

This brings up the subject of the essential role played by animals in the advancement of scientific knowledge. Scientists recognize that the use of animals is essential in certain types of experiments. Without animals, we would make almost no advances at all, or if we did it would be at great risk to humans. Imagine giving the first experimental polio vaccine to humans. Would you volunteer for such a study? What if it had *caused* polio in the test group? The fact that the polio vaccine was first proven safe in animals was an essential first step leading to its eventual use in humans. Scientists today are using animals to help determine the mechanisms of cancer and to search for potential treatments for AIDS, to name just a few examples. Although there has been controversy over animal research in recent years, scientists respect and value the contribution of animals to human (and animal) life and health. The use of animals in research is highly regulated by the federal government and most states to ensure that the animals are treated humanely.

Homeostasis

In the latter part of the nineteenth century, physiologist Claude Bernard developed the concept of **homeostasis.** The essence of homeostasis is that the body acts to maintain a stability of most physiological parameters. Central to this concept is the idea that the fluids bathing all the cells of our body constitute an internal environment. Physiological systems act to maintain the constancy of that internal environment. The many control systems you will read about in this book, such

as the regulation of blood pressure, regulation of respiration, and the control of body temperature, have as their end point the return of the internal environment of the body to normal physiological limits.

A simple but well-understood phenomenon occurs when you are exposed to a very cold environment while wearing inadequate clothing. You begin to shiver, move about by stamping your feet, and perhaps even assume a semi-fetal position with arms crossed. These responses increase heat production by the contracting muscles and conserve heat by minimizing the body surface exposed to the environment, preventing body temperature from falling below physiological limits. The homeostatic response is such that changes occur to oppose the environmental challenge and so maintain constancy of the internal environment. Unless the challenge is too great, body temperature will not be significantly altered. This homeostatic response is an example of a **negative feedback** control system. It is called negative feedback because an initial change in the sensed variable sets in motion a homeostatic response that tends to minimize, or negate, the initial change. Figure 1.3 shows the components of the negative feedback control system just described.

You see another example of homeostasis when you sweat heavily after running hard on a warm day.

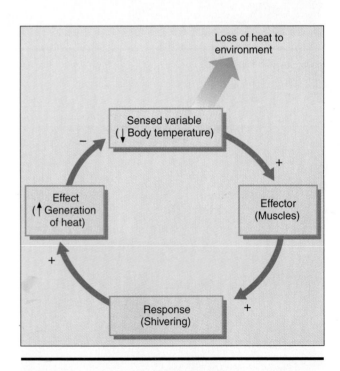

Figure 1.3 . A negative feedback control system. The negative sign associated with the arrow between *Effect* and *Sensed variable* indicates that the effect of the control system is to oppose the initial change in the sensed variable.

Your body must dissipate the heat generated by this muscular exercise, or your body temperature will rise to dangerous limits. Evaporation of the sweat cools your body, helping to stabilize body temperature. However, the increase in sweating removes water from the internal environment of the body. This triggers a series of changes in the internal environment that lead to the sensation of thirst. Runners then drink sufficient quantities of water to replace that lost in the sweat. Throughout this book you will read of homeostatic responses that regulate the hormonal output of the glands, cellular events that act to maintain constant energy supplies, and a host of other examples of homeostasis. It is important to understand that homeostasis is a concept, not a device. Any bodily response to a challenge that acts to reverse or minimize a change in the internal environment of the body is a homeostatic response.

You will also be introduced to the term *positive feedback*, a system in which the effect *enhances* the initial change in the sensed variable, causing an increasingly rapid movement *away* from homeostasis. This form of feedback is inherently unstable and is best exemplified by an atomic bomb. In this case, as each atom gives off subatomic particles it initiates the same response in more than one other nearby atom. Very quickly, the number of subatomic particles released is so large that an atomic explosion results. Positive feedback is a relatively rare event in the body, but it does operate in a few systems for a brief time. You will see this when we discuss blood clotting and the events leading to childbirth. These are not examples of homeostasis. Quite the opposite; they lead to change in the organism.

Throughout this book you will learn about the body's many mechanisms that prevent an alteration beyond the normal physiological limits of the various systems under discussion. You will see that the cardiovascular system reacts to different stresses in a way that maintains proper blood flow to all parts of the body. All other systems also have built-in mechanisms that function to maintain homeostatic control. It is our objective to present these control systems in an orderly and easily understandable way. We will use common experiences to illustrate our points. Although homeostatic control is evident in each organ system, humans are not perfect examples of homeostatic control. We are in dynamic equilibrium with our environment; we run, we take saunas, we go out in the cold, we do many things that strain normal constraints so that we often do alter our physiological state. If we did not do these things, we would be the ultimate organism in a state of homeostasis: the couch potato.

Summary

Informed decision-making is crucial to the preservation of personal health, as well as preservation of our complex society. A good knowledge of human physiology serves as a background for informed decisions relating to personal health. Your ability to make informed decisions will improve if you understand the scientific method, know how scientific research is performed and evaluated, and appreciate the role of experimental models, including animals, in advancing scientific knowledge and human health. A key concept in physiology is homeostasis: All of the organs and organ systems in the body act in concert to maintain the constancy of the internal environment. Homeostasis is often accomplished by the operation of negative feedback control systems, although there are also examples of positive feedback systems in the human body.

Questions

Conceptual and Factual Review

1. How do physiology and anatomy differ? How do they complement each other? (p. 2)

2. What steps constitute the scientific method? (p. 2)

3. What is a hypothesis? How does one test a hypothesis? (pp. 2–3)

4. Why is a control group an important component of a good experimental design? (pp. 5–6)

5. What is meant by homeostasis? (pp. 6–7)

6. What is the difference between positive feedback and negative feedback? (pp. 6–7)

Applying What You Know

1. If a set of data or facts is consistent with a hypothesis, does that mean that the hypothesis has been proved? Why or why not?

2. If someone tells you that they tried a new cure for a cold and it worked, does that constitute proof of efficacy of that cure? What questions should you ask?

3. If you run 100 yards as fast as you can, your heart rate will increase. Do you think that violates the concept of homeostasis?

7. Of what use are animals in scientific research? Why not use humans? (pp. 6–7)

4. Debate the following two statements: (a) "Numerous scientific studies have established that there is a close association between smoking and the incidence of cancer in humans"; (b) "Despite considerable effort, scientists have been unable to prove that smoking causes cancer in humans."

Chapter 2
The Cell

Objectives

By the time you complete this chapter you should be able to

◆ List some of the most important atoms that make up the human body.

◆ Understand the types of chemical bonds that hold atoms together.

◆ Define ions, acids, and bases.

◆ Describe the three classes of organic molecules: carbohydrates, lipids, and proteins.

◆ Know the general anatomy and function of a typical animal cell.

◆ Understand the structure of the cell membrane and be able to describe how it functions as a selectively permeable barrier to the movement of substances into and out of the cell.

◆ Explain how information is transmitted across the cell membrane.

◆ Know the structure and function of DNA and describe how it is replicated and how it is used to form RNA.

◆ Know the function of RNA.

◆ Describe the functions of the cytoskeleton, centrioles and spindle fibers, mitochondria, endoplasmic reticulum, ribosomes, and Golgi apparatus.

◆ Understand that cell growth, division, and differentiation are carefully controlled.

◆ List the levels of organization in a complex organism.

Introduction

What is life? This question has confounded biologists and philosophers for centuries. However, if one asks about an object, "Is it alive?," most of us have some practical criteria that help us to decide. If it moves on its own, reproduces (creates more objects like itself), and leaves a trail of little piles behind it on the ground, then most of us would agree that it is alive.

In this book we are going to describe the function of one particular organism: the human being. Humans are composed of countless smaller units of life called **cells.** Single cells possess all of the attributes of a living system: They eat, grow, eliminate wastes, reproduce, and sometimes even move about. In this chapter we describe the structure and functions of an animal cell, and then briefly describe how cells are joined together to form tissues, organs, and organ systems that make up the human body. We will focus on a typical animal cell, but you will learn in subsequent chapters that there are *many* different cell types. Indeed, trying to describe a typical cell is a lot like trying to describe a typical higher vertebrate by describing it as having a backbone, four appendages, a circulatory system, a digestive system, and skin. You might be describing a swan or a rhinoceros.

Although we begin this book with a chapter on single cells, you should not lose sight of the fact that this is a course in *human* physiology. It is not our intent to make you experts on the fine points of biochemistry, cell and molecular biology, or genetics. Rather, we hope that you gain an appreciation of the various functions common to many cell types, and of the various strategies the cell uses as it carries out these functions. Before we do, however, we need to briefly describe the nonliving building blocks of life itself.

The Chemistry of Life

All matter is composed of nonliving atoms. Indeed, one of the great fascinations for biologists is how nonliving atoms combine to form living cells, thereby creating the wonder of life. An understanding of how this occurs requires a basic knowledge of how atoms are held together.

Atoms, Molecules, and Chemical Bonds

An **atom** is the smallest functional unit of matter. The nucleus, or central core, of an atom is composed of positively charged particles, known as protons, and neutral particles, known as neutrons, tightly bound together. Surrounding the nucleus is a cloud of small, negatively charged particles, known as electrons, that neutralize the positive electrical charges of the protons of the atom. According to chemists, there are just over a hundred different types of atoms. Ninety-two of these occur naturally in the universe, and the rest are produced by humans. Table 2.1 shows a list of the most common atoms in living systems. Just three different atoms (carbon, oxygen, and hydrogen) account for over 95% of the atoms in the human body.

Although an atom is the smallest functional unit of matter, atoms rarely exist by themselves. Typically, two or more atoms join together to form a **molecule.** For example, although oxygen (O) is an atom, in the atmosphere oxygen exists as a molecule (O_2), which is formed by joining two oxygen atoms together. Another example is the simple sugar glucose, a molecule containing carbon, oxygen, and hydrogen.

Why exactly do atoms combine? What types of forces hold molecules together? The answers lie in the structure of atoms. Recall that atoms are composed of a nucleus and a cloud of electrons. These electrons are arranged around the nucleus in layers, or shells, like the layers of an onion (Figure 2.1).

Each shell can accommodate only a certain number of electrons (two in the first shell and eight in the second, for example). Atoms are most stable when each shell is completely filled. By forming a bond between them, two atoms with incompletely filled outermost shells can *share* electrons, thus filling their outermost shells completely. Such a bond is called a **covalent** (or electron-sharing) **bond.** Covalent bonds between atoms are so strong that they essentially do not break apart, and thus you do not find lone atoms

Table 2.1	**Some common atoms in living systems.**		
Atomic Name	**Symbol**	**Atomic Number**	**Atomic Mass**
Hydrogen	H	1	1.0
Carbon	C	6	12.0
Nitrogen	N	7	14.0
Oxygen	O	8	16.0
Sodium	Na	11	23.0
Phosphorus	P	15	31.0
Sulfur	S	16	32.1
Chlorine	Cl	17	35.5
Potassium	K	19	39.1
Calcium	Ca	20	40.1

The atomic number represents the number of protons in the atomic nucleus. Atomic mass is the total mass of the atom relative to the smallest atom (hydrogen), which is assigned a mass of 1.0. Because electrons weigh only about 1:2000 as much as a proton or neutron, almost all of the atomic mass is represented by the nucleus of the atom.

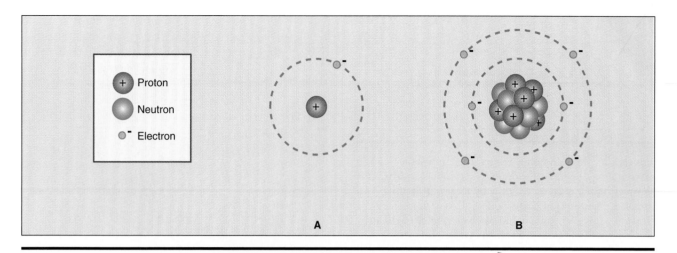

Figure 2.1 Atoms. (A) An atom of hydrogen (H), consisting of one proton and one electron. Hydrogen is the smallest of the atoms. (B) An atom of carbon (C), composed of six protons, six neutrons, and six electrons. Notice that the electrons of carbon are arranged in two distinct layers.

of hydrogen (H) in hydrogen gas, nor will you find single atoms of hydrogen or oxygen in water. Figure 2.2 shows covalently bonded molecules of hydrogen gas (H_2) and water (H_2O).

Sometimes an atom or molecule loses an electron to another atom or molecule that attracts the electron away, or gains an electron from an atom or molecule with a weaker attractant force. This loss or gain leaves the atom or molecule with a net electrical charge, because now there is an imbalance between the number of protons in the nucleus and the number of electrons surrounding it. The net electrical charge of an atom or

molecule is positive if an electron is lost, and negative if an electron is gained. A charged atom or molecule is called an **ion.** Ions can be either positively or negatively charged. Examples of ions are hydrogen (H^+), sodium (Na^+), chloride (Cl^-), calcium (Ca^{++}), potassium (K^+), and hydrogen phosphate (HPO_4^{--}). Notice that ions can have a shortage or a surplus of more than one electron.

Oppositely charged ions are attracted to each other and can join together to form neutral molecules. The attractive force holding these oppositely charged ions together is called an **ionic bond** (Figure 2.3). Ionic

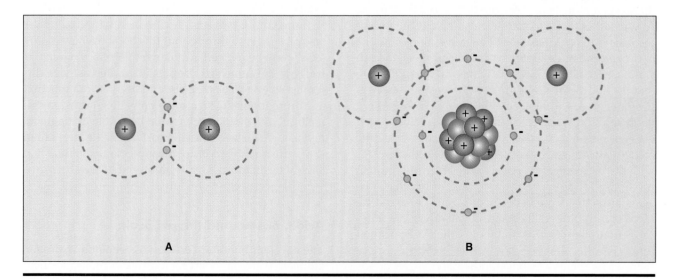

Figure 2.2 Covalent bonds. (A) A molecule of hydrogen gas (H_2) showing the covalent bonding, or sharing of electrons, between hydrogen atoms. (B) A molecule of water. Notice that the sharing of electrons between hydrogen and oxygen results in the optimum number of electrons in the outer shell of each atom (two for hydrogen, eight for oxygen).

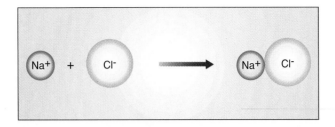

Figure 2.3 Ions and ionic bonds. To simplify the presentation, the ions of sodium (Na+) and chloride (Cl-) shown here are represented only by large balls with opposite charges to indicate their net loss or gain of an electron. However, you should be aware that these ions are composed of many protons, neutrons, and electrons.

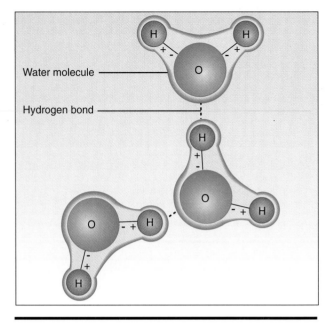

Figure 2.4 The polar nature of water molecules, showing how water molecules are attracted to each other by weak hydrogen bonds.

bonds are not as strong as covalent bonds, so molecules held together by ionic bonds tend to break apart rather easily, especially in aqueous (watery) solutions. Indeed, almost all of the sodium in the body is in the form of Na+, with very little actually forming an ionic bond with a Cl- to become molecules of common table salt (NaCl). Furthermore, there is far more ionic Ca++, K+, and Cl- in the body than calcium chloride (CaCl₂) or potassium chloride (KCl). Ions in aqueous solutions are sometimes called *electrolytes* because solutions of water containing ions are good conductors of electricity. As you will see, the ability of cells to control the movement of certain ions forms the basis for the electrical activity of nerves and for many of the transport processes of living cells.

In the water molecule depicted in Figure 2.2, the larger oxygen atom dominates the covalent bonds, so that the shared electrons spend more of the time near the oxygen atom than near the two hydrogen atoms. As a result of this unequal sharing, the oxygen end of the water molecule has a slight negative charge and the hydrogen end has a slight positive charge, even though water molecules are electrically neutral overall. (Note that the hydrogen atoms are not arranged at exact opposite ends of the molecule, but are close together). Because a water molecule has these oppositely charged ends, or poles, it is called a **polar** molecule. Opposite charges attract, so water molecules orient themselves with the more negative pole of one water molecule in proximity to the more positive pole of another. The weak attractive force between oppositely charged regions of molecules that contain covalently bonded hydrogen atoms (such as water) is called a **hydrogen bond.** Figure 2.4 shows how water molecules orient themselves relative to one another. Hydrogen bonds between water molecules are so weak that the bonds continually break and re-form, allowing water to flow.

Ions and polar molecules in aqueous solutions are attracted to the oppositely charged end of the polar

water molecules. Indeed, the degree of charge or polarity of a molecule or atom largely determines its water solubility (ability to become dissolved in water). Ions and polar molecules are soluble in water because they can mix with the water molecules. On the other hand, neutral nonpolar molecules such as fats are not very soluble in water because they are excluded from regions occupied by water.

It is often difficult to appreciate the relative sizes of objects such as ions and molecules, in comparison to more familiar objects. Textbooks are limited by the size of a page, so throughout this book we may depict ions, molecules, cells, and even organ systems at roughly the same size. To visualize the vast size differences between ions, cells, and organs, imagine that a sodium ion is the size of a penny. In contrast, a single human red blood cell would be ½ mile in diameter, and a normal-sized human heart would be larger than the entire earth! This is not the kind of information that should be memorized; it is presented here only to give you the perspective of relative sizes.

Acids, Bases, and the pH Scale

The covalent bonding between oxygen and hydrogen atoms in water is so strong that water molecules usually are quite stable. Very rarely, however, the electron from one hydrogen atom is transferred to an oxygen atom completely, and the water molecule dissociates or breaks apart into two ions: a hydrogen ion (H+) and a hydroxyl ion (OH⁻). Just as rarely, the two ions recombine to form a water molecule again. These two random events are so rare that only 1 out of 500 million water molecules exists in the ionized form at any one time.

Other molecules can generate or combine with H^+ in an aqueous solution more readily than water or OH^-. An **acid** is any molecule that can donate an H^+ (proton). When added to pure water, acids produce an acidic solution (one with a higher H^+ concentration than that of pure water). By definition, an aqueous solution with the same concentration of H^+ as that of pure water is a neutral solution. Conversely, a **base** is any molecule that can accept (combine with) an H^+. When added to pure water, bases produce a basic, or alkaline, solution (one with a lower H^+ concentration than that of pure water). Common acidic solutions are vinegar, cola, and orange juice; common basic solutions are baking soda in water, certain detergents, and drain cleaner. Because acids and bases have opposite effects on the H^+ concentration of solutions, they are said to neutralize each other in solution. You have probably heard that a spoonful of baking soda in water is a good home remedy for pain caused by too much acid in the stomach. Now you know that this remedy is based on sound chemical principles.

Because hydrogen ions are generally so rare in aqueous solutions, it is cumbersome to express their concentration in the usual units of moles per liter because the number would be so small. (A **mole,** abbreviated M, is 6.02×10^{23} particles of any substance). For example, a neutral solution such as pure water has an H^+ concentration of only 0.0000001 M (10^{-7} M). For convenience, therefore, the hydrogen ion concentration of a solution is indicated by the logarithm of the reciprocal of the hydrogen ion concentration, known as the **pH** scale:

$$pH = \log \frac{1}{[H^+]}$$

$$pH \text{ of pure water} = \log \frac{1}{0.0000001 \text{ M}} = 7$$

Why Salt and Sugar Dissolve in Water

Crystals of table salt are composed of a repeating lattice of sodium and chloride ions held together by relatively weak ionic bonds. When salt is placed in water, individual ions of sodium and chloride on the edge of the crystal break away from the lattice and are immediately surrounded by molecules of water. Because polar water molecules are attracted to charged ions, water molecules cluster around each ion, forming a cloud of water molecules known as a hydration shell. The hydration shell prevents the ions from reassociating back into the crystalline form. Only by removing the water molecules do the ions of sodium and chloride again become crystalline sodium chloride. Note that the water molecules in the hydration shell of sodium are oriented differently from those in the hydration shell of chloride.

Crystals of table sugar consist of a regular array of molecules of sucrose. You have not yet learned the structure of sucrose, but at this point it is enough to know that sucrose contains several polar groups. Thus, sucrose molecules in solution also become surrounded by a hydration shell that prevents their reassociation. Ions and polar compounds are water-soluble precisely because they can form weak bonds with molecules of water.

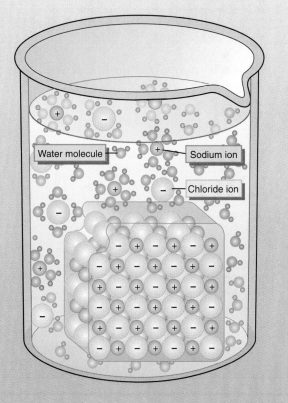

The pH of pure water is defined as a neutral pH. An *acidic* solution has a pH of *less* than 7 whereas a *basic* solution has a pH of *more* than 7, with each whole number change in pH representing a ten-fold change in the hydrogen ion concentration in the opposite direction. For example, an acidic solution with a pH of 6 (10^{-6} M H^+) has an H^+ concentration 10 times higher than that of water, whereas a solution with a pH of 8 (10^{-8} M H^+) has an H^+ concentration that is $\frac{1}{10}$ that of pure water. Figure 2.5 shows the pH scale and indicates the pHs of some common substances and body fluids.

The pH of normal body fluids such as blood plasma (7.4) is just slightly more alkaline than neutral water. Furthermore, the hydrogen ion concentration of blood plasma is very low relative to other ions (the hydrogen ion concentration of blood plasma is less than 1 millionth that of sodium ions, for example). If hydrogen ions are so rare, why do we spend so much time emphasizing them? The reason is that hydrogen ions are small, mobile, and positively charged. Therefore, they tend to be more reactive than larger ions, easily displacing other positive ions from their bonds with negatively charged regions of molecules. Hydrogen ions affect the transport of molecules across the cell membrane, the speed of certain chemical reactions, and even the shapes of proteins. In other words, a change in the hydrogen ion concentration

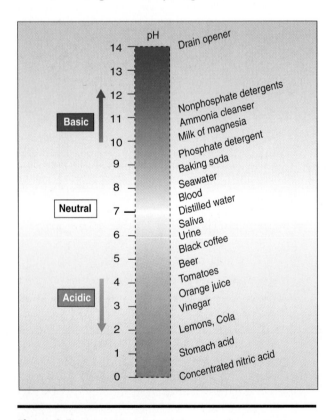

Figure 2.5 The pH scale and the pHs of some common substances and biological fluids.

of the body fluids can threaten the maintenance of homeostasis.

Carbohydrates, Lipids, and Proteins

Molecules that contain carbon atoms are called organic molecules. Organic molecules are important both as sources of energy and as the building blocks of living organisms. Three very important classes of organic molecules are **carbohydrates** (sugars and starches), **lipids** (fats and oils), and **proteins.**

Carbohydrates A clue to the basic structure of carbohydrates is found in its name: a "carbo-hydrate" has a backbone of carbon atoms with hydrogen and oxygen attached in the same proportion as they appear in water (2 to 1), hence the carbon is hydrated, or combined with water. **Monosaccharides** (meaning "one sugar") are the simplest carbohydrates, generally containing five or six carbon atoms arranged in a ring. More complicated carbohydrates are formed when monosaccharides combine into chains, creating **disaccharides** ("two sugars") or **polysaccharides** ("many sugars"). Because the process of connecting simple sugars together necessitates removing two atoms of hydrogen and one of oxygen, the process of forming carbohydrates from simple sugars is called **dehydration synthesis,** to reflect the fact that the equivalent of a water molecule is removed. Conversely, the breakdown of complex sugars and starches is a process known as **hydrolysis,** meaning that the products are formed by adding a water molecule (hydrated) as the bond between two monosaccharide molecules in the chain is broken (lysed). The process of dehydration synthesis is shown in Figure 2.6(A). If the process of dehydration synthesis continues, more complex carbohydrates such as **glycogen** are formed (Figure 2.6(B)). Glycogen is a polysaccharide in animals, whereas starch and cellulose are polysaccharides commonly found in plants.

An important feature of the synthesis and breakdown of complex molecules is that energy is required to form the bonds. Conversely, energy is released when the bonds are broken. For this reason complex carbohydrates are a good energy source for living organisms: The energy released during the breakdown of carbohydrates can be used to fuel biological processes (you will learn more about this in Chapter 14).

Lipids Like carbohydrates, lipids are composed primarily of carbon, hydrogen, and oxygen. However, the primary backbone of lipids is a chain of carbon atoms to which only hydrogen is attached: Oxygen is relatively rare. Lipids may take several forms, but the common feature is that they are nonpolar, and hence they are relatively insoluble in water. The most im-

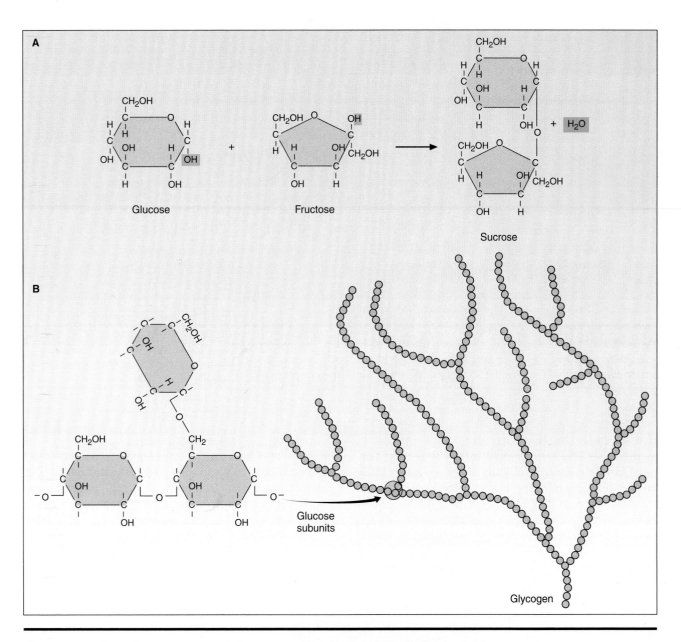

Figure 2.6 Dehydration synthesis. (A) The formation of ordinary table sugar (sucrose) by dehydration synthesis from glucose and fructose. Note that the equivalent of a water molecule is removed. (B) Glycogen, the storage form of carbohydrates in animals, is a highly branched polysaccharide formed from glucose subunits by dehydration synthesis.

portant classes of lipids are fatty acids, triglycerides, phospholipids, and steroids.

Fatty acids are chains of carbons and hydrogens with carboxylic acid groups (COOH) attached to them. Fatty acids are called **saturated** fatty acids if every carbon in the chain is bound to two hydrogen atoms; they are **unsaturated** if there are fewer than two hydrogens per carbon. **Triglycerides** are the primary storage form of lipids, just as glycogen is the primary storage form of carbohydrates. Each triglyceride molecule is composed of one molecule of **glycerol** covalently bonded to three fatty acids. The dehydration synthe-

sis of a triglyceride is shown in Figure 2.7. As is the case with carbohydrates, the synthesis of triglycerides requires energy; the breakdown of triglycerides releases energy that can be used by the body.

Phospholipids are similar to triglycerides in many ways. Like triglycerides they have a molecule of glycerol as the backbone, and they have two fatty acid tails. However, in place of the third fatty acid group is a negatively charged phosphate group (PO_4^{--}) and another group that varies depending on the phospholipid (Figure 2.8). The presence of charged groups on one end gives the phospholipid the special property

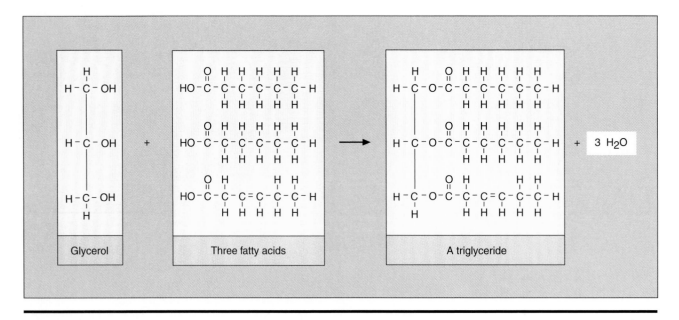

Figure 2.7 The dehydration synthesis of a triglyceride from glycerol and three fatty acids. Although they are shown here as the same length, fatty acid chains can vary in length considerably. Note that the third fatty acid molecule is an example of an unsaturated fatty acid.

that one end of the molecule is soluble in water, whereas the other end (represented by the two fatty acid tails) is relatively insoluble in water. Phospholipids are the primary building blocks of **cell membranes.**

Steroids do not look at all like the lipids described above, but they are classified as lipids because they are neutral compounds, relatively insoluble in water. Steroids consist of a backbone of three interlocking six-membered carbon rings and one five-membered carbon ring to which any number of different groups may be attached (Figure 2.9). The steroids, including the sex hormones estrogen and testosterone, are derived from **cholesterol.** Some of the cholesterol in our bodies is derived from the ingestion of cholesterol in the food we eat, especially butter, eggs, and fatty meats. Excessive ingestion of foods containing cholesterol should be avoided because cholesterol contributes to heart disease. However, cholesterol is also

Figure 2.8 Phospholipids. (A) The chemical structure of a typical phospholipid. In this particular phospholipid, one of the fatty acid chains is unsaturated, and the positively charged group attached to the phosphate group is choline. (B) Simplified representation of a phospholipid, emphasizing the polar head and two neutral tails.

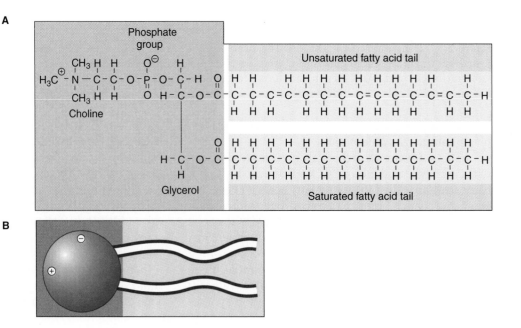

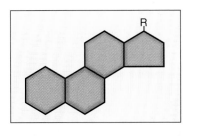

Figure 2.9   The structure of a steroid lipid. All steroids have the same basic structure of five- and six-sided rings of carbon; they vary only in the various attached groups (designated R) and in the attachments to the various carbons in the rings.

made by the body and is an important and natural component of the cell membrane.

Proteins The building blocks of proteins are called **amino acids.** Human proteins are built from 20 different amino acids, each with an amino group (NH_2) on one end and a carboxylic acid group (COOH) on the other (Figure 2.10). Other than a carbon and a hydrogen that join the amino and carboxylic acid groups together, the rest of the amino acid molecule (often designated *R* in diagrams of chemical structure) can vary considerably; a discussion of all their structures or names is beyond the scope of this book.

A **peptide** is formed when two or more amino acids are linked together as a result of dehydration synthesis. A chain of many amino acids is called a **polypeptide.** Once the length of the polypeptide chain exceeds about 100 amino acids it is generally called a protein. An important hallmark of proteins is their structural complexity, which is often described on four levels: their amino acid sequence (primary structure), their degree of twisting into a repeating spiral pattern known as a helix (secondary structure), their degree of bending or folding into a complex three-dimensional shape (tertiary structure), and the number and type of protein chains (quaternary structure). (See Figure 2.11.) The shape of a single protein chain (its secondary and tertiary structures) is determined in part by hydrogen bonds that form between different regions of the protein molecule itself.

Hydrogen bonds are relatively weak, so the tertiary structure of a protein may be influenced by the presence of other charged molecules nearby that are able to break the hydrogen bonds. This means that proteins may actually change their three-dimensional shape in the presence of charged or polar molecules. The ability to change shape is critical to the many functions of proteins, as we shall see in subsequent sections. Under certain conditions proteins may even be broken down and used as a source of energy for the cell (see Chapter 14).

Most proteins are soluble in aqueous solutions. There are exceptions, however; many of the proteins that are part of the cell membrane either are water-insoluble or have water-insoluble regions.

One of the most important functions of some proteins is to serve as **enzymes.** Enzymes are proteins that function as **catalysts,** helping chemical reactions to occur more efficiently in living cells.

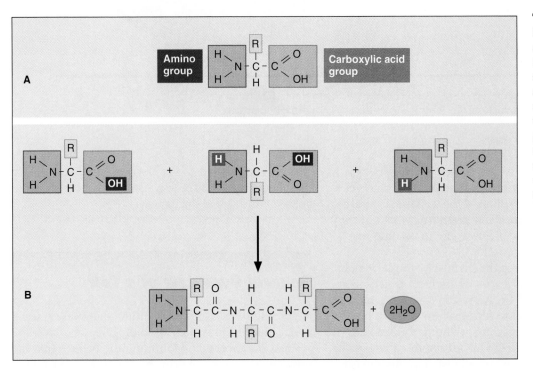

Figure 2.10   Amino acids and peptides. (A) The basic structure of an amino acid. There are 20 amino acids, each with a different R group. (B) The dehydration synthesis of a peptide chain from amino acids.

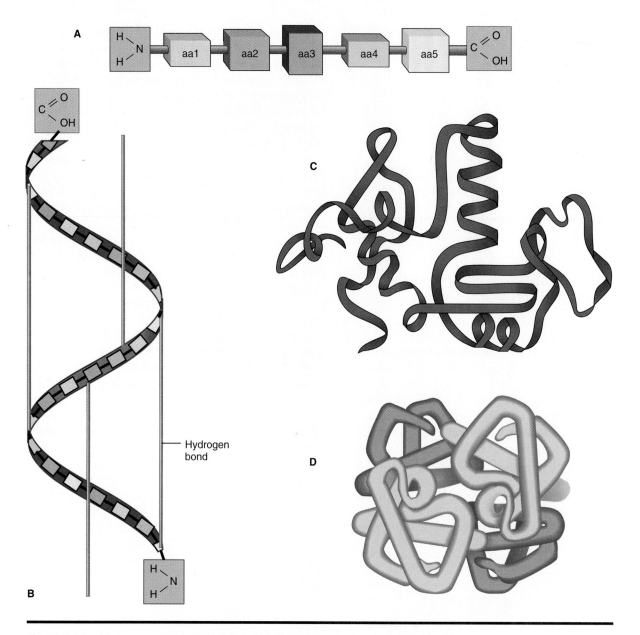

Figure 2.11 The different levels of structural complexity of a protein. (A) Primary structure. (B) Secondary structure. (C) Tertiary structure. (D) Quaternary structure.

The tertiary structure of a protein can also be permanently disrupted by high temperature or changes in pH. Permanent disruption of the tertiary structure of proteins, referred to as **denaturation,** is what causes the clear, viscous white of a raw egg to become hard when it is cooked.

We have tried to keep the chemistry simple by not introducing too many types of molecules at once. However, you should be aware that there are thousands of interesting molecules that we will not have time to mention in a brief course in human physiology. A few important molecules will be brought up in passing as we proceed. If you are interested in this subject

you are encouraged to take a course in biochemistry (the chemistry of life) or to consult a biochemistry textbook for additional details.

General Overview of a Cell

According to the cell theory of life, the smallest functional unit of life is the cell. Biologists who have examined the theory of evolution now believe that life probably began in the primordial seas over 3.8 billion

How Proteins Act as Enzymes

In chemistry, a catalyst is defined as a substance that facilitates a chemical reaction without itself being used up by the process. A human analogy would be a real estate agent who brings a home buyer and a home seller together, helping them to negotiate a deal that is satisfactory to them both. Catalysts speed up chemical reactions that would otherwise occur more slowly, but they do not change the final outcome of the reaction.

Many of the chemical reactions necessary for carrying out the function of a living cell would occur too slowly at ordinary body temperature to be of much use to the cell. In order to speed up these reactions, living cells (like sophisticated chemists) make use of catalysts. The cell's catalysts are special proteins called enzymes. Our bodies may have thousands of different enzymes, each highly specific for a certain chemical reaction.

to the enzyme is determined by the structure of a specific binding site on the enzyme. Binding of substrate to enzyme causes the enzyme to change shape slightly, imparting energy to the substrate molecule and facilitating the dissolution of the molecule into two products. Finally, the enzyme releases the products. Note that the enzyme, because it is unchanged by the process, is free to react with another substrate molecule again.

Another enzyme might bring two substrates together to form a single product, as depicted in Figure (B). This process might be envisioned as the binding of two substrate molecules to the enzyme, which would cause the enzyme to bring them into close proximity with sufficient energy that they combine. The binding of one substrate molecule by the enzyme might even change the shape of the substrate molecule so that it

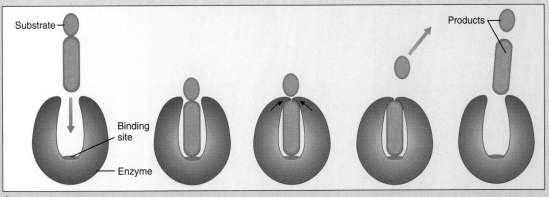

A

How do enzymes work? The ability of enzymes to facilitate a chemical reaction depends on their ability to bind weakly to other molecules and to change shape in the presence of charged or polar regions. The slight change in shape of the enzyme transfers energy to the molecules in the reaction, causing a chemical reaction to take place. One way that the process might occur is shown schematically in Figure (A). A molecule on which the enzyme is to act (the substrate) approaches the enzyme and binds to it. The ability of the substrate to bind

can combine more easily with another molecule. To make it even more complicated, some enzymes don't function unless they are assisted by certain other substances known as cofactors (generally atoms of certain metals such as iron, copper, and zinc) or coenzymes (certain organic molecules, the most common of which are vitamins). Regardless of whether a cofactor or coenzyme is required, an enzyme is a protein that facilitates a chemical reaction, without itself being used up in the process.

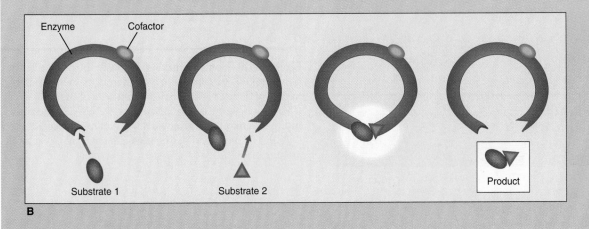

B

years ago and that it evolved from nonliving elements. But how nonliving chemicals and the physical forces of nature actually combined to create the first life-forms remains a mystery. All we know for certain is that at some time, the particular combination of chemicals present and the physical forces acting on them may have resulted in the formation of an entity that was able to sustain and reproduce itself, and life began. The fossil record suggests that the first life-form was a single cell, or life-unit, surrounded by an outer covering but devoid of much internal organization.

Today, 3.8 billion years later, modern cells are so complex and so varied in their shapes, sizes, and func-

tions that it is difficult to describe a typical animal cell. Some examples of single human cells are shown in Figure 2.12. Despite their apparent complexity and variety, there are some features shared by nearly all animal cells. A typical cell, shown schematically in Figure 2.13, is composed of three main parts:

◆ A cell membrane surrounds and covers the interior of the cell. The cell membrane is often called the plasma membrane because it keeps the liquid material of the cell, or **cytoplasm,** within the cell. The cell membrane is somewhat analogous to our own outer covering, the skin.

Osteocytes

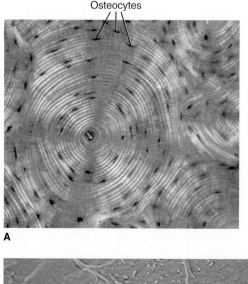

A

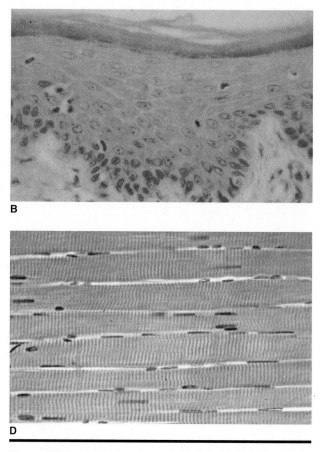

B

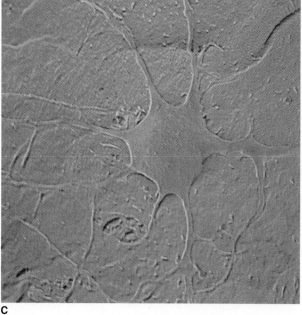

C

D

Figure 2.12 Some examples of cells in the human body. (A) Bone cells (osteocytes) embedded in an extracellular matrix of bone (×150). (B) Epithelial cells of the skin. (C) A single nerve cell, or neuron (×17,850). (D) A longitudinal section through several skeletal muscle cells (×150). The dark oval bodies along the margin of each cell are nuclei.

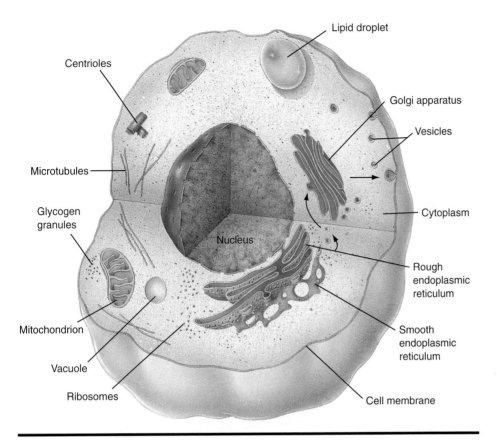

Centrioles

Lipid droplet

Golgi apparatus

Vesicles

Microtubules

Glycogen granules

Nucleus

Cytoplasm

Mitochondrion

Rough endoplasmic reticulum

Vacuole

Smooth endoplasmic reticulum

Ribosomes

Cell membrane

Figure 2.13   An artist's depiction of an animal cell.

◆ The **nucleus,** a membrane-covered, nearly spherical body in the cell, contains the genetic material that directs all of the cell's activities.

◆ The cytoplasm (*cyto-* is the prefix for *cell*) consists of all of the material inside the cell with the exception of the nucleus. The cytoplasm includes the **cytoskeleton,** a protein network that lends structural support to the cell membrane and cell contents, the **organelles** (quite literally, the "little organs" of the cell), and the **cytosol,** a jelly-like material that fills the remainder of the space. For the sake of clarity, the cytoskeleton is not shown in Figure 2.13; it will be described later. The cytosol is composed mainly of water in which the various molecular constituents of the cell, from small ions to large proteins, are dissolved. The cell's organelles (**mitochondria, ribosomes, endoplasmic reticulum, Golgi apparatus,** and other cellular ele-

ments) perform specialized functions in the cell, just as the major organs (such as the heart, liver, and lungs) have specialized functions in a higher organism composed of many cells.

The Cell Membrane

For a moment, imagine a house. The walls and roof of a house are composed of special materials that prevent rain and wind from entering. They also form a barrier that allows the temperature inside the house to be maintained at a level that can be higher or lower than the temperature outside. Windows allow light into the house. Doors open and close to allow people, pets, and other large items such as furniture and groceries in or out. Copper water lines permit the regulated entry of water and plastic sewer lines remove some of the waste. Power lines bring in energy, and information is transmitted from the outside world into the house via mail slots, telephone lines, and television cables.

Like the exterior of our hypothetical house, the cell membrane is the exterior of a living cell. It must permit the movement of some substances into and out of the cell, yet restrict the movement of others. What are the cellular equivalents of wood, brick, glass, copper, plastic and even fiber-optic cable? What special materials make up the cell membrane and give it its special properties?

Structure of the Cell Membrane

The cell membrane is constructed of lipids and proteins. However, it is by no means a simple structure. The lipids that form the structure of the cell membrane are primarily phospholipids. The lipid bilayer also contains some cholesterol. Cholesterol increases the structural strength of the cell membrane and makes it less permeable to small molecules.

Embedded in the lipid bilayer are numerous proteins. Some proteins, called **integral proteins,** span the entire width of the membrane. Other proteins extend through only one surface. The electrical charge characteristics of a protein determine its position in the membrane. Cell membrane proteins generally have a region that is electrically neutral and thus soluble in the lipid bilayer, and one or more charged regions that are insoluble in the phospholipid membrane but soluble in water. The charged regions of proteins tend to be those that extend out of the membrane, whereas the neutral portions tend to be embedded within it. The proteins in the cell membrane are analogous to the doors, windows, copper and plastic pipes, wire, and fiber-optic cable of the house, for they provide the means for transporting molecules and even information across the cell membrane. Some proteins create permanent water-filled channels through the lipid bilayer (like the copper pipes of the house) that allow substances such as water to enter the cell rather freely. Others open and close (like gates) only at specific times and only to allow passage of particular substances, such as sodium or calcium. Still others transmit information, as we shall see in a moment, or act as enzymes on the inner or outer surface of the cell. Finally, some proteins are sites of connection between cells and as points of attachment for the cell's cytoskeleton.

The cell membrane and its associated proteins are depicted in Figure 2.14(A). Note that some of the proteins have carbohydrate groups; these proteins, called **glycoproteins,** are involved in forming attachments between adjacent cells.

The phospholipid bilayer of the cell membrane is only about 3.5 nanometers thick (a nanometer, abbreviated nm, is 10^{-9} meters), too small to be seen in detail even with the most modern microscopes (Figure 2.14(B)). Imagine that you have been shrunk down to 20 nm in height. At this relative size, a single red blood cell would appear to be about half a mile in diameter. The phospholipid bilayer of the red blood cell membrane would appear to be about 13 inches thick, but cell membrane proteins with their amino acid and sugar chains sticking out would bring the total thickness of the cell membrane to about two or three feet. In contrast, the sodium ions that would be bouncing around randomly would be about the size of a penny. It is no wonder that many substances (including sodium) are restricted from passing freely through the cell membrane.

Although we have likened the cell membrane to the exterior of a house, the analogy is not entirely appropriate in several respects. First, unlike the walls of a house, the cell membrane of animal cells is not a rigid structure. If you could reach out and touch the cell membrane of the huge imaginary red blood cell described above, it would probably feel somewhat like gelatin, giving way under the force of your touch and springing back when you removed your hand. Most cells *do* seem to maintain a certain shape, but this is probably because of the cytoskeleton within the cell, the amount of fluid within the cell, and the limitations imposed by contact with other cells. Second, the phospholipids and proteins are not fixed or anchored to a particular position in the cell membrane, as are the doors and windows of a house. It is thought that the proteins drift in the constantly changing lipid bilayer, much as icebergs float about on the surface of the sea. Imagine if you were to get up in the morning and find that the front door to your house had moved three feet to the left! The cell membrane of animals is often described as a fluid mosaic to indicate that the membrane is not a rigid structure and that the pattern of proteins within it constantly changes.

Movement of Materials Across the Cell Membrane

Some materials cross the cell membrane freely, whereas others are partially or completely restricted from crossing. To understand how this comes about, we will first review what causes molecules to move in the first

HIGHLIGHT 2.3

Phospholipids: The Structure of the Cell Membrane

You probably have heard the phrase, "oil and water don't mix." If you doubt this, place cooking oil and water in a jar, shake it vigorously and then let it stand. Over time the oil moves to the top of the liquid, first as small droplets, then as larger drops that coalesce, and finally as a single layer of oil again. Why exactly do the oil and water separate? The reason is that the water molecules attract each other because they are polar. This excludes the neutral, nonpolar molecules of oil from regions occupied by water. Over time, the oil is forced into ever-larger droplets until it is completely separated from the water. The oil forms a layer on top of the water because it is less dense than water (Figure A).

hydro, meaning water + *philic*, meaning loving). The neutral fatty acid tails are insoluble in water (they are **hydrophobic,** from Greek *hydro*, meaning water + *phobic*, meaning fearing). When phospholipids are mixed with water they naturally tend to coalesce together to form a **lipid bilayer,** a double layer in which the hydrophobic tails of the two layers face each other, shielded from contact with water. The polar heads, on the other hand, remain in contact with water. Like pure lipids, some of the phospholipids will rise to the surface. But unlike pure lipids, phospholipids can form fairly stable spheres that enclose a core of water because there are polar heads on both sides of the lipid

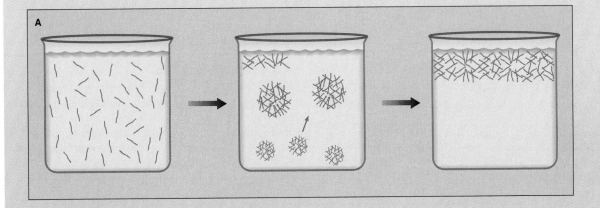

The cell membrane is composed primarily of phospholipids. As you now know, a phospholipid has several chains or tails of fatty acids attached to a polar head. Because it is polar, the head of the molecule is highly soluble in water (it is **hydrophilic,** from Greek

layer that prevent water from contacting the neutral fatty acid tails. These nonliving, water-filled spheres that we can create in a jar are similar to the basic structure of the living cell membrane: a phospholipid bilayer that encloses the cell cytoplasm (Figure B).

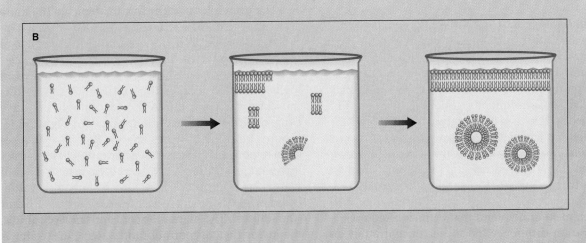

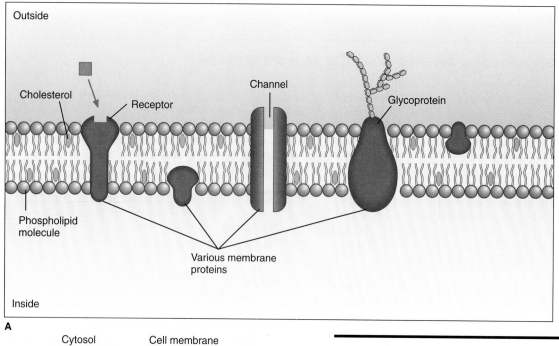

A

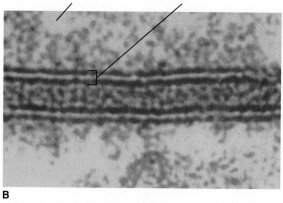

B

Figure 2.14 Cell membranes. (A) An artist's depiction shows that the basic structure of the cell membrane is a phospholipid bilayer strengthened with cholesterol. Note the variety of types of proteins in the membrane. Carbohydrate chains extend from some of the proteins (called glycoproteins), and only from the outer surface of the cell membrane. (B) An electron micrograph of the cell membranes of two adjacent cells.

place. Then we will describe the mechanisms available for getting molecules across the cell membrane.

Principles of Diffusion and Osmosis

Molecules in a gas or a liquid move about randomly as a result of thermal energy, colliding with other molecules and bouncing away with a change in direction. All particles have an equal chance of moving in any direction. Consequently, if a concentration difference exists between two regions in the solution, then strictly by chance molecules will tend to move away from the area of high concentration and toward the region of low concentration. The net movement of particles from one region to another as the result of this random thermal motion is known as **diffusion.** It is important to understand that diffusion requires a concentration difference; once the concentration of particles is the same throughout the solution, diffusion ceases. At this point a state of dynamic equilibrium exits in which random movement of particles

occurs at an equal rate in all directions. Diffusion is illustrated in Figure 2.15.

The principles of diffusion are depicted in Figure 2.16. Like Figure 2.15, Figure 2.16(A) shows what happens when placed at the bottom of a container of water; over time, the random movement of particles causes the diffusion of particles toward regions of lower particle concentration until the entire fluid is the same concentration. Although it might not have been obvious from the photograph in Figure 2.15, Figure 2.16(A) illustrates that water also diffuses from regions of its highest concentration to regions of lower concentration. However, the concentration of water (the liquid, or solvent) in an aqueous solution is *opposite* to that of the molecules other than water (the solutes). The higher the concentration of solutes, the lower the concentration of water. Pure water is the solution with the highest possible concentration of water. Thus, diffusion of water will always be toward regions with higher concentrations of solutes

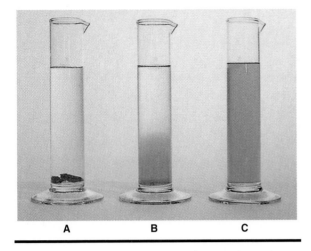

Figure 2.15 Diffusion. (A) Crystals of a blue dye have been placed at the base of a cylinder of water. (B) Over time the dye diffuses from its region of highest concentration (the bottom of cylinder) toward the top, where initially there is no dye at all. (C) Eventually an equilibrium is reached, and diffusion ceases.

and away from the more dilute (closer to pure water) regions.

The ultimate achievement of equal distributions of solutes and water by diffusion as shown in Figure 2.16(A) requires that there be no physical barriers to the movement of molecules. But now let us introduce such a barrier. Figure 2.16(B) illustrates that net diffusion between two regions of a solution separated by a barrier depends on two things: the presence of a concentration difference and the permeability properties of the barrier, which might range from very permeable (representing virtually no barrier to diffusion at all) to completely impermeable. In Figure 2.16(B), the barrier is permeable to the smaller solute but not to the larger one (nor to water).

Let us take a third case, one that demonstrates that water can be made to move not only by diffusion, but also by direct pressure applied to it (just as pure water moves through a hose). Figure 2.16(C) shows what would happen if a solution of a solute in water were separated from pure water by a barrier that was permeable to water but impermeable to the solute. In this example, the solute particles are prevented from diffusing, but the water diffuses toward its region of lower concentration, from right to left. The net diffusion of water across a selectively permeable membrane is called **osmosis**. You might think that osmosis will continue until all of the water ends up in the left-hand compartment, because as long as there is *any* pure water in the right compartment there would be a difference in water concentrations between the two sides. This does not happen, however. As water moves from right to left, the column of water becomes higher on the left than on the

right. This generates a dynamic, or hydrostatic, pressure difference between the two compartments, forcing some water to move from left to right. *Net* movement of water stops when the rate of water movement toward the left compartment (by osmosis) equals the rate of movement toward the right compartment (due to the differences in hydrostatic pressure).

We have been building up to a discussion of the cell membrane as a barrier to diffusion. The cell membrane is a *selectively permeable* membrane, allowing the free diffusion of water and certain solutes, but partially restricting or completely preventing the diffusion of other solutes. The permeability properties are determined by the structure of the phospholipid bilayer and by the presence of the integral proteins in the membrane.

Diffusion Through the Phospholipid Bilayer

Small, uncharged, nonpolar molecules can diffuse right through the lipid bilayer as if it did not even exist. Such molecules simply dissolve in the lipid bilayer, passing through it as one might imagine a ghost walking through a wall. Polar, or electrically charged, molecules cannot cross the lipid bilayer easily because they, unlike neutral molecules, are not very soluble in lipids. Thus, simply because of its lipid bilayer structure the cell membrane can restrict the passage of some molecules and allow the free passage of others. Two very important lipid-soluble molecules are oxygen (O_2), which diffuses into the cell and is used up in the process of metabolism, and carbon dioxide (CO_2), a waste product of metabolism that diffuses out of the cell and is removed from the body by the lungs. Other important molecules that cross the lipid bilayer by diffusion are urea (a small, neutral waste product that is removed from the body by the kidneys) and alcohol (or ethanol), which is not a normal substance produced by the body in any significant quantity. Other than the diffusion of neutral substances directly through the lipid bilayer, all other substances must rely primarily on channels or carrier molecules consisting of membrane proteins.

Diffusion Through Open Channels
One of the most rapidly diffusing substances to cross most cell membranes is water. The cell membrane is so freely permeable to water that the intracellular concentration of water (the concentration of water in the cytoplasm) is always essentially the same as that in the extracellular fluid (the fluid outside the cell). How can this be, given that water is a polar molecule that should not be able to cross the lipid bilayer? The answer is that some of the proteins spanning the lipid bilayer contain water-filled channels through which water and other small neutral molecules can diffuse (Figure 2.17). These channels might be considered analogous to the

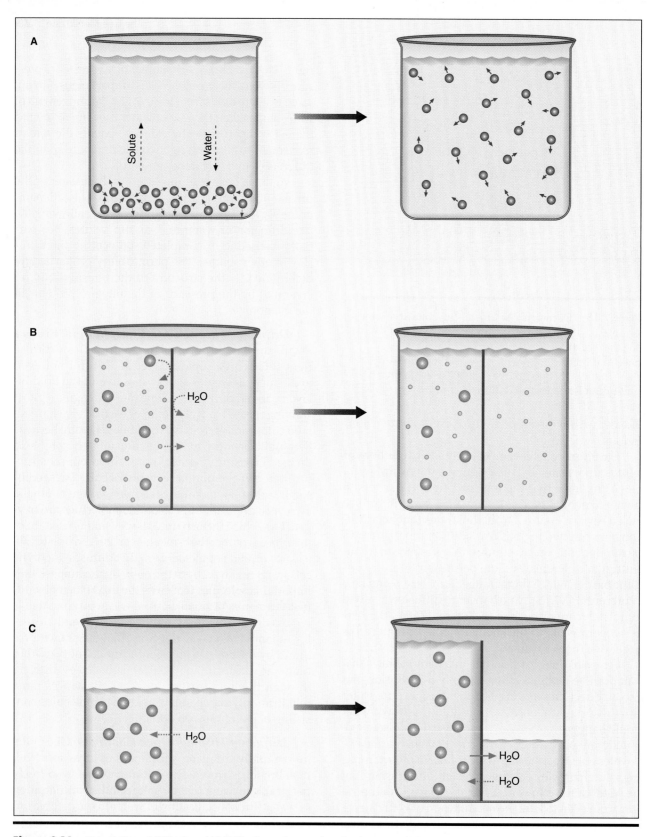

Figure 2.16 Principles of diffusion. (A) Diffusion when no barrier is present. Assume that space not occupied by solute molecules is occupied by water. Random movement of particles is indicated by the solid arrows and diffusion is indicated by the dashed arrows. (B) Diffusion across a semipermeable barrier, in this case, one that is permeable to the small solute but not to the large solute or to water. (C) Movement of water across a barrier that is permeable to water but not to the solute. Movement of water by diffusion is indicated by a dashed arrow; movement caused by the dynamic pressure generated by the column of fluid is indicated by the solid arrow.

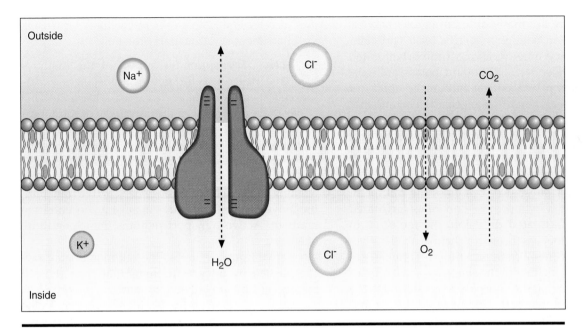

Figure 2.17 Diffusion of water through membrane proteins that serve as water channels. Water can move in either direction. Ions are restricted from using the water channels because they are too large, and particularly because they have a large shell of water molecules around them and because they are charged particles. Note that lipid-soluble substances such as oxygen and carbon dioxide can diffuse right through the lipid bilayer.

Panama canal between the Atlantic and the Pacific Oceans, except that they are always open and are so numerous that most cell membranes never restrict the osmotic movement of water. An exception is certain specialized cells in the kidney, as you will see later.

Diffusion Through Gated Channels Some channels are available for charged molecules and mol-ecules too large to pass through the water channels. Many of these channels can open and close, thereby acting as doors or gates (Figure 2.18). The opening and closing of the **gated channels** may be caused by the binding of a particular molecule to the channel in some way, or in response to a change in the electrical charge across the membrane. The details vary for different molecules, and at this point you should just be

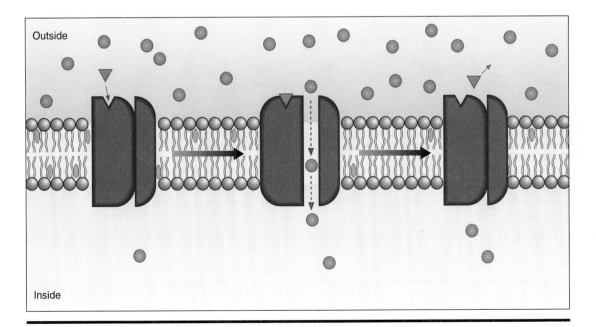

Figure 2.18 Diffusion through gated channels. Note that the molecule that triggers the opening of the channel is not the same one that diffuses through the channel.

aware that there are many different gated channels and that they can be highly specific for a particular molecule or ion. Some of the most important gated channels are those for sodium and potassium ions. These channels play a critical role in the ability of certain cells to generate an electrical current that can be transmitted over long distances as a mechanism of cell-to-cell communication. This is the subject of Chapter 3.

Facilitated Diffusion There is yet another way that hydrophilic substances can cross the cell membrane, and that is by a process known as **facilitated diffusion.** In facilitated diffusion, the molecule in question does not pass through a channel at all; rather, it attaches to a membrane protein. The attachment triggers a change either in the shape or orientation of the protein such that the molecule is transferred to the other side of the membrane and released there. Once the molecule is released, the protein returns to its original form. A model of how this process might occur is shown in Figure 2.19. The direction of net movement of molecules is always from a region of high concentration to a region of lower concentration, and it does not require a source of metabolic energy. Normal diffusion is simply helped by the presence of a particular protein, hence the name *facilitated diffusion.* Because the protein transports the molecule across the membrane rather than just opening a channel through it, the protein is sometimes called a **transport protein.** Glucose and amino acids, both essential to the activ-

ities of cells, enter most cells by facilitated diffusion via attachment to a transport protein.

Active Transport So far, all of the methods we have described for getting substances across the cell membrane have involved diffusion. That is, they have only allowed the movement of substances in the direction they would normally diffuse if there were no barrier. Some substances, however, must be transported across the cell membrane *against* their normal concentration gradient. This cannot happen by opening a channel. For this, we need **active transport,** the movement of substances against a concentration gradient. Active transport requires the expenditure of energy.

Active transport across the cell membrane is accomplished by integral proteins that span the cell membrane and are linked to a source of energy. Cells manufacture, store, and use a high-energy molecule called **ATP** (adenosine triphosphate; an adenosine molecule linked to three phosphate groups). Some transport proteins break ATP down to a lower-energy molecule of **ADP** (adenosine diphosphate) and use the released energy to transport one or more molecules across the cell membrane against their concentration gradients. This is called primary active transport because the energy source is derived directly from high-energy molecules stored within the cell. The process is depicted schematically in Figure 2.20. To picture this process, imagine that the active transport protein is a conveyor belt moving objects uphill,

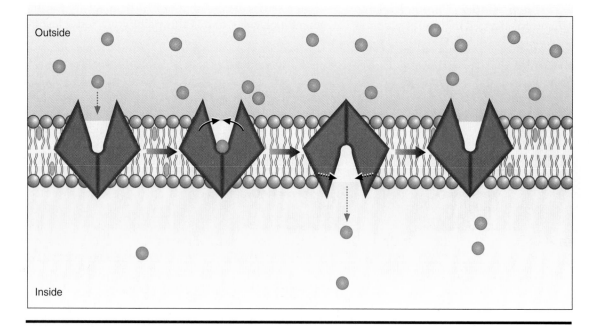

Figure 2.19 Facilitated diffusion. The binding of the molecule to be transferred triggers a change in conformation of the protein such that the molecule is transferred to the other side of the membrane. Once the molecule is released, the protein spontaneously returns to its original form.

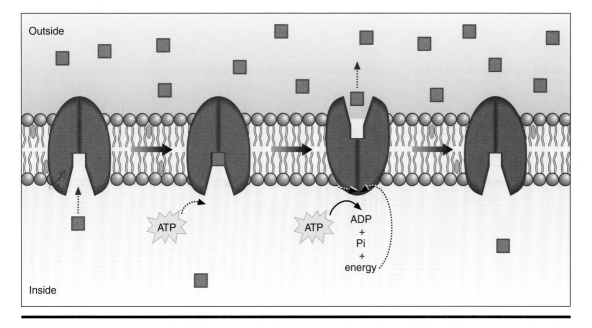

Figure 2.20   Primary active transport. The binding of the molecule to be transported activates the protein (an enzyme), which catalyzes the breakdown of ATP and thereby releases energy. The energy is used to cause a change of conformation of the protein. Again, the protein returns to its original conformation once transport is complete.

powered by a gasoline engine. In this analogy, ATP would be the gasoline and ADP the exhaust (with one very important modification; the ADP "exhaust" within the cell can be recycled to ATP).

Proteins that actively transport molecules across the cell membrane are sometimes called pumps, a word that describes their function very well. Some pumps are capable of transporting several different molecules at once, sometimes even in two directions at the same time. The best-known and one of the most important cell membrane pumps is **sodium-potassium ATPase** (or simply the Na$^+$-K$^+$ pump). The name aptly describes its function, for the suffix -*ase* means enzyme. The Na$^+$-K$^+$ pump is actually a membrane-integral protein that is also an enzyme; it catalyzes the breakdown of ATP and uses the liberated energy to transport sodium out of, and potassium into, the cell.

Not all active transport pumps use ATP directly. Some pumps use energy derived from the facilitated diffusion of one molecule to transport another, different molecule against its concentration gradient (Figure 2.21). This type of transport is called secondary active transport. The *active* part of the term comes from the fact that energy is required to move the actively transported molecule against its concentration gradient. The *secondary* refers to the fact that although the transport protein does not use ATP directly, it can transport a molecule actively only because energy *was* used to create the concentration gradient for the diffusing molecule in the first place. The great advantage of

secondary active transport is that it uses a source of energy that might otherwise be wasted by the cell. The facilitated diffusion process is used to provide the energy; the actively transported molecule just catches a ride.

Information Transfer Across the Cell Membrane

Because some proteins span the entire cell membrane and can easily be induced to change their shape in response to nearby charged groups, it is possible to transfer information across the cell membrane without any object crossing it. Proteins that transmit information are called **receptor proteins** because they receive, and then transmit, a signal across the cell membrane. Some hormones act by activating such receptor proteins on the cell membrane.

The principle of how a cell membrane receptor protein works is illustrated in Figure 2.22. A specific molecule capable of influencing the receptor approaches and, when it gets sufficiently close, binds to the receptor site. What happens next can be quite complex. Indeed, the actual sequence may involve one or more other proteins closely associated with the receptor, one of which may be an enzyme that causes a highly specific biochemical reaction to occur within the cell. At this point, you need not know exactly how the entire sequence of events occurs. The important principle is that some molecules restricted from entering the cell can still cause the cell to do something

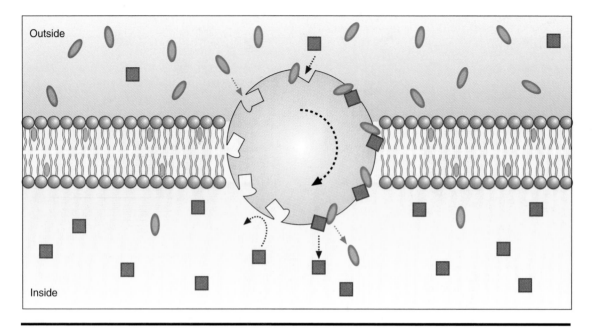

Figure 2.21   Secondary active transport. The energy for the secondary active transport of the square molecules is provided by the facilitated diffusion of the oval molecules, without direct expenditure of energy by the cell.

very specific merely by coming in contact with the appropriate cell membrane receptor. Receptors are highly specific for a particular molecule or group of similar molecules. For example, the receptor for the hormone insulin responds only to insulin and not to any other hormone. Furthermore, different cells have different sets of receptor proteins, which explains why some cells and tissues respond to a certain hormone and others do not. You will learn more about receptors and hormones in Chapter 5.

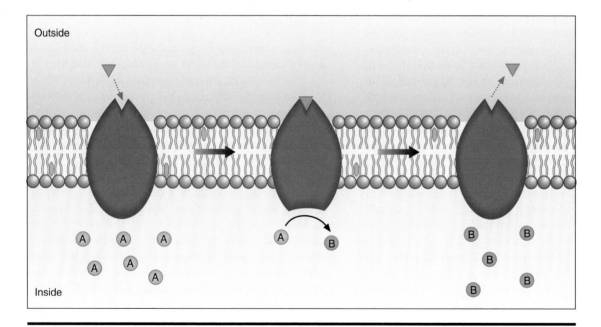

Figure 2.22   How receptor proteins work. The binding of the molecule to a receptor on the outside surface of the cell triggers some intracellular event; in this example, it catalyzes a chemical event in which a reactant A is turned into a product B.

The Nucleus

The nucleus is a membrane-bound, roughly spherical body located entirely within the cell. It is the command center of a living cell, much like the brain is the command center for the whole organism. The nucleus functions as a command center because it is the repository of the cell's genetic information; this genetic information directs all of the other activities of the cell. How does this work? Quite simply, the genetic information contained in the nucleus directs the synthesis of proteins within the cytoplasm, many of which are enzymes. Enzymes, in turn, are involved in the synthesis of most other biologically important molecules. The nucleus is bounded by a double membrane that keeps the genetic material contained in the nucleus, much as our skull serves as a barrier to protect and contain our brain.

DNA

The genetic information stored in the nucleus is contained in special molecules called **DNA** (deoxyribonucleic acid). Because of their size and the permeability properties of the double-layered nuclear membrane, DNA molecules are restricted to the nucleus. Thus, the information contained in the DNA must be copied and carried out of the nucleus in order for it to affect cell function.

The basic building blocks of DNA are smaller molecules called **nucleotides** (Figure 2.23). Each nucleotide consists of a sugar group, a phosphate group and one of four different nitrogenous bases. The chemical structures of the nitrogenous bases (guanine, cytosine, adenine, and thymine) are not terribly important for our understanding of human physiology.

Nucleotides are bonded together by strong covalent bonds between adjacent sugar and phosphate groups, forming a strand of nucleotides. A DNA molecule consists of two such strands of nucleotides, intertwined into what is called a double helix. The double-helical structure of DNA is somewhat like the single-helical structure of proteins, except that the DNA double helix is formed by multiple hydrogen bonds between certain base pairs of the two nucleotide strands, rather than by single hydrogen bonds between different amino acids in a protein. Guanine and cytosine can form three hydrogen bonds between them (but neither can bond to adenine or thymine), and adenine and thymine can form two hydrogen bonds. Thus guanine and cytosine are called complementary base pairs, as are adenine and thymine. The sequence of bases in *one* of the strands of the helical DNA molecule determines the entire genetic code. It might seem amazing to you that all of the genetic information making you unique is encoded by only four different code letters (nucleotides), when this sentence alone uses over twenty different letters of the alphabet. However, computers can accomplish information coding with only a binary, or two-digit, system.

Under most circumstances, DNA molecules cannot be seen clearly within the nucleus even with the aid of a microscope. However, during certain stages of cell division, DNA condenses with some associated proteins, becoming visible under the microscope as **chromosomes** (or "colored bodies"). Humans have 46 chromosomes (23 pairs), each consisting of a single long, double-stranded DNA molecule and associated proteins. Figure 2.24 shows the loose arrangement of DNA in the nucleus when a cell is not dividing, as well as the visibility of the chromosomes during cell division.

DNA Replication

With only a few exceptions, *every cell in the body contains an identical set of DNA molecules.* The main exceptions to this rule are red blood cells, which have no DNA and hence do not divide, and certain cells in the reproductive system. Thus virtually every cell, regardless of whether it is a cell in the heart or a cell in the brain, contains the same library of genetic information.

The process of making an exact copy of the cell's DNA before cell division is called **DNA replication.** The process can be envisioned as several sequential steps. First, a portion of the DNA molecule comes "unzipped"; that is, the two nucleotide strands partially separate from each other (Figure 2.25). Then other nucleotides with the appropriate complementary bases are joined to each of the nucleotide chains, forming a new complementary strand for each of the two original strands. This process continues until the entire DNA molecule has been replaced by two identical molecules. Eventually, the two new DNA molecules separate completely, one going to each of the two cells formed during cell division. The end result is a duplicate of the original DNA in each cell.

There are approximately 3 billion base pairs in human DNA, all of which must be copied every time a cell divides. The process of DNA replication is extremely accurate. Nevertheless, an occasional error in DNA replication may occur just by virtue of the sheer numbers of base pairs and the numbers of cells that are dividing in the human body (imagine retyping the entire contents of a library billions of times without making a single typographical error). Fortunately, however, cells have the ability to detect and fix most of the rare errors that do occur.

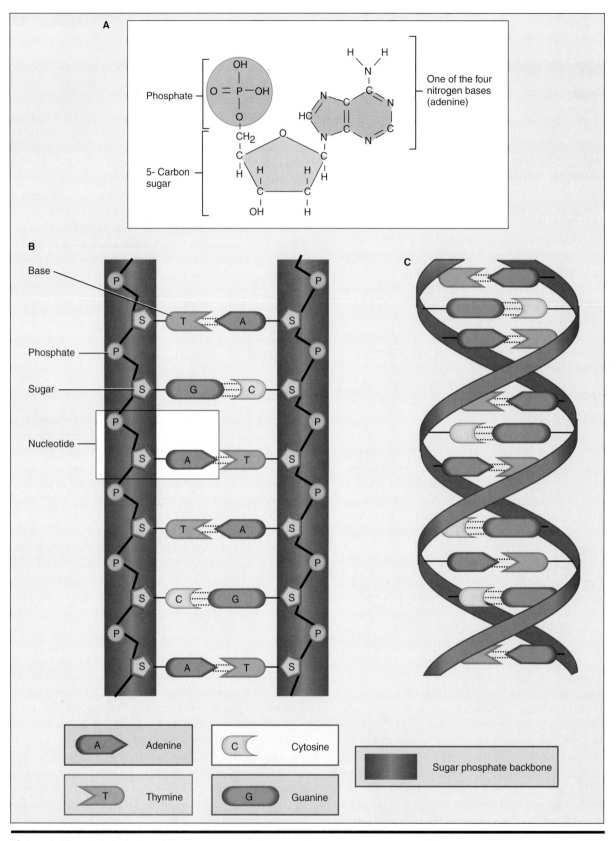

Figure 2.23   Nucleotides and the structure of DNA. (A) The structure of a single nucleotide, this one containing the nitrogenous base adenine. Only the nitrogenous bases are different between the four nucleotides. (B) Closeup of a portion of a DNA molecule showing that it is composed of two strands of complementary nucleotides. The backbone of each strand is held together by strong covalent bonds between the sugar and phosphate groups of adjacent nucleotides and the two strands are joined by weak hydrogen bonds between complementary base pairs. (C) A more distant view of a portion of a DNA molecule, showing the double-helical shape. The shape comes about because of the slight angles of the bonds between adjacent nucleotides.

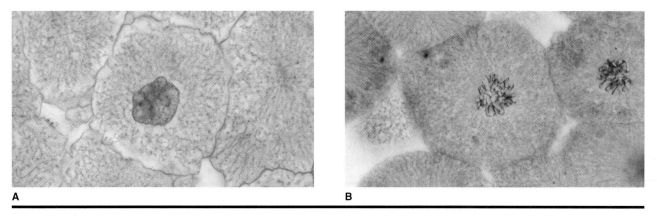

A   B

Figure 2.24   DNA in the nucleus. (A) The nucleus of a cell that is not dividing, in which the DNA is dispersed (×450). (B) The nucleus of a cell in the process of dividing, showing the now-visible chromosomes (×450).

DNA Transcription

For DNA to affect cellular function, the information contained within it must be read so that the information (or code, as it is sometimes called) can be taken out of the nucleus. Furthermore, like the information in a large library, only a portion of the information is being read, or translated, at any one time. Let us now introduce the concept of a gene, and then describe the

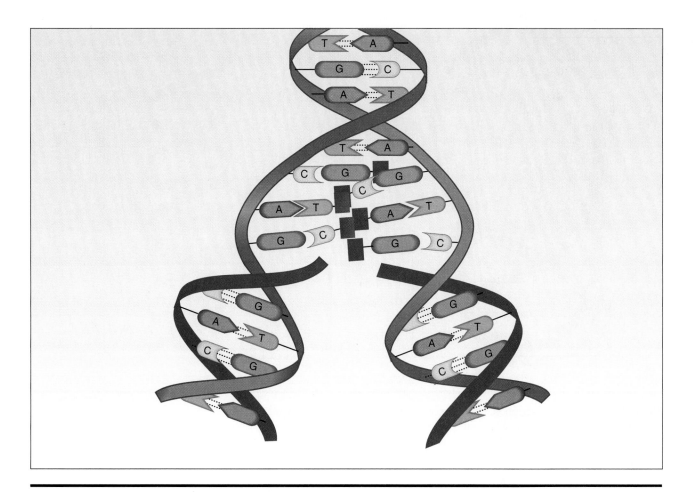

Figure 2.25   DNA replication. The new strands are indicated by the darker color. Notice that developing DNA molecules are identical.

process of how the genetic information is translated into action.

A **gene** is a segment of DNA that represents the recipe, if you will, for making a particular protein. A single gene may be represented by a sequence of several hundred base pairs in a DNA molecule. In order for the protein encoded by the gene to be made, the gene must first be copied so that the message can be taken out of the nucleus. **DNA transcription,** the process of making a copy of the gene, is somewhat akin to DNA replication, with several notable exceptions: Only a *portion* of the DNA is copied: the region represented by a single "recipe," or gene; the copy that is made of the gene is only a *single* strand of nucleotides; and the molecule that is made, called **RNA** (ribonucleic acid), is not quite an exact copy of the complementary strand of DNA nucleotides. In RNA, the sugar backbone is ribose rather than de-

oxyribose; ribose has one more oxygen molecule than deoxyribose. In addition, one of the four nitrogenous bases is different as well; in RNA, uracil replaces thymine (Figure 2.26). The processes of replication and transcription are summarized in Table 2.2. RNA translation, the process of using the RNA "recipe" to make a particular protein, will be described shortly.

The Cytoplasm

With the exception of the nucleus and the cell membrane, all other parts of a cell are referred to collectively as the cytoplasm. The cytoplasm includes the cytoskeleton, centrioles and spindle fibers, mitochondria, ribosomes, the endoplasmic reticulum, the Golgi

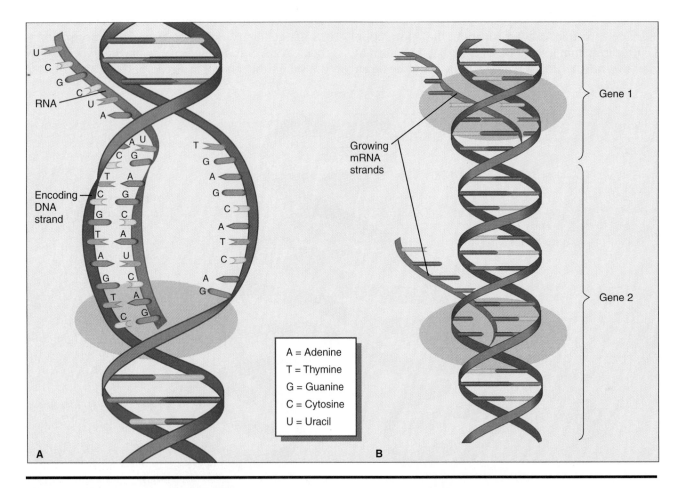

Figure 2.26 DNA transcription. (A) Transcription is proceeding downward in this example. Note that a single RNA strand is being made that is complementary to only one of the strands of DNA (the strand that carries the genetic code). Note also that in the RNA molecule, uracil replaces thymine as the complementary base pair to adenine. (B) A more distant view of DNA transcription showing that several genes may be read at the same time from different regions of the same DNA molecule.

Table 2.2 **A comparison of replication and transcription.**

	DNA Replication	DNA Transcription
When does it occur?	Just before cell division.	At any time.
What is copied?	All DNA.	Only a portion of the DNA, called a gene.
How does it happen?	Double-stranded DNA unzips; a complementary copy is made of each strand.	Only a portion of a double-stranded DNA unzips; a complementary copy is made of one strand only.
What is produced?	Two exact copies of the original DNA.	A single-stranded RNA molecule.
Where do the products go?	One complete set of DNA becomes part of each daughter cell during cell division.	RNA exits the cell nucleus, entering the cytoplasm.
What do these processes accomplish?	Replication provides the source of DNA for each daughter cell.	Transcription produces RNA, the recipe for manufacturing proteins in the cell.

HIGHLIGHT 2.4

The Human Genome Project

The human genome (the complete set of human genes) consists of roughly 100,000 different genes, each located in the same spot of the same chromosome for all humans. The reason that we do not all look exactly alike is because the *base pair sequence within each gene* varies slightly between individuals. Because humans all have the same gene sequence, it ought to be possible to create a map of the entire human gene sequence of each chromosome (just as one can construct a road map of the highways between New York City and Los Angeles) and ultimately to construct a map of the entire human genome contained in the 23 pairs of chromosomes. A major scientific effort is currently underway to do exactly that. This ambitious and technically difficult project involves hundreds of scientists around the world working together. The entire project is expected to cost hundreds of millions or perhaps even billions of dollars and should be completed sometime in the next decade.

As might be expected, any project of this magnitude is likely to be controversial. Currently, the genetic abnormalities that are responsible for a few genetically determined diseases (such as Down syndrome, Tay–Sachs disease, and Huntington's disease) have been identified, and it is now possible to safely test for these and other specific diseases while the fetus is still in the womb. But what if we knew the *entire* normal human genome? Proponents of the human genome project suggest that we have just barely scratched the surface in terms of being able to identify the genes responsible for other genetic disorders. It has even been suggested that in the future it might be possible to treat genetic disorders by repairing the faulty genome in an individual. On the other hand, opponents of the program point to its high cost and to the uncertainty that knowledge of gene location would necessarily produce any insights into how to repair the genome in disease states.

You should recognize that advances in science may lead to social, ethical, economic, or political dilemmas. Although one might assume that scientists should take a leadership role in dealing with these dilemmas because they are best equipped to understand the probable consequences of such advances, it is not necessarily the job of scientists alone to resolve these issues for us. Should a severely diseased fetus be aborted? Who would pay for procedures that are expensive and controversial? Should we be allowed to alter the human genome in individuals in order to repair "defects"? As a society, we may wish to consider such questions so that we may make informed choices.

apparatus, and various other vesicles, vacuoles, and cellular inclusions. The fluid that fills the remaining space in the cell is also considered part of the cytoplasm.

Cytoskeleton

As mentioned previously, an animal cell is not a particularly rigid structure. But neither is it just a fluid-filled bag, devoid of structural support. The cytoskeleton, depicted in Figure 2.27, consists of a loosely structured network of fibers called **microtubules** and **microfilaments.** Microtubules and microfilaments are, as their names imply, small tubes or solid elements composed of protein fibers. These structural elements are attached to the cell membrane and to each other, forming a framework for the fluid-like cell membrane, much as tent poles support a nylon tent. The cytoskeleton also supports and anchors the cell **organelles.**

Centrioles and Spindle Fibers

Centrioles are short, paired, rod-like microtubular structures located near the nucleus. Centrioles have no known function in a cell that is not dividing; their only role is in the process of cell division. Exactly what triggers the centrioles to participate in cell division is not well-understood. What we do know is that other microtubular elements, called **spindle fibers,** are pro-

duced as part of the preparation for the process of cell division, and that together, the centrioles and the spindle fibers are essential for the process of pulling the cell's replicated DNA into two parts, one for each of the daughter cells. The structure of a centriole is shown in Figure 2.28.

Mitochondria

Nearly all of the cells' activities require energy. Energy is available in the chemical bonds of the food you ingest, but this energy cannot be used directly by cells. Most of the energy in ingested nutrients must be converted to a more usable form before it can power the chemical and physical activities of living cells.

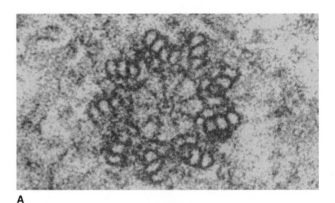

A

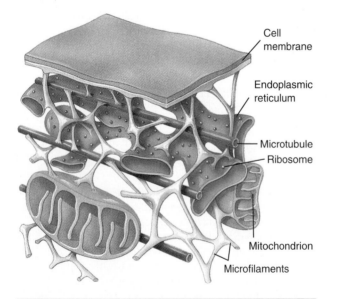

Figure 2.27 Cytoskeleton. Microtubules and microfilaments of the cytoskeleton support the cell membrane and anchor the organelles.

Cell membrane

Endoplasmic reticulum

Microtubule
Ribosome

Mitochondrion
Microfilaments

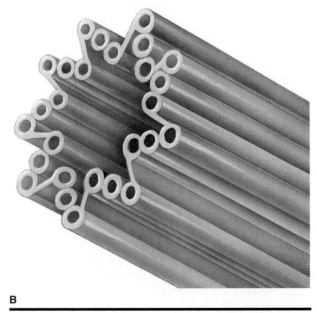

B

Figure 2.28 Centrioles. (A) Transmission electron micrograph of a centriole in cross-section, 300,000×. (B) A drawing of a centriole showing that it is composed of nine triplets of microtubules.

Mitochondria are the organelles responsible for producing most of the usable energy for cells; they are often called the power plants of cells. Not surprisingly, their number in different cells varies widely as a function of the requirement for energy. Cells with a high rate of energy use, such as skeletal muscle cells, may contain over 1000 mitochondria per cell.

A photograph of a single mitochondrion and a diagram of its structure and function are shown in Figure 2.29. A smooth outer membrane very similar to the cell membrane covers the entire surface. Contained entirely within the outer membrane is an inner membrane with numerous folds that increase its surface area. The inner membrane and the fluid contained within its folds contain hundreds of protein enzymes that facilitate the breakdown of chemical bonds in food, releasing the stored energy. The breaking of the chemical bonds of food is a process that uses oxygen and produces the metabolic waste product carbon dioxide (CO_2).

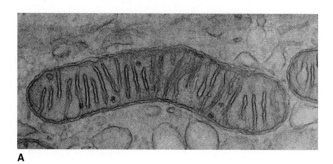

A

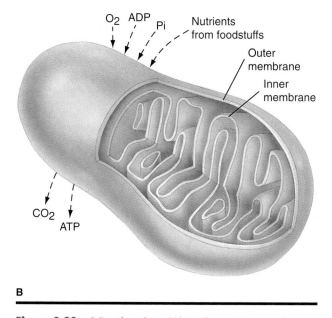

B

Figure 2.29 Mitochondria. (A) A photomicrograph (155,000×) of a mitochondrion. (B) Functional and structural overview of a mitochondrion.

The energy released by the breakdown of food in the mitochondria is not ready for export until it has been converted to a form that can be used by the cell. Within the mitochondria the energy released from food is used to produce ATP. The ATP is then exported from the mitochondria to the cytoplasm. You have already learned one use for ATP: It is the energy source for the active transport of sodium and potassium across the cell membrane.

Ribosomes

Ribosomes are small darkly stained bodies that are either attached to the endoplasmic reticulum (ER) or floating freely in the cytosol. It is the ribosomes that give certain regions of the endoplasmic reticulum a rough, granular appearance (review Figure 2.13). Ribosomes are the site of protein synthesis in the cell; indeed, ribosomes consist largely of RNA and the proteins being synthesized by RNA (remember, RNA was transcribed from a portion of the DNA in the nucleus, and a particular RNA carries the recipe for a particular protein). Within the ribosome the RNA code links an exact sequence of amino acids together, forming a particular protein. The process of reading the RNA code and translating it into the creation of a protein is called **RNA translation.**

The translation of RNA into a finished protein is a complicated process best left to specialized courses in cell or molecular biology. Briefly, three consecutive RNA bases in the RNA strand represent the code for one of the 20 different amino acids. With the RNA molecule as a template, the appropriate amino acids are strung together like beads on a string until the entire protein is complete.

As mentioned previously, some ribosomes are found in the cytosol and some are attached to the endoplasmic reticulum. The difference has to do with the function of the proteins produced. Proteins produced by ribosomes in the cytosol are generally for immediate use; an example is enzymes that catalyze chemical reactions in the cytosol. In contrast, proteins produced by ribosomes attached to the endoplasmic reticulum are generally destined for export from the cell or for the synthesis of new cellular membrane or other protein components of the cell.

The biochemical steps in the synthesis of the thousands of compounds made by a cell are complex and better studied in a biochemistry course. However, two points stand out. First, proteins are made by the ribosomes and then modified, if necessary, by the addition of nonprotein groups in the endoplasmic reticulum and the Golgi apparatus. Nonprotein compounds are made in the cytosol, the endoplasmic reticulum, and the Golgi apparatus. Second, the regulation and synthesis of compounds *other* than proteins is *directed* by proteins. This is because proteins serve as enzymes, directing the chemical reactions involved in synthesizing and breaking down other compounds. You have already learned that enzymes in the mitochondria are involved in *breaking down* foodstuffs so that energy can be liberated to build ATP. But enzymes can catalyze the synthesis of substances as well. Thus, the nucleus exerts control over the synthesis of nonproteins as well as proteins because it directs the synthesis of the enzymes necessary for other chemical reactions to occur. This can be represented by the following scheme, where the arrows refer only to the product of the activity of the preceding molecule:

DNA → RNA → protein enzymes → nonprotein compounds

Nonprotein compounds are made in the cytosol, the endoplasmic reticulum, and the Golgi apparatus, but not in the ribosomes; ribosomes make only proteins.

The Endoplasmic Reticulum

The **endoplasmic reticulum (ER)**, in conjunction with the ribosomes that are attached to it, synthesizes most of the chemical compounds made by the cell. If you were to consider a cell as analogous to an entire country, then the ER would be equivalent to all of the steel mills, sawmills, and chemical plants combined. It is important to note, however, that the materials manufactured by the ER are often not in their final form; refining and packaging is the role of the Golgi apparatus.

The structure of the ER and its role in the manufacture of proteins and other materials is shown schematically in Figure 2.30. The ER is an extensively folded, membranous system surrounding a fluid-filled space. A portion of the ER also forms the double-layered nuclear envelope, or outer membrane of the nucleus. The outer surface of some regions of the ER is dotted with numerous ribosomes, giving it a granular appearance, hence the name **rough ER.** Other regions devoid of ribosomes are called the **smooth ER.**

The rough ER is involved in the synthesis of proteins, as you might guess from the presence of associated ribosomes. Most of the proteins synthesized by the attached ribosomes are released into the fluid-filled space of the ER, and thus are isolated from the rest of the cytosol as soon as they are synthesized. Eventually the proteins enter the smooth ER, where they are packaged for transport to the Golgi apparatus. A few proteins may be released directly into the cytosol for use by the cell, however.

The smooth ER is not a site of protein synthesis. However, the smooth ER has a rich supply of enzymes on its membranes that are responsible for the synthesis of certain lipids, among them the steroid hormones. The smooth ER is also responsible for packaging the newly produced proteins and lipids for delivery to the Golgi apparatus. Newly synthesized proteins from the rough ER and lipids from the smooth ER are collected in the smooth ER. Small portions of the fluid-filled space are then surrounded by smooth ER membrane and pinched off, forming **vesicles** of smooth ER membrane containing the protein and lipid products. The vesicles migrate to the Golgi apparatus, where they fuse with the Golgi apparatus membrane, releasing their contents into the Golgi apparatus for further processing.

The Golgi Apparatus

If we think of the ER and its associated ribosomes as the cell's factory for producing proteins and lipids in a raw form, then the **Golgi apparatus** would represent the refining, packaging, and shipping center of the cell. The structure of the Golgi apparatus and the processes that occur there are diagrammed in Figure 2.31. In cross-section, the Golgi apparatus appears as a series of fluid-filled spaces surrounded by membrane, much like a stack of plates. Like the ER, the Golgi apparatus contains enzymes that further process the products of the ER into final form. It is here, for example, that sugar groups are added to proteins to form the glycoproteins of the cell membrane. The contents of each layer of the Golgi apparatus move toward the outermost layer in a slow but continuous process of vesicular formation and fusion. The contents of the outermost layer of the Golgi apparatus are ready for final packaging into vesicles.

Vesicles and Vacuoles

Vesicles are small spheres containing the products of the ER and the Golgi apparatus. You may think of them as the cell's storage and shipping containers. They are formed from a portion of ER or Golgi apparatus membrane, and are composed essentially of a lipid bilayer surrounding a fluid-filled space that contains a single product. Each vesicle contains only one product out of the thousands of products made by the ER and Golgi apparatus. If the contents of the vesicles are not immediately needed, they simply remain in the cell cytoplasm, like boxes in a warehouse awaiting shipment.

Vesicles containing products for export from the cell eventually move to the cell's surface, where they fuse with the cell membrane, releasing their contents outside of the cell. These vesicles are called *secretory vesicles* because the products are secreted by the cell. Other vesicles, called *storage vesicles*, contain products that are for the cell's own use; these vesicles generally are destined for the ER, the Golgi apparatus, or the cell

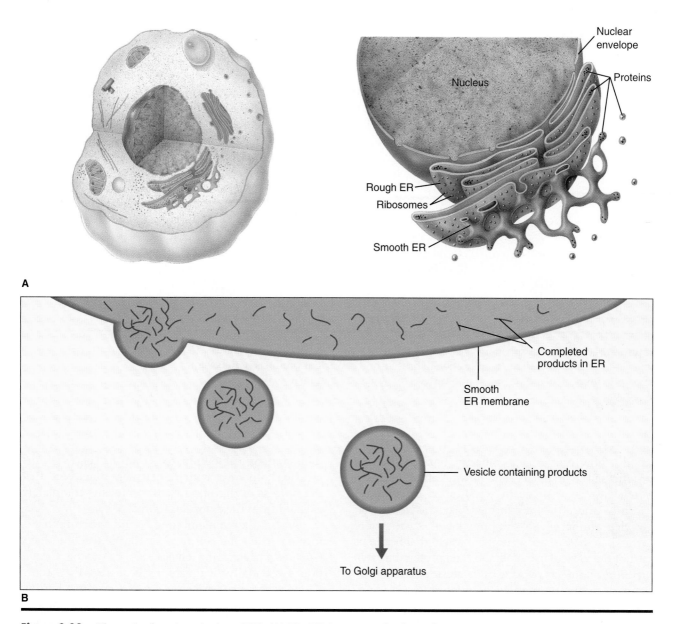

A

B

Figure 2.30 The endoplasmic reticulum (ER). (A) The ER is a network of membranes that are continuous with the nuclear membrane. The rough ER is dotted with ribosomes, the sites of protein synthesis. The smooth ER is devoid of ribosomes. (B) Vesicles containing the products of the ER are formed by pinching off portions of the smooth ER membrane.

membrane. Eventually, both secretory and storage vesicles migrate to their destination, where they fuse with the cell membrane or the organelle membrane (see Figure 2.30). In the case of storage vesicles, the important product may be an integral part of the vesicular wall itself. For example, certain cells of the kidneys have storage vesicles containing the specialized proteins that are the membrane's water channels.

When cell membrane water permeability needs to be increased, these pre-manufactured water channels can be quickly inserted into the cell membrane. This is why your kidneys can change the rate of urine formation so quickly, within a matter of minutes rather than hours. Notice that the vesicular membrane itself becomes part of the membrane with which it fuses. Nothing is wasted, not even the shipping container!

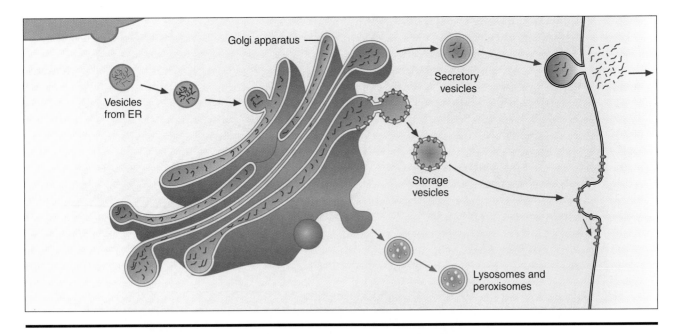

Figure 2.31 The Golgi apparatus. Vesicles from the ER fuse with the Golgi apparatus, where their contents undergo modification to final form. The various products of the Golgi apparatus are sorted and packaged into vesicles for storage or shipment to their final destinations.

But how do vesicles come to contain only one product in the first place, and how do all of these different vesicles manage to get shipped to the right "address" once they are formed? The answer is that vesicles are given a "shipping label" as they are formed, in the form of a specific protein that becomes part of the vesicular membrane. The protein has two functions; it directs packing of the vesicle so that only the right product is put into it, and it serves as an address label so that the final vesicle is sent only to the correct address. The concept of how this protein label might direct the packing and shipping of vesicles is shown in Figure 2.32.

Not all vesicles are used exclusively for storage. Certain vesicles, called **lysosomes** and **peroxisomes,** are small membrane-bound packages of powerful digestive and oxidative enzymes, respectively. Think of them as the cell's waste and toxic chemical treatment plants. The chemicals they contain are so powerful they must be kept in membrane-bound structures to avoid damaging the other components of the cell. Lysosomes (from the Greek words *lysis,* meaning dissolution, and *soma,* or body) actually fuse with and then digest entire objects that need to be removed, such as bacteria and old or damaged organelles. When their digestive task is complete, they become residual bodies, analogous to small bags of compacted waste. Residual bodies can be stored in the cell, but usually their contents are eliminated from the cell when the residual body fuses with the cell membrane. Note that the process of digestion by lysosomes is completely isolated from the cytosol (Figure 2.33).

A **vacuole** is a membrane-bound structure that is formed when the cell ingests a large foreign object (such as a bacterium). The process of vacuole formation is essentially the reverse of the process that occurs when a secretory vesicle secretes its product. During vacuole formation, the object becomes surrounded by cell membrane as it enters the cell, effectively isolating it from the cytosol. Ultimately it fuses with a lysosome and is digested.

Peroxisomes, the other type of waste-processing vesicles, contain powerful oxidative enzymes that could damage the cell if allowed access to the cytosol. The enzymes in peroxisomes destroy various chemical toxic wastes produced in the cell, perhaps the most important of which is hydrogen peroxide (H_2O_2). They also degrade compounds that have entered the cell from outside, such as alcohol. The process of detoxification of compounds by peroxisomes occurs entirely within the peroxisome (Figure 2.34).

Other Cell Inclusions

All of the vesicles described above are *membrane-bound* structures. The substances within them must

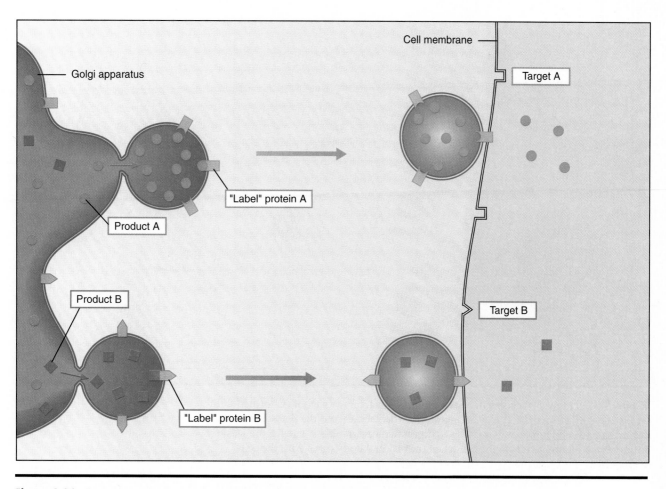

Figure 2.32 The concept of proteins as "address labels." As the vesicle forms at the membrane of the Golgi apparatus, the "label" protein directs the proper Golgi product into the vesicle. The label also directs the vesicle to the proper destination.

either be kept contained because they might damage the cell, or are simply being held (boxed and stored) until they are needed. But the cell can stockpile some materials in the cytoplasm without enclosing them in membranes. The raw materials for energy production, for example, are stored in the cell either as large droplets of lipids or as deposits of glycogen (the storage form of sugar). Such lipid droplets and glycogen granules, unbounded by membranes, are more analogous to large piles of coal on the ground next to a power plant than to boxed finished products. Our so-called fat cells are so specialized for the storage of lipids that most of their volume is actually occupied by large droplets of stored lipids. (Many of us try to reduce the amount of stored fat in our bodies by diet and exercise. You might be interested to learn that dieting and exercise reduce the volume of stored fat within the fat cells, but generally does not reduce the *number* of cells available for storing fat again should you stop dieting or exercising). In contrast to fat cells, muscle

cells generally store energy as glycogen granules rather than as fat.

Cell Growth and Division

From the moment your father's sperm penetrated your mother's egg, your particular set of genetic material (your DNA) was completely determined, and all of it was contained in a single cell, the fertilized egg. Today your adult body is composed of countless billions of individual cells. Obviously, that single cell must have reproduced to form other cells.

Single cells reproduce by dividing roughly in half. Each of the two new cells, called daughter cells, receives half of the cytosol and organelles. The daughter cells then grow, manufacturing more organelles, cell membrane, and other components until they are

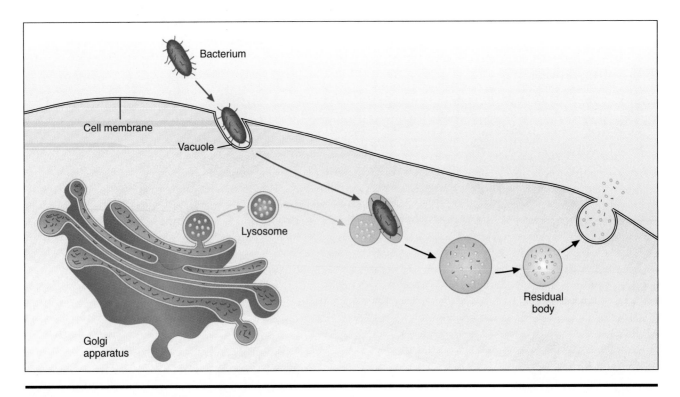

Figure 2.33 How a lysosome digests a foreign object. First a vacuole is formed as the object is engulfed by the cell. The membranes of the vacuole and the lysosome then fuse, releasing powerful lysosomal enzymes to digest the foreign object. Finally, the waste products are eliminated from the cell when the residual body fuses with the cell membrane.

again roughly the size of the original cell. Thus, the original single cell becomes two, two becomes four, four becomes eight, and so on.

An important feature of cell division is that the genetic material (DNA) is copied before the cell divides, so that each cell has a complete set of the original genetic material at the time of division. The process of the division of a cell into two daughter cells with identical sets of DNA is called **mitosis.** A more complete description of mitosis is found in Chapter 15.

If all cells were to continue to divide at the rate of the first several cells (one cell becomes two, two cells become four, four cells become eight, and so on), there would be no end to our growth. Because this doesn't happen, at some point the rate of cell division must begin to decline. The rate of decline is not uniform, however; some cells continue to divide throughout life, whereas others stop dividing altogether. Most notable among those that stop dividing are nerve cells; this may be part of the explanation for why one tends to *lose* cognitive ability later in life, rather than gain it. In contrast, the cells lining the intestinal tract and those forming the skin continue to divide, so that there is a constant replacement of dead cells with new cells.

Implicit in the finding that some cells continue to divide whereas others slow down or cease dividing altogether is that *cell growth and reproduction must be regulated.* Although these processes are not yet fully understood, there are some things we do know. First, cell growth and division are regulated by the genetic material in the cell. That is because the DNA ultimately directs the synthesis of all proteins, and proteins, in turn, serve as structural units and enzymes that determine the rate of most of life's chemical processes. Second, some genes **inhibit** processes involved in cell growth and division, whereas others have specific stimulatory roles. Cell growth and reproduction are the result of the action of many genes, not just one.

Cell Differentiation and Embryonic Development

Consider again the first cell formed by the union of sperm and egg. As that first cell became two, then

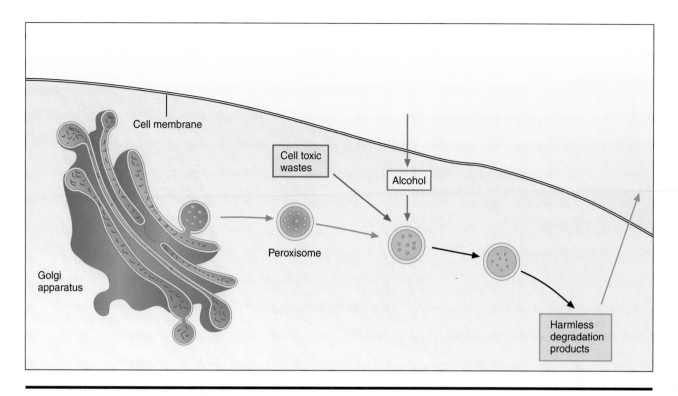

Figure 2.34 How peroxisomes degrade toxic wastes.

four, and so on, each daughter cell was essentially identical to its parent. But the cells of our adult bodies are *not* all alike. This is one of the great mysteries in developmental biology; how is it that a small ball of cells differentiates into a very complex adult organism with many different cell types, when *all* of the cells share exactly the same set of DNA?

At some time during early development, certain cells begin to **differentiate;** that is, they begin to take on very different forms and functions. What causes this differentiation? How is it decided that one or more cells will begin to change so that a head, a spinal cord, and backbone develop from the previously formless mass of cells? The answers are not yet fully understood, but again, we do have some clues. First, only a very small fraction of the genes that make up the DNA are actually being expressed (or "read") at any one time in any given cell. Second, differentiation occurs because precisely which genes are being expressed by a cell can change over time.

We do not yet know all of the details concerning how gene expression is regulated; further research will undoubtedly continue to improve our understanding. We do know, however, that a change in gene expression results in a change in the specific RNA molecules produced by the cell, thereby directing the cell to produce very specific proteins. This in turn alters the course of the cell's form and function because proteins serve as structural units and as the enzymes that

direct so much of the cell's biochemical machinery. Apparently, very early during embryonic differentiation, the addition of only a single protein can have profound effects on the course of development and perhaps on the course of gene expression in subsequent cells. Researchers have found, for example, that in the frog embryo a single protein appears to signal the embryo to start developing the head. Later in development, changes in gene expression result in even finer tuning of the developmental process, determining which muscle cells in the heart become smooth muscle cells (the cells of the heart's blood vessels) and which become heart (cardiac) muscle cells. The student who is interested in the regulation of gene expression might wish to take a course in genetics or molecular biology.

Levels of Organization in Multicellular Organisms

The final outcome of cell division and cell differentiation is the complex multicellular organism that we know as the human body. Indeed, the ways in which cells are organized together makes sense based on the functions of the individual cells. For example, you will learn that muscle cells are specialized to

HIGHLIGHT 2.5

Cancer

Cancer is second only to heart disease as the leading cause of death in this country. It is a complex disease that has proven difficult to prevent or treat. What we do know is that cancer represents a *failure to control cell division properly.*

In all living organisms it is important that cells have the ability to divide, not only so that the immature organism may grow, but also so damaged or worn out cells can be replaced periodically. The normal rate of cell division and differentiation is controlled by numerous different genes, each of which affects a stimulatory or inhibitory process in the regulation of cell growth and division. If some of these regulatory genes become damaged, then the ability to control cell division may be lost and cell division may proceed unchecked, ultimately interfering with normal organ function. You may liken the situation to an old automobile in which the brakes have failed, the accelerator pedal is beginning to stick on occasion, the steering mechanism is sloppy, and the tires are completely bald. One of these defects alone might not be enough to cause a problem, but together they equal disaster. This is what makes cancer such a complex disease. Because cell division is controlled by multiple genes, no one single genetic defect can be identified that "causes" cancer all by itself.

A change in the DNA sequence is known as a **mutation.** The list of agents that can cause mutations and that therefore can cause cancer includes certain viruses, numerous forms of radiation (including UV [ultraviolet] light of sunlight), benzene (an organic solvent), and certain chemicals found in tobacco smoke, to name just a few. However, because the genetic damage must be fairly severe, it may take years of exposure to carcinogens before cancer occurs. Some mutations occur during DNA replication just by random chance alone, although as mentioned before, they are rather rare.

The fact that multiple genes control cell division and that damage is cumulative also explains why certain forms of cancer (and other diseases) tend to run in families even though seemingly a few family members may get the disease. If your ancestors received only minor genetic damage to the line of cells that produced sperm or eggs, they would have passed these defects on to you without any signs of the cancer itself (analogous to a fairly new car with nearly bald tires). But if other defects are later added to those you inherited, cancer may occur. This is why it is in our best interest to reduce our exposure to risk factors in those areas in which our ancestors have a predisposition to disease, including cancer.

---

contract; when this occurs, they are performing work. But no single muscle cell can lift a 100-lb barbell; it takes countless cells working together to do so. In this section, we describe the levels of organization in a complex organism. An overview, which includes the two levels of organization we have already discussed (chemical and cellular), is shown in Figure 2.35.

Tissues

Tissues are groups of cells with similar (though not necessarily identical) functions. We can generally categorize all cells as falling into one of four broad types of tissues: muscle, connective, nervous, and epithelial tissues. Several tissue types may form an organ.

Muscle Tissue Muscle tissue is composed of cells that are specialized for contraction. Chapter 6 deals with muscle tissue in more detail. At this point, you need only recognize that muscle tissue is further categorized into three types: skeletal muscle, responsible for movement of the skeleton; cardiac muscle, responsible for the pumping of blood by the heart; and smooth muscle, which surrounds hollow organs and

tubes such as the blood vessels, the gastrointestinal tract, and the bladder.

Connective Tissue Connective tissue supports and connects various other tissues and organs. It represents a broad group that includes bones, tendons, and ligaments (also discussed in Chapter 6), as well as the tissue that connects soft organs to each other. A distinguishing feature of all of these subcategories of connective tissue is that the cells synthesize and release into the spaces between them certain molecules that give the connective tissue its strength. You will learn, for example, that bone is composed mostly of mineral deposits surrounding a few living connective tissue cells.

Paradoxically, blood cells are also categorized as connective tissue. Although this may seem strange to you at first, the explanation is that blood cells originate from progenitor cells located in the marrow of bones.

Nervous Tissue Nervous tissue is composed of cells specialized for generating and transmitting electrical impulses, thereby creating an important communication network for the organism. Nervous tissue is found in the brain and the spinal cord, as well as in the nerves that transmit information to and from the

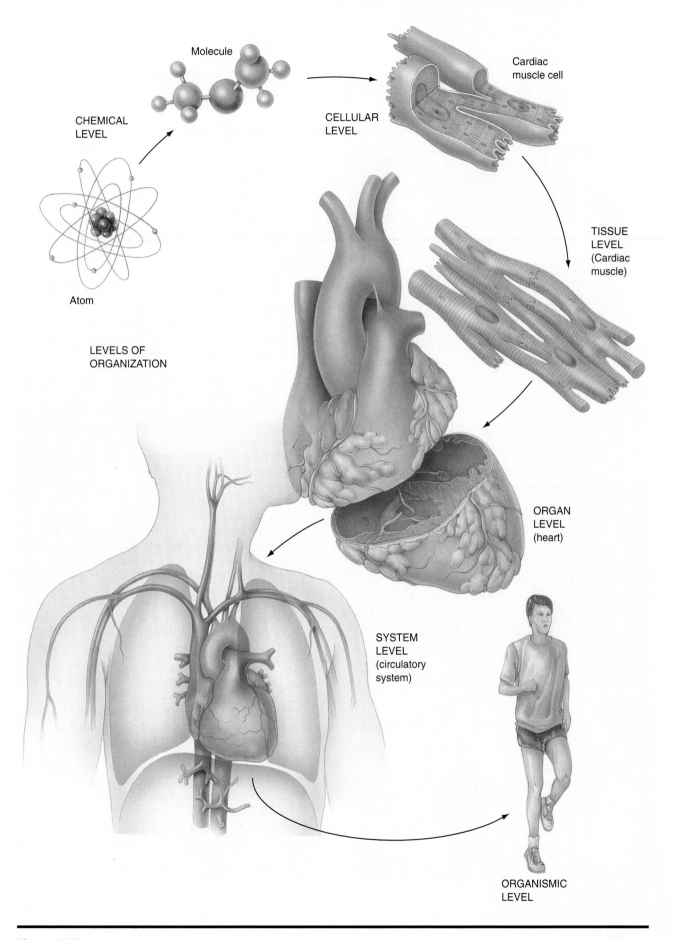

Molecule

CHEMICAL
LEVEL

CELLULAR
LEVEL

Cardiac
muscle cell

Atom

TISSUE
LEVEL
(Cardiac
muscle)

LEVELS OF
ORGANIZATION

ORGAN
LEVEL
(heart)

SYSTEM
LEVEL
(circulatory
system)

ORGANISMIC
LEVEL

Figure 2.35   Levels of organization in a higher organism.

HeLa Cells: 40 Years of Cell Research

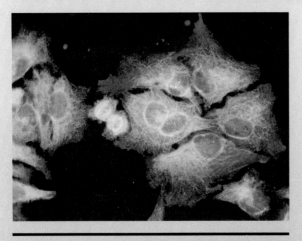

Immunofluorescent light micrograph (×400) of HeLa cells in culture.

In the early 1950s, researchers at Johns Hopkins University led by Dr. G.O. Gey obtained some living cancer cells from a 31-year-old patient with cancer of the cervix, in the hope of growing the cells *ex vivo* (outside the body) so that they could study cancer. Using tissue culture techniques that were still rather new at the time, Dr. Gey and his associates were able to keep these cells and their progeny alive in the laboratory, thereby establishing the very first culture of human cells. Eventually he made them available to other researchers. These cultured human cancer cells have proven invaluable for our understanding not only of the mechanisms underlying the development of cancer, but also of subjects as basic as cell nutrition, viral growth, protein synthesis, and drug effects on cells. They are called *HeLa* cells in honor of *He*nrietta *La*cks, the patient from whom they were originally obtained

(the pseudonym Helen Lane was initially used to protect her identity, but today it seems appropriate to honor her by using her true name). Ms. Lacks died a few years after the discovery of her cancer, but her cells live on. Indeed, HeLa cells have become so widely used in human cell research that there are over 1000 references to HeLa cells in the medical scientific literature each year.

The story of HeLa cells illustrates that scientific knowledge often advances slowly, as a result of years or even decades of work by hundreds of scientists around the world. As scientists make progress, they share their findings with others by publishing their results in the scientific literature so that the latest information eventually becomes common knowledge. It is not unusual that they sometimes disagree on the meaning or interpretation of some of the more recent discoveries, at least at first. These apparent disagreements give scientists the opportunity to design and test new hypotheses to determine why they initially disagreed. Ultimately the process leads scientists toward the truth, but scientists can never be absolutely certain that they have truly arrived at the truth. This may seem like a minor distinction, but it is an important one when we consider how science is brought to our attention by the media. By giving wide publicity to a recent scientific discovery before it has been reexamined and confirmed, the news media may inadvertently make it appear as if we have finally uncovered the whole truth. This is not to say that new scientific discoveries should not be made available as soon as possible for fear of misinterpretation, but only that we all (scientists and nonscientists alike) need to use our judgment before we accept each new public announcement about science. Good science, like the many advances made with HeLa cells, will stand the test of time.

various organs throughout the body. In Chapter 3, you will learn how the cells that make up nervous tissue generate and send electrical impulses.

Epithelial Tissue Epithelial tissue is a rather diverse category of groups of cells that are joined together to form sheets of tissues that line or cover various organs. Examples are the layers of cells that line the inside of the stomach, the intestines, and the lungs; the cells that cover the outside of our bodies (our skin); and the cells that make up the hollow tubules in the kidney. Many of the cells that form epithelial tissues are highly specialized for transporting materials across the entire cell layer, such as the transfer of nutrients from the intestines into the bloodstream and of waste products from the bloodstream into the urine.

Epithelial tissues also form glands, which are tissues specialized for secretion of substances into either the blood or a hollow organ or duct. The subject of glands and their secretions is covered in more detail in future chapters.

Organs

Organs are composed of two or more tissue types joined together that perform a specific function or functions. For example, the heart is an organ comprised primarily of cardiac muscle tissue, but it also contains some smooth muscle tissue (forming the blood vessels to the heart), nervous tissue (the electrical conducting system of the heart), and some connective tissue (most notably in the heart valves). The specific function of the heart is to pump blood. Some organs have several specific functions; for example, the kidneys remove wastes and contribute to the control of blood pressure.

Organ Systems

Organ systems are groups of organs that together serve a broad function important to the survival of the whole organism. A good example is the organ system responsible for the digestion of food. The digestive system includes the mouth, the pharynx (throat), the esophagus, the stomach, the intestines, and even the liver and the pancreas. Organs that make up an organ system often must interact in some coordinated way to accomplish their function.

In the final analysis, the higher organism that we know as a human being consists of billions of cells, all linked together in several levels of organization into one amazing living unit. How does it all work? That is what the rest of this book is all about.

Summary

Living cells are composed of nonliving units of matter. The smallest functional unit of matter is an atom. Atoms are joined together to form molecules by several types of chemical bonds, including covalent or electron-sharing bonds (the strongest), ionic bonds (moderate strength), and hydrogen bonds (the weakest). An acid is a compound that can donate a hydrogen ion to an aqueous solution, whereas a base is a compound that can accept a hydrogen ion. Thus, acids and bases tend to neutralize each other in aqueous solutions. The hydrogen ion concentration in biologic fluids is so low that it is generally expressed by a negative logarithmic scale called the pH scale, in which a pH of 7 is considered a neutral solution.

Cells use the three classes of organic molecules (carbohydrates, lipids, and proteins) as building blocks for cellular structures and as sources of energy. Carbohydrates are chains of carbon atoms with hydrogen and oxygen atoms attached to the carbons in the same proportion as they are found in water (2:1). Lipids are a backbone of carbon with only hydrogens attached; a hallmark of lipids is that they are relatively insoluble in water. Proteins consist of chains of amino acids linked together. Proteins in particular may have very complex shapes that can be altered by nearby charged molecules. Because proteins can change shape so easily, they have many different kinds of functions within the cell. Proteins are the receptors, transport molecules, channels, and gates of the cell membrane, as well as the contractile elements of muscle cells. In addition, the enzymes responsible for setting the rates of specific chemical reactions in the cell are also proteins.

The cell membrane is composed of cholesterol and proteins embedded in a bilayer of phospholipids. Lipid-soluble substances easily cross the lipid bilayer by diffusion, but charged or polar molecules cannot. Proteins serve as channels and active transport sites for the movement of many of these charged and polar substances, and also as receptor sites for transmission of information.

The nucleus of the cell contains all of the genetic material in the form of DNA. The genetic code in DNA contains the recipes for all of the proteins, and thus, it ultimately directs all of the cell's activities. Because DNA is restricted from leaving the nucleus, portions of the DNA code are first transcribed into smaller molecules called RNA. It is RNA that leaves the nucleus and is responsible for the actual synthesis of proteins.

Numerous other identifiable structures each have their own specific function in the cell. A cytoskeleton of flexible fibers helps to maintain the cell's shape and anchors the other organelles. Centrioles and spindle fibers are short tubular elements involved in cell division. Mitochondria are the cell's energy factories, using raw energy supplies to produce the high-energy molecule ATP. Ribosomes are the site of protein synthesis. The endoplasmic reticulum (ER) and the Golgi apparatus, both appearing as somewhat flattened stacks of connected membrane, are the cell's factories; it is here that the proteins made by the ribosomes are modified and most of the other molecules are produced.

Vesicles perform several functions in a cell. Some vesicles store and transport the products of the Golgi apparatus. Other vesicles, called lysosomes and peroxisomes, store powerful chemicals that digest foreign objects such as bacteria and remove toxic wastes. Many cells also contain stockpiles of raw energy in the form of lipid droplets or glycogen granules.

Early in development, all cells divide and grow, and many continue to do so throughout life. Cell division and growth are regulated by certain genes. Cells also differentiate; that is, they begin to take on a variety of shapes and functions. Even though all cells share the exact same genetic code, differentiation can occur because the portion of the genetic code that is actually being read, or used, can change.

Multicellular organisms such as human beings are composed of billions of cells, all working together in a coordinated fashion. Cells with a similar function form a tissue. The four broad types of tissues are muscle, connective, nervous, and epithelial. An organ is made up of several tissues joined to perform a specific function, and an organ system is a group of organs that serve a broad survival function.

Questions

Conceptual and Factual Review

1. What is an atom? What two types of particles make up the atomic nucleus? (p. 10)
2. What is a molecule? What is an ion? (p. 10, p. 11)
3. What are the three principal types of chemical bonds that hold atoms, molecules or ions together? Which is the strongest and which is the weakest? (pp. 10–12)
4. What is the name of the scale used to describe the hydrogen ion concentration of a solution? (p. 13)
5. Define acidic and basic (alkaline) solutions. (p. 14)
6. What are the three principal classes of organic molecules (molecules containing carbon atoms)? (p. 14)
7. Describe the processes of dehydration synthesis and hydrolysis. (p. 14)
8. Why are lipids relatively insoluble in water? (p. 14)
9. What special type of lipid makes up the bulk of the cell membrane, and what are its special properties? (p. 15, p. 16)
10. What are proteins composed of? (p. 17)
11. What functions do proteins serve in the cell membrane? (p. 22)
12. What is diffusion? What is the special term used for the diffusion of water? (p. 24, p. 25)
13. Why can oxygen diffuse right through the phospholipid bilayer of the cell membrane, whereas sodium ions cannot? (p. 25)
14. How does water cross the cell membrane? (p. 25)

15. What distinguishes active transport from facilitated diffusion? (p. 28)
16. What is the difference between primary and secondary active transport? (p. 28, p. 29)
17. Describe how information can be transmitted across the cell membrane, even in the absence of molecular movement across the cell membrane. (p. 29, p. 30)
18. What is DNA? (p. 31)
19. What determines the exact genetic code carried in the DNA? (p. 31)
20. Describe DNA replication. (p. 31)
21. What is a gene? (p. 34)
22. What is the purpose of copying the information contained in a gene into the form of RNA? (p. 34)
23. What is the function of mitochondria? (p. 36, p. 37)
24. What is ATP and what is its function in the cell? (p. 37)
25. Where are proteins made in the cell? (p. 37)
26. What is the function of the endoplasmic reticulum? (p. 38)
27. How do materials get from the endoplasmic reticulum to the Golgi apparatus? (p. 38)
28. Where are secretory and storage vesicles formed, and how are they addressed so that they are shipped to the correct destination? (pp. 38–40)

29. What is a peroxisome? (p. 40)

30. What do we mean when we say that cells differentiate in a multicellular organism? In general terms, how does differentiation come about? (p. 42, p. 43)

31. Distinguish between a tissue and an organ. (pp. 44–47)

Applying What You Know

1. What do you think would happen if you placed an animal cell (such as a red blood cell) in pure water? Explain.

2. Why do you suppose that glucose and amino acids cannot diffuse directly through the phospholipid bilayer of the cell membrane, as does oxygen?

3. Explain why agents that are capable of damaging a cell's DNA, such as radiation or cancer-causing chemicals, are generally more dangerous to the healthy development of a fetus than they are to the health of an adult.

4. Discuss the possible ways that a cell might get rid of waste products.

5. What might be the advantages and disadvantages to a cell of being part of a multicellular organism, as opposed to remaining as a single cell?

Chapter 3
Cell Communication

Objectives

By the time you complete this chapter you should be able to

◆ Appreciate the two methods (electrical and chemical) that cells use to communicate with each other.

◆ Define diffusion potential and understand how it is generated.

◆ Understand that charged particles (ions) move from one place to another in response to an electrochemical gradient.

◆ Know what membrane potential is and how it comes about.

◆ Explain how the Na^+-K^+ pump contributes to the control of cell volume.

◆ Describe the anatomy of a neuron.

◆ Understand how a graded potential is generated in an excitable cell.

◆ Explain how an action potential is generated and propagated, and how the way that it is generated differs from that of a graded potential.

◆ Describe synaptic transmission.

◆ List four neurotransmitter substances.

◆ Know that hormones are chemical messengers.

◆ Define *neurohormone* and *paracrine factor*.

Introduction

The human body is composed of countless billions of cells. Although all cells perform many of the basic functions described in the previous chapter, most cells have specialized functions as well. For example, cells in the eye are specialized to receive light input, muscle cells are specialized to contract, certain cells in the bone marrow are specialized to produce red blood cells, and so on. Specialized cell functions benefit the organism as a whole only if cells communicate with each other. For example, the visual cells of the eye are not useful unless they communicate with cells of the brain, the sensory cells responsive to postural changes are useful only if they communicate with postural muscles, and so forth. If you have ever observed a 10-month-old child just at the stage of learning to stand upright and walk, you can appreciate the tremendous coordination of bones and muscles that is required. Such seemingly simple tasks as walking demand accurate and rapid communication between sensory cells, brain cells, and muscle cells.

Not all communication between cells is as rapid as the mere fractions of a second required for maintaining upright posture or for playing basketball. Some forms of communication are much slower, on the order of many seconds to several minutes. For example, recall a situation in which you have just avoided a potentially life-threatening situation, such as a severe auto accident. Perhaps 15–30 seconds later you may have felt weak and shaky, even though the emergency had already passed. What you felt was a phenomenon called an adrenaline rush; it was due to a slow form of communication that takes 15–30 seconds to develop (in the context of cellular communication, times greater than a second can be considered slow).

In addition, the final *effect* of communication between cells may not be fully observed for months or even years, especially if it requires the growth or differentiation of cells, tissues, and organs. An example is the coordination of sexual maturation. Although maturation of reproductive organs evolves through a series of events that starts before an individual is born, reproductive organs do not reach full maturity for another 12–14 years. Heightened sexual desires and development of secondary sexual characteristics, such as facial hair in males and breasts in females, signal sexual maturity. It is no accident that the timing of these events is coordinated quite precisely. In this chapter, we will explore the major principles of how cells communicate with each other, but if you look closely you will find additional examples of cell communication in nearly every chapter that follows.

In complex organisms, communication between cells is accomplished either by electrical currents or by chemicals. Most electrical currents are transmitted by specialized cells called **neurons,** or nerve cells. Electrical signals can be transmitted very rapidly (within a fraction of a second for the distance from your head to your toe, for example); in fact, the ability to maintain one's balance on a ski slope is controlled by electrical currents. In general, information in the form of electrical currents goes only to locations that are "wired," or served by nerve cells. A notable exception is the transmission of electrical current between muscle cells of the heart. The ability of heart muscle cells to transmit electrical currents directly from one cell to the next is, in part, responsible for the heart's ability to pump blood in a coordinated fashion.

Chemical forms of communication are considerably slower than electrical communication. Consequently, chemical forms of communication developed in the course of evolution only where speed was not of the essence. However, chemical communication is ideally suited for situations where communication must be sustained for long periods of time and for communicating with many different tissues and organs at once. Chemical communication is used to regulate growth and sexual maturation, for example (see Chapter 15).

In this chapter we review the basic principles of how electrical and chemical communication systems work at the cellular level. To understand how the electrical communication system works we must first understand how an electrical charge is generated and maintained across the cell membrane of *all* cells. Thus, it may seem as though we are placing most of our emphasis on electrical communication. Do not lose sight of the fact that chemical communication is equally important, even if seemingly more straightforward. The next two chapters cover how electrical (neural) and chemical (endocrine) systems are used to control bodily functions.

Membrane Potential

An important feature of living cells is the difference in electrical charge between the inside and the outside of the cell membrane. This difference in charge is called the **membrane potential** because energy was required to separate these charges; thus, these separated charges have the potential to do work. In this section we will first learn how the physical properties of a membrane can cause a separation of charge (a diffusion potential) to occur, and then we will learn how active transport of ions contributes to the development of the membrane potential of living cells.

Diffusion Potential

In the previous chapter we showed that because of random motion, molecules tend to diffuse across a membrane from an area of high concentration toward an area of lower concentration, provided that the membrane is permeable to the molecule in question. For neutral molecules, diffusion ceases when the concentrations of the particles are equal on each side of the membrane (Figure 3.1(A)). For ions (which carry a net positive or negative charge), the situation is somewhat more complex. Movement of ions is influenced by the *electrical* properties of the two fluids, as well as by differences in concentration. Take the example of two fluids separated by a semipermeable membrane (Figure 3.1(B)). On one side is pure water. On the other side is a solution of Na^+, which can cross the membrane, and an equal number of negatively charged ions, which cannot diffuse. Both solutions are electrically neutral, meaning that they contain an equal number of positive and negative charges. Initially, Na^+ diffuses across the membrane because of a difference in concentration, but as Na^+ diffuses a difference in electrical charge develops. This electrical charge begins to *oppose* further diffusion of positively charged Na^+ ions to the positive side of the membrane, because like charges repel each other and unlike charges attract. The electrical charge that develops across a selectively permeable membrane when ions diffuse is called a **diffusion potential** because it comes about as the result of diffusion.

Because movement of ions is influenced by both concentration and charge, an ion can actually be made to move *against* a concentration gradient if there is sufficient electrical charge across the membrane. This situation is depicted in Figure 3.1(C). At rest the membrane separates two solutions containing equal concentrations of Na^+ and a negatively charged molecule that cannot diffuse. An electrical charge is then introduced across the membrane. Although Na^+ is initially at equal concentrations on both sides of the membrane, the introduction of a charge produces an *electrical* force that causes net movement of Na^+ toward the negatively charged side. The net movement of Na^+ will stop when the concentration of Na^+ becomes suf-

ficiently high on the negatively charged side of the membrane that diffusion to the right due to the concentration difference exactly equals the movement to the left caused by the electrical attraction.

The concentrations and voltages at which chemical and electrical forces exactly counterbalance each other can be calculated, but rely on formulas that are beyond the scope of this text. The important point is that ions move down their **electrochemical gradient,** which takes *both* chemical concentration and electrical charge into account. The principles depicted in Figure 3.1 for movement of positively charged ions also apply to negatively charged ions, which move toward regions of positive charge and away from regions of negative charge.

Membrane Potential

In living cells, the cell membrane separates the fluid inside the cell (the cytosol) from the fluid outside (the extracellular fluid). The cytosol and the extracellular fluid are dilute solutions primarily composed of ions, and ions carry an electrical charge. Compared to the extracellular fluid, the cytosol has slightly fewer positive charges and more negative ones; thus, an electrical charge exists across the membrane. This difference in charge is called the membrane potential. Membrane potentials are expressed with the fluid outside the cell defined as the reference point, having zero electrical potential (Figure 3.2). In most cells the membrane potential is about – 70 millivolts (a millivolt, abbreviated mV, is a thousandth of a volt). The negative sign indicates that the inside of the cell has more negative charges than positive ones. The actual number of ions that are in imbalance is relatively small compared to the total ionic concentrations on both sides of the membrane. Most of the imbalance in charges is found right next to the membrane because the unbalanced opposite charges attract each other; most of the remaining fluid both inside and outside the cell is electrically neutral.

But how does the membrane potential come about, and how is it maintained? In order to understand the membrane potential, we need to focus on three facts:

◆   As mentioned above, the inside of the cell contains more negative charges than the extra-

---

Figure 3.1 (*Opposite*)   Forces governing the diffusion of neutral and charged substances. (A) Diffusion of neutral substances. (B) Diffusion of a small positively charged ion; note that a charge develops across the membrane. (C) Diffusion of a charged ion against a concentration difference, caused by introducing an electrical charge across the membrane.

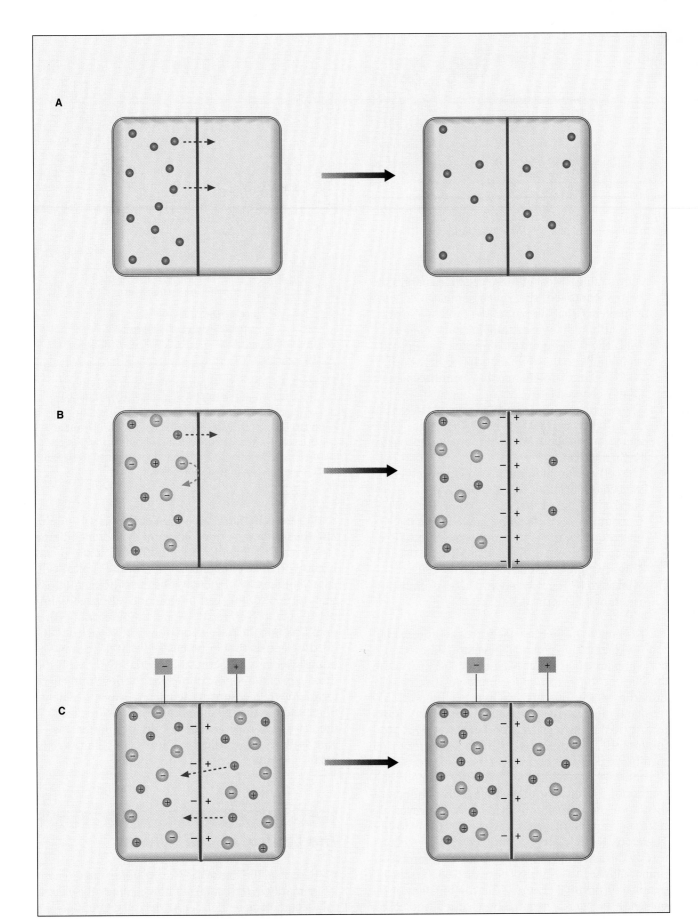

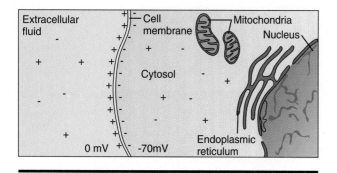

Figure 3.2 Membrane potential. By convention, the electrical potential of the extracellular fluid is zero.

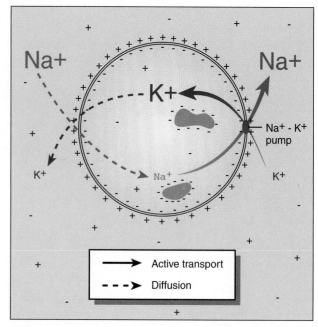

Figure 3.3 The Na⁺-K⁺ pump. The pump effectively excludes Na⁺ from the cell because it actively transports out of the cell any Na⁺ that leaks in by diffusion. In contrast, the inwardly transported K⁺ can diffuse out of the cell readily because the cell membrane is much more permeable to K⁺ than to Na⁺. The red structures represent large, negatively charged molecules.

cellular fluid. Most of these negative charges are permanent (fixed) sites on large molecules, such as proteins, that are part of the protein's structure, although there are other small anions as well. The proteins are too large to cross the cell membrane, so they remain trapped in the cell. A single large protein may have multiple negatively charged sites.

◆ The cell membrane is fairly permeable to K⁺, but only slightly permeable to Na⁺. In fact, the cell membrane is about 50 times more permeable to K⁺ than to Na⁺.

◆ The cell membrane contains transport protein molecules, introduced in Chapter 2, called Na⁺-K⁺ ATPase, or the Na⁺-K⁺ pump for short. The pump uses the energy derived from the breakdown of the high-energy molecule ATP to transport Na⁺ out of the cell and K⁺ into the cell. Three Na⁺ are transported out of the cell for every two K⁺ transported in. These relationships are shown in Figure 3.3.

Keeping in mind these facts, we can now explain how the membrane potential develops. First, because the pump transports three Na⁺ out for every two K⁺ in, its *direct* action is to remove more positive charges from the cell than it adds. In a living cell, a portion of the membrane potential, perhaps about –10 mV, is due directly to this unequal transport of Na⁺ and K⁺. More importantly, the pump contributes *indirectly* to the membrane potential because it raises the intracellular K⁺ concentration and lowers the intracellular Na⁺ concentration. The effect is to produce concentration gradients for the diffusion of these ions. However, the cell membrane is fairly permeable to K⁺ but not to Na⁺. Therefore, the K⁺ that was pumped into the cell can readily diffuse out, but the Na⁺ that was pumped out cannot diffuse in as readily. The overall effect is the net removal of positive charges. The combination of the presence of large, negatively charged molecules

within the cell, the direct action of the Na⁺-K⁺ pump, and the different permeabilities of the cell membrane to Na⁺ and K⁺ together account for the actual membrane potential of about –70 mV in the typical cell.

Another way of viewing the effect of the pump is that it makes the cell membrane effectively impermeable to Na⁺, for the small amount of Na⁺ that does leak into the cell is soon transported back out. Indeed, scientists first hypothesized that the cell membrane was very impermeable to Na⁺. Only with the discovery of the Na⁺-K⁺ pump and the development of techniques to measure the actual one-way movements of ions across the cell membrane did they realize that the cell membrane is actually slightly permeable to Na⁺, with the inward diffusion of Na⁺ being exactly balanced by outward active transport.

The Control of Cell Volume

We described the role of the Na⁺-K⁺ pump in this chapter primarily in the context of how it affects the electrical voltage, or potential, across the cell membrane. We did so because it is necessary for you to understand how the membrane potential is created and maintained before we discuss how some cells use changes in voltage as a way of communicating. Most cells have little need for the membrane potential; red blood cells, for example, survive and thrive with membrane potentials of only about −10 to −20 mV. The primary function of the pump in most cells is to control the total number of solutes within the cell. This in turn is the primary determinant of cell volume. How does the pump help to maintain homeostasis of cell volume?

Recall that the cell membrane is a soft, pliable, selectively permeable membrane. In fact, it is highly permeable to water, relatively permeable to K^+, and considerably less permeable to Na^+. As you have just learned, there are many large protein molecules (serving as enzymes, structural elements, etc.) and other molecules and organelles such as DNA, RNA, mitochondria, ribosomes, and vesicles inside the cell. These cannot cross the cell membrane; therefore, they are restricted to the inside once they are manufactured within the cell. Many of these cellular elements are negatively charged, so they attract positive ions such as Na^+ and K^+. The end result is that in the absence of membrane pumps, cells have a natural tendency toward a much higher total solute concentration inside than outside. Considering that this would lead to a diffusion gradient for the inward diffusion of water and that the cell membrane is highly permeable to water, water would enter the cell. Eventually the cell would swell, burst, and die.

There are at least three ways that cells could handle this potential water overload problem. One would be to create a strong, rigid cell wall that could withstand fluid pressures sufficiently high to oppose the diffusion of water. That is what plant cells do, but not animal cells. The second would be to transport water out of the cell at the same rate it diffuses in. However, as far as we know, there are no active water transport molecules on animal or plant cell membranes. The third solution would be to transport solutes out of the cell (it really wouldn't matter which solute) in order to lower the intracellular solute concentration until it is the same as the total solute concentration outside the cell; then there would be no diffusional force for the entry of water. This is exactly what the Na^+-K^+ pump does; it effectively keeps Na^+ out of the cell, limiting the total number of solute particles in the cell and thereby controlling cell volume. As it does so it generates the membrane potential.

There is another indirect benefit of the Na^+-K^+ pump. Because it pumps Na^+ out of the cell, it indirectly produces the electrochemical gradient for the inward movement of Na^+. Some of this inward movement occurs via facilitated diffusion transport proteins (review Figure 2.19). The energy derived from the facilitated diffusion of Na^+ can be used to transport *another* substance *against* its concentration gradient via the same transport protein; this is an example of secondary active transport (Review Figure 2.21). In effect, the cell uses energy derived from the facilitated diffusion of sodium to do something else important to the cell. This is how glucose is absorbed in the intestines and in the kidneys; it is actively transported into the cells of the intestines and kidneys coupled to the inward diffusion of Na^+.

Electrical Activity of Nerve Cells

Nerve cells (neurons) are called excitable cells because they have evolved the capacity to alter their membrane potential under physiological conditions. Muscle cells are also excitable, and this excitability plays a key role in the development of muscle contraction, as you will see in the next chapter. Neurons have evolved specifically for the purpose of communication. In effect, neurons function as tiny living electrical cables, actually producing electrical impulses and transmitting these impulses over long distances. In order to understand how this is done we need to examine the specialized structural and functional properties of a typical neuron.

Anatomy

A representative neuron might appear as depicted in Figure 3.4. The major portion of a neuron is the **cell body,** which contains the nucleus and other organelles typical of all cells. Projecting from the cell body are extensions of the cell body, known as **dendrites,** and at least one much longer projection, the nerve **axon,** or nerve fiber. It is not always possible to tell a dendrite from an axon just by looking at them; the difference between a dendrite and an axon has more to do with function than form. As you will see shortly, the axon is specialized for transmitting electrical information over long distances, whereas the dendrites are specialized for receiving incoming information from other neurons and transmitting it only short distances, generally to the cell body. The region where the cell body narrows to form the axon is called the **axon hillock.** The axon may be quite long compared to the dimensions of most other single cells; some neurons have axons that are over a meter in length. Side branches, or **axon collaterals,** may extend from the main axon. At the end of the axon and its collaterals are the **axon terminals,** which are regions of the neuron specialized for transmitting information to

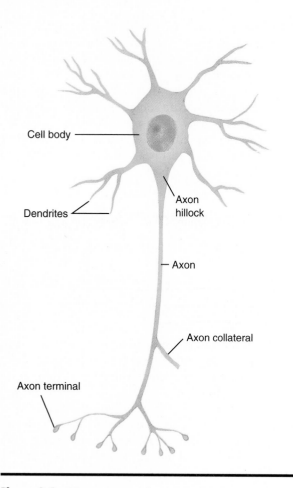

Figure 3.4 The anatomy of a representative neuron.

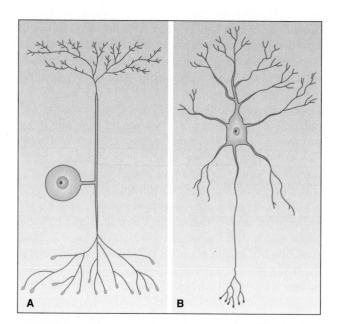

Figure 3.5 Different types of neurons. (A) A sensory neuron with a branched axon. (B) A neuron in the brain with extensive dendritic branching.

another cell. The functions of the parts of a neuron are summarized in Table 3.1 (opposite page).

Not all neurons look like the "typical" cell depicted in Figure 3.4. Although they all share the ability to transmit information electrically, they come in a variety of shapes and sizes depending on their location and function. Some neurons, particularly those specialized for receiving sensory information (such as pain from the skin) appear to have two axons joined to the cell body by a short extension (Figure 3.5). In reality, this is a single axon that sends branches in opposite directions. The action potential in such a sensory neuron is initiated at one end of the axon and travels directly to the other end. Still other neurons, particularly those in the brain, have highly branched dendrites that permit them to receive information from many different neurons at once.

General Function

The transmission of information by a neuron can be divided into three phases. Phase one is the generation of a **graded potential** in the dendrites and the cell body. A graded potential is a small change in the membrane potential in a localized region of a dendrite or the cell body. Its magnitude varies depending on the strength of the signal that generated it. Phase two is the generation of an **action potential** at the axon hillock, which travels down the axon to the axon terminals. An action potential is a large and rapid change in the membrane potential that, once initiated, cannot be stopped, sweeping down the entire length of the axon. Phase three is **synaptic transmission,** the process whereby information provided by the action potential is transmitted from the neuron to another cell.

At this point, you know the basic structure of the cell membrane and the principles of how the membrane potential in a normal cell is generated and maintained. What is different about neurons, such that their membrane potentials can be changed quickly? The answer is that neurons have a lot of gated channels in the cell membrane. Gated channels are defined both by the signal that causes them to open and close (chemically gated or voltage gated) and by the primary ion that they let pass through (Na^+, K^+, or in some cases Ca^{++}). The opening and closing of gated channels for a particular ion alters the cell membrane permeability to that ion, thereby either allowing the ions to diffuse rapidly or restricting their diffusion. Rapid changes in ion movements alter the membrane potential.

Table 3.1 **Summary of the parts of a neuron and their functions.**

Name	Description	Function
Cell body	The main part of the neuron.	Contains the cell's nucleus and organelles. Neurotransmitters are synthesized here, then transported down the axon to the axon terminals.
Dendrites	Finger- or thread-like extensions of the cell body.	Have chemically gated ion channels that respond to certain neurotransmitters with the generation of graded potentials.
Axon hillock	The point where the cell body narrows to become the axon. The first part of the axon.	The first region of the neuron to have voltage-gated ion channels. The action potential is initiated here.
Axon	A long projection from the cell body.	This region propagates the action potential in an all-or-none fashion, from the axon hillock toward the axon terminals. Also transports neurotransmitter from the cell body to the axon terminals.
Axon collaterals	Side branches of an axon.	Function the same as the main axon, providing information to other neurons in the form of action potentials.
Axon terminals	The rounded tip, or terminal, of an axon.	Store neurotransmitter and release it into the synaptic cleft in response to action potentials.

Graded Potentials

The dendrites and the cell body of a typical neuron (but not its axon) have the chemically gated types of ion channels. The binding of a particular chemical to a receptor associated with the channel causes the channel to open, thereby making the membrane more permeable to a particular ion. A typical channel in the dendrite of a neuron would be a chemically gated Na^+ channel, such as that depicted in Figure 3.6. Because there is a substantial electrochemical gradient favoring diffusion into the cell, Na^+ enters through the temporarily opened channels, making the membrane potential less negative in that local region. The magnitude and even the duration of a graded potential can vary depending on how many gated channels are opened and how long they remain open. This, in turn, depends on the quantity and longevity of the chemical substances that bind to the channels. There may be hundreds of these local regions of altered permeability on a single neuron at any given time, each covering only a small local region and lasting only a short time. However, these small potentials can undergo **summation;** that is, they can add onto each other in both space and time. Depending on the strength, location, and timing of the chemical stimulus, regions of the dendrites and the cell body may undergo slight, transitory changes in membrane potential. Graded potentials are short-lived, disappearing over time and with distance from the stimulus. Thus, the membrane potential returns rather quickly to the normal resting membrane potential once the stimulus is removed (the membrane potential in an excitable cell is called the resting membrane potential when the cell is not being excited).

A point of clarification about the measurement of membrane potentials may be helpful. When Na^+ enters the cell and the inside of the cell becomes *less* negative, we say that the membrane potential has been reduced, or that partial **depolarization** has occurred (the membrane has become *less* polarized).

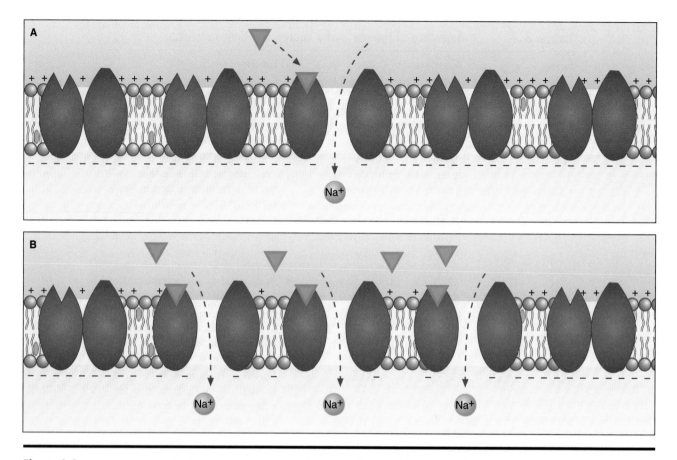

Figure 3.6 How graded potentials are produced. (A) A chemically gated Na+ channel opens when a particular chemical binds to the receptor. This allows Na+ to enter, producing a graded potential in the region of the channel. (B) An increase in the quantity of the chemical causes more channels to open and a greater graded potential.

Conversely, if the membrane potential were to become even more negative than usual, we would say that the cell membrane has become **hyperpolarized.** Finally, return of the membrane potential to the normal resting level is known as **repolarization.** Experimentally, the devices used to record membrane potentials plot a *decrease* in membrane potential (or depolarization) as an upward deflection on the recorded graph, as depicted in Figure 3.7. This figure illustrates a slight graded depolarization, as might be seen with a single localized change in Na+ permeability; the summation that can occur if several such graded depolarizations occur in rapid succession; and a hyperpolarizing event. The hyperpolarization could be caused by activation of a chemically gated K+ channel because increasing K+ permeability would increase the diffusion of K+ out of the cell and thereby make the membrane potential more negative. Notice the speed with which these graded potentials disappear; a single depolarizing event completely dissipates within a fraction of a second. This is because the chemical substances that cause the change in membrane permeability are short-lived; once they are gone the gates close and the membrane permeability quickly returns to normal.

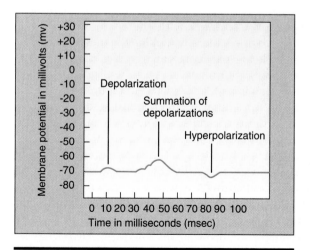

Figure 3.7 Graded potentials in an excitable cell.

Action Potentials

Graded potentials in the region of the dendrites and cell body are local events that are not transmitted over great distances. But something different happens at the axon hillock. If the graded potentials summate sufficiently to produce a depolarization to about −55 mV in the region of the axon hillock, an action potential may be generated that then sweeps down the axon at a constant speed. Once it is initiated it cannot be stopped; that is, it proceeds until it reaches the axon terminal. In more common terms, it is known as a **nerve impulse**. An action potential comes about because the axon hillock and the rest of the axon have *voltage*-gated channels rather than chemically gated ones. Thus, the opening and closing of ion channels is affected by the membrane potential itself. The membrane potential at which an action potential is generated is called the **threshold**. (See Figure 3.8.) −55 mV

The axon hillock is the first place these voltage-gated channels are found (they are not present on the cell body or on the dendrites). Once voltage-gated channels are activated at the axon hillock, the change in membrane potential spreads to adjacent areas of the axon, activating voltage-gated channels there as well. For this reason, the activation of voltage-gated channels tends to be self-propagating as it proceeds down the axon. Note that the propagation of an action potential down the axon is an example of a positive feedback system; the opening of voltage-gated channels causes the membrane potential to change even further, causing the opening of voltage-gated channels in adjacent regions, and so on. The process is terminated only when the action potential reaches the axon terminal and thus has nowhere else to go. Figure 3.9 depicts the regions of the neuron with respect to their differing membrane excitabilities.

Under normal circumstances action potentials start at the axon hillock and spread down the axon to the terminals because the axon hillock is the first region to have voltage-gated channels. However, there is nothing inherently one-way about the axon. For example, if an axon were stimulated electrically in the middle with electrodes, two action potentials would be produced, one traveling in each direction from the point of origin. However, the action potential traveling toward the cell body would die out at the axon hillock because the cell body and dendrites do not have voltage-gated channels.

Getting the action potential started is only the first phase of an action potential, however, for eventually the membrane potential must be returned to its original resting state. Two additional events occur to make this happen. First, the voltage-gated Na^+ channels do not stay open indefinitely even if the membrane were to stay depolarized; they automatically close after a time delay. Second, the axon hillock and axon also contain voltage-gated K^+ channels. These were also activated by depolarization of the membrane to threshold, but they are so slow to open that they are just beginning to open as the Na^+ channels are closing. The combination of closing the Na^+ channels and opening the K^+ channels repolarizes the membrane, because now there is a net movement of positively charged ions out of the cell (Na^+ stops diffusing in and K^+ begins diffusing out). Finally, the K^+ channels close when the membrane potential returns to near the resting level. Figure 3.10 illustrates the phases of the action potential and the accompanying changes in axon membrane permeabilities to Na^+ and K^+.

The number of Na^+ and K^+ ions that cross the membrane with each action potential is so small that a neuron can generate many action potentials in succession without significantly altering the intracellular concentration of Na^+ and K^+. The small gain of Na^+ and loss of K^+ from the cell that may occur after prolonged periods of nerve stimulation are easily reversed by the normal pumping activity of the Na^+-K^+ pump. In other words, the primary role of the Na^+-K^+ pump in a neuron is the same as it is in any cell; to balance the normal small gain of Na^+ and loss of K^+ by the cell, regardless of whether those losses or gains come from passive leaks, facilitated diffusion, or movement through channels during the generation of an action potential.

So far, we have looked at an action potential as it would appear if viewed from one location on the axon as the action potential changes over time at that one

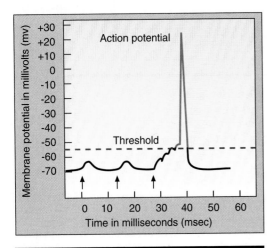

Figure 3.8 An action potential generated at the axon hillock. This action potential was preceded by two graded potentials, but was not initiated until several graded potentials summated (the third arrow) so that the threshold was reached.

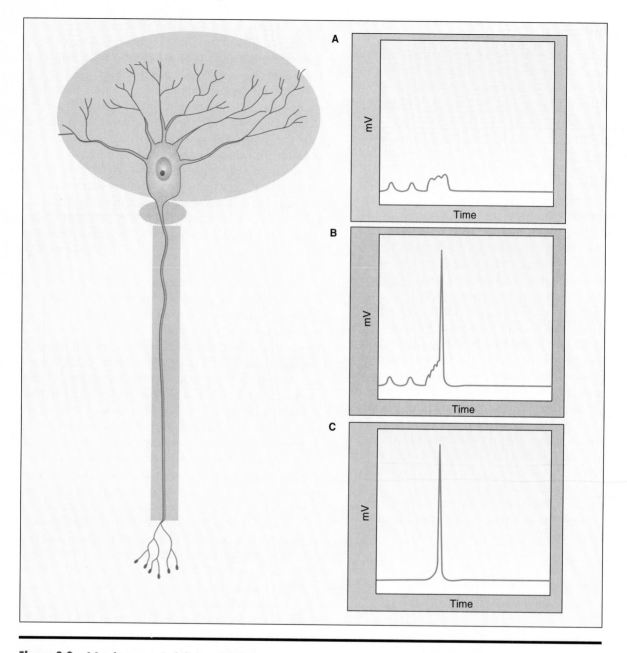

Figure 3.9 Membrane excitabilities of different regions of the neuron. (A) Dendrites and the cell body produce brief, local graded potentials. (B) The axon hillock is the first region to have electrically gated ion channels; here, graded potentials may be converted to an action potential. (C) Self-propagated action potentials proceed down the length of the axon.

location. In actuality, however, the action potential sweeps down the axon, from the hillock to the axon terminals. Action potentials travel at a constant speed and amplitude and under normal circumstances cannot be stopped once they are started. Thus, they are said to be self-propagating, like ripples on a pond (although a ripple is not perfectly self-propagating; it will decline in amplitude over long distances). The constant amplitude and rate of travel is maintained by

diffusion of charged particles in the region of the action potential (Figure 3.11). Because opposite charges attract, the positive charges in the local region of the nerve impulse can spread by diffusion to more negative regions nearby, slightly depolarizing the adjacent region and bringing it closer to threshold. These charges are not crossing the axon membrane, but are simply diffusing to a nearby region either inside or outside the axon. Once this local diffusion causes

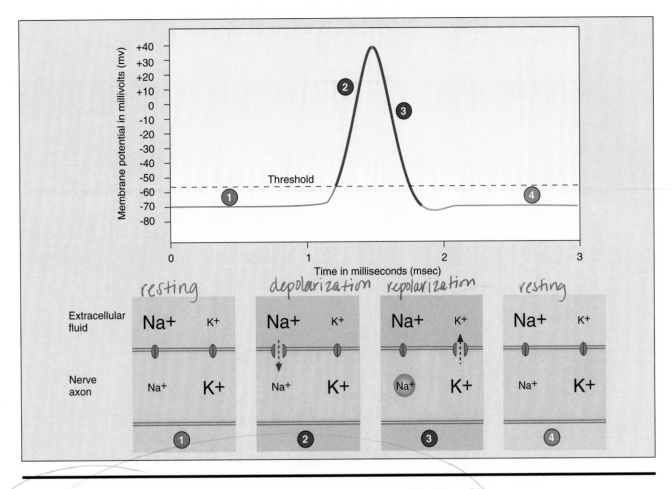

Figure 3.10 The phases of an action potential. The opening and closing of the Na⁺ and K⁺ channels is shown for each phase. (1) The resting membrane potential. (2) The depolarization phase of the action potential, characterized by opening of the Na⁺ channels. (3) The repolarization phase, characterized by closing of the Na⁺ channels and opening of the K⁺ channels. (4) Return to the resting membrane potential after the K⁺ channels close. This action potential has a different shape from that of Figure 3.8 because the time scale has been greatly expanded and because it is not preceded by graded potentials.

threshold to be reached in the nearby region, the Na⁺ channels open and Na⁺ diffuses inward, perpetuating the action potential. This whole process is a continuous one; thus, you can think of the nerve impulse shown in Figure 3.11 as sliding smoothly down the axon toward the left as each adjacent region is brought to threshold by local diffusion of charged particles.

Refractory Period

If a neuron is stimulated weakly, only a few action potentials are generated. A more intense stimulus generates more action potentials per unit of time, and a very large stimulus generates action potentials in rapid succession (Figure 3.12). Nevertheless, each action potential is a separate event, alike in amplitude and shape. You either get an action potential in re-

sponse to the graded potential, or you don't. There are no "wimpy" little action potentials and no giants either, regardless of how large the initial stimulus was. The identical nature of each action potential is known as the **all-or-none rule.**

Why doesn't a larger stimulus generate a larger action potential? When there are a lot of action potentials in a row traveling in an axon, how do they manage to stay at discrete distances from each other instead of running up on top of each other and creating one big action potential pileup? The explanation is that during the time that a segment of axon is producing one action potential, it is incapable of producing another. The period of time during which the axon cannot respond at all is known as the **absolute refractory period** (Figure 3.13). The absolute refractory period comes about because of how the action

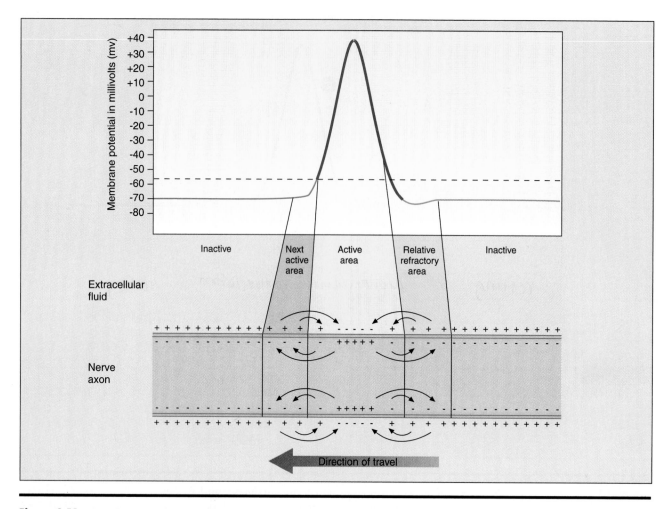

Figure 3.11 Local current flow during an action potential. Changes in membrane potential brought about by the opening and closing of ion channels in the membrane induce charged ions to move toward oppositely charged regions. The arrows indicate the direction of movement of positive charges only; negatively charged ions move in the opposite direction.

potential is created in the first place; a change in voltage causes voltage-gated Na⁺ channels to open, further depolarizing the membrane. But if the channels are *already* open because an action potential is in progress, they cannot open any further and, therefore, they cannot create a second action potential on top of the first. Furthermore, because each action potential results in the opening of essentially all of the Na⁺ channels in the first place, a bigger action potential cannot be created just by the presence of a bigger initial stimulus at the axon hillock. Toward the end of the action potential, when the membrane potential has returned almost to the resting level and the Na⁺ channels are again closed, there is a **relative refractory period** during which another action potential can be initiated, but only if the stimulus is stronger than usual. These refractory periods ensure that each action potential is a discrete event.

Speed of Nerve Impulse Conduction

Although nerve impulses in any given neuron travel at a constant speed under normal physiological conditions, the speed of conduction differs widely among neurons. Conduction velocity depends on the temperature, the diameter of the nerve axon, and also on whether the axon is coated with a sheath of an insulating material called **myelin;** such axons are called **myelinated axons.**

Nerve impulse speed increases if a tissue is warmed above normal temperature and decreases if the tissue is cooled. In a warm-blooded animal internal temperature generally remains fairly constant and thus temperature is not normally an important factor. Experimentally, however, local cooling of a nerve can completely interrupt transmission of action potentials. In humans, local cooling of an injured tissue can reduce

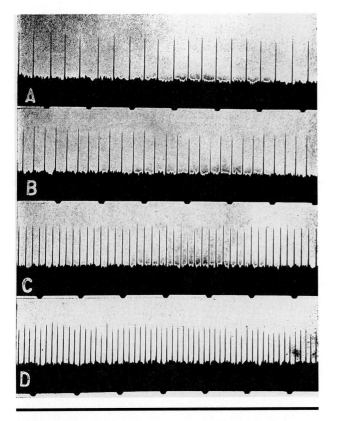

Figure 3.12 Recordings of action potentials from a single sensory neuron at different stimulus intensities (stimulus intensity increases from top to bottom). The time markers at the bottom of each tracing represent 1/5 of a second. These recordings were made in 1934, when such recordings were considered state of the art.

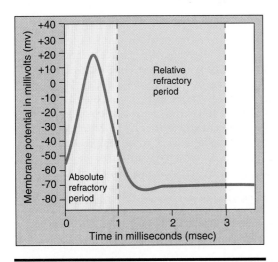

Figure 3.13 Refractory period. During the absolute refractory period a neuron cannot initiate another action potential under any circumstances; during the relative refractory period another action potential can be generated, but it takes a greater than normal stimulus.

pain in part because nerve impulse conduction may be blocked and in part because nerve receptors for pain may be inhibited. This is one of the reasons that it may be helpful to apply an ice-pack to an injury such as a twisted ankle. Although at first it may seem that the ice makes the pain worse (by activating pain receptors near the surface of the skin), if the ice can be held in place long enough the area may become numb. Nerve impulse speed also increases with increasing diameter of the axon. Speeds of conduction range from about 1/2 meter per second in the very thinnest axons up to about 130 meters per second in the thickest axons. In everyday terms, nerve impulse speed varies from about 1 mile per hour, a slow walk, to over 260 miles per hour.

The longest nerve cells in the human body can be over a meter in length, extending from the tip of the toe to the spinal cord. For these cells, the speed of conduction can be of critical importance. Conduction speed increases if the axon is insulated with myelin. This is because myelin does not conduct electric current and thus acts as an insulator similar to the insulation surrounding an electrical wire. Myelin is produced by non-neural cells that surround the axon. The myelin sheath is not continuous, but is interrupted at regular intervals. The points where the sheath is interrupted are called **nodes of Ranvier.** Because myelin prevents the inward flow of current across the axon membrane during an action potential at a nearby node, the only flow of current in the region of the myelin sheath is by local current flow either inside or outside the cell. At the next node, however, this local current flow causes threshold to be reached, generating another action potential, which in turn causes local current to flow beneath the myelin sheath to yet another node, and so on. These principles are illustrated in Figure 3.14. In effect, a new action potential is created at each node. Impulse conduction from node to node in this fashion is called **saltatory conduction** (from the Latin word *saltare*, meaning *to jump*).

Saltatory conduction in myelinated nerves is much faster than conduction of nerve impulses in unmyelinated fibers because conduction by local current flow is much faster than conduction that requires the opening and closing of ion channels. The advantage of myelination of a nerve fiber is that it allows an increased speed of information transmittal without having to increase axon diameter.

Synaptic Transmission

By now you understand how an electrical signal, the nerve impulse, is generated in a neuron and how the impulse is transmitted down the axon to the axon

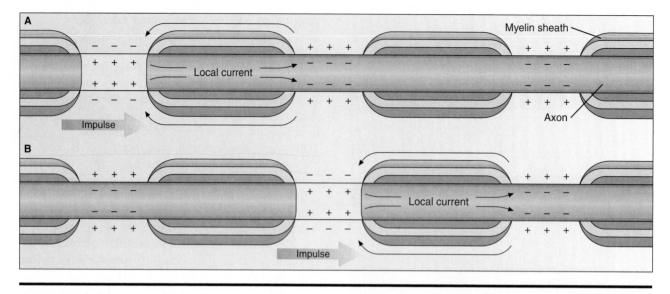

Figure 3.14 Saltatory conduction. A new action potential is generated at each successive node of Ranvier. The mechanism for bringing each node of Ranvier to threshold is the local movement of current between the nodes.

terminals. However, the nerve impulse cannot pass from one neuron to the next because there is no direct connection between the two neurons. How, then, is the information transmitted? Communication between a neuron and another cell is the function of a specialized junction called the **synapse.** The transfer of informa-

tion from a neuron to another cell by chemical means is called synaptic transmission. In this section we will see how an action potential produces the release of a chemical from the axon terminal of a **presynaptic neuron.** The released chemical is called a **neurotransmitter** because of its function as a transmitter of in-

MILESTONE 3.1

The Discovery of How Action Potentials are Generated

Our basic understanding of how action potentials are generated and conducted in nerve cells was not obtained by studying live nerve cells in human beings. Most of our information came from innovative studies conducted in the late 1930s and 1940s using "giant" axons from the squid. These axons initiate rapid propulsion during the squid's escape behavior; thus, it was advantageous for these axons to evolve with a fast conduction velocity. However, squid axons do not have a myelin sheath to increase conduction velocity. Fur-

thermore, squid live in very cold (near freezing) sea water, which would tend to produce a rather slow conduction velocity. However, conduction velocity can be improved by increasing axon diameter. Perhaps for this reason, squid axons evolved to have a diameter approximately 10 times larger than that of the axons of human nerve cells. It was their large size, coupled with the ease of maintenance of the nerves in the laboratory (squid are cold-blooded, hence room-temperature sea water will do), that convinced scientists that they would be a useful experimental animal model for studying the cellular mechanisms underlying the process of nerve conduction. Indeed, giant squid axons were the first neurons in which intracellular electrical potentials were measured. Thus, they were the first axons in which scientists could demonstrate that the depolarization and repolarization phases of an action potential were due to transient, voltage-dependent changes in the permeability of the membrane to Na^+ and K^+, respectively. Two British scientists, A. L. Hodgkin and A. F. Huxley, shared a Nobel prize for this work in 1963.

HIGHLIGHT 3.1

Diseases of Nerve Impulse Conduction

Several human diseases are caused by destruction of the myelin sheath around myelinated nerve axons, leading to disorders of nerve impulse conduction. **Multiple sclerosis (MS)** is a disease typically first seen in adults between the ages of 20 and 40 in which the myelin sheath of nerves in the brain and spinal cord becomes progressively damaged. The disease has a confusing number and variety of symptoms because of the many neural pathways that may be affected, and although it is progressive, it may undergo periods of remission. The disease begins as lesions (areas of damage) of the myelin that eventually become hardened (sclerotic) scar tissue. These sclerotic areas no longer serve as effective insulators of the nerve; thus, impulse conduction velocity may be slowed. In addition, the sclerotic area may damage the

axon itself, further interfering with impulse conduction.

A somewhat similar but rare disease is **amyotrophic lateral sclerosis** (ALS), or Lou Gehrig's disease, so named for the famous baseball player who was afflicted with it. In this disease, the sclerotic areas typically begin in the lateral columns of the spinal cord. The primary symptom of ALS is progressive weakening and wasting of skeletal muscle tissue.

The underlying causes of these diseases are not well understood, and at the moment there are no effective cures. Patients with ALS generally do not live beyond two to five years once they are diagnosed with the disease. The prognosis for those with MS is generally better; roughly 85% are able to lead productive lives.

formation and because it is released by a neuron. The neurotransmitter, in turn, acts on a **postsynaptic neuron.** Typically, the axon terminal of the presynaptic neuron synapses with a dendrite or the cell body of the postsynaptic neuron. Bear in mind, however, that although we are describing how information is passed from one neuron to another, neurons also form

synapses with other cell types, including muscle cells, glands, and other tissues.

The Synapse

Figure 3.15 shows the structure of typical synapse and the events that occur during synaptic transmission.

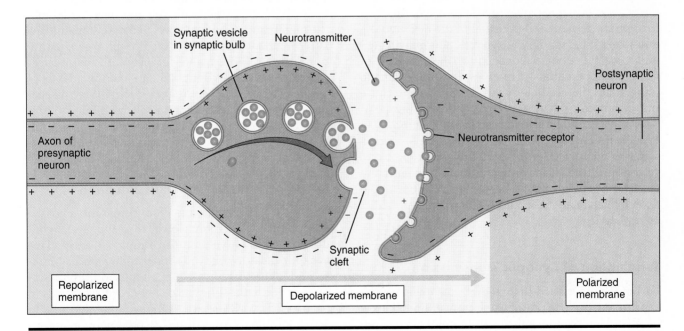

Figure 3.15 Synaptic transmission. The presynaptic neuron releases neurotransmitter into the synaptic cleft in response to an action potential. The neurotransmitter diffuses across the synaptic cleft to bind to specific receptors on the cell membrane of the postsynaptic cell. The binding of neurotransmitter permits the opening of chemically sensitive ion channels in the cell membrane of the postsynaptic neuron, altering the permeability of the cell membrane in that local region and producing a graded potential.

The axon terminal of a presynaptic neuron ends in a rounded structure called the **synaptic bulb.** The synaptic bulb contains packets of chemical neurotransmitter packaged in small membrane-bound vesicles. When an action potential arrives at the synaptic bulb, a series of events is initiated that ultimately result in the movement of these vesicles to the cell membrane. There the vesicles fuse with the cell membrane, releasing their contents into the space between the pre- and postsynaptic cells; this space is known as the **synaptic cleft.** Because the synaptic cleft is rather narrow, some of the neurotransmitter can reach the postsynaptic cell membrane by diffusion. At the postsynaptic cell membrane the neurotransmitter binds to a receptor, which results in changes in membrane permeability to certain ions, such as Na^+. This, in turn, produces a small graded depolarization of the postsynaptic cell membrane in the area of the synapse.

You may have noticed that the production of a graded potential by synaptic transmission is very similar to how we described the production of a graded potential earlier in this chapter. In fact, the chemical substances that we alluded to earlier are the neurotransmitters released by presynaptic neurons. Indeed, we have come full circle; the development of graded potentials at the synapses on the dendrites and cell body of a neuron may, if sufficiently summated at the region of the axon hillock, result in an action potential. The action potential travels down the axon, producing the release of neurotransmitter substance, which in turn produces a graded potential in the next neuron, and so on. Whether the postsynaptic neuron will generate an action potential depends on the sum of all of the graded potentials it receives. It is important to understand that a single synapse is unlikely to cause the generation of an action potential in the postsynaptic neuron. Indeed, a postsynaptic neuron may have hundreds of synapses with dozens of presynaptic neurons (Figure 3.16). Only when the entire sum of positive and negative influences from many synapses produces a net graded depolarization at the axon hillock of about –55 mV will an action potential be initiated.

Termination of Synaptic Transmission

Earlier we mentioned that graded potentials may be short-lived, decaying rather quickly over time. This is because the neurotransmitter substances remain in the synaptic cleft for only a short period of time, and the increase in postsynaptic cell membrane permeability to Na^+ is directly related to the concentration of neurotransmitter. There are three possible fates of neurotransmitter that is released into the synaptic cleft: it may diffuse out of the synaptic cleft into the general

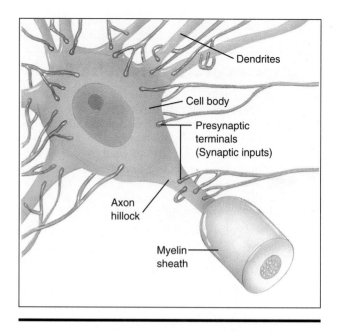

Figure 3.16 Multiple presynaptic inputs to a single nerve cell.

circulation, where it ultimately will be destroyed; it may be degraded by special enzymes produced by either the pre- or postsynaptic cell; or it may be taken back up by the presynaptic cell and repackaged into the membrane-bound vesicles, to be used again. This latter method of removal, called re-uptake, is an important one. Indeed, interference with neurotransmitter re-uptake accounts for most of the actions of cocaine, as we shall see. Together, these three mechanisms of neurotransmitter removal ensure that a stimulus can be terminated nearly as quickly as it can be initiated.

Speed of Synaptic Transmission

The rate of transmission of information across a synapse is considerably slower than the rate of conduction of a nerve impulse down an axon. This synaptic delay, on the order of ½ millisecond (a millisecond, abbreviated msec, is a thousandth of a second), is due to the time it takes for neurotransmitter to be released, to diffuse across the synaptic cleft, and to activate the postsynaptic cell to become more permeable to Na^+, as well as the time required for Na^+ to enter the cell and produce a graded potential and then a nerve impulse. Given these steps, even ½₀₀₀ of a second seems relatively fast! Nevertheless, it is considerably slower than the speed of conduction of action potentials. If an action potential is analogous to a car traveling on a major highway, then the synaptic delay would be analogous to stopping at a toll booth to pay a toll. Because this delay is fairly consistent for most synapses, research scientists have used the measurement of the time it takes between applying a stimulus and recording an action potential at another location to determine how many synapses may be present in that pathway.

Neurotransmitter Substances and Receptors

The human body actually uses several dozen different neurotransmitter substances, each with its own particular mode of action and effect on the postsynaptic cell. As mentioned earlier, most neurotransmitters produce a depolarizing graded potential, increasing the chance that the threshold will be passed and that an action potential will be generated by the postsynaptic neuron or cell. Others produce hyperpolarizing graded potentials in the postsynaptic cell, reducing the chance that the threshold for an action potential will be reached. In general, most neurons produce primarily one neurotransmitter substance at their axon terminals.

Some neurotransmitter substances exert an excitatory influence on one postsynaptic cell type and an inhibitory influence on another. They can do this because different receptor subtypes can exist for a single chemical substance (in this case the neurotransmitter). Activation of different receptor subtypes can produce very different postsynaptic events, a subject beyond the scope of this book.

Four of the most common neurotransmitter substances are **acetylcholine** (ACh), **norepinephrine** (NE), **epinephrine** (E), and **dopamine** (DA). Acetylcholine is most commonly found outside the brain. It is the neurotransmitter at the junction between a neuron and a skeletal muscle cell, where it stimulates the muscle to contract. It also inhibits the rate at which the heart beats. ACh is an example of one neurotransmitter producing both stimulatory and inhibitory actions, depending on the type of receptor it activates. Norepinephrine, epinephrine, and dopamine are

HIGHLIGHT 3.2

Cocaine

Cocaine and its more potent form, crack cocaine, are among the most abused drugs in the United States today. Cocaine has a number of different actions, but one of the most important is that it stimulates the area of the brain (the limbic system) responsible for pleasant and unpleasant feelings such as rage, pain, sorrow, and especially pleasure. How does it do this?

As you have learned, one of the ways in which synaptic transmission between neurons is terminated is by re-uptake of the neurotransmitter into the presynaptic neuron, followed by repackaging of the transmitter into secretory vesicles so that it can be used again. Cocaine blocks the re-uptake of dopamine, one of most important neurotransmitter substances in the limbic system. The effect is that more dopamine remains in the synaptic cleft for a longer period of time. Ultimately this results in excessive stimulation of neurons in areas of the brain associated with pleasure, which could account for the euphoria or "high" during the initial use of cocaine.

What happens, though, with prolonged use? Biological systems have an astounding capacity to adapt. In this case, the body adapts to a perceived oversupply of dopamine in the synaptic cleft by reducing the production of dopamine. As the rate of dopamine production declines, more cocaine is required to maintain even a normal amount of dopamine in the synaptic cleft. This adaptation may account for **tolerance** (the need to have more and more cocaine to achieve the same effect). It would also account for **withdrawal** (the symptoms that occur when the drug is not available) because the absence of cocaine combined with an inadequate supply of dopamine would result in understimulation of these brain centers.

Cocaine also produces overstimulation of pathways involving the neurotransmitter norepinephrine, again by blocking the re-uptake of norepinephrine back into the presynaptic neuron. This can be particularly dangerous because norepinephrine is an important neurotransmitter of the nerves to the heart and blood vessels. Cocaine may produce severe fluctuations of blood pressure and an overstimulation of the heart, ultimately leading to cardiac arrest and death. Finally, cocaine use during pregnancy can cause undernourishment of the fetus and produce numerous neurological defects.

Although scientists understand the basic mechanism of cocaine's action at the cellular level, what they do not yet understand is why cocaine and especially crack cocaine are so addictive. Why does a strong physical dependency develop (despite many user's good intentions not to become dependent) in which the user's lifestyle begins to center around obtaining and using the drug? The euphoric effect may play a role, but is probably not the entire answer, for there are other drugs that are addictive but do not have the same pleasurable component as cocaine. For obvious reasons, the mechanisms underlying addiction are the focus of considerable research today.

widely distributed in the brain. Norepinephrine is also the neurotransmitter for a branch of the nervous system that regulates many of the automatic functions, such as blood pressure and the rate of breathing.

Neural Information Processing

So far we have described how an action potential, or nerve impulse, is generated in a nerve cell and transmitted from the nerve cell to another cell. These nerve impulses form the basis of rapid communication throughout the body. But if all the information is in the form of identical electrical impulses, how one can perceive all of the colors in a sunset, feel emotions, or guide a basketball into the hoop from a distance of 20 feet? At this point, a full and adequate explanation escapes us. However, we can list some of the biological phenomena that probably contribute to information processing and integration.

Convergence and Divergence of Neural Information

Information is not transmitted strictly from one nerve cell to another in a straight line. Recall that many different presynaptic cells may form synapses with a single neuron; this joining of the output of many presynaptic cells onto one postsynaptic cell is called **convergence.** In human terms, an example might be that the sight of a steak, the sizzling sound it makes on the grill, and its smell may all provide information (via neurons and their synapses) to one neuron that contributes to the feeling of hunger. Furthermore, one nerve cell can provide information to a variety of postsynaptic cells, a process called **divergence.** In the example of the 20-foot jump shot, information from a single visual cell in the eye may provide information to different areas of the brain involved in vision, control of muscle movement, memory, and even emotions. The principles of convergence and divergence at the cellular level are depicted in Figure 3.17. In this figure, output from neurons numbered 1–3 converges on neuron 4. The output of this neuron, in turn, diverges to four other cells numbered 5–8.

Coding for Stimulus Intensity

The coding for stimulus intensity is in part determined by how many neurons in a given pathway are active and by the number of action potentials per unit time

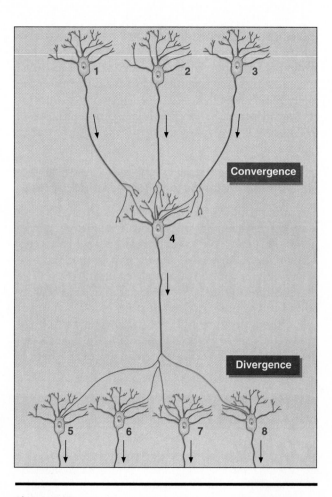

Figure 3.17   Convergence and divergence of neural networks. Neurons 1–3 all provide input to neuron 4 (convergence). Neuron 4 provides input to neurons 5–8 (divergence).

in a given neuron. The latter can vary from zero in an inactive neuron up to several hundred impulses per second. Neurons vary considerably both in the threshold membrane potential required to initiate an action potential and in the maximum number of action potentials per unit time. Note that because action potentials are conducted in an all-or-none fashion and do not vary in amplitude as they travel down a neuron, intensity is *not* encoded by action potential amplitude, but by frequency (review Figure 3.12).

Inhibitory Versus Stimulatory Input

As mentioned previously, synaptic input to a neuron can either increase or decrease the likelihood that the neuron will initiate an action potential. This depends on both the type of neurotransmitter released and the type of receptor present on the postsynaptic membrane.

Direct Electrical Communication

Electrical impulses generated by neurons are usually transmitted to adjacent neurons by a chemically mediated step because the distance between the two cells is too great for direct electrical transmission. Although this is true for most excitable cells, there are important exceptions. A few excitable cell types are capable of direct electrical connections. The most important of these include certain muscle cells, particularly those of the heart and digestive tract. Cells in these muscles are joined together by special connections known as **gap junctions** (Figure 3.18). Gap junctions are composed of membrane proteins that span the gap between two adjacent cells, forming a direct channel between them. Muscle cells joined by gap junctions can transmit electrical information in the form of ions directly from cell to cell, so that an action potential proceeds not just down the length of a single cell, but through all of the muscle cells in the tissue. This is the basis for the coordinated contraction of the heart, and for the process used to move food down the digestive tract. Some nerve cells also have gap junctions (sometimes called electrical synapses), but the chemical synapse is far more common.

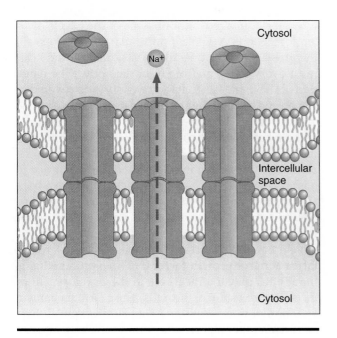

Figure 3.18 Gap junctions. Because an electrical charge (in the form of ions) can be transmitted through gap junctions, an action potential in one cell will result in an action potential in the next cell.

Chemical Communication

There are many different ways that cells use chemicals to communicate, and thus the chemical messengers defy easy classification. The only common thread seems to be that they represent an almost universal way of sending and receiving information from one cell to the next. Indeed, every form of communication except the direct electrical connections between certain muscle cells relies, at least in part, on chemical messengers. Even the synapse between two nerve cells is a form of chemical communication. The classification of chemical messengers into subgroups relies primarily on the type of cell that releases the chemical and on how the chemical messenger reaches its site of action. Table 3.2 is a summary of the chemical messengers, showing their sites of origin and purpose, and some examples.

Hormones

When scientists first discovered that cells could communicate by means of chemical messengers carried by the blood, they called the chemicals hormones. According to the classical definition, a **hormone** is a chemical that is secreted by a gland into the bloodstream, traveling to other sites in the body to exert its effect. The site at which the hormone acts is called the target for the hormone (Figure 3.19). A single hormone may have only one target tissue or organ, or it may have many diverse targets and thus many different regulatory effects. Hormones control processes as diverse as the rate of urine formation, the amount of sugar in the blood, and the rate of sexual maturation, to name just a few. The hormonal regulatory systems are so important that an entire chapter is devoted to them (Chapter 5).

Because the blood carries hormones to all organs and tissues, in theory it would be possible for every cell in the body to be a target for a particular hormone. In general, however, hormones act only on specific cells. This is because a target cell needs to have a specific receptor for that hormone or it cannot respond to it at all. To make it more complicated, different cells may have very different receptor types for the same hormone. It is the special properties of these receptors that determine which cells are targets for a particular hormone and which are not, and what effect is elicited by the hormone.

Paracrine and Autocrine Factors

From the time of the discovery of the first hormone, numerous other chemical messengers have been discovered that do not fit the classical definition of a hor-

Table 3.2 Comparison of types of chemical messengers.

Chemical Messenger	Sites of Origin and Purpose	Examples
Hormone	From a gland specialized for secretion into the blood. For communication with many distant sites.	Insulin, testosterone, estrogen.
Paracrine factor	From a cell into the nearby extracellular space. For communication with nearby cells.	Histamine, prostaglandins, growth factors.
Autocrine factor	From a cell into the nearby extracellular space. For communication with self.	Growth inhibitory factors (can also act on nearby cells as a paracrine factor).
Neurotransmitter	From a presynaptic neuron into the synaptic cleft. For communication with another neuron.	Norepinephrine, acetylcholine, dopamine.
Neurohormone	From a neuroendocrine cell into the blood. For communication with cells at distant sites.	Oxytocin, antidiuretic hormone.

mone. Some cells communicate with nearby cells by releasing chemicals into the extracellular space. These chemicals function as messengers even though they do not circulate in the blood. Such local chemical mes-

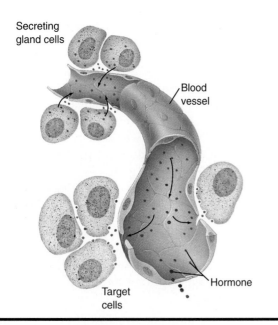

Figure 3.19 Hormones as chemical messengers. A hormone is a chemical messenger that is secreted into the bloodstream by cells of certain glands. Hormones are available to all cells in the body, but they act only on target cells that have the appropriate receptors.

sengers are called **paracrine factors.** The word *factor* reflects the fact that these are not hormones in the classical sense and that we do not always know the exact structure of some of these substances. A common paracrine factor with which you may be familiar is **histamine,** a chemical released from specialized cells called **mast cells** that are present in most tissues. Histamine is responsible for the local swelling of a tissue when it is damaged, and for some of the effects of allergic reactions. A whole group of paracrine factors are the **prostaglandins,** a family of fatty acids found in almost every tissue and organ in the body. (The name derives from the fact that they were first discovered in semen and were thought to originate from the prostate gland.) Prostaglandins are not hormones, but they seem to have a variety of roles in local tissue function, including constriction or dilation of blood vessels, blood clotting, and promotion of inflammation, just to name a few. The various growth factors that have been identified in a number of tissues and organs are also paracrine factors. Growth factors are peptides or proteins that participate in the regulation of growth and development of a particular tissue. One such growth factor is **nerve growth factor** (NGF), a protein that is thought to play a critical role in the development of the nervous system of the fetus. The exact mechanism of action of most growth factors is incompletely understood at the present time.

In general, paracrine factors are synthesized and released in the vicinity of their target cells. Although paracrine factors can sometimes be found in the blood, they are still called paracrine factors rather than hormones because the tissues that produce them are not specialized for secretion (i.e. are not glands).

Closely related to paracrine factors are **autocrine factors,** which are chemical messengers that act directly on the cells that produce them. A schematic representation of paracrine and autocrine factors is depicted in Figure 3.20. Note that the distinction between a paracrine and an autocrine factor is in whether the target tissue is the same tissue that produced the messenger. Indeed, some chemical messengers are both paracrine and autocrine factors. Examples of autocrine (and paracrine) factors are the inhibitors of cell reproduction that are produced by cells specifically to regulate their own (and nearby cells') division. Inhibitors of cell reproduction are thought to be one of the major mechanisms for controlling the rate of cell reproduction in most tissues. Indeed, one of the defects in cancer cells may be that they lack these inhibitors of cell division or do not respond to them, so that cell division continues unchecked.

Neurohormones

At the beginning of this chapter, we referred to electrical and chemical forms of communication as the two forms of communication between cells. In fact, the boundaries between these two forms are often blurred. The nervous system, for example, is composed of neurons that are highly specialized for transmitting information over long distances via electrical impulses, yet even in the nervous system there is a chemical step (synaptic transmission) in which a chemical messenger (neurotransmitter) links the electrical activity of two nerve cells.

Some neurons, called **neuroendocrine cells,** do not synapse with another neuron; instead, their axon terminals are located near blood vessels. Neuroendocrine cells release their chemical messengers, called **neurohormones,** directly into the bloodstream (Figure 3.21). The name *neurohormone* reflects the fact that these chemical messengers are released from nerves, but travel in the blood to act on distant target sites, as do the hormones. Neurohormones can be released rather quickly into the bloodstream because their release is triggered by an action potential in the neuroendocrine cell itself. An example of a neurohormone is **oxytocin.** Neuroendocrine cells in the brain release oxytocin into the bloodstream in response to stimulation of sensory nerves by suckling. The oxytocin is carried in the blood to the mammary gland, where it is responsible for the movement of milk toward the nipple, a process known as milk ejection. This process ensures that milk is available to an infant when it is suckling, but not at other times.

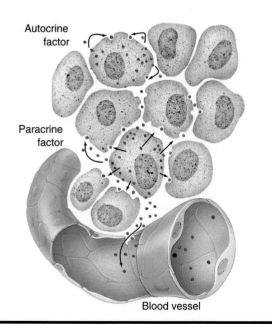

Figure 3.20 Paracrine and autocrine factors. Paracrine and autocrine factors are local chemical messengers acting on nearby cells (paracrine action) or on the cell that produced the messenger (autocrine action). Some of these factors do escape into the circulation, but dispersion by the blood is not the norm.

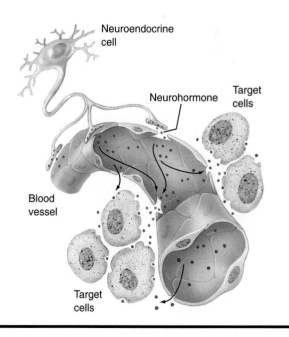

Figure 3.21 Neurohormones. Neurohormones are released into the blood in response to an action potential in modified neurons called neuroendocrine cells. The neurohormone then circulates in the blood, acting in the same way as a hormone.

Summary

Complex organisms have developed a number of strategies for communication between cells. The strategy used depends, among other things, on the distance between the cells, whether the speed of communication is important, and whether the information needs to be transmitted to a broad range of tissues or organs or only to a very specific cell or tissue. In general, cells communicate by either electrical or chemical means.

All cells have an inherent electrical activity, in that the inside of the cell has a slight negative charge compared to the outside. This difference in charge is called the membrane potential, and is generally about -70 mV. It comes about because of the active transport of Na^+ out of the cell and K^+ into the cell by a cell membrane transport protein known as Na^+-K^+ ATPase (the Na^+-K^+ pump), coupled with the more complete leak, or diffusion, of K^+ out of the cell than of Na^+ into the cell.

Although all cells have an electrical potential across the cell membrane, only certain cells are excitable; that is, they use *changes* in electrical potential to transmit information. The most important of these are neurons, or nerve cells. Neurons can generate a rapid change in membrane potential, called the action potential, that is propagated down the length of the nerve axon. The action potential is brought about by sequential opening and closing of voltage-sensitive cell membrane channels for Na^+ and K^+. Communication *between* nerve cells involves a short chemical step called synaptic transmission. In synaptic transmission the action potential in a presynaptic neuron causes the release of a chemical messenger called the neurotransmitter, which in turn affects the membrane potential of a postsynaptic neuron. Four important neurotransmitters are acetylcholine, norepinephrine, epinephrine, and dopamine. Neurotransmitters act by activating chemically sensitive ion channels, which in turn causes a slight excitation (depolarization) or inhibition (hyperpolarization) of the postsynaptic neuron. The specific neurotransmitter released at these synapses, the convergent and divergent ways in which neurons connect to each other, and the number of action potentials traveling down a nerve cell per unit time all contribute to how neural information is processed as it is transmitted.

Some specialized excitable cells, most notably the muscle cells of the heart and gastrointestinal tract, can transmit electrical currents directly between them without the necessity of a synapse.

Chemical communication is widespread throughout the body as a general strategy for communication between cells. Chemical communication is considerably slower than electrical communication. Hormones are chemical messengers released from glands into the bloodstream that act on target cells at distant sites. Hormone action is determined by the presence of, and type of, receptors on the target cell. Paracrine factors are chemical messengers that are released locally, in the vicinity of the target cells. Examples of paracrine factors include histamine, prostaglandins, and various growth-promoting factors. Closely related to paracrine factors are autocrine factors, chemical messengers that affect the cell that produces them. Most cells produce inhibitory autocrine factors that limit their own reproduction.

Some forms of chemical communication seem to bridge the gap between electrical communication and chemical communication. The neurotransmitter released into the synaptic cleft of a synapse is a form of chemical communication. In addition, neuroendocrine cells release chemical messengers called neurohormones into the bloodstream. Oxytocin is one example of a neurohormone.

Questions

Conceptual and Factual Review

1. What are the two general methods cells use to communicate? (p. 51)
2. What determines the rate of net movement of *neutral* particles across a semipermeable membrane? What additional factor influences the movement of *charged* particles? (p. 52)
3. Under what circumstances can a charged particle undergo net diffusion *against* a concentration gradient across a semipermeable membrane? (p. 52)
4. Under resting conditions, to which charged particle are cells more permeable: Na^+ or K^+? (p. 54)
5. What is a graded potential? What causes it? (p. 56, p. 57)

6. In what region of a neuron can graded potentials first produce an action potential? Why does it first occur in this region? (p. 59)
7. What factors determine the speed of conduction of an action potential? (p. 62, p. 63)
8. What is a neurotransmitter, and what causes its release? (pp. 64–66)
9. What is convergence? Divergence? (p. 68)
10. What types of cells have gap junctions, and what is their function? (p. 69)
11. Why is paracrine communication not effective over long distances? (p. 70)

Applying What You Know

1. Why do cells in a multicellular organism need to communicate?
2. Under what circumstances would an organism such as a human use chemical forms of communication, and when would electrical forms of communication be more appropriate? Consider the relative advantages and disadvantages of each method.
3. If you could block the action of the Na^+-K^+ pump, what do you think would happen to the resting membrane potential of all cells? Could an excitable cell generate an action potential? Why or why not?
4. A biology student is growing cells in the laboratory. She notes that the cells grow and divide until they are almost touching each other, and then they seem to stop dividing. What could explain this phenomenon?

Chapter 4
Neurophysiology

Objectives

By the time you complete this chapter you should be able to

◆ Describe many of the primary characteristics of brain function.

◆ Understand some of the electrical circuitry providing those functions.

◆ Know the functional divisions of the central nervous system.

◆ Describe the overall anatomical organization of the major brain structures and of the spinal cord.

◆ Know the different areas of the brain that are responsible for making you aware of your environment and regulating behavioral responses.

◆ Understand how reflex activity can be controlled by both spinal and brain centers.

◆ Appreciate the difference between consciousness and sleep.

◆ Understand the effect of some psychoactive drugs on those states.

Introduction

Of all the organs in the body, your brain is the most complex and the most resistant to scientific investigation. It is the organ that makes us uniquely human by adding a dimension to all the attributes that are in relatively rudimentary form in other animals. We communicate, think, and respond to the world both qualitatively and quantitatively; this is different from other animals, although this is not to deny that animals can communicate and have the ability to synthesize new thoughts. But no animal, other than the human one, has the ability to read this book, or to write poetry, or to travel to the moon.

The brain has been called the knapsack of intelligence, an apt but limited description. Certainly the brain is central to our human condition. In common jargon, when someone describes another person as being brain dead, the person is referring to someone who is not quite bright. In a medical sense however, this term refers to one who has functioning organs, yet is in a coma, with no hope of recovery. Why is this state called brain death? It is the brain that initiates movement, thought, speech, and all that makes us uniquely human. How does the brain accomplish all these things?

First consider common perceptions that are conditioned, in large part, by current science fiction. Movies present walking, talking robots, controlled by a central computer. The computer allows the robot to sense the external world; it can see it, hear it, touch it, and respond appropriately. It also senses its own mechanical parts, for how else could it move, pick up items, and deposit them accurately? The central computer must be able to integrate many inputs (such as vision, sound, touch, and position), then act in response. In order to move from point to point, it must sense obstacles so that it knows to move around them. It must also perceive the position of the floor on which it moves and adjust its balance to slight inclines and declines in floor level. Is this an accurate and complete description of the human brain? Not quite.

Science fiction shows robots as talking, and in fact today some do. But although robotic "thought" is limited, human thought is nearly boundless. A robot must be programmed by a designer, limiting the central computer to only the concerns the programmer herself has considered. Her approach is as follows: "Robot, if X happens then you will do Y." A series of finite if/then commands may be programmed into the computer. Although computers can be programmed to respond differently to the same stimulus based on previous experience, a condition we call learning, the range of responses is quite limited relative to human

responses. The computer does not think in human terms. For the robot, each stimulus provokes a relatively stereotypical response, and only programmed stimuli will lead to a response.

Humans respond differently to similar stimuli under different conditions, and we also respond to novel stimuli, something a computer is incapable of doing. At birth, many "hard-wired," or stereotypical, responses are evident: A limb will be withdrawn from a painful stimulus, the baby will cry when he or she is hungry or suffering discomfort. Although there are some similarities between the human brain and the science fiction robot, the differences are far more substantial than are the similarities.

Think for a moment of all the things you do that may be the result of a brain function. As you read this sentence, your eyes transfer the different alphabetic symbols to your brain. In turn, your brain rapidly integrates those symbols, making them into words and sentences. The thoughts expressed by this writer in symbolic form are now made known to you. While reading, you may have shifted position, moved the book, stretched your legs, or made other movements. You did not fall over, lose sight of the printed page, or lose track of the thoughts expressed. All this was done without conscious effort on your part. Your brain takes care that proper muscular adjustments are made without your thinking about them. Sight, hearing, smell, and all senses of your environment are impressed on your consciousness when required, but suppressed from the conscious level when appropriate. The brain receives input from both the external world and your internal world and integrates those signals so as to adjust output in order to maintain homeostatic balance. The brain is also the organ that controls that most elusive property, consciousness.

History of Neurophysiology

If asked, most people would be able to give an approximate answer to the question of what the brain does. Whereas we now accept that the brain is the organ of thought, movement, consciousness, sight, and all our senses, the early Greeks believed that thought originated within the heart. Hebrew writings of 5000 years ago assigned the capacity of emotions to the kidneys, and Aristotle, in the fourth century B.C., believed that the brain functioned only to cool the blood. Very little attention was directed toward the brain until relatively recently; the heart was seen as the foremost organ of the body. Our language retains the vestiges of this focus as we still speak of hard-hearted or soft-hearted people, of losing heart, of being heartbroken, and of something being at the heart of the matter. How then did our knowledge of brain function come about?

Our understanding of brain function came slowly from careful anatomical observations and from many animal experiments in which parts of the brain were removed or stimulated electrically. Early anatomists recognized that the brain is not a homogeneous organ, but rather is composed of discrete parts, many of which could be easily identified. Furthermore, the brain is attached to the spinal cord, from which numerous nerves exit and travel to the muscles and organs of the body. Indeed, no part of our body is free of innervation. Thus, the brain seemed to be an organ centrally placed and well-designed to coordinate body functions.

By the middle of the nineteenth century, anatomists performed ablation (surgical removal) experiments in an attempt to understand the brain's function. Among the first were studies in which they removed small parts of the brain from hens. Depending on which part of the brain was surgically removed, they observed a different deficit. Removing parts of the lower brain, the medulla, interrupted normal respiration. If parts of the outermost brain, or cerebrum, were ablated the hen "lost instincts," meaning it would no longer eat or perform other normal activities. These types of crude studies clearly indicated that the brain controls many functions, and that different areas of the brain serve specific functions.

Electrical studies of the brain were an outgrowth of the realization that specific brain areas have specific functions. As you learned in Chapter 3, a weak electrical current applied to a nerve causes that nerve to generate an action potential that is then propagated down the nerve. This led researchers to test the results of stimulating the brain surface with a weak electrical current. They demonstrated that electrical stimulation of a discrete area of the brain of an anesthetized animal might cause a limb to move in a particular direction. Stimulation of another spot on the surface of the brain might cause a digit to contract. Stimulation of a specific brain area always led to the same effect. We shall talk more of that later. These early workers made it clear that a good anatomical map of the brain was necessary for an understanding of its function. Anatomical studies, coupled with electrophysiological ones, detailed much of what we know about brain function.

Structure of the Nervous System

The nervous system is composed of all the neural elements of the body. However a clear division may be made between the elements contained in the bony structures of the body, the skull, and spinal column, and the neural elements outside of the bony case. The former is composed of the brain and its associated spinal cord and is called the **central nervous system (CNS).** The latter, which includes all the neural elements not contained in the CNS, is called the **peripheral nervous system (PNS). The PNS is composed of two branches. The first is the somatic nervous system,** which carries motor information from the CNS to the skeletal (voluntary) muscles via the **efferent nerves;** the **afferent nerves** carry sensory information from muscles, joints, skin, and bones back to the CNS. The other branch is the **autonomic nervous system (ANS),** which carries information to and from the glands, smooth muscles, and heart, via the **sympathetic** and **parasympathetic** subdivisions, and is involved in the regulation of homeostatic control. Schematically, this is depicted in Figure 4.1.

The human brain is pictured in Figure 4.2. As viewed from the top (superior view), it is evident that the brain is divided into two similar parts, the left and the right hemispheres. Running down the middle of the brain is the longitudinal fissure, making the separation visible. Below this fissure is a large tract of nerve fibers connecting the two hemispheres, called the **corpus collosum.** The outer portion of the brain, called the **cerebral cortex** or **gray matter,** as it lacks the white myelin coating, is composed of a series of folded convolutions. The elevated parts are called **gyri,** and the depressions between the gyri are called **sulci.** (The singular of *gyri* and *sulci* are *gyrus* and *sulcus.*) Gray matter contains mainly cell bodies and dendrites of the neurons.

As viewed from a lateral, or side view, the cerebral cortex is divided into four separate areas, or lobes: the **occipital lobe,** the **parietal lobe,** the **temporal lobe,** and the **frontal lobe.** The central sulcus separates the frontal lobe from the parietal lobe and the lateral sulcus separates the temporal lobe from the parietal lobe. A less evident separation exists between the parietal and occipital lobes. Each of these lobes has a separate and unique function that will be described below. Lying just below the cortex is a thicker mass of brain tissue composed of myelinated fibers having a whiter cast, appropriately called the **white matter.** Both gray matter and white matter are also seen in the spinal cord.

Surrounding the brain is a composite of four tough membranes called the **meninges,** which contain the jelly-like substance of the brain. (Three meninges surround the spinal cord.) The meninges, along with the bony skull, help to protect the brain from injury.

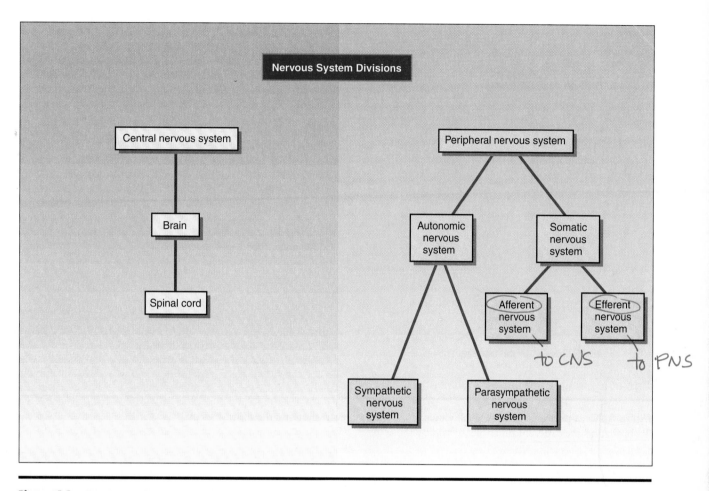

Figure 4.1 Divisions of the nervous system.

Cellular Elements

The CNS is composed of nerve cells and associated cells, which will be described in detail. The changes of the electrical charges that occur in the membranes of these cells have been described, and you are encouraged to review the section on action potentials in Chapter 3. The neurons can be considered the major working elements of the brain and spinal cord, despite the fact that many other non-neural cell types are present. Each has a unique function, and the proper functioning of each is a requirement for normal brain activity. First, let us describe the non-neural elements of the brain.

Glia

Approximately half of the volume of the brain cavity is filled with a set of cells called **glia.** The precise function of these elements is not completely known, but we do have some knowledge of a few of the glial cell types.

Astrocytes As the name implies, these cells are star-shaped and are thought to perform what might best be described as housekeeping duties. They seem to be able to regulate the ionic environment in which the neuronal elements reside. Because precise ionic concentrations of K^+ and Na^+ in the fluids that bathe neurons are necessary for normal activity, the **astrocyte** is an important cell. It also may regulate blood flow to certain areas of the brain as neural activity is altered.

Oligodendrocytes In Chapter 3 you learned that the myelin sheath surrounding axons increases the speed of conduction of the action potential along that axon. **Oligodendrocytes** produce the myelin that wraps around the myelinated fibers. In the periphery, Schwann cells supply the axons with myelin. This is an extremely important function, as impairment in myelination results in severe neuromuscular deficits. For example, in the disease **multiple sclerosis,** the insulating myelin becomes impaired, resulting in a severe deficit in muscular control along with other

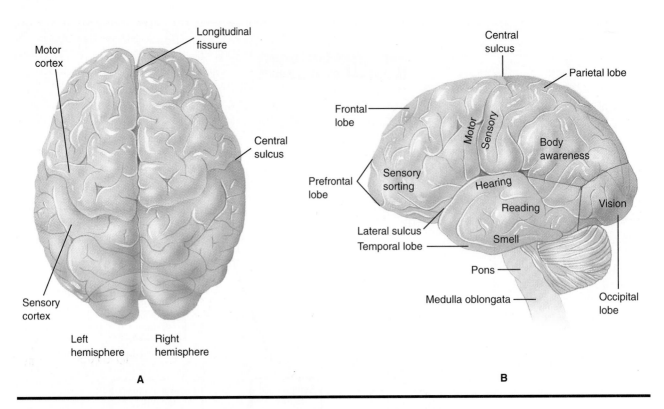

Figure 4.2 Superior (A) and side (B) views of the human brain showing the four lobes of the brain and the different functions served by those lobes.

neurological symptoms, such as alterations in personality, memory loss, emotional instability, and pain. All these disorders may be traced to the impairment of electrical conduction along the central neurons.

Vascular Cells

As you will learn in Chapter 8, most capillaries throughout the body are highly permeable, allowing most blood solutes to pass freely across them into the interstitial space. The capillaries of the brain are an exception. Oxygen, carbon dioxide, and glucose may freely cross the capillary membrane; however, many other bloodborne solutes that can easily cross other capillaries are restricted in their movement across brain capillaries. This limitation of the brain capillaries to the movement of certain solutes is called the **blood–brain barrier.** Figure 4.3 illustrates the unique anatomy of brain capillaries that is responsible for the blood–brain barrier. Adjacent endothelial cells of the brain capillaries are joined very tightly, forming what are called tight junctions, so that solutes are unable to pass between the adjacent cells. In addition, astrocytes attach to the capillaries with foot-like processes. This anatomical arrangement causes the capillary system of the brain to be less permeable to bloodborne solutes than the other capillary networks of the body. There

are small areas of the brain that do not possess the blood–brain barrier, but they account for no more than 1% of the total capillaries of the brain.

The blood–brain barrier most likely serves to protect the brain from sudden changes in the environment surrounding the neural elements. This protection presents a serious problem, though, when physicians try to treat infections of the brain. Many drugs are unable to cross the blood–brain barrier and cannot reach the site of infection, requiring the use of special drugs that are able to cross the blood–brain barrier.

Figure 4.4 shows the relationship between the glia and neural elements in the brain. It points out how the glia act as structural as well as functional units. The upper part of the figure is a single neuronal cell body showing the numerous connections with other nerves and with glial processes.

Organization of the Brain

In general, the brain is organized in a hierarchical manner. Large divisions of the brain are more easily seen than the many smaller ones that exist, and it is the larger divisions that concern us at this juncture. Regardless of

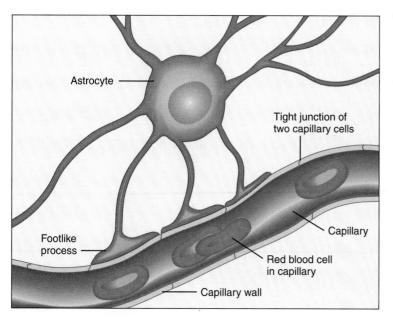

Figure 4.3 Diagram of a brain capillary showing the blood–brain barrier. The combination of the capillary tight junctions and the foot processes of the astrocytes reduces capillary permeability.

the ability to distinguish anatomical divisions of the brain, many parts act in concert to allow smooth integrated function. Figure 4.5 shows the major divisions and some of the structures contained within them.

Telencephalon (Cerebrum)

The brain structures credited with the higher levels of function such as thought, intelligence, and ability to plan and solve problems are located in the **telencephalon,** or cerebrum. It is composed of the **cerebral hemispheres,** which are the foremost part of the brain, and a series of internal structures formed into nuclei or groupings of densely packed nerve cells. It is important to note the distinction between a nucleus as it refers to the tight grouping of cells in a brain

structure and the nucleus of a single cell that was described in Chapter 2.

Diencephalon

The **diencephalon** is the smallest and least distinctive division of the brain, containing the fewest groups of nerve cells. In general, it serves as a relay station conducting information from sense organs to other parts of the brain. Two of its major structures are the thalamus and the hypothalamus.

Mesencephalon (Midbrain)

In general, the **mesencephalon** serves as a relay station for both visual and auditory input to the

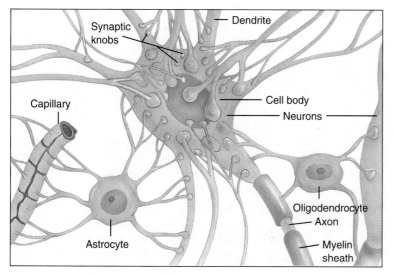

Figure 4.4 Relationship of glial tissue to neural tissue in the brain. Each neural cell body is in contact with many glial cell processes.

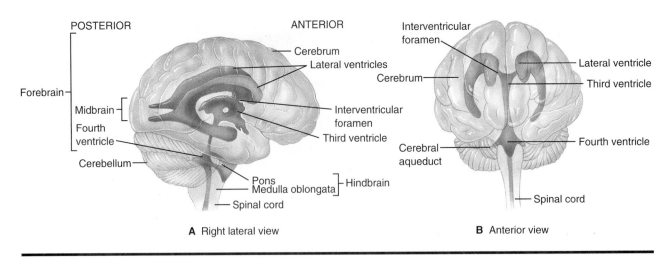

Figure 4.5 Major divisions of the brain, showing the ventricular system. The center of the brain contains a series of hollow chambers, the brain ventricles, filled with cerebrospinal fluid.

cerebrum. It also contains nuclei that aid in the integration of motor output from the cerebrum.

Metencephalon

The two major areas of the **metencephalon** are the cerebellum and the pons. The cerebellum coordinates both conscious and unconscious movements and acts as a processing center for postural reflexes. The pons aids in connecting the two hemispheres of the cerebellum, and links the cerebellum to medullary nuclei.

Myelencephalon

The **myelencephalon** contains many of the structures that are responsible for the function of primary physiological systems. Both the respiratory and cardiovascular control centers are located within it.

Ventricular System

If the brain is sectioned through the central sulcus, a lateral view of the internal structures is exposed. Figure 4.5 shows a side and top view of the complex anatomy of the ventricular system. If the CNS is cut in cross-section, the white and gray matter may be seen, and new structures become evident. At the center of the brain is a series of hollow, liquid-filled chambers called the **ventricles.** The fluid filling the ventricles is called **cerebrospinal fluid (CSF).** Although the CSF bathes the interior of the brain, spinal cord, and the outer surface of the brain, it is not clear exactly what function it performs. It may play a protective role in acting as a shock absorber, as well as acting to support the brain within the skull as the brain floats on the surrounding CSF, reducing its effective weight.

Because the CSF is the source of the fluids that make up the local environment of the neurons, it is also possible that the CSF provides an avenue for nutrients and biologically active chemicals to reach different parts of the brain. Figure 4.6 shows the ventricular system of the brain and its connection to the spinal fluid. You can see that the CSF circulates throughout the entire CNS. CSF is secreted by a highly vascular, specialized tissue along parts of the ventricular system called the **choroid plexus** and flows as the arrows in Figure 4.6 indicate. The fluid bathes the ventricular system and the outer cortical regions, then flows along the spinal cord and returns to the brain, where it is absorbed back into the blood from the area called the arachnoid villi.

Spinal Cord

Lying just below the temporal lobes at the base of the brain is the **brain stem,** below which is the spinal cord, which passes through the vertebrae of the spine. Figure 4.7 presents an overall view of the brain and the spinal cord as they appear in the body. Note that spinal nerves exit the spinal cord at the juncture of each pair of vertebrae. These spinal nerves are grouped into the **cervical, thoracic, lumbar,** and **sacral** nerves according to their location along the spine. An enlarged view of several vertebrae in relation to the spinal cord and spinal nerves is also shown. Spinal nerves exit the cord through a small hole between each pair of vertebrae. The anatomy is such that no matter how the vertebrae are moved, the spinal nerves are unaffected. No normal movements of the spine can cause the vertebrae to impinge on the spinal nerves. They are wonderfully protected from the twisting and turning movements to which the spine is subjected. However, the intervertebral discs, which separate

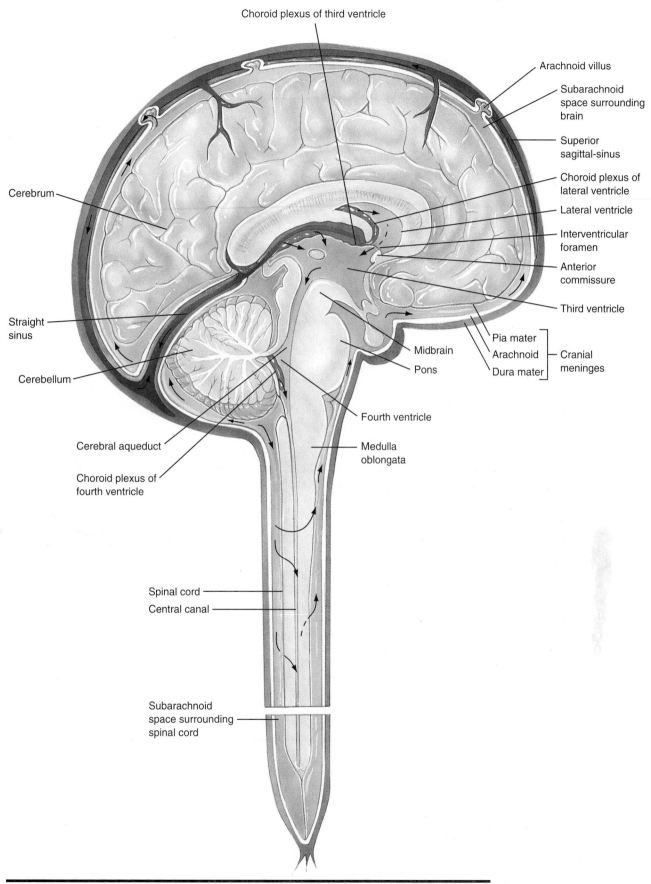

Figure 4.6 Diagram of the CNS showing the circulation of the CSF through the ventricular system, spinal cord, and the surface of the brain. CSF is formed by the choroid plexus.

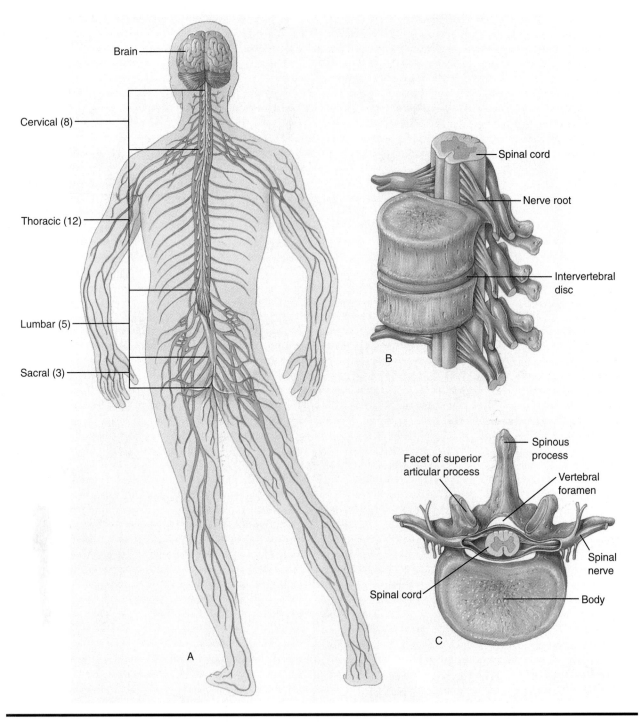

Figure 4.7 (A) Relationship of the spinal column, spinal cord, and spinal nerves. (B) An enlarged view of the junction of two vertebral bodies. (C) View of a cross-section through one vertebra.

adjacent vertebrae and provide a cushion between them, are sometimes the source of difficulty. On occasion one may bulge into the space occupied by the spinal nerve of that segment and compress it, causing pain and even loss of some muscle control. Surgical removal of the offending disc may be required in order to relieve pressure on the spinal nerve.

In addition to the spinal nerves, 12 pairs of **cranial nerves** exit the CNS directly from the brain through small holes in the skull. The cranial nerves and the spinal nerves make up the peripheral nervous system, carrying information to and from the CNS. Table 4.1 lists the cranial nerves and the major function of each nerve. Most of the cranial nerves are sensory nerves;

Table 4.1   **Cranial nerves, name, composition, and major functions.**

Cranial Nerve	Type of Nerve	Function
I Olfactory	Sensory	Smell ✓
II Optic	Sensory	Vision ✓
III Oculomotor	Motor and sensory	Muscles that control movement of eye and lens
IV Troclear	Motor and sensory	Movement of eye
V Trigeminal	Motor and sensory	Sensory from nose, scalp, lips, forehead, teeth, gums, and jaw; motor to chewing muscles
VI Abducens	Motor and sensory	Eye muscles
VII Facial	Motor and sensory	Facial muscles, salivary and tear glands, taste buds
VIII Vestibulocochlear	Sensory	Hearing and equilibrium ✓
IX Glossopharyngeal	Motor and sensory	Swallowing muscles, taste buds, middle ear, salivary gland, carotid sinus
X Vagus	Motor and sensory	Muscles of pharynx and larynx, taste buds, visceral sensations ✓
XI Accessory	Motor	Muscles of larynx, head, neck, and shoulders
XII Hypoglossal	Motor and sensory	Tongue

(handwritten margin note: "know these")

that is, they carry information from the periphery to the brain. As noted in Table 4.1, many also contain motor fibers carrying information away from the brain, activating muscle groups.

If a cross-section is made across the spinal cord where a spinal nerve exits the cord, it can be seen that the spinal nerve is actually composed of two nerves that join a short distance from the spinal column. Figure 4.8 shows such a cross-section. Two nerves

leave the cord: a ventral nerve (**ventral root**) and a dorsal nerve (**dorsal root**), which contains a small swelling a short distance from the point of exit. The two nerves then join to form the spinal nerve. (*Dorsal* refers to the nerve's position toward the back; *ventral* refers to the front.) You will see later that the two parts of the spinal nerves send information in different directions. Figure 4.8 shows that a sensory neuron entering the spinal cord may synapse directly with a

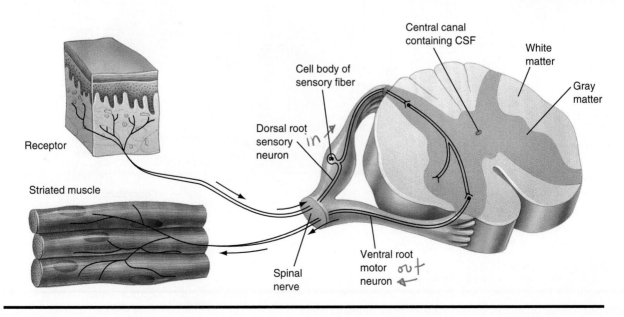

Figure 4.8   Cross-section of the spinal cord showing the dorsal and ventral roots.

motor neuron in the same segment of the cord or do so via an intermediate neuron called an association neuron or interneuron. Note that the spinal cord, like the brain, is also divided into white and gray matter, and also has a central hollow canal containing CSF. The spinal nerves and the cranial nerves make up the peripheral nervous system.

Cortical Function

Two experimental methods using anesthetized animals have allowed researchers to construct functional maps of the cortex. In the first method, they expose a portion of the cortex and apply a very weak electrical current to the exposed area. Any motor activity is noted. For example, if a small area of the cortical surface just in front of the central sulcus is stimulated electrically, voluntary muscle groups move. By carefully noting which muscles contract in response to well-localized electrical stimuli, it is possible to construct a detailed map of the areas of the cortex that control specific motor activities.

In the second method, researchers place recording electrodes on the brain surface and stimulate some distant part of the body, either mechanically or with a weak electrical current. Any evoked electrical activity on the brain surface is recorded. In this way, it is possible to map the sensory input to the cortex from all parts of the body. These maps have also been constructed in conscious patients undergoing brain surgery. It is possible to stimulate the cortex of unanesthetized humans, as no sensation of pain is experienced when the cortex is stimulated electrically or even cut with a knife. Figure 4.9 shows a cortical map obtained from such experiments. Note that specific functions reside in specific brain areas rather than being intermingled, although elaborate connections exist between essentially all areas of the brain.

Frontal Lobes

The area of the frontal cortex that controls muscle movement is called the **primary motor cortex.** Immediately in front of the primary motor area is the **premotor cortex,** which is also concerned with voluntary motor activity. The larger part of the frontal lobes

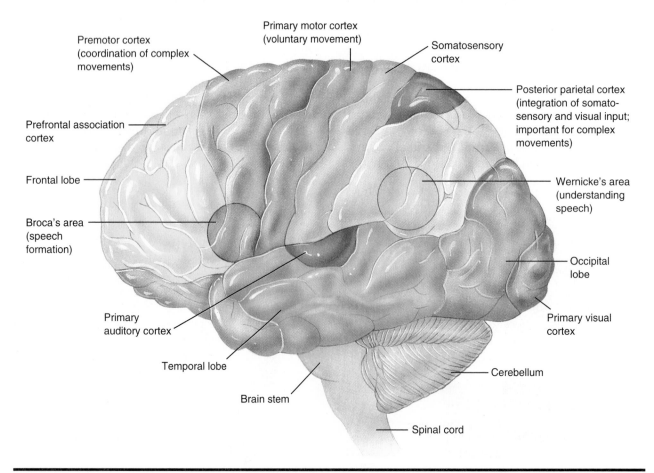

Figure 4.9 Map of the functional divisions of the human cortex.

MILESTONE 4.1

Prefrontal Lobotomy

In 1848 an accident occurred that was to have long-lasting and serious repercussions in the treatment of some mental disorders. Phineas Gage was then the foreman of a work gang preparing right-of-way for railroad track. His job was to tamp blasting powder into place in preparation for removing rock and soil. As he was tamping it down with a crowbar the powder exploded, driving the crowbar through his skull, carrying away his left frontal cortex and part of his upper face. Miraculously, he survived, but he exhibited a profound personality change. Before the accident Mr. Gage was a model citizen, highly respected for his probity. His employer found him to be dependable, hard working, and well-liked by all his associates. Following recovery Mr. Gage was a different person. He took to drink, was profane in language and behavior, and had trouble holding down a job. His physician described him as "manifesting little deference for his fellows, impatient of restraint when it conflicts with his desires, . . . obstinate, yet capricious." He seemed to have lost the inhibition imposed on us by society, and he acted freely, although not always in his best interests.

It is difficult to tell at this distance from the accident the exact cause of the profound behavioral change. It may have been due in part to the severe facial disfigurement caused by the explosion, a result of losing a large part of the frontal lobes, or a combination of both. Whatever the cause, it led neurobiologists to recognize that some aspects of behavior have their roots in brain substance. Researchers later came to accept that the prefrontal lobes act as a kind of brake or restraint from totally free behavior. It followed that if a person suffered from a severe behavioral disorder that caused depres-

sion, perhaps suicidal thoughts, or almost any mental disorder that made life difficult, severing the connections of the prefrontal lobes from the rest of the brain would free the patient from those restraints and resolve the behavioral problems.

In 1935 Dr. Antonio Moniz performed the first prefrontal lobotomy. He cut two openings in the skull and, by inserting a thin knife through the holes, cut the nerve fibers connecting the prefrontal lobes with the thalamus. He reported that his patients who previously suffered chronic distress, depression, phobias, and hallucinations were relieved of the symptoms. He believed the surgery improved the quality of their life and allowed them to resume a moderate amount of normal activity. For many decades this surgery was performed on tens of thousands of people. However, when more skeptical and analytical researchers studied the patients more thoroughly, a very different picture emerged.

Prefrontal lobotomy certainly did alter behavior, but not as benignly as Moniz believed. The lobotomized patient was more likely to suffer very severe deficits of normal behavior, often becoming grossly incapacitated. Many complained that they were no longer human and could do nothing about it. We now recognize that prefrontal lobotomy is not the panacea Moniz and others once thought it to be, and it is no longer part of the treatment for psychological ills. This sad episode of psychosurgery does point out the important fact that behavioral and intellectual capacities seem to reside in certain brain areas. Removing the prefrontal lobes from contact with much of the rest of the brain causes permanent impairment of psychological function.

seems to be required for the programming and coordination of complex voluntary movements. Coordination between the primary motor cortex and the premotor cortex is necessary for the almost infinite variety of the fine complex control we have over our muscles. As anyone who has tried it knows, threading a needle is not an easy task. Extremely fine coordinated muscular control is necessary to accomplish the task.

A conspicuous feature of the cortical map is that the respective brain areas responsible for different parts of the body do not correspond in size to those body parts. For example, the size of the cortical area controlling hand movement is proportionally much greater than the cortical area controlling arm movement. This makes sense because the range and fine motor control of hand movements, particularly of the thumb, are far greater than those of the arm.

Another function assigned to the frontal lobes is the control of certain thought and behavioral traits. For some time it was believed that separating the most anterior portion of the frontal lobes, the prefrontal cortex, from the rest of the brain was a proper cure for many behavioral abnormalities (see Milestone 4.1). Although we know that the frontal lobes play a crucial role in coordinating and planning complex movement and speech, it is not entirely clear what the relation of the frontal lobes is to personality and behavior.

Parietal Lobes

The parietal lobes receive sensory input, and a map similar to that of the motor cortex may be constructed. If the area just behind the central sulcus of the parietal lobe is stimulated in an awake patient, the patient will tell the surgeon that he or she feels something in some localized area of the body. Similar data can be obtained from mammals, mainly cats, although they cannot speak. If electrodes are placed on the brain of an anesthetized cat and a part of the body is stimulated mechanically, an electrical signal will be recorded from the corresponding area of the cortex. Thus, both the nonhuman mammalian cortex and the human cortex are anatomically similar. Although this is not the main subject of this book, it is well to point out that many similarities exist between the human brain and the brains of nonhuman mammalian species. The similarities afford an excellent chance to study brain function in nonhuman mammals and learn much about human brain function.

Again, the relative size of the brain areas does not correspond to the size of the body part from which they receive input. Consider how sensitive your tongue is. The smallest piece of food is keenly felt, and so the tongue requires many sensory nerves and a correspondingly large cortical projection area. Your arm, in contrast, is far less sensitive to touch, and has far fewer sensory nerves, requiring a relatively small cortical area to which impulses are projected. Figure 4.10 shows a graphical representation of the relative areas of the cortex devoted to each part of the body, illustrating the disparity in the size of the sensory cortex relative to the size of the anatomical part it serves. Note that the distortion of the figure corresponds to the relative sensitivity of each body part. The hand, lips, and tongue occupy far greater cortical areas than their individual sizes represent in proportion to the body's overall surface area.

Cerebral Lateralization The two hemispheres control both the motor and sensory reception of the opposite, or contralateral, side of the body. If the motor cortex of the left side of the brain corresponding to the arm is stimulated, the right arm will move. If the right cerebral cortex is stimulated, motion will occur on the left side. Similarly, if recording electrodes are placed on the sensory cortex, evoked potentials are recorded on the opposite side of the brain from which the stimulus was applied. A cortical evoked potential is one that is produced by a sensory nerve activating a cortical neuron. Thus, the right hemisphere controls movement of the left side of the body and receives sensory

Figure 4.10 Cortical distribution of sensory input of the human brain. The distorted figure of a person whose body parts are sized in proportion to the surface area occupied by those parts on the human cortex is shown above.

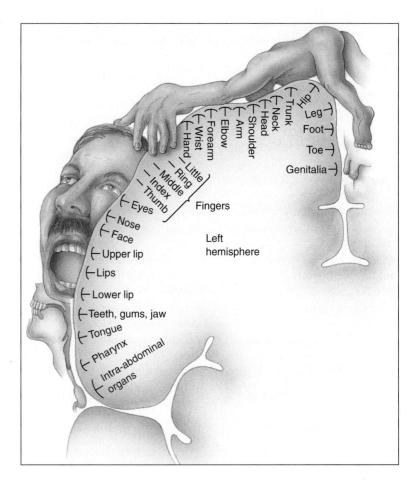

input from the left side. However, the two hemispheres are connected by the large bundle of fibers running in the corpus collosum, whose nerves allow the two hemispheres to share information and coordinate integrated function.

Our understanding of the function of the corpus collosum and hemispheric lateralization was enhanced by studying patients who had their corpus collosum surgically severed to alleviate severe symptoms of epilepsy. These "split-brain" patients are able to perform most tasks with no noticeable deficit. However, if an object is placed so that its visual projection is only to the right side of the brain, the person will see it perfectly well, but may not be able to name it, even though it is a common object. This experiment demonstrates that the two hemispheres are functionally different, each having some strengths and weaknesses not shared by the other hemisphere. It also demonstrates that information flow between the hemispheres is necessary for the full range of human responses to our environment.

Experiments have demonstrated that for most people the left hemisphere is largely associated with verbal and written language, as well as analytical abilities. The right hemisphere, in contrast, processes information having to do with perception of wholes, such as face recognition, and emotional behavior. An individual with destruction of some part of the right hemisphere might be able to identify a picture of a nose or an eye, but not be able to recognize the photograph of a person with whom he is acquainted. Latest research into the functional aspects of the two hemispheres indicates that each half of the brain is a functional complement of the other half. Rather than one hemisphere being a dominant one, they act in concert, sharing information to allow well-coordinated movement and thoughts. This is not to say that specific brain areas never control specific functions. For most people, the left hemisphere controls speech. In the mid-nineteenth century, Paul Broca noticed that some people lost their ability to talk intelligibly, despite the fact that they had no impairment of motor control of their tongue and vocal cords. However, they could read and write normally and understand all that was said to them. He pointed out that they had lesions in an area of the frontal lobe of the brain, now called **Broca's area** (refer to Figure 4.9). We now know that Broca's area is a primary speech area, and lesions in this part of the brain interfere with speech.

Aphasia People with deficits in their ability to speak are said to be **aphasic.** Aphasia is a disturbance of language that may take many forms, depending on the exact location and size of the lesion. Some patients may be unable to recall specific words or names of objects; others may speak in meaningless phrases and even be unable to repeat short phrases. The exact deficits experienced depend on the extent of the damage to the frontal lobe. Careful analysis of aphasics has shown that control of language resides in the left hemisphere in over 95% of all people, and in the remaining 5% of people speech resides in the right hemisphere. About 70% of left-handed people have left hemispheric control of speech; the remainder have right hemispheric control. It is not yet known why this variation in the position exists.

Another type of aphasia indicates that the left frontal lobe is intimately concerned with comprehension of language. This is not to be confused with a deficit in the ability to interpret sounds. For example, many deaf people use American Sign Language (ASL) as their means of communication. ASL is a separate language that uses visual signs rather than auditory ones as a means of communicating thoughts. ASL has its own visual vocabulary, grammar, and structural elements. Some people who have used ASL for communication have become aphasic in that they are unable to understand or communicate in ASL, even though ASL does not require any vocalization, but rather depends on visual input. Those people have been shown to have suffered damage to their left hemispheres, indicating that language skills reside in that hemisphere. It also indicates most clearly that it is language that is affected, not simply a deficit in hearing or vision.

Temporal Lobes The temporal lobe is also involved in speech comprehension. Patients with damage to Wernicke's area, in the posterior part of the temporal lobe, have the ability to speak well-formed words and syllables, yet their speech is not intelligible. The structure of their speech may best be described as a "word salad." Words are used in almost random manner, conveying no meaning. Some words may be in correct order, but they may be followed by meaningless sounds or words that do not fit with the preceding ones. Furthermore, these people have difficulty understanding both written and verbal language. Perhaps the strangest part of this syndrome is that many of these patients do not seem to know that their use of language is unintelligible.

Also located in the temporal lobe is the primary center for hearing. Lesions to the posterior part of the temporal lobe may cause word deafness. The person may be able to read and write with little or no abnormalities. However, spoken language may be severely disturbed. The temporal lobe is closely associated with language, both spoken and written, yet a deficit in one may occur without a deficit in the other. In some manner not understood, the temporal lobes integrate written and oral symbols, using different neural circuits, so as to make those symbols meaningful.

Occipital Lobes

Lying in the rearmost portion of the cortex is the occipital lobe. It is here that the primary visual area resides. If this area is stimulated in an awake patient, the response is a visual one. The patient will report seeing some type of light, but not necessarily an object or scene. This part of the CNS is discussed in more detail in Chapter 7, with a more complete description of the visual system.

Hindbrain

The hindbrain, also called the **brain stem,** is located just above the spinal cord and comprises three major divisions: the medulla oblongata, or medulla, the pons, and the cerebellum. Structures in this part of the brain also play an important role in emotional responses, sleep patterns, and some pleasurable sensations.

Medulla

Although this area of the brain is small compared to the ones described above, the functions that reside in the **medulla** are most basic to survival. The medulla contains groups of neurons, or nuclei, that control cardiovascular and respiratory function. Information from the lungs, heart, and vascular receptors is relayed to these centers, where they are integrated with other neural and hormonal stimuli, as is described in Chapters 8 and 10. (The receptors mentioned above are not to be confused with receptor molecules located on cell membranes, as discussed in the preceding chapter. You will see in Chapters 8 and 10 that receptor organs exist in the circulatory and respiratory systems that are able to sense alterations in pressure and gas concentrations of blood, and signal those changes to the brain. Specialized nerve endings in those organs respond to the parameter being sensed.) The result is to keep blood pressure at the proper level and respiration at an appropriate rate and depth. All this is done at the subconscious level. You have no sensation that allows you to become aware of your own blood pressure, nor do you often consciously regulate the rate and depth of your breathing. Should those nuclei suffer damage, control of blood pressure and respiration would be severely compromised, possibly leading to death. Figure 4.11 shows this part of the brain with the other areas cut away for clarity. The more important structures are labeled along with the functions they control.

Pons

The **pons** (a name that means bridge) is located immediately above the medulla and contains nerve tracts

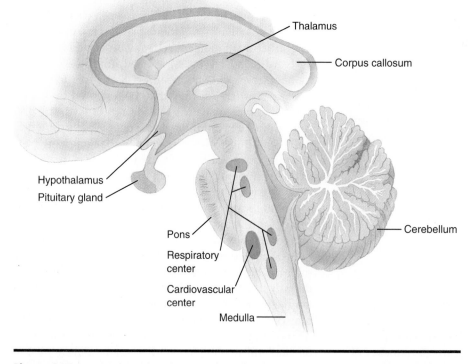

Thalamus

Corpus callosum

Hypothalamus
Pituitary gland

Pons

Respiratory center

Cardiovascular center

Medulla

Cerebellum

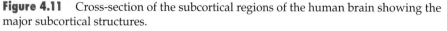

Figure 4.11 Cross-section of the subcortical regions of the human brain showing the major subcortical structures.

connecting the spinal cord with other areas of the brain. Within the pons are the origins of the V, VI, and VII spinal nerves, as well as nuclei that control, in part, the rhythmicity of respiration.

Cerebellum

The **cerebellum,** or little brain, lies just below and behind the occipital and frontal lobes of the cerebrum. Its structure is irregular, composed of what appear to be many folded leaves layered one on the other. Despite the fact that the neuronal connections of the cerebellum are well-understood, little is known about the cerebellum's exact role in muscular function. What is clear from some ablation experiments and from human patients who have suffered damage to the cerebellum is that it receives information from the spinal cord relating to the position of muscle groups in many parts of the body. This suggests that the cerebellum is an important controller of muscle tone, particularly of the muscles responsible for maintaining posture.

In addition, fine control over repetitive movements seems to rely on a well-functioning cerebellum. The drunk person who cannot touch his nose or walk a straight line has impaired cerebellar function. A pigeon with a cerebellar lesion may fly perfectly well, but not be able to land on a branch of a tree, either missing it completely or barely managing to grasp the limb. The cerebellum also appears to be necessary for the rapid, coordinated movements required of a piano player or a typist.

Major Subcortical Structures

Beneath the cortex of the brain are many structures that receive and send information to other parts of the brain. Electrophysiological experiments have allowed researchers to map the connections and determine some of the functions of those structures.

Thalamus

A large oval mass of neural tissue lying alongside the third ventricle is the **thalamus.** The largest portion of the thalamus is composed of cell bodies that receive input from sensory fibers. For this reason it is often called the relay station of the brain. It integrates neural activity from both the cerebrum and the periphery, relaying that information to the cerebellum, the cortex, and other brain areas. It also connects with the cerebral cortex and the hypothalamus and is involved with motor function and emotional processing.

Hypothalamus

Lying just below and forming part of the floor and sides of the third ventricle is the **hypothalamus.** It is an area of densely packed cell bodies that regulate many of the homeostatic mechanisms of the body. The hypothalamus plays an important role in the homeostatic control of your body. For our purposes, however, it is necessary only to point out that it helps regulate biological rhythms such as sleeping and waking, body temperature, blood pressure, salt and water balance of the body, heart rate, and behavioral drives such as thirst, hunger, and sex.

Limbic System

The **limbic system** is not a separate, well-defined structure, but a ring of structures that includes part of the thalamus and hypothalamus. This complex set of interconnecting pathways plays an important role in emotion, behavior, and motivation. Stimulating specific parts of the limbic system with electrodes implanted in the brain of conscious animals may elicit emotional changes. For example, a charging bull may change almost immediately to a passive, nonaggressive animal when stimulated in the proper area of the limbic system. Conversely, a quiet, passive animal may exhibit rage and aggressive behavior when stimulated in a different area. Conscious humans may report feelings of great pleasure if the electrode is placed appropriately.

Human experimentation is often possible during brain surgery. It is sometimes necessary to open the skull of a patient, exposing the brain surface. The surgeon cuts the bone while the patient is completely anesthetized with a general anesthetic. The patient is then allowed to become fully conscious and experiences no pain when the brain tissue is cut or even stimulated electrically. The patient can respond verbally to any electrical stimulation applied to his or her brain. Although these experiments are naturally limited, they have produced excellent information concerning localization of brain function, and in general support findings derived from animal experiments.

Experiments such as these indicate that facial expressions characteristic of emotion are not learned responses but are hard-wired in the limbic system. At birth, the emotional cues such as happiness, fear, and sadness are already set in place in the deep structures of the brain. A strong piece of evidence that facial signals are not learned is that people who are born blind smile when happy and display other facial signs of emotion when appropriate, yet they have never seen those expressions.

Reticular Formation

A network of fibers running from the spinal cord up through the brain stem is called the **reticular formation.** The reticular formation is composed of widely spread neurons that signal the cortex and act as a wake-up call, promoting arousal, alertness, and attention. When researchers selectively destroy some areas of this network, the animal exhibits hyperarousal. Destruction of other areas of the reticular system results in coma.

Brain Neurotransmitters

As discussed in the preceding chapter, nerves communicate with each other by releasing chemical messengers at their terminal ends. They diffuse across the space between the terminal end of one nerve and the cell body of a neighboring nerve, combining with a specific receptor molecule on the postsynaptic membrane, causing it to depolarize, or even hyperpolarize. You learned in Chapter 3 that some of the major neurotransmitters are acetylcholine, epinephrine, norepinephrine, and dopamine. These are also significant within the brain. Many of the neurons of the CNS release or respond to those chemicals. However, in addition to these three chemicals, others play important roles in transferring information from one neuron to another. They may be divided into two general categories.

Monoamines

These neurotransmitters are modified amino acids including epinephrine, norepinephrine, **serotonin,** hista-

mine, and dopamine. Of these, dopamine is of particular interest. Patients with **Parkinson's disease,** a chronic, progressive condition of rigidity and progressive paralysis, exhibit deterioration of the neurons that produce dopamine. If some medical treatment can supply dopamine to the proper part of the brain, it may alleviate considerably the symptoms of Parkinsonism.

Neuropeptides

This more recently discovered class of neurotransmitters is composed of short chains of amino acids. Some of these neuropeptides, such as arginine, vasopressin, and angiotensin II, are also hormones. Within this class of neurotransmitters are the **enkephlins** and **endorphins,** which are sometimes called **opioid neurotransmitters.** They bind to the same receptor sites as does morphine. Endorphins create a general sense of well-being ranging from reduced perception of pain to euphoria. The well-known runner's high is believed to result from the release of endogenous opioid neurotransmitters. In addition to these neuropeptides, many others have been isolated and shown to effect neurotransmission. The number of compounds identified gets larger each year, and for our purposes it is not necessary to list them all. You should note, however, that the brain makes many polypeptide neurotransmitters that play some still unknown role in the transmission and storage of information in the brain.

Nitric Oxide

A new neuroactive compound has been discovered that seems to have wide biological effects. **Nitric oxide (NO)** is a very reactive gas, and oddly enough is a highly toxic one. NO is one of a class of highly react-

HIGHLIGHT 4.1

Alzheimer's Disease

Alzheimer's disease is one of the most common degenerative diseases of the brain, named for Alois Alzheimer, who first described it. For reasons that are unknown, large areas of the brain in some people may atrophy. The weight of their brains decreases, as does the number of functioning neurons. The most consistent finding in these patients is that the activity of an enzyme required for the formation of acetylcholine is severely reduced. The implication of this finding is that cholinergic neurons of the brain are selectively lost. The loss appears greatest in the cortex and in some of the deeper nuclei in the brain.

People who have this disease exhibit progressive mental deterioration, memory loss, and disorientation of time and place. As the disease progresses the patients may no longer recognize their closest relatives. They may completely lose contact with the world about them, despite the fact that they appear to be able to see, hear, and talk normally. Researchers have shown that the degree of loss of cognitive ability is proportional to the loss of the enzyme required for the synthesis of acetylcholine.

MILESTONE 4.2

Fetal Tissue Implants and Parkinson's Disease

In 1817, Dr. James Parkinson recognized that a similar constellation of symptoms was present in a number of his patients. The first sign was a reduction in the strength of their muscles. Over time a rhythmic tremor developed and gradually progressed to paralysis. Despite the generalized paralysis and debility, there was no loss of mental abilities. This degenerative disease was given the name Parkinson's disease, in recognition of the person who first reported it as being a separate disease entity. No hint of its cause was found until the mid-twentieth century.

Dr. Oleh Hornykiewicz noted as he performed autopsies that patients with Parkinson's disease exhibited lesions in an area of the brain called the basal ganglia. It was known at that time that the basal ganglia monitor voluntary motor movements. The basal ganglia receive signals from active motor fibers and so regulate motor neuron output to ensure smooth and coordinated movement. But more important to an understanding of Parkinson's disease was the knowledge that neurons in the basal ganglia (as well as some others associated with it) synthesize dopamine, and that the autopsied brain areas of patients with Parkinson's disease showed lower than normal dopamine content.

These findings led to the hypothesis that a lack of dopamine was at the root of the disease, and finally to the conclusion that Parkinsonism is a dopamine deficiency state that results from disease or injury to the neurons that produce dopamine. Thus, if dopamine could be supplied to the patients, their condition might improve. Fortunately a precursor to dopamine, L-DOPA, can be given orally and is able to cross the blood–brain barrier. Early treatment with L-DOPA pro-

duced dramatic improvement in many patients, but was unable to stop the progression of the disease. Unfortunately, L-DOPA also causes some serious side effects at doses high enough to improve the symptoms. What then?

Animal research led to the discovery of an animal model for Parkinson's disease. These animals had a small piece of their adrenal medulla surgically implanted in the area of the brain that requires dopamine. Because the adrenal medulla makes a compound related to dopamine, it was a way of increasing the dopamine content of the basal ganglia without subjecting the entire body to high levels of dopamine. Animals so treated showed marked improvement. This method is unsuitable for humans, as adrenal tissue is difficult to obtain and it does not secrete dopamine. But it did show that if researchers could find a method to supply dopamine to the proper brain area, the symptoms of Parkinson's disease might be alleviated. The results ultimately led to another technique that holds promise in humans.

Human fetal tissue may be obtained from spontaneously aborted fetuses. This tissue can then be surgically implanted in the brain of patients so that the dopamine it produces will increase the level in the basal ganglia. Unlike tissue taken from an adult, fetal tissue has the advantage that it can be introduced into a recipient and not be rejected. The research involving fetal tissue implants holds promise not only for patients with Parkinson's disease, but also for patients who suffer from a lack of certain other hormones and neurotransmitters. As yet this is an experimental procedure, but it may have great value in treating many diseases.

ive compounds known as free radicals. These chemicals, generated normally, are thought to be damaging to cellular molecules. Fortunately, each cell possesses other chemicals or enzymes that rapidly destroy or inactivate the free radicals.

NO is not a chemical that one would have thought to be of biological importance, yet we now know that many neurons of the CNS, as well as other cell types, contain receptors for NO, and that these receptors control many important functions. (See Milestone 4.3 for a more detailed description of the gas.) Among its actions, NO has been shown to play an important role in regulating blood pressure by its action on the cardiovascular center of the brain. If a drug that inhibits the production of NO is injected

into the ventricular system of the brain and reaches the brain's cardiovascular center, which controls blood pressure, the animal's blood pressure increases dramatically. This is powerful evidence that normal blood pressure is maintained, at least in part, by continual release of NO from nerve terminals in the medullary cardiovascular center.

The neurotransmitters described above are only a few of the many neurotransmitters known and thought to be active in conducting impulses across synapses. A more complete list appears in Table 4.2; it is not all-inclusive, but should impress on you the great number and chemical variability of neurotransmitters. Undoubtedly, many more neurotransmitters will be discovered in the future.

Table 4.2	Some chemicals known or thought to act as neurotransmitters.

Peptides	Nonpeptides
ACTH	**Acetylcholine***
Angiotensin II	**Dopamine**
Bradykinin	**Epinephrine**
Endorphins	**Gamma-aminobutyric acid (GABA)**
Gastrin	**Glutamate**
Glucagon	Glycine
Growth hormone releasing factor	**Histamine**
Insulin	**Nitric oxide**
Oxytocin	**Norepinephrine**
Somatostatin	**Serotonin**
Substance P	
Thyrotropin releasing factor	
Vasopressin	

* Boldface type denotes major known neurotransmitters.

Peripheral Nervous System

The extensive network of nerves exiting from the brain and spinal cord makes up the peripheral nervous system, which is further subdivided into the somatic nervous system and the autonomic nervous system, each of which has a different functional role. The somatic nervous system is the *voluntary* nervous system. Sensory nerves supply information that is generally, but not always, experienced at the conscious level through vision, hearing, and smell, as well as through muscles, joints, and skin. The output of the somatic nervous system is primarily to skeletal muscle that is under voluntary control.

The autonomic nervous system, in contrast, innervates glands, blood vessels and other structures *not* under voluntary control. Sensory fibers in the autonomic nervous system monitor blood pressure, gastrointestinal motility, heart distension, and other homeostatic systems, none of which are generally sensed by a person, except in rare circumstances.

MILESTONE 4.3

Discovery of the Role of Nitric Oxide

For many years it has been known that acetylcholine dilates blood vessels. Investigators removed small arteries from animals and measured their contractile properties by adding drugs to the bathing solutions in which they were submerged. If acetylcholine was added to the fluid, the arterial smooth muscle relaxed. This simple experiment seemed to indicate that the nerves controlling the arterial smooth muscle released acetylcholine, which directly caused relaxation. However, in 1980 Robert Furchgott removed the endothelium, the internal layer of cells from arteries, and showed that when the endothelium was removed acetylcholine had no effect on relaxation. He proposed that acetylcholine receptors in the endothelial cells responded by releasing a relaxing factor, which caused the muscle to relax. He called this as yet undetermined factor **endothelium-derived relaxing factor** (EDRF).

Extensive efforts were directed at identifying EDRF, and in 1987 it was discovered that EDRF was the simple gas nitric oxide (NO). Further investigations revealed many unusual characteristics of NO. Instead of being manufactured in advance and stored in granules or vesicles, as are other neurotransmitters, it is produced only on demand and is very quickly metabolized compared to other neurotransmitters. Within a short time, scientists demonstrated that NO was formed from the amino acid arginine in the presence of an enzyme called nitric oxide synthase. NO rapidly escapes from the cell by simple diffusion and exerts its effect on the nearby cell. In the above case, it acts on the arterial smooth muscle and causes relaxation. Further studies showed that the distribution of nitric oxide synthase is widespread, occurring in many parts of the brain. But how can we measure its action in the brain?

Once researchers discovered the nature of the chemical reaction that forms NO, it became possible to synthesize an inhibitor of nitric oxide synthase. If the inhibitor is injected or infused into special brain areas, those areas no longer produce NO. Any resulting physiological difference noted can then be ascribed to the loss of action of NO. In this way, we see that NO in the brain tissue plays important roles in blood pressure control, hormone release by the pituitary gland, and perhaps memory processing. Among the very exciting features of this new knowledge is evidence that a simple gas is a neurotransmitter. There is even highly suggestive evidence that carbon monoxide may also be a neurotransmitter. If so, it will open up a new field of investigation and perhaps bring us a better understanding of brain function.

Somatic Nervous System

The cranial nerves, as noted previously, exit the brain directly through openings in the skull. They are immune from any type of compression originating from damage to the vertebrae. Even transection of the uppermost levels of the spinal cord will leave intact all efferent and afferent pathways of the cranial nerves. With the exception of the vagus nerve, they are all part of the somatic nervous system. The other portion of the somatic nervous system is made up of spinal nerves that leave the spinal cord at each junction between adjacent vertebrae.

Figure 4.12 shows the arrangement in greater detail, adding two adjacent spinal segments. The possible flow of information along the spinal nerves is more complex than in Figure 4.8. The dorsal roots carry information *to* the CNS via nerves called **afferent** nerves, whereas the ventral roots carry information *away* from the CNS via **efferent** nerves. In addition, within the spinal cord branches of the afferent and efferent nerves may travel up or down the spinal cord, activating other neurons at higher or lower levels.

Sensory information from receptors that are consciously appreciated, such as pain and temperature, travel along a spinal nerve, but shortly before entering the spinal cord diverge from that nerve to form the dorsal root. There the fibers enter a small swelling in the dorsal root called the **dorsal root ganglion.** The ganglion contains the cell bodies of the sensory fibers, which cause the swelling in the nerve root. The axon of the sensory fiber continues out of the ganglion into the spinal cord, where it may take one of many routes. It might synapse directly with a motor fiber, or

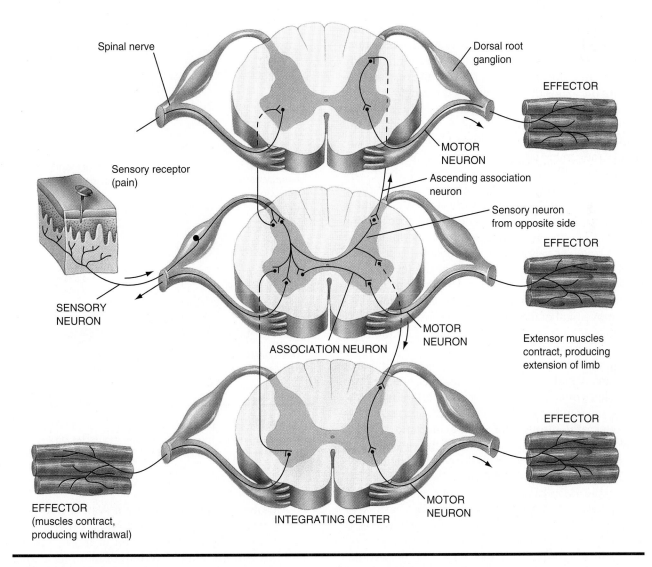

Figure 4.12 Representation of the complexity of reflex arcs. Although only three spinal levels are presented, a reflex may involve more than three spinal segments.

synapse with another fiber called an interneuron, which in turn synapses with a motor fiber. The interneuron, in turn, may synapse with a motor fiber at the same level as entry, or travel up or down the cord to one in another level of the spinal cord. Alternatively, the sensory fiber might travel up the spinal cord to the brain, where a particular sensation may be registered.

Cell bodies of the motor fibers lie within the spinal cord, and their axons leave the cord at the same level as the cell body. They then travel through the ventral roots, joining with the dorsal roots to form a spinal nerve. The axon continues along the spinal nerve to specific muscle groups served by that spinal nerve. From this description it is evident that spinal nerves are mixed nerves containing both efferent fibers and afferent fibers, as indicated by the arrows in Figure 4.12.

The division in the spinal cord between the areas containing cell bodies, the gray area, and the area containing axons is easily seen. The innermost part of the spinal cord is butterfly-shaped, containing a high density of cell bodies, which are not covered with myelin. Surrounding the gray matter is the white matter, containing the axons of the myelinated nerve fibers.

A final set of axon bundles must be added to complete the wiring diagram of the motor nerves of the spinal cord. These are diagrammed in Figure 4.13. Voluntary movement begins with a conscious effort originating in cell bodies in the cortex of the brain, signaling cells in the motor cortex. Axons from these cells travel through the lower parts of the brain, where they cross to the opposite side at the level of the medulla. Another system travels down the spinal cord to the level at which they synapse with a motor neuron. They exit the spinal cord at the level of this synapse. It is this crossing, or **decussation,** that accounts for the fact that the right side of the brain controls movement of the left side of the body and the left motor cortex controls muscle movement on the right side of the body.

Figure 4.13 shows only two descending pathways of the many pathways for motor neurons, but these are enough for our purposes. What is important to understand is that motor fibers originate in the cortex, follow discrete paths through the spinal cord, and synapse with the final motor neuron called the **lower motor neuron.** Note that the axons of the pyramidal fibers extend from the brain to the level of the spinal cord at which the postsynaptic motor fiber exits the cord. Thus, the axons of the pyramidal cells may be three or more feet in length. If the axon of the motor fiber is severed, it will regenerate. However, should the axon of the nerves contained in the spinal cord be severed, no regeneration takes place. The muscles below the point of cleavage will be permanently paralyzed.

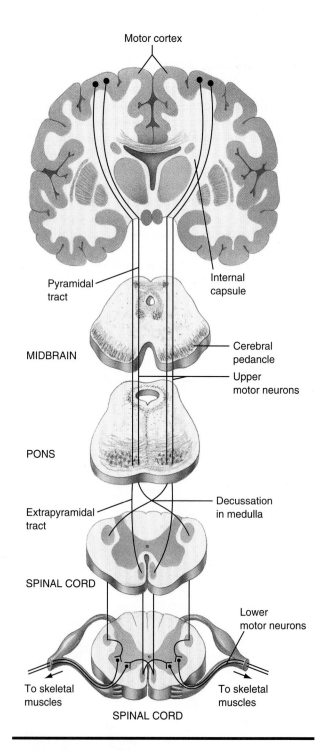

Figure 4.13   Pyramidal tracts showing cross-over, or decussation, in the medulla. Voluntary muscles on one side of the body are controlled by the opposite side of the brain.

Although the pyramidal motor system is responsible for innervating muscle groups, other motor neurons also contribute to control of muscle movement. An extrapyramidal system of neurons arising from the cerebellum and other structures in the brain monitors the activity of the pyramidal system and coordinates overall muscle movement so that it is smooth and well-directed.

Almost every muscle attaches to two bones so that contraction of the muscle causes a bone to move away from or toward the other bone. An opposing muscle group attaches to the two bones so that contraction of that muscle will oppose the movement of the first muscle group. Figure 4.14 shows this arrangement for the leg and two of its major muscles. Contraction of the biceps femoris causes the leg to flex. Contraction of the quadriceps muscle has the opposite effect, causing the leg to extend. When the leg flexes, not only does the biceps femoris contract, but the quadriceps muscle is also slightly contracted. Thus, as one muscle group contracts the other does not relax completely; rather, as one muscle group is stimulated to increase its contractile state, the opposing muscle group undergoes a relative relaxation, remaining partially contracted.

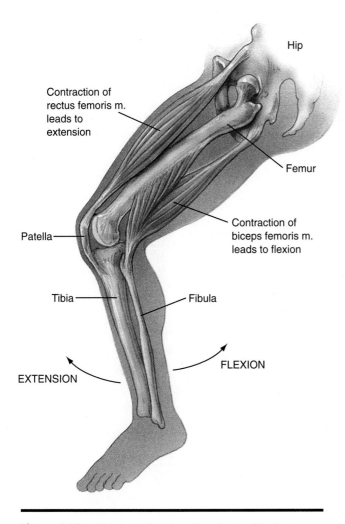

Figure 4.14 Diagram showing the relationship between flexor and extensor muscles of the leg.

Contraction of the two opposing muscle groups, with one in a greater contractile state, serves to stabilize the contraction and position of the limb. Contraction of only a flexor muscle would result in a poorly controlled movement that would be unstable at each point in the movement. Action of opposing pairs of extensors and flexors is critical for the maintenance of erect posture. Contraction of opposing muscles serves to strengthen the position of any limb. This dynamic is analogous to the guy wires that pull in opposite directions to maintain a tent pole in the erect position. You shall see when we discuss reflex activity that other pathways and control systems stimulate and inhibit muscle groups without any conscious input.

A normal person is able to close his or her eyes and easily touch the tip of his or her nose without difficulty. To do so requires precise knowledge of the spatial position of the finger being moved and the static position of the nose. Information relating to the position of limbs comes from receptors in the joints and muscles that inform the brain of position. Without such knowledge, precise movement is not possible. Anyone who has bitten her lip following a dental appointment during which a local anesthetic was injected understands this. The local anesthetic not only prevents the sensation of pain, but it also prevents knowledge of the precise position of the anesthetized area. Receptors that signal position are called **proprioceptors.**

Autonomic Nervous System

In the previous section you saw that the somatic nervous system receives sensory input from skin, muscle, viscera, and sense organs, all of which are perceived at the conscious level. The efferent side of the somatic system is also under conscious control and sends fibers to voluntary muscles. However, there is another set of nerves that are called autonomic, as they are generally involved in activity that does not reach the conscious level. In general, the nerve fibers of the autonomic nervous system regulate the internal environment. They convey sensory (afferent) input from internal organs such as the glands and the heart and detectors throughout the body that sense blood pressure, gas concentrations in the blood, ion concentrations in blood, and a host of other detectors that monitor the internal environment. The efferent nerves of the autonomic nervous system regulate glandular and visceral activity as well as the heart and smooth muscle of the circulatory system.

Although the autonomic nervous system functions as a unit in that it controls the homeostatic balance of the body, two distinct divisions of the autonomic nervous system exist: the sympathetic and the parasympathetic divisions. The anatomy of these divisions is different from that of the somatic nervous system, as some of the cell bodies of the sympathetic and parasympathetic nervous systems are located *outside* of the spinal column and brain. Small areas containing the cell bodies are called **ganglia.** In the autonomic nerves located outside the brain they are seen as small swellings (see Figure 4.15).

The autonomic nerves that exit the spinal cord and enter ganglia synapse with cell bodies within the ganglia, and so are called **preganglionic** fibers. Almost all preganglionic fibers release the neurotransmitter acetylcholine from their nerve endings and thus are called cholinergic fibers. The axons running from the ganglia to the organs of innervation are called **postganglionic** fibers. The relationship of the ganglia to the target organ, as well as the type of neurotransmitter released, separates the sympathetic division from the parasympathetic division. The sympathetic nervous system also is functionally different from the parasympathetic nervous system. Those differences are discussed below. Figure 4.16 illustrates the anatomical distribution of both the sympathetic and parasympathetic divisions of the autonomic nervous system.

Sympathetic Nervous System Preganglionic sympathetic fibers exit the spinal cord and enter ganglia near the cord, but relatively far from the target organ. Most sympathetic ganglia lie in a series along the spinal column called the **sympathetic chain.** Figure 4.16(A) shows that the sympathetic fibers exit the spinal cord from the posterior gray horn and travel in the nerves of the posterior root. They then enter a ganglion at the same spinal level, or the nerve may travel up or down the sympathetic chain and enter a ganglion at another location. Within the ganglion the fiber synapses with a postganglionic neuron, which in turn may travel some distance before terminating in **adrenergic** endings that release the neurotransmitter norepinephrine. General excitation of the sympathetic nerves occurs during situations that require flight-or-fight reactions. This is the experience you have all encountered when suddenly startled, or when you have some reason to fear for your safety. Your heart rate speeds up, your pupils dilate so that more light reaches the inner eye, and blood flows away from the skin, making it look pale. Nonvital functions slow down, such as intestinal digestion and flow of blood to internal organs. It is as if the sympathetic nervous system were designed to prepare an individual either to flee from some threatening situation or to fight for survival.

Parasympathetic Nervous System Preganglionic fibers of the parasympathetic division exit the CNS from some of the cranial nerves and from the lowest level of the spinal cord, the sacral segments. The nerve endings of the preganglionic fibers release acetylcholine and so are also called cholinergic fibers. However, the ganglia containing the cell bodies of the postganglionic fibers are far removed from the CNS, and are close to the organ they innervate. Postganglionic parasympathetic fibers release acetylcholine at their nerve endings and, in general, have

Figure 4.15 Diagram of the anatomical relationships of the autonomic nervous system. The sympathetic ganglia are located at a distance from the organ of innervation, whereas parasympathetic ganglia are near their organ of innervation.

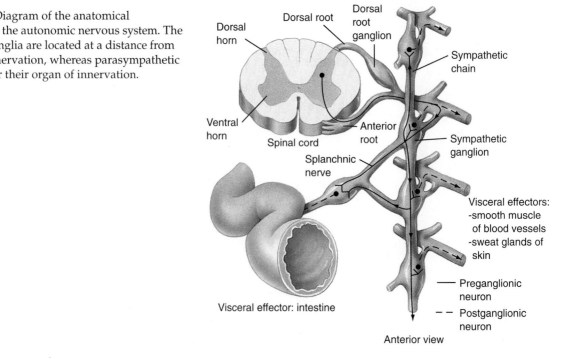

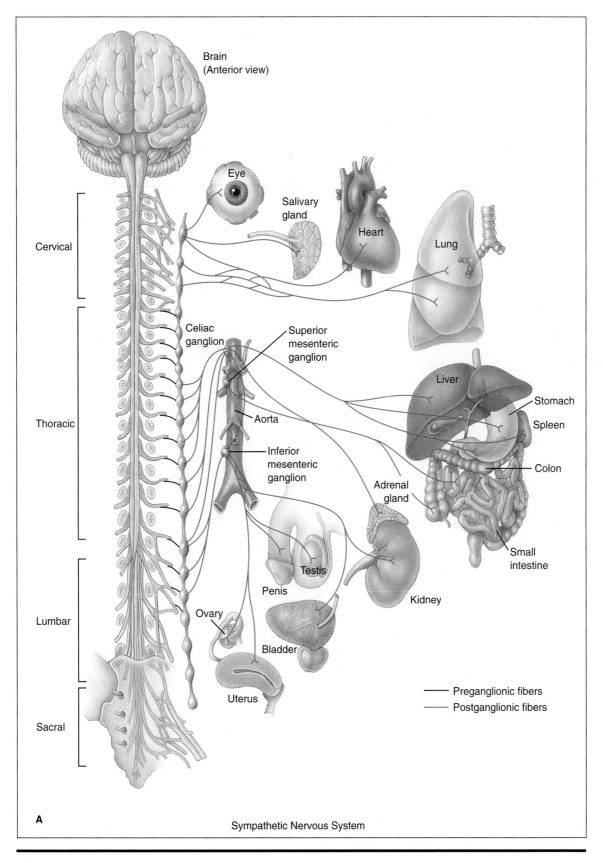

Brain
(Anterior view)

Cervical

Eye

Salivary
gland

Heart

Lung

Thoracic

Celiac
ganglion

Superior
mesenteric
ganglion

Aorta

Inferior
mesenteric
ganglion

Liver

Stomach

Spleen

Colon

Adrenal
gland

Small
intestine

Testis

Penis

Kidney

Lumbar

Ovary

Bladder

Uterus

—— Preganglionic fibers
—— Postganglionic fibers

Sacral

A

Sympathetic Nervous System

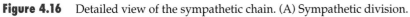

Figure 4.16   Detailed view of the sympathetic chain. (A) Sympathetic division.

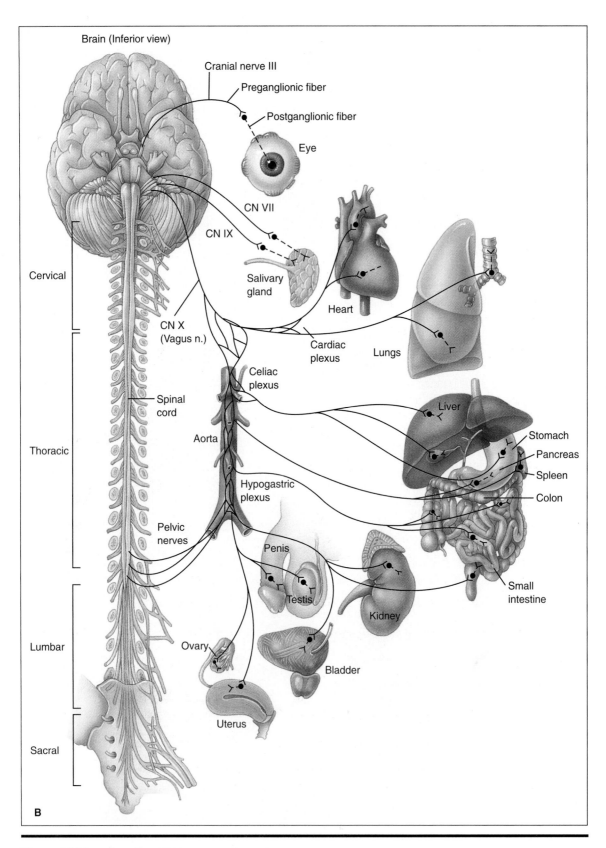

Brain (Inferior view)

Cranial nerve III

Preganglionic fiber

Postganglionic fiber

Eye

CN VII

CN IX

Cervical

Salivary gland

Heart

CN X (Vagus n.)

Cardiac plexus

Lungs

Celiac plexus

Spinal cord

Liver

Aorta

Stomach

Thoracic

Pancreas

Spleen

Hypogastric plexus

Colon

Pelvic nerves

Penis

Small intestine

Testis

Kidney

Lumbar

Ovary

Bladder

Uterus

Sacral

B

Figure 4.16 (continued)   (B) Parasympathetic division.

physiological effects that are opposed to those of the sympathetic fibers. For example, stimulation of the parasympathetic nerve to the heart, the vagus nerve, causes the heart to slow down.

From the above description you can see that many organs are dually innervated, receiving both sympathetic and parasympathetic postganglionic fibers. This may give the impression that the systems work in an alternate manner; that is, only one system would be active at a time. However, that is not the case. Generally both are active simultaneously, and depending on the circumstances one becomes more active as the other is inhibited. This may be demonstrated in an anesthetized animal. If the heart rate is measured and the parasympathetic supply to the heart is severed, the heart rate increases, indicating some tonic parasympathetic inhibition of the heart rate. On the other hand, if only the sympathetic supply to the heart is interrupted, the heart rate decreases, indicating tonic sympathetic stimulation of the heart rate. Both divisions seem to act simultaneously in opposing directions. This dual, tonic influence allows for a more finely regulated control system. It is as if both a brake and an accelerator are applied to the heart simultaneously, allowing for a much finer control and more rapid response than if only one system were operative. Table 4.3 presents a functional comparison of both systems, showing how stimulation of each system causes very different effects in the body.

Reflexes

Although humans are thinking beings, possessing the largest cortical brain area in the animal kingdom, clearly not all of our actions are accomplished at the conscious level. Actions of an involuntary nature are called **reflexes.** Many of them are obvious, such as blinking an eye when an object suddenly approaches it, withdrawal of a hand that touches a hot object, and extension of a leg when the patella tendon of the knee is lightly hit in a doctor's office. None of these responses to a stimulus requires conscious effort. They are hard-wired into our nervous system and cause movement without conscious intervention. In addition, many reflexes occur constantly within our bodies and we are never even conscious of them. Autonomic reflexes control our heart rate, blood pressure, and many other functions that are beyond our ability to detect at the conscious level, although at times we may be aware of a racing heart.

Skeletal Muscle Reflexes

Of all skeletal muscle reflexes, the simplest ones are those that are **monosynaptic.** That is, in the neural pathway of the reflex only a single synapse is involved. The patellar reflex referred to above is an excellent example.

Table 4.3	Comparison of some of the contrasting effects of stimulation of the parasympathetic and sympathetic divisions of the autonomic nervous system.	
End Organ	**Parasympathetic Stimulation**	**Sympathetic Stimulation**
Blood pressure	Decrease mean pressure	Increase mean pressure
Digestive secretions	Increase	Decrease
Eye	Pupillary constriction	Pupillary dilation
GI tract activity	Increase	Decrease
Heart	Decrease force and rate of contraction	Increase force and rate of contraction
Penis	Erection	Inhibition of erection
Respiratory passages	Decrease diameter	Increase diameter
Respiratory rate	Decrease	Increase
Skeletal muscle	Decrease blood flow	Increase blood flow
Sweat glands	Not innervated	Increase sweating

Figure 4.17 illustrates the details. A tap on the patella tendon causes the extensor muscle to stretch suddenly. Receptors in the muscle sense the sudden, passive stretch and set up an action potential in the sensory neuron. The sensory neuron enters the spinal cord, where it synapses directly with a motor neuron in the ventral horn. Activation of the motor neuron that innervates the stretched extensor muscle causes it to contract, resulting in the knee-jerk response. This pathway, which an impulse follows from receptor to effector, is called a **reflex arc.** Note that in this simple reflex only one sensory neuron and one effector or motor neuron is involved. The entire reflex arc is contained in the spinal cord, without any involvement of the brain. This is not to say that the individual is unaware of being tapped on the knee with a rubber hammer. Rather, it means that the movement at the knee joint does not require nervous input from higher centers. In fact, the knee-jerk reflex is present in people who have had their spinal cord severed above the level required for the reflex.

More complex spinal reflexes also control many of our responses to various stimuli. A good example is the crossed extensor reflex withdrawal of a leg when one steps on a tack. You may not be aware of it, but simultaneously the opposite leg is caused to straighten to allow you to take weight off the injured leg. Figure 4.18 diagrams the neural and motor pathways concerned in this complex response. The painful stimulus to one foot sends the message to the spinal cord via the sensory nerve. The sensory neuron sends one branch to the same side of the cord and another to the opposite side of the cord, where they synapse with associ-

ation neurons in the gray matter. The association neurons, in turn, synapse with motor neurons, innervating both flexor and extensor muscles of both legs. The flexor muscles of the injured side are stimulated as the extensor muscles are inhibited, causing withdrawal of the leg. On the opposite side, the extensor muscles are stimulated and the flexors inhibited, causing straightening of the opposite leg.

In addition, pain fibers also convey information to the brain and you become consciously aware of the thorn and the pain it causes. But that information is not necessary for the reflex withdrawal of the injured leg, which is controlled entirely at the level of the spinal cord; the reflex occurs even when the spinal cord is severed. This may be demonstrated in an individual with a complete section of the spinal cord cut above the level of the nerves supplying the legs. That person is unable to move his legs voluntarily because all nervous connections between the brain and the final efferent somatic nerves are severed. However, a tap on the patella will still elicit the knee-jerk reflex, withdrawal of a limb when stuck with a pin, and extension of the opposite limb.

Learned Reflexes

The reflexes discussed above are built into the nervous system and require no learning. Certainly nothing in our consciousness prepares us to extend a leg when the patella tendon is struck lightly with a hammer. However, there are other reflexes that may be developed with practice, such as those learned by

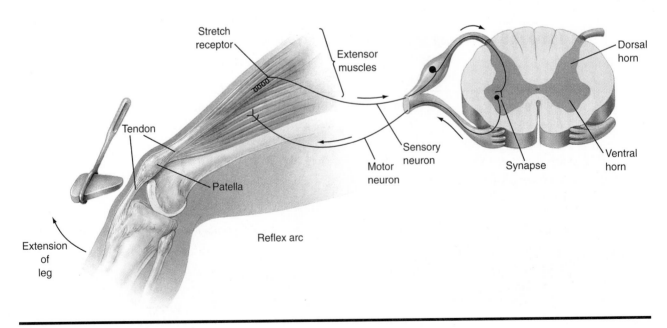

Figure 4.17 Diagram of the patella tendon reflex (knee-jerk reflex). This is the simplest of reflex arcs, involving only one synapse.

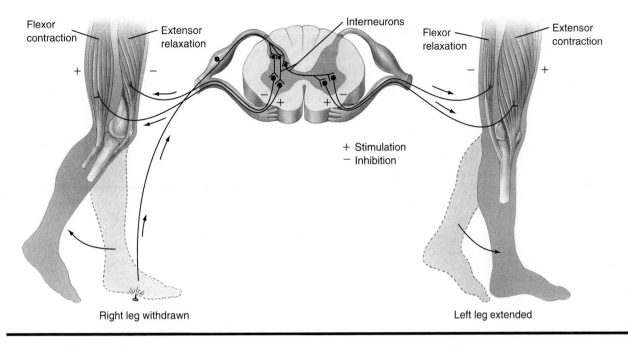

Flexor contraction

Extensor relaxation

Interneurons

Flexor relaxation

Extensor contraction

+ Stimulation
− Inhibition

Right leg withdrawn

Left leg extended

Figure 4.18 Crossed extensor reflex involving both legs. This reflex comprises multiple synapses and muscle groups.

athletes. A basketball player who is able to dribble a ball down court while avoiding defensive players actually moves reflexly much of the time. He or she learns to switch hands or spin to get around an opposing player. The player does not need to do a lot of thinking while making these movements. Rather, long hours of practice and repetitive movements set up neural circuitry that allows the athlete to perform complex, rapid movements without conscious effort.

Another example of reflex learning is seen as a competent typist types a transcript. A fast typist might type as rapidly as 75–100 words per minute. At an average of 5 letters per word, that equals 6–8 keystrokes each second. The sound or sight of each word sets off a series of very rapid hand and finger movements requiring no conscious effort on the part of the typist. A beginner, in contrast, must think of the position of each key and make the appropriate movements following some thought process. Sufficient repetition will allow the beginner to become an accomplished typist as the neuromuscular reflex pathways become established.

Consciousness

No simple all-inclusive definition of consciousness exists, although most people feel they know what it is. Certainly it must be an awake state in which we are aware of our thoughts, our surroundings, and our own physical and mental condition. If this definition comes close to defining consciousness, does it mean that it is something separate from our body? Is it the sum of some chemical and electrical activity of our brain, or is it separate from that, having a nonmaterial component? Although philosophers and scientists may argue the matter, we take here the more material view that consciousness resides in well-structured electrochemical happenings in the brain. We are able to measure alterations in brain activity that correlate well with states we would call consciousness, and we can explain our ability to see, hear, feel, and respond to stimuli in terms of altered neuronal activity.

Electroencephalogram

In the late nineteenth century, electrical activity of nerve and muscles could be studied, as refinements in recording apparatus moved at great speed. An English physiologist placed recording electrodes on the skull and was able to measure very small cyclical alterations in the electrical voltages at the surface of the skull. The voltage changes reflect the activity of many millions of brain cells. These changes in the voltage may be recorded and are called an **electroencephalogram**, or **EEG**. Figure 4.19 shows some typical recordings. An alert person with eyes open generates electrical waves, as shown in the first EEG tracing. Closing the eyes alters the pattern, as shown in the second EEG tracing. If the individual is allowed to fall asleep, EEG activity

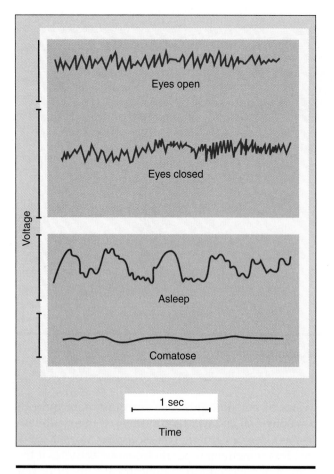

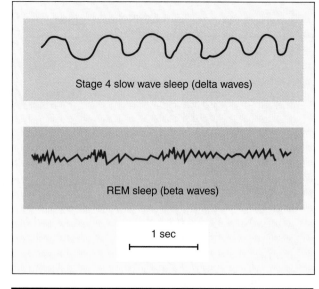

Figure 4.20 EEG waves during slow wave sleep and REM sleep.

Figure 4.19 Electrical recordings (EEG) from the skull of a person during different states of consciousness. Each state of consciousness is easily recognized by its unique EEG tracing.

changes in a well-ordered way. The third tracing in the figure presents a typical pattern of a person sleeping peacefully. The bottom tracing is one from an individual who has suffered permanent brain damage and is comatose. This tracing is sometimes called a flat-line EEG.

Sleep

The EEG tracings shown in Figure 4.19 indicate that brain activity correlates with the state of consciousness. As a person moves from an alert state to a less alert state, the overall electrical activity of the brain changes in a well-defined manner. Thus, there is an electrical correlate of the conscious state. Additionally, the EEG indicates that **sleep** is not an absence of neural activity. On the contrary, the brain is active even during deep sleep. Furthermore, the pattern of the EEG changes during sleep. As an individual progresses from drowsiness to deep sleep, the EEG slowly

changes into the pattern seen in Figure 4.19 labeled *asleep.* However, periodically the electrical activity changes into the pattern seen in the lower tracing in Figure 4.20. This period of sleep is characterized by several well-defined features, including dreaming, increased respiration and brain metabolism, and its defining characteristic, rapid eye movements. Although the eyes remain closed, they move rapidly, almost as if the person were watching a fast tennis match. Because of the rapid eye movements associated with this stage of sleep, it is called **REM sleep.** It is also called paradoxical sleep, as it is a paradox that at a time when there seems to be great brain activity, the person is sound asleep.

A fairly common error of students studying for an exam is to stay up all night cramming for the exam given in the morning. At the time of the exam, the student finds that his or her brain isn't functioning as well as it should. Information studied the night before is difficult to retrieve. From experiences such as this it is easy to understand how we have come to accept that sleep is necessary for proper brain function, although the mechanism by which this happens is not understood.

Perception of Pain

Although perception of pain is a common experience, the precise nature of that perception is not well understood. As you will learn in Chapter 7, the body contains specific receptors for pain, and pain fibers may be traced anatomically. However, in contrast to other senses, no cortical localization may be assigned to

HIGHLIGHT 4.2

Physiological Function of Sleep

Although it is evident that sleep deprivation interferes with rapid mental responses, most recent research indicates that sleep may play a more general role in normal physiology. For years we have known that animals deprived of sleep for days or weeks will die, although the cause of death was not understood because at autopsy all the animal's organs and organ systems appeared normal. Recently, however, researchers discovered that the cause of death was widespread bacterial infections. It seems that the immune system of the animals was severely compromised by sleep deprivation, allowing bacteria not normally harmful to multiply to lethal levels. Similar (but short-term) experiments were done with human volunteers. In the short term the immune system was boosted by sleep deprivation, in that some of the white blood cells and chemical messengers associated with a well-functioning immune system actually increased in the blood. (You will learn in Chapters 9 and 10 that the white blood cells play an important role in warding off infection.)

Although the data obtained thus far do not clearly define the function of sleep, it does appear to be required for more than just restoration of mental abilities. Sleep may also play an important role in maintaining the immune system in a functional state.

---

pain. Regardless of where the brain is stimulated electrically, pain is not experienced.

Unlike other systems, the brain also appears to have a built-in system for the suppression of pain. CNS neurons may release opioid neurotransmitters, called endogenous opioids. They have the ability to reduce the perception of pain, apparently by blocking a pain-producing substance called **substance P.**

The intensity of pain felt is a function not only of the strength of the stimulus but also of a host of psychological factors. Should one sustain an injury during a football game, the pain may not be felt until long after the injury, when the person is no longer intensely involved in the game. It is also evident that there is a large placebo effect. About 30% of people with chronic pain report a lessening of pain if told they are getting morphine but are given an injection of salt water instead. Perhaps this is analogous to the crying child who finds the pain of a scraped knee disappears when kissed by his or her mother.

Learning and Memory

A newborn child might seem to be an individual with a clean slate to be filled with all that he or she can learn in a lifetime. But this would be incorrect. Even at birth many stimulus-response actions are built-in. The newborn does not need to be taught to suckle at the breast. That reflex is built into the nervous system, a result of thousands of years of evolution. If the child is placed in a cold environment, goose bumps will appear on the skin. But the child has much to learn, and will learn many things in his or her life.

All **learning** depends on the train of neurochemical events called **memory.** You may meet someone and sometime later you remember his name when you see him again. Your memory of his name, however, may fade with time, and at a much later date you might not be able to recall it. At a still later date it may not even be possible for you to remember the circumstances under which you met, or even that you did meet. This description of one simple example of learning is not too different from all other learning experiences, whether it is learning physiology as it appears in this book, or learning that an acquaintance is trustworthy, or learning the facts that are necessary for you to perform your job. Learning is the acquisition of facts, or knowledge, and it is less than perfect.

We all forget things, some things more easily and more quickly than others. Memory is spoken of as having two subtypes: short-term memory and long-term memory. Although there is no precise definition of those two terms, it is clear that remembering a name for a day is different from remembering your mother's birthday. One you might forget easily and quickly. The other you should remember for your entire life! We must emphasize at this point that the neural circuits accounting for memory must be very large. Multiple synapses must be involved in both short-term memory and long-term memory, perhaps even thousands of synapses. The important point is that there is a difference between them, and that the difference seems to reside in the ease with which the neural circuits may be activated.

Short-Term Memory It is not possible to define the neurological events that make up any individual memory, but it is possible to understand some of the events that must take place in the CNS for a memory to be stored. In brief, a new event imposes itself on the

CNS by causing a particular set of neurons to interact in a unique manner. The event begins when receptors (such as visual, auditory, and pain) are stimulated and the neural elements connected to the receptors carry that information to the brain. Stated differently, a new series of synapses is stimulated to release the neurotransmitter substance of the synapses. Immediately following such an event, recall is easy. If you look quickly at the series of numbers 2,7,8,3,6,6,3, you will be able to remember them for a time and recite their order easily. By tomorrow, or even a few minutes from now, you probably won't be able to recall those numbers, yet you can recall other sets of seven numbers readily. You must store in your head many telephone numbers you use regularly, including those of people you care about, over time. Those long-remembered numbers are evidence of long-term memory. The difference between which numbers you remember and which ones you forget must reside in the permanence of the synaptic circuits involved in wiring those numbers into your brain. Evidence from simple animals having very few brain cells has shown that as a stimulus is repeatedly presented to the animal, the synapses affected by the stimulus may fire more readily in response to later stimuli. Repetition changes some characteristic of synaptic transmission.

Short-term memory is not simply a result of repeated exposures to the same experience. Rather, there seems to be a requirement for a neural circuit that allows recording of each new experience. A sad but excellent example of this is the record of a man who suffered a stroke in both temporal lobes. He recovered quite well physically and his long-term memory was intact. However, his short-term memory was essentially gone. He could keep a conversation in his memory for about 15 minutes and then it would disappear. The patient was unable to recount any of the events that took place during the day. In essence, he was living only in the immediate present, having no recall of the most recent events. It was as if the neural circuit necessary for recording and storing new events was missing.

Long-Term Memory You hold many **long-term memories** in your CNS, as mentioned above. How do the changes in synaptic transmission differ between short-term memory and long-term memory? In the example of telephone numbers, the neural circuitry must be the same or very similar, in that the neurons involved are the same for both long-term and short-term memory. What, then, is the difference? Short-term memory seems to affect the sensitivity of only some synapses in releasing its neurotransmitter, perhaps increasing the number of receptors on a given cell. Long-term memory traces appear to be of a different character, and may actually be visualized with the aid of electron microscopy. Again, researchers

using simple animals and repeatedly stimulating a nerve so that long-term memory takes place are able to demonstrate anatomical changes in the affected synapse. The total area at the presynaptic terminal of the vesicles containing the neurotransmitter increases. If that synapse is made to remain inactive for a time, the area of the vesicles decreases.

Continued stimulation of the same set of synapses causes an increase not only in vesicular area, but also in the number of vesicles in the presynaptic terminal. The increase in the size and number of synaptic vesicles indicates that more neurotransmitter is released for each action potential and so makes the receptor cell fire more easily. If we carry this information over to the newborn, it is reasonable to consider long-term memory, or learning, the continued accumulation of new synaptic circuits that fire more easily than non-stimulated synapses.

Storage of Memories

At the beginning of this chapter we made an analogy between the CNS and the central computer of a fictional robot. Computers may also store information and might be considered repositories of memory. As new information is put into a computer, it may be stored in a random manner wherever an empty place exists on the hard disc. Regardless of the type of information entered, the place of storage can be random, although computers may be designed to store in a nonrandom manner. The human brain, however, does not work like the computer model. The quality of the information to be stored determines the location to which it is assigned. You have already seen that verbal memories have their neural circuits located in the temporal and parietal lobes. If either Broca's area or Wernicke's area sustains trauma, memories having to do with speech and language are impaired. Our brains have memory traces assigned to specific areas, depending on the character of the memory. For example, our ability to recognize faces depends on the integrity of a portion of the brain located on the underside of the occipital and temporal lobes.

The filing system of memory traces in the brain is such that localized damage does not wipe out memory traces having to do with all learning in a time-related fashion, as would be the case if each succeeding fact were stored in the order of learning in the same location. Instead, localized damage may leave an individual deprived of one type of memory but in possession of all other types.

False Memories

You have seen that memories may be fragile; often, forgetting is easy, yet sometimes memories may be evoked of events that either did not occur or did so in a very different context than that remembered. Human memories are not simply the retrieval of a single fact. Rather, they are complex, consisting of auditory, visual, tactile, and even emotional events. You have seen that each of the above experiences is recorded in different parts of the cortex: visual events in the occipital lobe and tactile events in the parietal lobe. Memories are stored in different parts of the brain; recall of those memories requires accurate retrieval of those traces from many parts of the brain. With the passage of time some of the memory bits may be lost, leaving others intact. This means that parts of the memory may be retrieved, but with some of the context lost. The time or the place of the memory may not be remembered as it actually happened.

Motivational Systems

As with memory, specific areas of the brain are associated with specific motivations. Many of our drives arise from small, well-defined areas in the brain, rather than from some widespread network of neuronal connections.

Deep in the brain is a small area called the hypothalamus (refer to Figure 4.11). Although it is only about 1% of the total brain weight, it is a very important group of neurons as it controls many of the endocrine and vital functions of our body, including thirst, pleasure, and sexual drives. If stimulating electrodes are placed in appropriate positions in the hypothalamus, any one of the above drives may be stimulated. Conversely, if other discrete, well-defined areas of the hypothalamus are destroyed, the opposite drive may result. For example, if the ventromedial nucleus of the hypothalamus is destroyed, an animal will go on eating almost continuously. It is as if the animal no longer feels the fullness associated with eating large amounts of food. Its hunger cannot be satisfied, and the animal may grow to grotesque proportions. In this case, that nucleus is called the satiety center. Once it is removed from the CNS the major drive of the animal is to consume food. The animal seems never to be sated with food.

Similar centers reside in the hypothalamus for other life-sustaining functions. The sensation of thirst appears to arise from a hypothalamic nucleus. Stimulation of that nucleus either by certain neurotransmitters or by an increase in the concentration of blood solutes bathing the nucleus causes an animal to drink. Should this nucleus be damaged, the animal may not drink at all, regardless of its state of dehydration. Despite the everyday experience that causes us to quench our thirst, drinking is determined more by hypothalamic stimulation than by a feeling of a dry mouth.

Pleasure Center

Neuropsychologist James Olds placed electrodes in an area of the hypothalamus of rats and arranged an apparatus so that the rat was able to stimulate itself by pressing a bar in the cage. Each time the rat pressed the bar, a short train of weak electrical shocks was delivered to the hypothalamus. Figure 4.21 illustrates the device used to accomplish this. Soon the rat learned that it could stimulate itself by pressing the bar in the cage. Although we are unable to say precisely what the rat felt when it did so, it is presumed that the sensation was a very pleasurable one. If allowed, the rat would press the bar continually for hours on end, often as many as 1000 times in an hour. In fact, it would press the bar without thought of food or water, and unless the apparatus was turned off the animal would actually be in danger of dying from starvation and dehydration. It is difficult to interpret these experiments other than by assuming that the experimental animals experienced such a high degree of pleasure from the stimulating electrodes that they would forgo the most basic needs of food and water in order to continue the pleasurable sensation.

Figure 4.21   Self-stimulating apparatus for rats. When the bar is pressed, the rat receives a very mild electric current applied to the pleasure center.

Rage Center

Other emotional patterns may be elicited by stimulation of different hypothalamic areas. Electrical stimulation of discrete areas of the hypothalamus may cause an animal to assume all the appearances of rage. It will develop a posture indicative of rage, the claws will extend, it may hiss or present some other vocalization associated with rage, it will assume a defensive stance, and present all the signs of a very angry animal. Under normal circumstances animals do not exhibit such behavior, so the pattern of rage must be held in check by other centers, perhaps the pleasure center. But whatever the relationship between hypothalamic centers, such experiments indicate that emotive behavior has much of its origin in a very small portion of the brain. Damage to those nuclei, or nonphysiological stimulation of them, may lead to seriously inappropriate behavior in humans as well as in other animals.

Drug Effects

Drugs are used by many societies, and all of them have some effect on the CNS. These drugs may be used in a variety of settings. Some are used for medicinal purposes, some are used as social accessories, some are used more casually (as is caffeine), and some are used for what might be called recreational purposes (as is alcohol). The latter are usually highly psychoactive and often addictive. Most drugs that affect the CNS have a common mode of action: They resemble in some way a normally occurring compound and attach to the receptor for that compound, either on the cell membrane or within the cell. In this way, the drug mimics the action of a normal neuroactive compound, but does so in an exaggerated manner or by modulating normal function so as to produce an unusual effect.

Nicotine The only drug isolated from tobacco that acts on the CNS is **nicotine,** which stimulates receptors sensitive to the neurotransmitter acetylcholine. At high doses it is a dangerous drug, producing tremors and even convulsions. Fortunately, those effects are not noted at the doses usually achieved by smoking; rather, the smoker experiences more mild effects, such as a feeling of relaxation, some increase in alertness, and a general feeling of satisfaction or easing of tension. Not consciously noted are other physiological changes: increased heart rate and blood pressure and a decrease in appetite. Smokers, on average, weigh as much as 5–10 pounds less than nonsmokers. However, more serious physiological effects come from the tars contained in the smoke, which increase the incidence of lung cancer and heart attacks. A one-pack-a-day smoker may increase his or her risk of contracting lung cancer tenfold.

Long-term smokers show both psychological and physiological dependence on nicotine. Cessation of cigarette smoking leads to a strong craving for "just another puff," anxiety, difficulty in concentration, and increased hunger. These are true symptoms of addiction. For most, the symptoms of withdrawal last about one month. After that time most who are addicted to smoking feel no adverse effects of withdrawal and may never return to smoking.

Alcohol The most widespread and commonly used drug is **alcohol.** It is a simple molecule that is rapidly absorbed from the stomach and the upper part of the intestines. However, the rate of absorption may be slowed if the stomach is full. As is discussed in Chapter 13, a full stomach slows the rate at which it empties its contents into the intestines, which in turn reduces the rate of absorption of alcohol into the blood. Because alcohol is metabolized by the liver at a reasonably constant rate, it is possible to predict blood alcohol levels if the amount of alcohol consumed is known, as well as the time over which it was consumed. An adult will metabolize about one-third of an ounce of pure alcohol per hour. Thus, if one drinks more than one-third ounce an hour (one ounce of 80 proof whiskey contains about ⅓ ounce of pure alcohol), the blood alcohol level will continue to rise.

Some people may show personality changes under the influence of alcohol, and the nature of these changes depends on the individual. It may be a sense of well-being, mild euphoria, or even the opposite. The person may become argumentative or morose, or exhibit impairment of concentration and a reduction of sexual inhibitions. Alcohol levels that cause such changes in behavior account for many antisocial acts. As many as 50% of crimes such as robbery, rape, and other sexual assaults are committed by people showing alcohol intoxication, indicating that it diminishes social restraints.

As the dose of alcohol increases, behavior changes even more radically: muscular coordination, memory, and social inhibitions all become progressively dulled. With increasing doses, the incapacitation becomes more and more severe, finally resulting in coma or death. Figure 4.22 shows the relationship between blood alcohol level and the symptoms of intoxication. The shading between the categories of induced behavior indicates that there is no one specific value for any given effect. Rather, the precise blood level of alcohol required for a given effect is variable between people, and may also vary within a single person in different circumstances. What should be clear is that as the alcohol level in blood increases the behavioral changes become progressively more severe, in the direction shown by Figure 4.22. The bracket indicates the

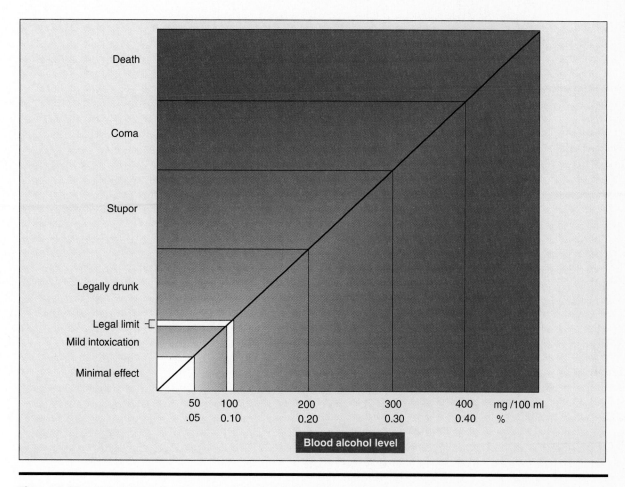

Figure 4.22 Relationship between blood alcohol level and behavior.

legal limiting values indicating intoxication, as they vary from state to state.

Figure 4.23 presents a guide to the blood alcohol level that would be achieved by a 150-pound person who consumes the number of drinks shown on the horizontal axis over the time noted at the end of each line. If a person weighing 150 pounds were to chug-a-lug 20 ounces of whiskey in just a few minutes, his or her blood alcohol level would be expected to reach 400 mg/100 ml, or 0.4%. If he or she drank the same quantity over an eight-hour period the blood alcohol level would be about 0.3%. Ten drinks in eight hours would result in a blood alcohol level of about 0.14%. In order to estimate the blood alcohol levels for people of different weights, find the value for the 150-pound person and multiply that value by 150/weight of the person.

Long-term ingestion of large amounts of alcohol has many deleterious effects, often resulting in a physiological dependence on the drug. If alcohol is suddenly stopped, some people may experience confusion, hallucinations, convulsions, and even death. The hallucinations are generally quite frightening and often consist of seeing nonexistent insects of great size. The person experiencing the hallucinations may insist on trying to kill these insects or escape from them. This syndrome is called **delirium tremens** (DTs), but the neural mechanisms responsible for dependence are unknown.

Opioid Drugs An **opioid** drug is closely related to morphine, and may be a naturally occurring or synthetic drug. Highly addictive, morphine is a naturally occurring chemical found in the opium poppy, and has been in use for thousands of years. Opium has been used medicinally for treatment of cough, headache, colic, and many other common complaints. Until 1914 opium was sold over the counter without restriction, but in that year its adverse societal effects were recognized and a law was passed banning use of opium products except for medical reasons.

Morphine is the most active drug contained in the crude extract of the opium poppy. As it is poorly absorbed from the intestine, it is generally taken intravenously, as a direct route into the blood. The drug produces a feeling of euphoria, well-being, and contentment. It also is a powerful analgesic, reducing

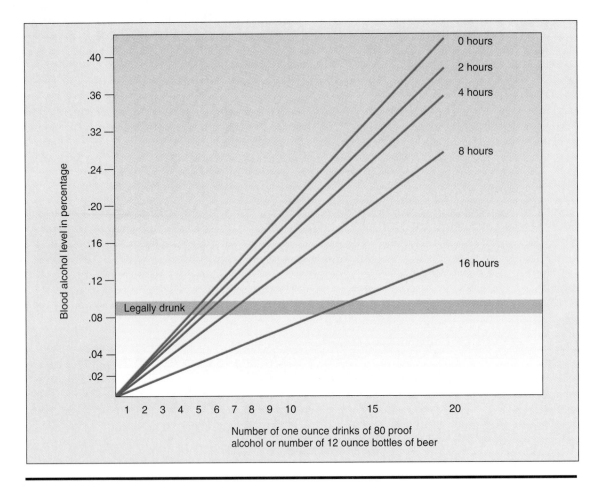

Figure 4.23 Relationship between the number of alcoholic drinks and blood alcohol level.

the perception of pain by blocking release of pain-producing substances. Higher doses of morphine cause sedation and depress the respiratory center profoundly, which may result in death from respiratory failure.

Continued use of morphine and other opioid drugs (such as heroin, codeine, and Demerol) produces a physical dependence characterized by severe symptoms upon withdrawal of the opioid: fever, chills, vomiting, cramps, intense aches and pains, and an intense desire for the drug. It is this powerful drive for more drug that creates so much of the antisocial activity of the addict. The cost of the drug through illicit channels is so high that criminal activity is often required to keep the addict supplied with enough drug to avoid withdrawal symptoms. A variety of methods have been used to break the addictive behavior in addicts, and though there is no universal cure, many programs report success in weaning addicts away from opioid drugs.

Cocaine Cocaine has been used by Andean natives as a stimulant for many centuries. They chew the leaves of the coca plant to release the active ingredient cocaine, enjoying a sense of well-being and staving off fatigue. The dose of cocaine that can be achieved in this way is very small and the natives do not suffer the severe consequences of people in more technologically advanced societies who use the pure drug.

In the latter part of the nineteenth century, purified cocaine was used as a very effective local anesthetic for eye surgery. As it can be absorbed rapidly not only from the cornea but also from the nasal mucosa, it has come to be used as a recreational drug. Cocaine causes an intense euphoric experience and powerful psychostimulation. These properties account for compulsive use of the drug. At low doses it increases mental alertness and reduces fatigue. At higher doses muscular coordination is progressively lost and tremors appear, followed by seizures. Death may occur as a result of heart failure (see Chapter 8) or respiratory failure. Cocaine is a highly addictive and dangerous drug.

Summary

You have seen that the brain is the most complex organ in the body. Although it may be described anatomically with some precision, its functions are not so easy to understand. The brain controls most of the vital functions of the body, including blood pressure, heart rate, respiration, and endocrine function. It also accounts for the aspects of our intelligence that make us human, such as our ability to learn, think, speak, understand written and spoken language, respond to novel stimuli, and exhibit emotion.

Careful anatomical studies allow us to divide the nervous system into component parts: the central nervous system, which is made up of the brain and spinal cord, and the peripheral nervous system, which is composed of the somatic and autonomic nervous systems. The autonomic nervous system is also divided into two divisions: the sympathetic and parasympathetic nervous systems.

The brain may be divided into four major lobes, each of which serves different functions. The cortex of each lobe is a major receptive area for different sensory input and for motor output. Destruction of any small area of cortex will leave a deficit in some sensory or motor ability. Deeper structures of the brain control other functions, such as motivational behavior, pleasure, rage, and aggression.

The somatic nervous system is composed of nerves that exit the spinal cord between the vertebrae and contain both motor and sensory fibers. These are the conduits that carry information to the CNS and send information back to the periphery, stimulating muscles to contract or relax. Conscious muscular movement originates with cells in the cortex. The neural activity set up in the axons of those cells is monitored by other neurons so that constant feedback is sensed by the cerebellum and other brain structures. Smooth, well-coordinated muscle control requires constant monitoring of efferent neural activity. Sensory input also reaches the brain and makes us aware of the world about us. We experience that world with all our senses, including skin receptors that tell us if something is hot or cold, or painful.

The autonomic nervous system governs many of the functions not under conscious control. In general, the sympathetic nervous system prepares the individual for fight-or-flight responses. Activation of the sympathetic nervous system causes increased heart rate, decreased blood flow to the gastrointestinal tract and decreased motility, reduced blood flow to the skin, and other responses that prepare an individual for a potentially traumatic experience. The parasympathetic nervous system in many ways acts in the reverse fashion of the sympathetic nervous system.

The seat of consciousness is the brain, yet the exact mechanisms that account for consciousness are as yet unknown. We do know that different nuclei in the brain serve specific functions we assign to consciousness. Short-term and long-term memory are stored in different brain structures, as are different motivational systems, such as pleasure, rage, and aggression. In addition, many behavioral aspects of our being may be altered by drugs. Most of the psychoactive drugs are addictive, requiring higher and higher doses to maintain the desired effect. The cost of addiction rises with continued use, and as addicts increase their dosage they encounter toxic effects that sometimes lead to death.

Questions

Conceptual and Factual Review

1. What are the two major divisions of the nervous system? (pp. 76–77)
2. Distinguish between the gray matter and the white matter of the brain and the spinal cord. (p. 76, p. 83)
3. Glia perform a few different functions within the brain. What are they? (pp. 77–78)
4. What are the major subdivisions of the brain? (pp. 78–80)
5. How is CSF produced, and what might its functions be? (pp. 80–81)
6. What is the major difference between the spinal nerves of the dorsal and ventral horns? (p. 83)
7. What cortical areas are the primary sites for vision, hearing, speech, motor activity, and sensation? (pp. 84–86)
8. What are the major functions of the different lobes of the brain? (pp. 84–88)
9. What is cerebral lateralization? (pp. 86–87)
10. What is a neurotransmitter? Name three types. (pp. 90–91)
11. Describe the anatomical and functional differences between the sympathetic and parasympathetic nervous systems. (pp. 95–99)
12. Define a reflex. Are reflexes inborn or learned? (pp. 99–101)
13. What are the neural pathways involved in a simple reflex arc? In a crossed extensor reflex? (pp. 99–101)
14. What neural alterations occur as an individual learns something? (pp. 103–104)
15. What are some of the anatomical correlates of learning? (p. 104)

Applying What You Know

1. Will the patella tendon reflex be present or absent in a person whose spinal cord has been severed at the thoracic level?
2. Will the eye blink reflex be present in the same person as above?
3. What is learning?
4. A person sustains a severe blow on the head and complains that he has difficulty moving his right arm and leg. Where do you think the injury exists?
5. Some people have difficulty in naming objects following a stroke. Where do you believe the damage is probably located?
6. Why does a drunk person have difficulty walking a straight line?
7. What property of psychoactive drugs makes them socially dangerous?
8. Each year thousands of people die from an overdose of drugs such as morphine, heroin, and cocaine. What is the most likely cause of death in those people?

Chapter 5
Endocrine System

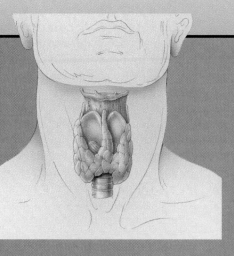

Objectives

By the time you complete this chapter you should be able to

◆ Understand that endocrine glands secrete hormones, which act slowly and are long lasting in comparison to neural effects.

◆ Define the three general types of hormones that are secreted: peptides, amines, and steroid hormones.

◆ Comprehend that all hormones require a receptor molecule for their action to be expressed.

◆ Appreciate that some hormones bind to cellular receptors, causing the liberation of intracellular chemicals, which result in the hormone effect.

◆ Understand that some hormones, such as steroids and thyroxine, are able to cross the cellular membrane without the intervention of a membrane receptor.

◆ Understand the intracellular effect of these hormones in regulating the synthesis of specific proteins by nuclear DNA.

◆ Understand that although hormones enter the blood and are carried to all parts of the body, they act only on their target cells, which have receptor molecules for those hormones.

◆ Comprehend that the action of any hormone depends on the presence of receptor molecules for that hormone and is the reason that hormones affect only their target cells.

◆ Recognize that pathophysiological conditions may arise from oversecretion or undersecretion of a hormone.

Introduction

We have discussed some of the ways in which cells communicate with each other. Recall that the nervous system accomplishes rapid communication between cells. Neural impulses may be conducted by a nerve as fast as 200 miles per hour. In less than 0.01 seconds, the action potential can travel from the cortex to an effector neuron in the lower part of the spinal cord and initiate contraction of a muscle group. Such rapid communication is necessary if one is ever to hit a baseball thrown by a good pitcher, or to jump back rapidly onto the curb as a speeding car suddenly appears. However, not all of life's processes require such rapid intercellular communication. Despite wide variations in the intake of food and salts, the concentrations of glucose and salts vary only slightly from day to day. That regulation requires a longer-acting system that responds more slowly than that described for the nervous system. Many of the slower working processes are carried out by the **endocrine system** (from Greek *endon*, meaning *within* and *krino*, meaning *to separate*).

An example of a slow communicating system in operation is that which occurs during a long distance run. The runner requires a great deal of energy to sustain herself during the run. The major energy source for the exercising muscles is glucose, obtained from the blood. Should that glucose enter the exercising muscle cells without being replenished, the glucose concentration in her blood would fall with each passing minute. Soon the glucose level of blood would be so low that she could no longer continue. But this does not happen; vigorous exercise may continue for many hours in a fit person. What keeps the level of glucose in blood high enough to allow exercise to continue despite the increased use of glucose? As you shall see, a variety of hormones are released into the circulation that increase both the blood concentration of glucose as well as the uptake and use of glucose by the skeletal muscle cells. A subset of body cells is signaled to produce glucose so that another subset of cells may use it. Figure 5.1 is a representation of this homeostatic response.

The endocrine system acts slowly, as it takes time for hormones to be synthesized and released and to travel to some distant organ via the blood. More time is required for the hormone to diffuse out of the blood to the cells it acts on. Still more time is needed for the receptor cell to respond to the hormonal signal, which often requires synthesis of some other protein molecule. The effects of endocrine secretions are also longer-acting than the effects of neural action. Hormones do not disappear from the blood rapidly, nor do the intracellular effects turn off immediately, as do the effects of neural stimulation when the action potentials no longer reach their effector organ.

Every day, life requires many adjustments to a changing environment. The environmental temperature may vary widely, yet your body temperature remains relatively constant. Your intake of foods such as glucose, fats, and proteins, as well as salts of potassium and sodium also may vary widely. Yet the concentration of all those substances in your plasma changes only slightly during the day and even during your lifetime. Stability of this type does not require rapid cellular communication; instead it requires slow, generalized dissemination of information. The endocrine system meets these requirements by releasing hormones from endocrine glands in response to a specific signal. Endocrine glands are those that release chemicals called hormones directly into the bloodstream, where they are carried to their target organs.

Despite the importance of the endocrine system, little was known about it until this century, when modern scientific methods were applied to the study of the endocrine glands. It was not that earlier physicians were unaware of the presence of endocrine glands; rather it was that no one knew how they functioned. Clues came slowly at first, and then more rapidly. It became evident that the mystery of the glands could be unravelled, in large part, by removing them from test animals and noting any changes in physiological function. Extracts were then made of the same glands (taken from other animals) and injected into those with the gland removed. If the extract reversed the specific effects of surgical removal of the endocrine gland, this was evidence that a hormone was produced by that gland. It should be stressed that this second part of the experiment is crucial. These evidential data led to treatment of individuals with inadequate endocrine function; such treatment is known as **replacement therapy** and this experiment showed that this process could be used to restore function. We will discuss that further when considering individual hormones.

First let us discuss the role of the endocrine gland and how it is distinguished from other glands, known as **exocrine glands.** The cells of exocrine glands empty their secretions into ducts that empty to the outside of the body. Some examples of exocrine glands are the salivary glands, the sweat glands, and the digestive glands of the pancreas. All of these glands have ducts into which the secretions of the gland are directed. The ducts, in turn, empty their contents into the mouth, the skin, and the intestines, all of which are considered to lie outside of the body.

Unlike exocrine glands, endocrine glands do not contain a duct system; however, they produce hormones. The cells of the endocrine glands secrete their

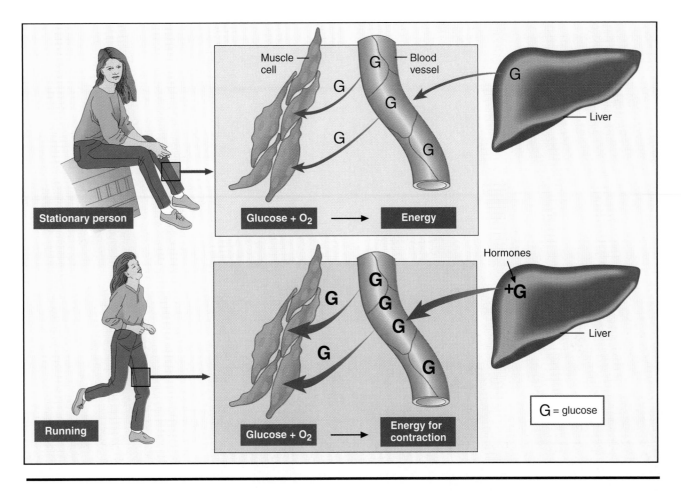

Figure 5.1 Representation of the hormonal effect that stimulates glucose production and glucose entry into muscle cells during running. This allows the muscle cells to produce the extra energy required for the muscular effort.

hormones directly into the interstitial space rather than into ducts. These chemicals then diffuse into the bloodstream. Once in the blood, the hormones are carried to all parts of the body.

Hormone Action

Despite the far-reaching effects of many of the hormones, most have no *direct* action on cellular chemistry. Hormones that cannot cross the cell membrane combine with a specific receptor molecule on the cell surface. Following this combination of hormone and receptor, a second chemical is released in the interior of the cell, promoting a cascade of cellular events. This will become clear when the specific reactions are discussed. However, at this juncture, it is important to note that the hormone is considered the **first messenger** in a series of events. The first messenger causes the

release of another chemical, called the **second messenger,** which in turn is responsible for a series of actions within the **target cell.** Other hormones are able to cross the cell membrane and enter the cell. There they serve as their own second messenger. However, not all cells contain receptors for all hormones. If that were the case, each hormone would exert its effect on all cells of the body. Instead, hormones exert their effect only on cells that contain a receptor for that particular hormone.

Chemical Nature of Hormones

Although the body secretes a large number of hormones, they may be divided into three main classes based on their chemical structure. It is not important to go into the details of that chemical nature at this point. It is important to understand the classification. Figure 5.2 illustrates the molecular pattern of the three classes, as exemplified by a few representative hormones.

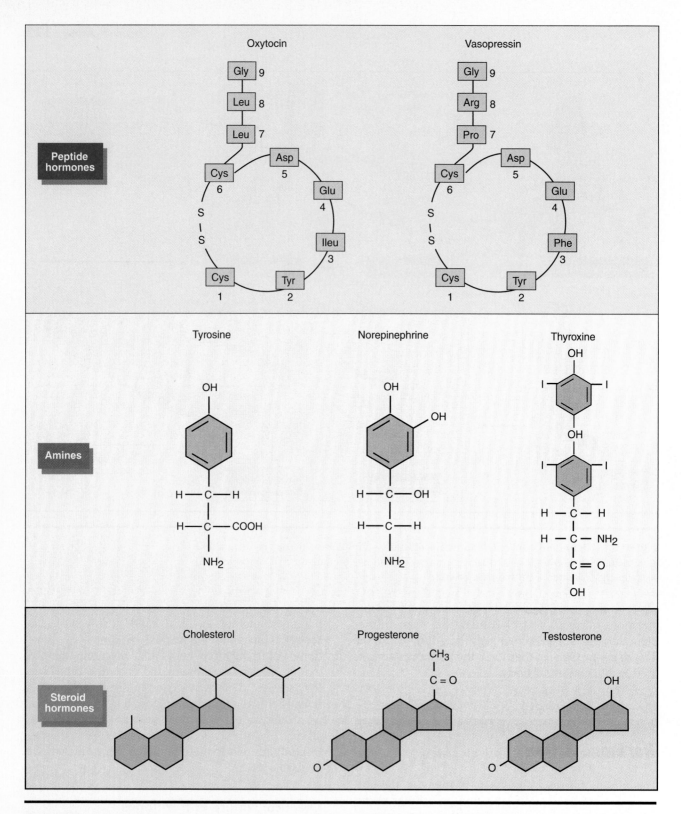

Figure 5.2 Representations of the three categories of hormones. The letters in each box of the peptide hormones stand for a single type of amino acid. The numbers represent the numerical position along the peptide chain where the amino acid is attached. Oxytocin is a peptide containing nine amino acids; vasopressin also contains nine amino acids. (The abbreviations represent different amino acids whose structures are not important at this juncture.) Note that the difference between vasopressin and oxytocin resides in two different amino acids at positions 3 and 8. The structure of the amine hormones is based on the tyrosine molecule shown at the left of the diagram. Steroid hormones are based on the cholesterol molecule shown on the left.

Peptide hormones are composed of amino acids, the building blocks of proteins, and may contain as few as three or four amino acids or as many as several hundred. The two hormones shown, oxytocin and vasopressin, or antidiuretic hormone (ADH), are very similar in chemical structure. Only two of the amino acids of ADH are different from those of oxytocin, in positions 3 and 8, yet the two hormones have very different functions.

Amines are derivatives, as the name implies, of the amino acid tyrosine, and are relatively small molecules. Some were discussed in Chapter 4. Both norepinephrine and thyroxine, the hormone of the thyroid gland, are easily seen to be derivatives of the amino acid tyrosine.

Steroid hormones are derived from the cholesterol molecule and are secreted by many of the reproductive endocrine glands and the adrenal gland. Note that the difference in structure between progesterone and testosterone is quite small; however, the effects of the two hormones are quite different.

The three categories of hormones, in general, attach to receptor molecules in different parts of their target cells. Peptide hormones are soluble in water and circulate in the blood plasma. These hormones are incapable of penetrating the cell membrane and gaining access to the cell interior. For this reason, they must first attach to a special receptor molecule on the surface of the cell in order to be transported into the cell interior. Steroid hormones and thyroid hormones are lipid soluble and must be carried in plasma by a carrier molecule, as their chemical nature is incompatible with dissolving in plasma. Thus, they bind to carrier proteins in the plasma and in some fashion are released at the cell surface, where they are able to penetrate the cell membrane. Once inside the cell, the steroid hormone attaches to a receptor molecule and is carried into the nucleus of the cell. There it acts on cellular DNA to increase or decrease the production of a specific protein that is the agent for the hormonal effect. Hormones such as steroid hormones do not require a cytoplasmic receptor to enter the nucleus; however, upon entering, they also combine with a specific nuclear receptor to cause synthesis of a specific protein.

Cell Membrane Effects

Figure 5.3 illustrates how a peptide hormone acts. The endocrine gland releases its hormone into the extracellular fluid, which diffuses into a blood vessel within that gland (1). The hormone then travels through the circulation, arriving at the capillaries of the target organ (2). The hormone diffuses out of the capillary and binds to cells having specific receptors for that hormone embedded on the cell membrane (3). This initiates the first effect of the hormone and is the rea-

son the hormone is called the first messenger. Often the hormone's first reaction is to activate an enzyme that is an integral membrane protein, called **adenylate cyclase** (4), which in turn causes the release of the second messenger, called **cyclic AMP (cAMP)** into the

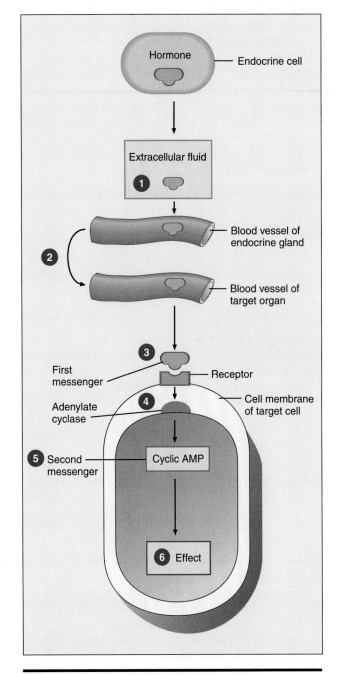

Figure 5.3 Mechanism of action of a hormone that binds to a surface receptor such as peptide hormones. The hormone is represented by the green figure, whose structure allows it to bind to a specific membrane-bound receptor, shown in blue. The hormone–receptor complex then activates adenylate cyclase in the membrane to cause release of the second messenger, cyclic AMP. The second messenger then causes an effect.

cell interior (5). The release of cAMP in target cells is responsible for a cascade of reactions terminating in the specific effect (6). Because cAMP is the second chemical responsible for the cascade effect, it is called the second messenger. Although cAMP was used in this example as the second messenger, it is not the only one. For example, calcium plays that role in other cells.

When the second messengers are liberated, this can cause a cellular effect. The precise effect depends on both the specific hormone bound to the cell and the second messenger liberated in the cell. It is important to realize that hormone effects are specific to their target cells. There they regulate specific processes already present in the cell.

Steroid Action

Some hormones, such as steroid hormones and thyroid hormone, are able to penetrate cell membranes and enter the cytoplasm of the target cells without the aid of a membrane receptor. However, once inside the cell, they too are bound by a specific hormone receptor. The sequence of effects for this type of hormone is illustrated in Figure 5.4. (It is important to realize that these hormone effects are also specific to their target cells. They do not do anything new to a cell.) Processes 1 and 2 in the figure are similar to those for hormones that bind to surface receptors. Steroid hormones and thyroid hormones penetrate the cell membrane of their target cells without first binding to a cell surface receptor (3 and 4). Once within the cell, steroid hormones can enter the nucleus and bind to its specific receptor. The thyroid hormone combines with a receptor on the surface of the nuclear membrane and enters the nucleus (5). Within the nucleus, the hormones act on a particular segment of chromosomal DNA (6), resulting in the release of mRNA, which in turn causes the formation of a specific protein (7). The effector protein formed by the action of the two hormones is different, as the hormones act on different segments of DNA. Formation of the protein is responsible for the cellular effect of the hormone on the target cell.

The sequence of events following release of the hormone from the endocrine gland is complex. Most hormones require either a membrane receptor or a cytoplasmic receptor for their effects to be established. The effects of the hormone–receptor interaction may be many, including enzyme regulation, protein synthesis, altered permeability of the cell membrane, and secretion of another substance.

Synthesis and Storage

Thus far, hormone action in general has been discussed. In this section discussion centers on how cells go about making, storing, and releasing hormones. A brief ex-

planation of those processes is necessary for a solid understanding of how hormones exert their homeostatic control on so many of our metabolic processes. Because hormones exert such powerful effects on the body, their release must be carefully controlled.

Many hormones are stored in the cell as granules (small solid grains); others are stored in liquid-filled sacs called vesicles. However, some hormones, particularly those with long actions such as steroid hormones, are made by the cells only when the stimulus

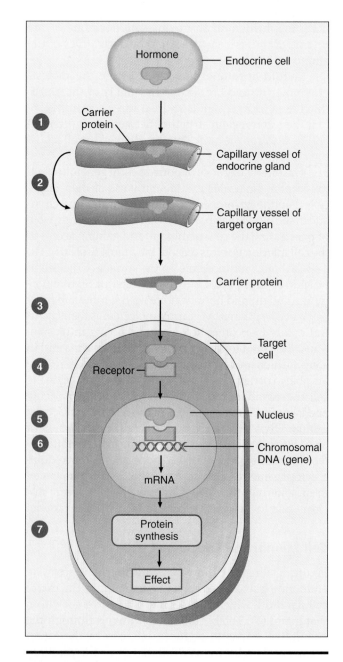

Figure 5.4 Mechanism of action of a hormone that binds to a cytoplasmic receptor and acts on the nuclear DNA, such as steroid hormones.

for secretion arrives at the cell. In general, hormones that are long-acting and are able to diffuse across cell membranes are not stored in granules. Peptide hormones that cannot cross cell membranes and are usually shorter-acting hormones generally are stored in granular form in the cells. An electron micrograph of an endocrine cell extruding a hormone-containing granule is shown in Figure 5.5.

Control of Hormone Secretion

All hormones regulate physiological processes and are released in response to a change in the physiological state of the animal. Most act to maintain the constancy of the internal environment, such as keeping the glucose concentration of plasma constant in the face of physiological challenges, or in regulating the ionic concentrations of plasma. Some hormones respond very rapidly to a stimulus, acting within seconds, whereas others take many minutes to hours before increased plasma levels of the hormone may be achieved.

Stimuli

Despite the wide variations in the rate at which hormones are secreted, all endocrine glands have a few properties in common. All can be stimulated to increase the secretion of their hormones. The stimulus may be neural or hormonal. Many endocrine glands receive nerve fibers that end on their secretory cells; the rate of secretion of the hormone made by those cells may be regulated by alterations in the rate of action potentials flowing along the nerve. Other endocrine glands respond to chemical stimuli. Those chemicals may be hormones produced by still other endocrine glands, or other blood borne chemicals. The nature of those chemicals is discussed in relation to each gland. What is important at this point is to recognize that endocrine glands respond to a variety of stimuli that act to increase or decrease the output of particular hormones.

Feedback Control

The secretion of all hormones is well-controlled by a number of mechanisms. Their concentration in blood is set so that they increase or decrease as the demands of the environment change. The regulaton of hormone concentrations is modulated by both nervous signals and hormonal signals, often working together so as to maintain homeostasis in the face of a challange.

Negative Feedback   Most control systems function to maintain constancy of a regulated variable. A control system familiar to all is the thermostatically controlled furnace (see Figure 5.6). The thermostat controlls the internal temperature of the house. The setting on the thermostat, set by the occupant, represents the preferred temperature, or the **set point.** When room temperature falls below the set point, a

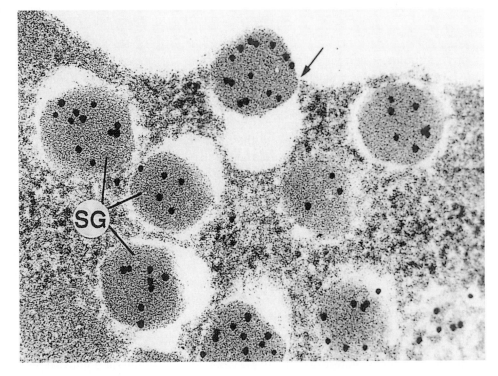

Figure 5.5 Photomicrograph of an endocrine cell showing granules in different stages during the secretory process. The label SG stands for secretory granule.

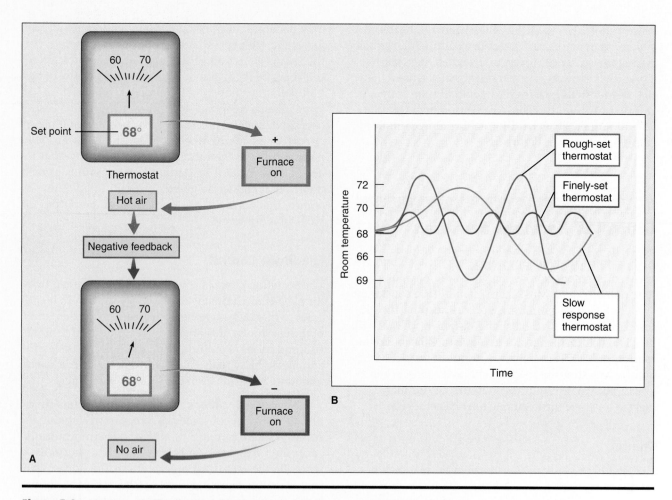

Figure 5.6 (A) Negative feedback system, as illustrated by the workings of a thermostatically controlled furnace. (B) Variation about a mean for a negative feedback system.

positive stimulus is sent via an electric signal to the target, which in this case is the furnace. The furnace responds by blowing hot air into the room, and the temperature rises. When the temperature exceeds the setting on the thermostat, the signal to the furnace stops, reducing the flow of hot air into the room to zero. The thermostat feeds back to the furnace a negative signal (i.e., removal of a positive signal). This is called a **negative feedback** system. It is the negative feedback arm of the system that maintains a near constant temperature.

If one were to set the thermostat at 68°F and plot the room temperature during a few on/off cycles of the furnace, it would not be a perfectly horizontal line at 68°F. The furnace can be turned on only when the thermostat detects a negative deviation from the set point. The furnace will turn off again only when the thermostat detects a positive deviation from the set point. Thus, in this example, room temperature will oscillate around the mean temperature of 68°F, but the variation

depends on how finely the thermostat is tuned. If it is made so that a very small deviation can be detected, the room temperature will vary as depicted by the purple line in Figure 5.6(B). If the thermostat is less sensitive to temperature variations, the room temperature may vary as depicted by the green line in Figure 5.6(B). This demonstrates two other properties of control systems. First, the variable being regulated is never absolutely stable. It will always vary to some degree about a mean determined by the set point. Second, the magnitude of the variation about the mean is a reflection of the sensitivity of the detector.

A third property of a feedback system is determined by the rate at which the active component of the system can operate. Using the hot air analogy, if the blower fan is not forceful, the room temperature will rise slowly when the furnace is signaled to turn on. The rate at which room temperature rises will be slow, and the oscillations about the mean temperature of the set point will be as shown by the yellow line in

Figure 5.6(B). For most physiologic systems, the detectors are quite sensitive and the variation about the mean is small. Body temperature is a good example. Yearly environmental temperatures in temperate climates vary as much as 140°F, yet the temperature of a healthy body is maintained within about one degree at approximately 98.6°F.

Positive Feedback A feedback system as described above provides for stability of the regulated variable. If that is the case, what would be the consequence of **positive feedback?** Using the example of a thermostat and furnace, as the temperature in the room increases, the signal to the furnace would be positive rather than negative. This would cause the furnace to supply more heat to the room rather than causing it to turn off. The rise in temperature leads to a signal for even more heat. This increase in temperature leads to still more heat. If allowed to continue, the result would be an unstable, disastrous system, yet this seemingly unstable feature is occasionally encountered in the body. An example of positive feedback was presented in Chapter 3, in the discussion of action potentials. Once the threshold voltage of a nerve fiber is exceeded, an irreversible action potential results. This occurs because the rising membrane voltage causes voltage-sensitive sodium channels to open, allowing the voltage to rise even more rapidly. Such a positive feedback cycle is responsible for the very swift rise in membrane voltage called the action potential. Other examples of positive feedback are ovulation and coagulation of blood, which are discussed in later chapters.

Endocrine Systems

A number of glands exist throughout the body that secrete hormones into the blood. Each gland secretes one or more hormones unique for that gland. However, all have the effect of regulating a physiological process, such that homeostasis is maintained. Table 5.1 lists the major endocrine glands and their primary functions.

Pancreas and Blood Glucose

One of the most exciting topics in endocrinological research is that dealing with the discovery of **insulin,** a protein hormone secreted by the pancreas, which controls the concentration of sugar (glucose) in blood plasma. Regulation of plasma glucose is important as glucose is the only energy source for the brain, which cannot store glucose, yet needs a constant supply. It is the lack of insulin or the inability of its target cells to respond to insulin that is the cause of **diabetes mellitus** (*mellitus* is Latin for *sweetened with honey.*)

As long ago as the first century A.D., the Greek physician Aretaeos of Kappadokia recognized and

Table 5.1	**List of major endocrine glands and their primary effects.**	
Gland	**Major Hormone**	**Primary Effects**
Adrenal		
Cortex	Aldosterone	Regulates renal Na and K excretion
Medulla	Epinephrine	Mimics sympathetic nervous system stimulation
Gonads		
Ovaries	Estrogen, progesterone	Reproductive hormones, stimulate secondary sexual characteristics
Testes	Testosterone	
Kidneys	Renin-angiotensin	Regulates salt and water balance
Pancreas	Insulin, glucogon	Regulates glucose uptake by cells and glucose metabolism
Parathyroid	PTH	Regulates Ca and PO4 metabolism
Pituitary		
Anterior	Tropic hormones	Stimulate growth and secretion of other endocrine glands
Posterior	Vasopressin (ADH)	Renal water conservation
	Oxytocin	Promotes uterine contractions
Thyroid	T4	Regulates metabolic rate

described the symptoms of diabetes. The name he gave it still stands; *diabetes* means *to pass through* or *siphon.* According to Aretaeos, "Diabetes is a wonderful affliction, not very frequent among men, being a melting down of the flesh and limbs into urine. The patients never stop making water. The flow is incessant, as in the opening of aqueducts. Life is short, disgusting and painful; thirst unquenchable; excessive drinking, which, however, is disproportionate to the large quantity of urine, for more urine is passed; and one cannot stop them from drinking or making water; . . . and at no distant term they expire."

Metabolic Effects of Pancreatic Hormones

You will see in Milestone 5.1 the effects of a lack of insulin. But what is it about insulin that prevents these severe symptoms? What metabolic processes does it regulate? Is insulin the only hormone secreted by the islet cells of the pancreas?

Following the work of Best and Banting, it became possible to inject pure insulin into any muscle and study its effects on the body. Insulin injection resulted in a rapid decrease in the concentration of glucose in blood plasma. This was the expected result as Von Mehring and Minkowsky showed that the lack of a pancreas (insulin) resulted in an increase of blood glucose. Normal glucose levels following a 12-hour fast are about 80–120 mg per 100 ml of plasma, averaging approximately 90 mg per 100 ml. If excessive insulin is administered, the glucose level of plasma may fall so low that the brain is deprived of sufficient glucose for normal function. In such cases the individual would lose consciousness and possibly go into shock. This clearly demonstrates that insulin is a potent hormone regulating, in part, the glucose concentration of blood.

Insulin is not the only hormone secreted by the pancreas; some extracts obtained using a slightly different method from that of Banting and Best actually increase plasma glucose. This was taken to mean that the pancreas secreted at least two hormones controlling the glucose level of blood. The second hormone was found to be secreted by the alpha cells of the pancreas, and was named **glucagon.** As the investigations of pancreatic extracts became more sophisticated, still another pancreatic hormone was identified as **somatostatin,** which is secreted by the delta cells of the pancreas. Although it is known that somatostatin inhibits the secretion of both insulin and glucagon, its physiological role in maintaining blood glucose levels is not yet well-understood. But it is fair to state that these three hormones act in concert to maintain the glucose concentration of plasma within the normal range.

Hyperglycemia As stated above, the normal concentration of glucose in blood plasma is approximately 80–120 mg per 100 ml. Concentrations above this level are considered **hyperglycemic** (*hyper = too much, gly = glucose, emic = concentration in blood*). Following the ingestion of a large quantity of glucose, the concentration of glucose in plasma increases beyond the fasting level. Figure 5.7 shows the results of feeding a normal person a high-carbohydrate meal.

Shortly after the meal, the plasma glucose concentration increases, as does the insulin concentration. At the same time, the glucagon concentration of plasma falls. The increasing concentration of plasma glucose resulting from intestinal absorption of carbohydrates signals the pancreas to increase insulin output at the same time glucagon release is reduced. These changes in hormone concentration serve to minimize changes in plasma glucose.

If blood is obtained from an untreated diabetic person, you would see that the starting level of glucose would be higher than normal, and following a meal the glucose level of plasma would rise even higher, perhaps as high as 1200 mg per 100 ml.

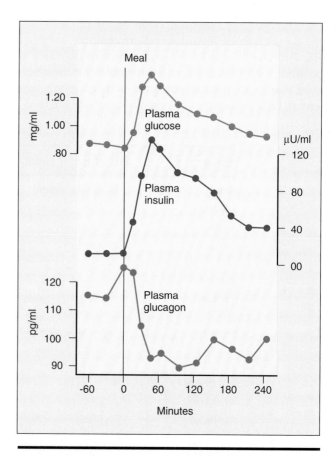

Figure 5.7 Results of an experiment showing stimulation of insulin secretion and suppression of glucagon secretion following a high-carbohydrate meal.

Discovery of Insulin

Aretaeos' accurate description of untreated diabetes mellitus, or sugar diabetes, profiled the symptoms, which were known for almost 2000 years. Only in the seventh century did Chinese physician Chen Chuan record that the urine of diabetics was sweet. Eleven more centuries passed before the pancreas was suspected of being involved in this disease. The direct cause of diabetes remained unknown until the twentieth century.

German pathologist Paul Langerhans discovered small islands of cells in the gland that seemed to be different from those supplying digestive juices. They are now called **islets of Langerhans,** or simply pancreatic islets. Figure 1(A) is a diagram of a section of the pancreas. The circular area of lighter-staining cells embedded in the exocrine pancreas is shown. Furthermore, three distinct cell types are seen in the islets of Langerhans: **alpha cells, beta cells,** and **delta cells.** The cytoplasm of those cells contains many granules, suggesting their endocrine function, each secreting a different hormone, as will be described below. Figure 1(B) is a modern photomicrograph of those cells.

Not until 21 years after Langerhans' original description of the pancreas in 1890 were the first definitive experiments done. Joseph Von Mehring and Oscar Minkowski surgically removed the pancreas of a dog and discovered that the dog exhibited all the symptoms of diabetes mellitus: increased thirst, high rate of urine flow, increased blood glucose, presence of glucose in the urine, and generalized wasting of body weight. This single experimental demonstration led to the treatment of diabetes mellitus.

It wasn't until 1921 that Frederick Banting and Charles Best ended the search for the active substance from the pancreas. Like Von Mehring and Minkowski, they first removed the pancreas from dogs, creating diabetes mellitus. The dogs were then treated with purified extracts from the pancreas obtained from other dogs, which made the diabetic symptoms disappear. Figure 2 shows Banting and Best with one of their dogs successfully treated for the disease. Biochemical separation of the pancreatic extracts revealed the active component to be a polypeptide called insulin. Insulin was then shown to be secreted by the beta cells of the pancreas. (continued on next page)

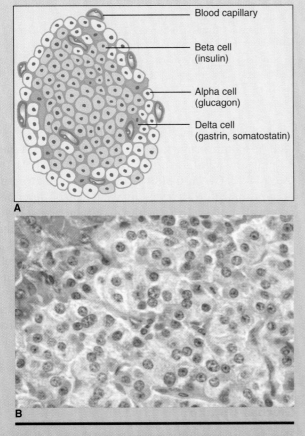

Figure 1(A) Diagram illustrating the presence of islet of Langerhans in the body of the exocrine pancreas, showing alpha, beta, and delta cells in the islet tissue. (B) Photomicrograph showing the presence of the islets of Langerhans within the body of the pancreas.

Figure 2 Photograph of Banting and Best standing with their dog after successful treatment of the animal's diabetes mellitus.

Almost immediately the world changed for diabetics. Until Banting and Best, the only treatment was to place the diabetic on a very restrictive, low-calorie diet. The treatment simply slowed the inevitable. With commercial manufacture of insulin, the picture changed dramatically. Figure 3(A) shows a photograph of a

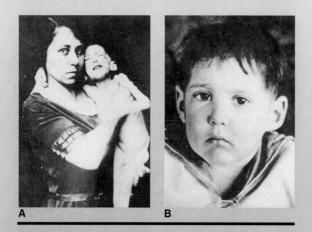

A   **B**

Figure 3 (A) Photograph of a three-year-old girl with untreated diabetes mellitus, December 1922. Note that the young girl fits Greek physician Aretaeos' description of the disease. (B) A photograph of the same girl in Figure 3(A) taken three months after treatment by Banting and Best with insulin.

three-year-old girl with untreated diabetes mellitus. She was fortunate, as she was one of the first patients to be treated with the newly purified insulin. In Figure 3(B) she is shown three months later. A revolution had taken place in the treatment of diabetes mellitus.

Although all the steps in insulin action have not yet been uncovered, the treatment of diabetes mellitus illustrates a common scenario in the discovery of the mechanisms of hormone action and the treatment for many endocrine diseases. Researchers first remove a gland from an experimental animal and note symptoms of the disease. Next they isolate an extract of that gland and inject it into the affected animal. Reversal of the symptoms indicates that the disease is due to the lack of some active factor of the gland. This factor is then isolated, purified, and used to treat the disease in humans. In addition to purifying insulin from animals, researchers using modern techniques of molecular biology have been able to modify the genetic make-up of bacteria so that the bacteria produce insulin. This insulin may be obtained in a highly purified form for use in human diabetics.

Since the experimental work of Best and Banting, insulin has been purified from animals and bacteria and used to treat human diabetics. Untold millions of diabetics receiving insulin have been able to live long and productive lives rather than dying prematurely.

Researchers have been able to study the rate at which glucose can enter the different cells of the body. This rate of entry in the presence and absence of insulin may be compared. Experiments such as these show that insulin acts on cellular membranes, particularly of striated muscle cells, to facilitate glucose entry into those cells. As more glucose leaves the blood and enters cells, glucose levels fall. Although the exact mechanisms responsible for increasing the cellular transport of glucose are still unknown, we are able to state that the facilitated diffusion of glucose into cells is stimulated by insulin.

One exception to the action of insulin noted above is found in brain cells, which are permeable to glucose without the intervention of insulin. Because the brain uses glucose as its metabolic fuel, it is essential that plasma glucose be maintained above some minimum level of about 40–50 mg/100 ml. Very low plasma glucose levels may lead to irritability, fainting, and even shock.

Hypoglycemia A relatively rare event for normal people is **hypoglycemia** (low plasma glucose), which is a fall in plasma glucose to levels significantly below normal. If this were to occur for any reason, the pancreas would reduce its output of insulin and in-

crease the secretion of glucagon. Both changes in secretion would bring glucose levels back to normal, helping to maintain homeostatic balance. It is the interplay between insulin and glucagon that maintains normal glucose levels, by what may be described as a push–pull system. Insulin allows increased cellular entry of glucose from the plasma, causing a reduction of plasma glucose levels. Glucagon, however, has the opposite effect of increasing plasma glucose levels.

Although hypoglycemia is rarely seen in normal people, diabetics may suffer this condition as a result of administering excess insulin. It is crucial that diabetics monitor their blood levels of insulin as frequently as possible and adjust the dosage of insulin accordingly.

Physiological Consequences of Impaired Glucose Uptake

Thus far, the discussion suggests that the only defect incurred by insulin deficiency is an inability to keep glucose levels relatively constant. Other derangements in physiology follow as a consequence of that deficiency.

Acidosis A deficiency of insulin secretion has effects that go beyond the inability to regulate plasma glucose levels. Because cellular entry of glucose is reduced, cellular metabolism is altered, decreasing glucose metabolism and increasing the metabolism of fats. Increased fat metabolism leads to an increase in the production of acidic products called ketone bodies, which create a condition known as **acidosis.** Before the advent of insulin treatment, severe acidosis was a major contributing cause of death in diabetics.

Water Loss Untreated diabetics have very high levels of glucose in their blood. The concentration of glucose may be so high that the amount of glucose presented to the kidneys exceeds their ability to reabsorb all that is presented (see Chapter 12), so large amounts of glucose enter the urine. This excess urinary solute necessitates the excretion of large volumes of water to carry off the excess glucose. Excess excretion of urine due to excess osmotic particles is called osmotic diuresis. The loss of water may be so great that the individual becomes severely dehydrated; despite near constant thirst, he or she can not drink sufficient water to replace the urinary loss.

Summary of Blood Glucose Homeostasis

Figure 5.8 summarizes the control elements that serve to maintain relatively constant blood glucose concentrations. Following a meal, absorption of glucose from the small intestines increases the concentration of glucose in plasma. The beta cells of the islets of Langerhans respond to the higher blood glucose levels by increasing the secretion of insulin and inhibiting glucagon secretion. These events are diagrammed on the right side of Figure 5.8.

A decrease in glucagon secretion reduces the stimulus on the liver to release glucose; therefore, less glucose is released into the blood. Simultaneously, insulin levels of plasma increase and act on both the liver and muscle cells. The liver is signaled to allow an increase in the entry of glucose, which is then converted into the storage form of glucose, glycogen. At the same time, muscle cells respond to the increase in insulin by transporting glucose more rapidly into the cell interior. All these actions serve to lower the concentration of glucose back toward normal.

The left side of Figure 5.8 shows the responses that occur when plasma glucose levels fall, such as during fasting. The reduction in plasma glucose acts to lower the secretory rate of insulin and simultaneously increases the secretory rate of glucagon. The liver responds by breaking glycogen stores down into glucose and releasing the newly formed glucose into the blood. In addition, the rate at which glucose enters the liver and muscle cells will be reduced, thus stabilizing plasma glucose concentrations.

During moderate exercise, the use of glucose by the exercising muscles may increase as much as twenty- to thirty-fold. Unless extra glucose is supplied to the blood, glucose levels would fall, as the glucose

HIGHLIGHT 5.1

Diabetes

Thus far, diabetes has been discussed as if it were one malady. Actually, it can be divided into two classes: Type I diabetes and Type II diabetes.

Type I diabetes, known as **insulin-dependent diabetes** or **juvenile onset diabetes,** is most often seen in people less than 40 years of age, and it often strikes children. It is characterized by a severe deficit in the ability of the pancreas to secrete normal amounts of insulin. The reasons for this are varied. In part, a genetic disposition may be responsible. Environmental factors may also be involved, particularly viral infections that affect the pancreas. Individuals afflicted with Type I diabetes must be supplied with replacement doses of insulin to make up for the deficiency of the pancreas. The actual dose and frequency of administration depends on the severity of the disease. Simple methods for measuring blood glucose levels may be done quickly at home. This allows the patients to monitor their own blood glucose levels and adjust the daily dose to meet needs.

Type II diabetes, or **non-insulin-dependent diabetes,** is the most common form of diabetes. It accounts for about 85% of all diabetics and is called **adult-onset diabetes.** In this form of the disease, the pancreas is able to secrete normal amounts of insulin. The defect resides not in the pancreas, but in the response to insulin. For reasons not well-understood, the individual does not respond normally to insulin levels in blood. The target tissues for insulin are unresponsive to the hormone due to a defective receptor for insulin or due to some postreceptor defect. The symptoms of the disease, however, are the same as in Type I diabetes. Fortunately for those with Type II diabetes, this disease may often be treated successfully by changing the diet. Obesity appears to be a predisposing factor in Type II diabetes. Thus, patients with this form of diabetes are advised to lose weight and increase their physical activity.

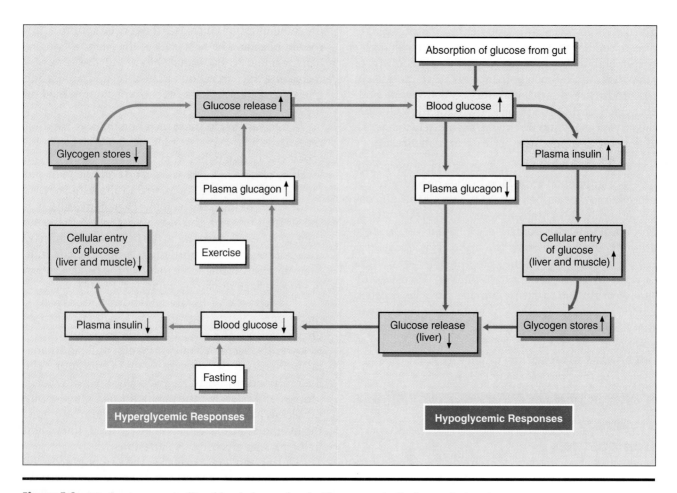

Figure 5.8 Mechanisms controlling blood glucose levels. The arrows in the boxes designate either an increase or decrease in the substances or actions in the box. On the right side of the figure are the responses that lower blood glucose levels (hypoglycemic responses). On the left side of the figure are the responses that serve to increase glucose levels (hyperglycemic responses).

leaves the blood and enters the exercising muscle. However, exercise or stress is a stimulus to the sympathetic nervous system to release epinephrine into the general circulation, which has multiple actions on glucose metabolism. It helps to maintain glucose levels within the normal range by acting on liver cells to break down glycogen into glucose, act on the pancreas to stimulate glucagon secretion, and inhibit insulin secretion. These actions prevent plasma glucose concentrations from deviating from normal during exercise, as well as during periods of stress.

As outlined in Figure 5.8, the control of blood glucose is an example of an excellent negative feedback system. Any stimulus that would raise blood glucose levels sets up a series of events that act to oppose a deviation from normal. As the plasma glucose concentration falls, the original signal is reduced. If plasma glucose falls below some minimum, a new set of responses occurs to increase plasma glucose.

Adrenal Glands

The adrenal gland is a small triangular gland lying just above the kidney. Until 1855 its function was unknown. In that year, physician Thomas Addison described a disease in which the patient loses appetite, becomes weak and lethargic, and shows an increase of skin pigmentation. At that time no treatment was available, and the patient usually died within a short time. Addison attributed the symptoms to a lack of adrenal function, as he found that all of his patients who died with these symptoms showed diseased **adrenal glands** on autopsy. Figure 5.9 shows the position and structure of the adrenals.

Since Addison gave the first description of the disease, the illness was given the name **Addison's disease,** although Addison had no idea of the nature of

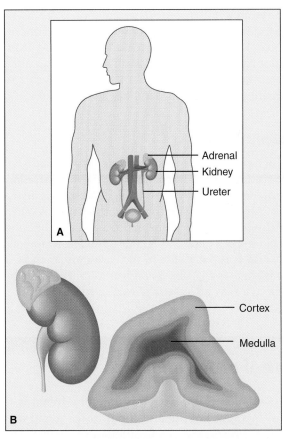

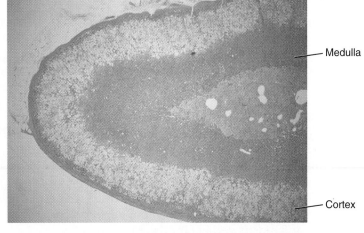

Figure 5.9 Location and structure of the adrenal glands. In the human body, each adrenal is located immediately above one kidney. The adrenal itself is divided into two distinct parts: an inner portion (the medulla) and an outer portion (the cortex).

the missing ingredients supplied by the adrenal gland. The adrenal glands are small endocrine glands lying just above the upper pole of the kidneys. Researchers later isolated and purified the adrenal factors necessary for sustaining life.

If an adrenal gland is cut in half, two distinct parts are seen. There is the inner (medullary) portion and an outer (cortical) portion, called the **adrenal medulla** and the **adrenal cortex,** respectively. Extracts from the adrenal medulla contain the hormones epinephrine and norepinephrine that are secreted in response to sympathetic stimulation (see Chapters 3 and 4).

Extracts from the adrenal cortex yield a set of hormones having a very different chemical make-up from the catecholamines. They are all known as **steroids.** These adrenal cortical hormones may be divided into three classes, depending on their actions on the body. The two major categories are the mineralocorticoids and the glucocorticoids. Additionally, a much smaller supply of sex hormones is secreted.

Glucocorticoids

The group of hormones known as **glucocorticoids** consists of **cortisol, cortisone,** and **corticosterone.** Cortisol is the major hormone, representing over 90% of the total glucocorticoids secreted by the adrenal glands.

All these hormones act on cells to control, in part, metabolic activity, particularly of protein synthesis. For example, glucocorticoids signal the liver to convert amino acids into glucose and also signal adipose tissue to release fatty acids into the circulation. These hormones also have another function that causes them to be considered **permissive hormones.** In order for epinephrine to constrict blood vessels, the presence of cortisol is necessary. However, cortisol by itself has no vasoconstrictor properties. Thus, it is permissive because it permits another hormone to exert its effect. Although much is still unknown about the exact mechanisms of action of cortisol, we do know that the general effects of the hormone aid in maintaining homeostasis when external pressures (such as stress and inflammation) are brought to bear on the body.

Stress Exposure to any noxious stimulus causes a generalized response known as **stress.** The stimulus may be exposure to injury, cold, or disease; it may even be severe anxiety. Whatever the stimulus, the body responds by increasing the secretion of adrenal cortical hormones. This was first demonstrated by Canadian physiologist Hans Selye, who noted that animals subjected to stress show enlarged adrenal glands. We now know that any stress causes the central nervous system to signal the adrenal glands to secrete glucocorticoids. The increased plasma levels of these hormones allow the individual to withstand stress more easily than in the absence of these hormones. Some of the effects of increased plasma levels of glucocorticoids include an increased ability of blood vessels to respond to catecholamines, stimulation of protein breakdown, inhibition of glucose uptake by many cell types other than brain cells, increased release of fatty acids from adipose tissue, and an increase in the readily available energy source glucose during the period of stress.

An animal with nonfunctioning adrenal glands may die as a result of applied stress that would have little effect on a normal animal. Although the precise mechanism of the protective action of adrenal hormones is unknown, we do know that they play an important role. For example, glucocorticoids are required for survival of fasting animals. Yet during the fast, glucocorticoid levels of blood do not rise. It is the presence of glucocorticoids that is required for survival.

Inflammation Most tissues of the body react to injury by swelling, turning red, causing changes in the walls of the small vessels of the injured area, and causing pain. Humans are subject to many forms of inflammatory diseases. Joints may become inflamed acutely or chronically, as in **arthritis.** Inflammation of the gastrointestinal tract may result in severe bowel problems, such as **colitis.** Whatever the cause, it has been found that glucocorticoids are excellent anti-inflammatory agents. Cortisol acts to reduce the inflammation produced by these conditions. Although it would seem that cortisol could be used to cure conditions such as rheumatoid arthritis, a chronic and often debilitating inflammatory disease, that is not the case. High doses of cortisol are required to reduce the inflammation. High doses have adverse side effects that prohibit its long-term use, but for short-term use, cortisol is very helpful; it may also be used topically to treat certain inflammatory diseases of the skin.

Mineralocorticoids

Mineralocorticoids are steroid hormones that regulate the sodium and potassium concentrations in the body fluids. The major mineralocorticoid hormone is **aldosterone.** Its site of action is the kidney, where it increases Na^+ retention and simultaneously increases K^+ secretion by the kidney and its loss into the urine. This is an extremely important action as the mass of salt (NaCl) and the volume of water that makes up the body determines, in large part, the blood volume. Should an individual lack sufficient aldosterone, inappropriate amounts of salt and water will be lost in the urine and blood volume will fall, decreasing blood pressure. Low secretion (hyposecretion) of aldos-

terone results in inappropriate retention of K^+ as K^+ secretion is reduced. K^+ retention causes the concentration of K^+ in plasma to increase to toxic levels, leading to muscular weakness. Recall that the ratio of K^+ concentrations across the cell membrane determines the membrane potential (see Chapters 2 and 3).

Anabolic Steroids

Another class of hormones secreted by the adrenal cortex is the **anabolic steroids,** of which **testosterone,** a hormone that has masculinizing activities, is an example. Any hormone that acts as testosterone is called an **androgen.** This hormone is also a sex hormone and is found in both male and female adrenals. Among its many effects is the ability to increase muscle mass. If secreted in inappropriately high amounts or taken by injection, it may also cause masculinization.

Control of Secretion

The regulation of the secretory rate of adrenal hormones involves both neural input to the gland and special hormones that act to stimulate secretion. The scheme of control is diagrammed in Figure 5.10.

Adrenal Cortical Hormones

A common feature of many endocrine systems is that the secretion of one hormone is controlled by another. The hormone that stimulates the secretion of the second hormone is known as a **tropic** hormone; (*tropic* is from the Greek word meaning *to nourish.*) In the case of adrenal cortical hormones, the sequence of control involves two different tropic hormones. Thus, **adreno-corticotropic hormone (ACTH)** stimulates the adrenal cortex to secrete its hormones. ACTH is released from a pea-sized gland at the base of the brain, called the anterior pituitary gland, and reaches the adrenals via the blood. However, the control of ACTH secretion is dependent on still another tropic hormone, **cortico-tropin-releasing hormone (CRH).** CRH is secreted by the hypothalamus and is carried to the anterior pituitary by the venous blood draining the hypothalamus. There it induces an increase in the secretory rate of ACTH, which in turn causes the adrenal glands to increase the release of cortical hormones.

As noted above, stress increases the circulating levels of adrenal cortical hormones. But, as indicated in Figure 5.10, stress acts on the cortex of the brain to signal the hypothalamus to release CRH, which then

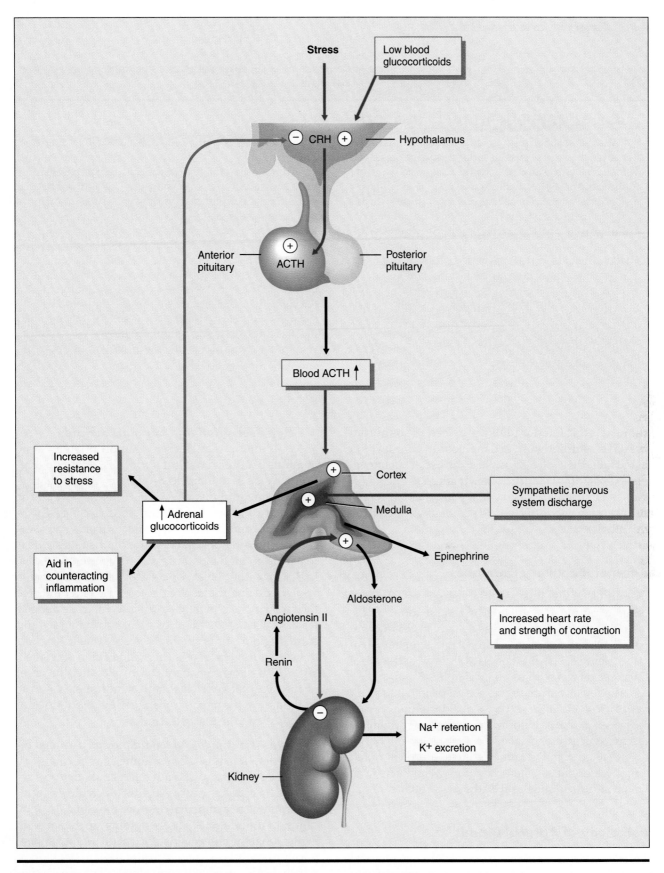

Figure 5.10 Summary of the control of secretions from the adrenal gland. Blue boxes leading to blue arrows represent a positive stimulus. Red arrows represent inhibition of the designated hormone (negative feedback). Notations in yellow boxes represent the physiological effect resulting from an increase in the level of the circulating hormone.

Use of Anabolic Steroids in Sports

The ability of anabolic steroids to increase muscle mass has led to their use by athletes to enhance performance. In recent years, powerful synthetic steroids that have the properties of anabolic steroids have been produced. These are often taken in large doses by athletes who believe it is necessary for them to "bulk-up" in order to improve their performance. It is not at all clear that their use aids performance. The only data indicating that anabolic steroids increase strength come from their use in castrated males. These people have very little circulating testosterone, and exhibit a lower muscle mass than normal men. Testosterone and other anabolic steroids do increase their muscle mass and masculinize them. However, to date little evidence exists showing that normal men who are trained athletes will benefit from taking large doses of anabolic steroids. Conversely, they are at great risk of doing irreparable damage to themselves.

Chronic use of high doses of anabolic steroids has been shown to cause liver damage. This is a serious side effect that can threaten the life of the user. In addition, nonmedical use of steroids may have adverse effects on behavior, as it leads to aggression and hostility toward others. Anabolic steroids may also cause hallucinations or delusions. In a study of 41 men and women using anabolic steroids to enhance body-building skills, 9 of the 41 subjects experienced psychotic symptoms during periods of exposure. Steroid use that was classified as moderate in these studies was 10 to 100 times the dose prescribed for medical conditions. Whatever the positive effects of steroids, if there are any, one thing is certain: The risk is real.

---

stimulates the anterior pituitary to increase its output of ACTH. Additionally, if the glucocorticoid level in blood falls, that will also increase the hypothalamic secretion of CRH. Thus, secretion of adrenocortical hormones is stimulated directly by another hormone, ACTH, and indirectly by nerves that stimulate the hypothalamus.

Adrenal Medullary Hormones

Output of the medullary hormones epinephrine and norepinephrine is regulated by neural input to the medullary portion of the adrenal glands. Increased sympathetic stimulation causes adrenal release of those two catecholamines. Their action is widespread and some of their actions were discussed in Chapters 3 and 4.

The lower part of Figure 5.10 shows that the kidney is involved in the response to aldosterone, as well as the regulation of aldosterone secretion. This will be discussed in greater detail in Chapter 12. It is important to stress here that the kidney secretes a protein called renin, which causes the generation of angiotensin II, which in turn stimulates the adrenal cortex.

Pathology of Adrenal Glands

Two types of adrenal malfunction may be encountered in patients. Some patients have an inappropriately increased secretory rate of adrenal hormones (hypersecretion); others may secrete too little of the hormones (hyposecretion). Both abnormalities result in severe symptoms that may be life-threatening.

Hypersecretion of Adrenal Steroids An adrenal tumor may secrete inappropriately large amounts of adrenal cortical hormones, or hypersecretion may be secondary to a pituitary tumor that secretes large amounts of ACTH. In either event, the results are similar, and the condition is called **Cushing's syndrome.** Excess mineralocorticoids cause excess retention of NaCl and water by the kidney. This may lead to high blood pressure (hypertension). The high circulating levels of cortisol stimulate appetite while reducing the use of glucose. This accounts for the fact that such patients exhibit an increase in body fat, as well as a redistribution of that fat in the face and trunk regions. Also, the high levels of male sex hormones (androgens) cause a masculinization of women. Figure 5.11 shows a patient before and after the onset of the symptoms of Cushing's syndrome. Note the roundness of the face, the change in fat distribution of the body, and the masculinization.

Hyposecretion of Adrenal Steroids Adrenal insufficiency, or **Addison's syndrome,** may be caused by a number of factors. The adrenal glands may be destroyed by certain diseases such as tuberculosis, by an autoimmune response that destroys the adrenals, or by cancer of the adrenals. The symptoms are the same no matter what the ailment. Insufficient quantities of these hormones result in salt and water loss by the kidney, which leads to severe low blood pressure (hypotension). Because cortisol levels are also very low, the permissive action of cortisol on catecholamine action is minimal. This acts to make the hypotension even more severe. The decrease in cortisol also causes loss

Figure 5.11   A patient before (A) and following (B) the onset of Cushing's syndrome. Note the masculinization resulting from the high rate of adrenal steroid secretion.

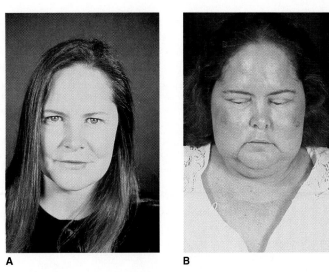

A   B

of appetite, weight loss, and digestive symptoms such as nausea, vomiting, and diarrhea. If left untreated, a sudden adrenal crisis with severe low blood pressure may lead to sudden death. Fortunately, daily replacement therapy is now available and patients with nonfunctioning adrenals may live normal lives. Many people who are unable to secrete normal amounts of hormone may take daily doses of those hormones in appropriate amounts to keep them healthy.

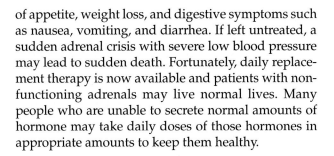

Pituitary Gland

Although the brain is usually considered to be solely concerned with nervous activity, it also serves as one of the major endocrine glands of the body. Located at the base of the brain is a pea-sized bit of tissue lying in a small depression of the skull that supports the brain. The entire endocrine system has been called the orchestra that keeps the body in tune. If that is an apt description, the **pituitary gland** may be considered the conductor of the orchestra.

Anatomy of the Pituitary Gland

Figure 5.12 shows the position of the pituitary gland, with its attachment to the hypothalamic nuclei. The tissue that attaches the pituitary to the base of the brain is called the **infundibulum.** The pituitary itself is divided into two distinct lobes, the **anterior lobe** and the **posterior lobe,** both of which secrete different hormones.

In order to understand the workings of the pituitary gland, it is necessary to understand its unusual anatomy and its relation to the hypothalamic region of the brain. In Figure 5.12 you see that the pituitary gland is attached to the hypothalamus by a narrow

stalk of the infundibulum, which contains neural elements innervating the posterior pituitary. These neural elements serve as conduits for the movement of hormones from their site of synthesis in the hypothalamic nuclei to the site of storage in the pituitary. ADH, a major hormone affecting kidney function, is synthesized in the hypothalamus and transferred to its storage site, the posterior pituitary, along the neurons connecting those two anatomical sites. ADH is then released from the posterior pituitary by nervous signals arriving from the hypothalamus.

The stalk also contains an unusual blood supply. Rather than venous blood being returned directly to the heart, it first travels down the stalk and enters the anterior pituitary. There it breaks up into another capillary network supplying the anterior pituitary with blood that contains secretions from the hypothalamic area. Hormones secreted by the hypothalamus reach the anterior pituitary directly and alter the rate of release of anterior pituitary hormones into the general circulation. As noted in Figure 5.12, many of those hormones are tropic hormones that stimulate other endocrine glands. In addition, some of the anterior pituitary hormones act directly on target cells to change their metabolic rates.

Gonadotropins In Chapter 15 we will discuss in detail the control of the hormones that affect both primary and secondary sexual characteristics and sexual function. At this point, however, it is necessary only to point out that the anterior pituitary plays a pivotal role in regulating secretions of sex hormones by the male and female sex organs. The **gonadotropins** are tropic hormones that regulate secretion of other hormones produced by the ovaries and testicles.

Human Growth Hormone Human growth hormone (hGH) regulates the rate of growth of body cells and bone. (Animal growth hormone is inactive when

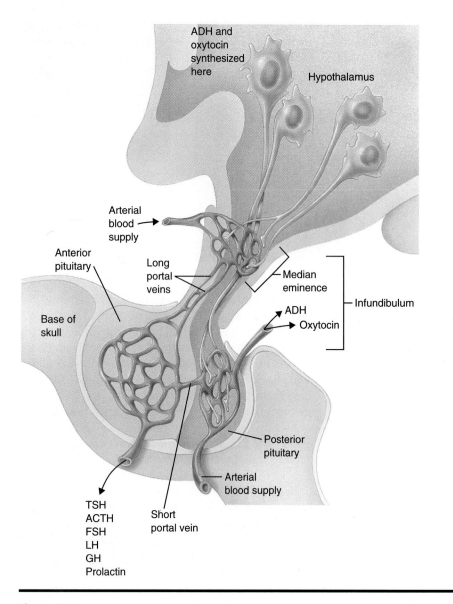

ADH and oxytocin synthesized here

Hypothalamus

Arterial blood supply

Anterior pituitary

Long portal veins

Median eminence

Infundibulum

ADH

Oxytocin

Base of skull

Posterior pituitary

Arterial blood supply

TSH
ACTH
FSH
LH
GH
Prolactin

Short portal vein

Figure 5.12 Anatomy of the pituitary gland, showing both the anterior and posterior pituitary and the connection to the hypothalamus. Note the unusual blood supply to the anterior pituitary. Venous blood from the hypothalamus does not enter the general venous pool, but rather flows to the anterior pituitary. There it again breaks up into capillaries. In this way, tropic hormones secreted by the hypothalamus are carried directly to the anterior pituitary.

injected into humans, indicating a different chemistry. For this reason, we use the term *human growth hormone*.) As a baby develops and increases in size, hGH plays a major role in maintaining that growth. At the beginning of puberty, hGH levels fall about 25%, growth slows, and growth finally ceases when adult stature is attained. However, hGH continues to be secreted throughout life, indicating that there is more to the control of growth than simply the presence or absence of hGH. For example, no matter what the level of hGH in plasma, if a child is undernourished, she will not attain her optimal height. The fact that average height of populations has increased over the years reflects the role of improved nutrition in growth. Anyone who has seen suits of medieval armor is struck by their small size. This is evidence that a few hundred years ago the average height of European males was considerably less than the present height. The reason for this change over the years is most likely due to an improved diet. This is not to imply that hGH is not a regulator of growth, but to point out that growth is a very complex process that is still less than perfectly understood.

However, we do understand that hGH is required for normal growth and that an excess or deficiency of this hormone will result in either excessive growth (giantism) or in deficient growth (dwarfism). Figure 5.13 shows two people who illustrate the extreme effects of excessive and deficient hGH. Both were born with significantly different rates of hGH secretion. Should a pituitary tumor in which hGH is secreted in large amounts develop after puberty, when bone growth is no longer possible, that person cannot increase his or her height. However, this excess secretion may cause the soft tissue to continue growing and the bones may thicken. This condition is known as **acromegaly.** The internal organs and the hands, feet, and face enlarge. The appearance of the person is so affected that changes are easily distinguished. Figure 5.14 shows a person before and after the onset of acromegaly.

Although it had been recognized for many years that a deficiency of growth hormone resulted in severely stunted growth, little could be done to help affected children, as it was very difficult to obtain pure hGH from human pituitary tissue. The little hGH available was insufficient to treat all children who were unable to attain near normal height. However, recent research in the field of molecular biology has developed techniques that cause bacteria to produce peptides, which lead to the production of hGH. Much larger quantities of hGH are now available and may be used to prevent dwarfism in children who lack sufficient levels of hGH.

Reproductive Hormones Among the many hormones released by the anterior pituitary is a set of reproductive hormones. They act on the testes of the male and the ovaries of the female. All of these hormones are

discussed in detail in Chapter 15. At this point, it is necessary only to recognize the remarkably wide variety of hormones released by the pituitary gland.

Thyroid Gland

The **thyroid gland** is located in the neck, as shown in Figure 5.15 on p. 133, and is relatively small, weighing about an ounce. It is richly supplied with blood and is

A

Figure 5.13 (A) Photograph of a person exposed to high levels of growth hormone during her early years. (B) A person whose pituitary secreted inadequate amounts of growth hormone during his early life.

B

AGE 9   AGE 16

AGE 33   AGE 52

Figure 5.14 Progression of acromegaly. This condition is caused by an excess of growth hormone secreted after puberty. There is an enlargement of the bones of the hands and face so that the facial characteristics change dramatically.

composed of many sac-like structures called follicles. Within the follicles is a thick fluid containing the active components of the thyroid gland.

Thyroid Hormone Synthesis and Storage

Experiments with radioactive iodine allowed researchers to trace the course of iodine from blood to thyroid hormones, and so determine the pathway of production of those hormones. The thyroid gland avidly takes up iodine from the blood and synthesizes iodine-containing thyroid hormones, the two most active compounds of which are triiodothyronine **(T3)** and tetraiodathryonine, or **thyroxine (T4).** The biochemical scheme of thyroxine production, storage, and secretion is outlined in Figure 5.16 on p. 134.

MILESTONE 5.2

A Medical-Ethical Dilemma

Human growth hormone is unique to human beings, so it is not possible to extract growth hormone from animals for human use. Until recently, this resulted in a very small supply of hGH, as it had to be isolated from human tissue. This required strict rationing of the hormone, necessitating careful allocation of hGH to children destined to be severely stunted in growth because they lacked sufficient hGH of their own. However, it is now possible to produce hGH using molecular techniques that insert the hGH gene into bacteria. The supply of hGH is now much greater and the hormone is

cheaper. The ethical dilemma stems from the question of how to respond to parents who have reason to believe that their normal child may not grow to the height the parents would prefer, and want the child treated with growth hormone. Should hGH be given to anyone just to ensure an adult stature of 6 feet? Of 6 feet, 6 inches? Is it appropriate for physicians to prescribe a hormone on the basis of a height preference alone, even if the child can be expected to grow normally to a stature well within the normal range? What criteria should be used for allocating hGH?

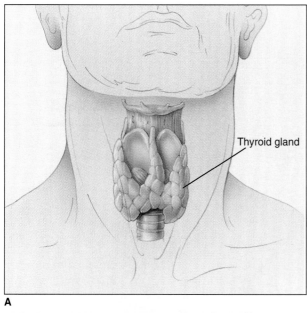

A

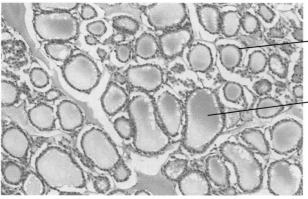

B

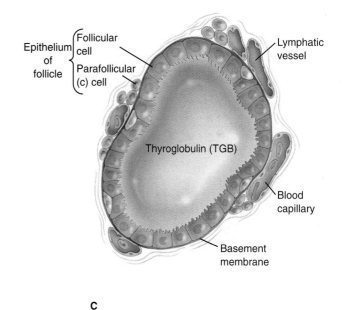

C

Figure 5.15   (A) The thyroid gland is located on either side of the neck in the region of the larynx. When normal it is not seen externally. (Compare with Figure 5.18, showing a person with severe goiter, an enlarged thyroid gland.) (B) Microscopic anatomy of the thyroid gland, showing the follicular nature of the gland. (C) Diagram illustrating the structure of a single thyroid follicle.

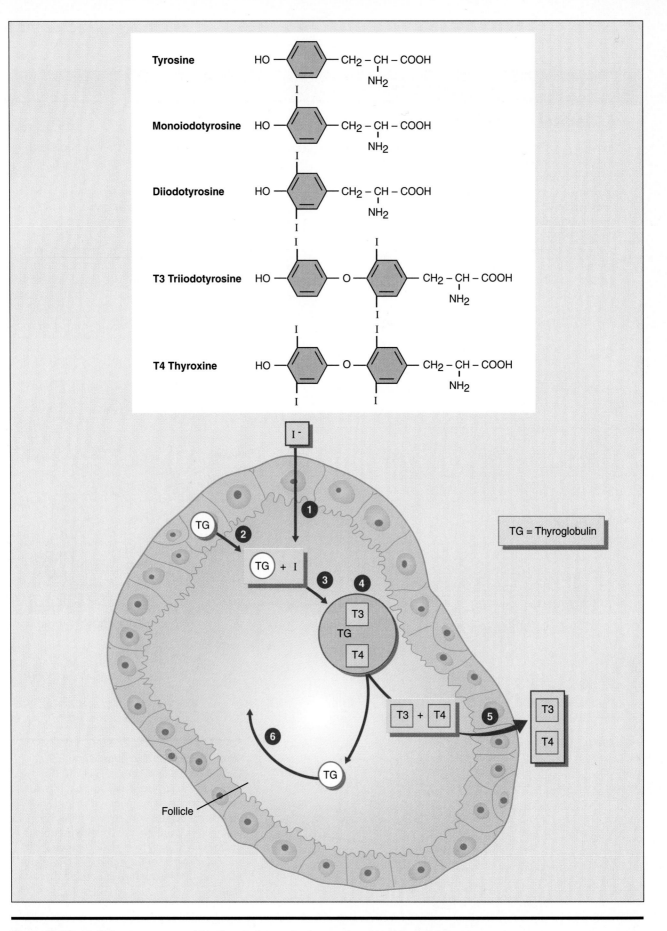

Figure 5.16 Cellular transport and biochemical steps in the uptake of iodide and formation and secretion of thyroxine.

MILESTONE 5.3

Historical Aspects of Thyroid Function

As noted in the introduction to this section, endocrine physiology was an essentially unknown science until early in this century. However, for more than a millennium people were aware of at least one endocrine disorder, known as **goiter.** Individuals so affected exhibit a swelling of the neck that may be so large it appears to hang from the chin to the chest. This swelling is due entirely to an enlargement of the thyroid gland located in the neck. The prevalence of this condition was so great that in many localities it was not considered an abnormality. In Switzerland, some districts reported that as many as 50% of the men of military age showed the presence of goiter. In 1883, in one district of Switzerland, over 25% of the children were reported to have goiter, as shown in the figure below. Why was goiter so

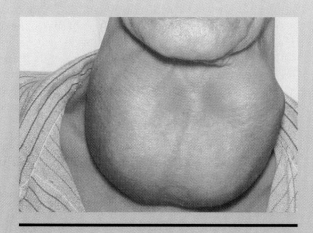

A young girl showing the presence of a large goiter. The swelling on the neck is due entirely to an enlargement of the thyroid gland.

prevalent in many parts of the world, but relatively rare in other areas? Why is it now a rarity in all industrial countries?

Until the middle of the nineteenth century, little was known about what causes diseases, as they were simply blamed on noxious "miasmas" or some such vague term. However, with the new knowledge obtained from studies of microscopic organisms, researchers suspected that a microbial infection was the causative agent of goiter. Soon this idea had to be discarded as it became clear that goiters were confined to well-defined geographical areas and visitors to those areas did not contract goiter. It was not a communicable disease. Improvement of chemical techniques in the early part of the twentieth century led to the finding that the thyroid gland of animals contained small quantities of **iodine,** a substance found in trace amounts in other glands or tissues. This turned medical thinking away from the theory of goiter as an infectious disease to the theory that it was a manifestation of a dietary deficiency, probably of iodine. Two findings supported this hypothesis. First, soil in areas in which goiter was endemic was found to have a low iodine content. Second, if iodine was given to people with goiter, it caused the goiter to decrease in size and often to disappear completely.

In the early part of the 1920s, some countries began adding iodine to common table salt; this is known as iodized salt. Very low doses were added, amounting to only 1 mg of NaI (sodium iodide) per day, compared to the average human intake of salt (NaCl) of approximately 5000–7000 mg per day. This small amount of iodide was sufficient to prevent new cases of goiter. These findings, in turn, led to a search to determine the nature of the protective effect of iodine.

The first step in the process is the active uptake of iodide from plasma into the follicular cells of the thyroid gland. Next, the iodide ion moves across the opposite cell membrane into the follicle (1). The follicular cells synthesize a large glycoprotein molecule called **thyroglobulin** (TG), which is also transported into the follicle (2) and which contains the amino acid tyrosine. Within the follicle, TG combines (3) with iodide that has been oxidized to iodine, forming two different complexes with tyrosine. One complex combines with one iodine atom and is called monoiodotyrosine **(MIT)**; the other complex is a combination of TG with two atoms of iodide and is called diiodotyrosine **(DIT)**, not shown in the figure. A further coupling oc-

curs (4) so that thyroglobulin combines with tyrosine that has three or four iodine attachments (T3 or T4). Within the follicle is a colloid that can cleave T3 and T4 from TG, allowing them to be secreted into the blood and tissues (5). After cleavage of T3 and T4, the remaining iodine is removed from the thyroglobulin molecule and recycled for the resynthesis of TG (6).

Although this is a long, complex system for synthesis, storage, and secretion of a hormone, there is some advantage to its complexity. Iodine is a relatively rare but essential element, so we need a good method for stockpiling it. The follicular fluid, with its high concentration of TG, forms a pool of precursor molecules for storage and thyroxine secretion. As much as a two-month supply of thyroxine may be stored, to be released in response to appropriate stimuli.

Function of Thyroxine

Thyroxine is an unusual hormone in that it has no single specific target cell. Thyroxine enters most cells of the body and stimulates the metabolic rate of each cell it enters. In this way, thyroxine determines not only the basal metabolic rate of an individual, but also the rate of growth of a developing child and, perhaps most importantly, the development of the nervous system. Infants born with a thyroxine deficiency do not develop normally; their physical and mental growth are retarded. Unless thyroxine is given to them within a few weeks of birth, they will suffer severe mental retardation.

Control of Secretion of Thyroxine

Figure 5.17 is a representation of the control system for thyroxine secretion. It is a negative feedback system similar to that seen for other hormones, except that a two-step process is involved. Thyroxine acts on a center in the hypothalamus to inhibit secretion of **thyrotropin-releasing hormone (TRH).** If thyroid hormone levels in blood fall, the inhibition of TRH release is reduced and more is secreted. TRH is carried to the anterior pituitary via the venous portal system, stimulating the anterior pituitary to secrete **thyroid-stimulating hormone (TSH).** This tropic hormone stimulates the thyroid gland to synthesize, store, and release thyroxine into the general circulation. The negative feedback arm is thyroxine itself when it inhibits hypothalamic release of TRH.

Figure 5.17 points out how an iodine deficiency can lead to goiter. No matter how hard the thyroid may be stimulated to work by increasing levels of TSH, it will not be able to produce sufficient thyroxine if sufficient iodine is not available. An insufficient synthesis of thyroxine will reduce the secretion of that hormone into the blood, so that inhibition of TRH release from the hypothalamus is reduced.

The thyroid gland will be continually stimulated, but will be unable to produce thyroxine. However, it will be able to produce thyroglobulin and to store it in the follicular lumen as colloid. The thyroid gland will continue to enlarge, often to grotesque proportions, until iodine is supplied to the diet (see photo on p. 135). This is an example of negative feedback control gone awry.

We now have the clue that solves the mystery of the sharp geographical distribution of goiter. The soil of many areas of the world has a very low iodine content; food grown in those areas will also have a very low content of iodine. Thus, people whose food comes entirely from that district will be iodine deficient and exhibit goiter. The cure is simple. In our culture sodium iodide is added to table salt, protecting the population from an iodine deficiency. Modern transportation has also helped reduce the incidence of goiter, as food may be shipped great distances rapidly. In particular, seafood is rich in iodine and is now available to people living far inland.

Thyroxine Deficiency and Excess

Occasionally the thyroid gland may secrete too little hormone, resulting in a condition known as **hypothyroidism.** In contrast, a tumor may form in the thyroid, causing the gland to secrete extra hormone, a condition known as **hyperthyroidism.** Both conditions are injurious.

Adult Hypothyroidism Thyroid deficiency in adults gives rise to a wide constellation of symptoms stemming from the generalized effect of thyroxine on the metabolic rate. The lower metabolic causes the lowered body temperature and heart rate, increased sensitivity to cold, and general lethargy. In addition, it causes the facial features to be altered as the tissues swell, causing puffiness of the face (Figure 5.18). In severe cases, a dulling of mental processes occurs.

Fetal Hypothyroidism If a pregnant woman suffers an iodine deficiency or the thyroid of the fetus does not develop normally, the child will be severely affected. There will be abnormal and stunted physical growth as well as abnormal brain growth, resulting in mental retardation. This condition is known as **cretinism.** It is vital that a pregnant woman have normal levels of thyroxine and iodine so that cretinism may be avoided. Although hypothyroidism may be corrected in an adult by supplying thyroxine, it may not be corrected in a child born with the syndrome of cretinism. Figure 5.19 on p. 139 shows a child with cretinism.

Another form of hypothyroidism has been described in which the gene for the cellular receptor for thyroxine is defective, and the receptor reacts poorly with thyroxine. This genetic defect has been related to hyperactivity and attention disorders in some children.

Excess Secretion of Thyroxine

Occasionally an individual may suffer a tumor of the thyroid gland, in which the gland secretes excess thyroxine. This condition causes the metabolic rate, the body temperature, and the heart rate to increase. The individual may lose weight and become overactive or nervous. In long-lasting cases, excess fluid and tissue are deposited behind the eyes, causing them to

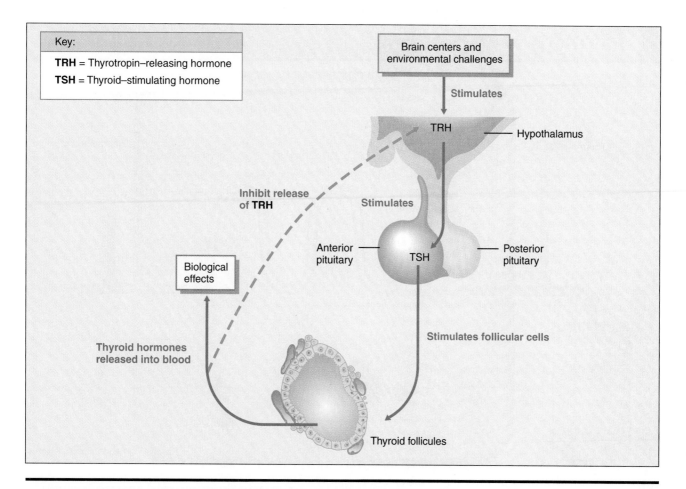

Key:

TRH = Thyrotropin–releasing hormone
TSH = Thyroid–stimulating hormone

Brain centers and environmental challenges

Stimulates

TRH — Hypothalamus

Inhibit release of **TRH**

Stimulates

Anterior pituitary — TSH — Posterior pituitary

Biological effects

Stimulates follicular cells

Thyroid hormones released into blood

Thyroid follicules

Figure 5.17   The thyroid gland regulates thyroxine secretion. Note that if the negative feedback arm represented by the dotted line is broken, a continuous stimulus will be applied to the thyroid gland. If iodine intake is too low to allow sufficient quantities of thyroxine to be produced, negative feedback is lost. The gland will continually enlarge in an attempt to supply more thyroxine, causing goiter formation.

HIGHLIGHT 5.3

Metabolic Rate

Every cell in the body obtains its energy from reactions that break down chemical bonds and thus release energy. At the same time, cells also take up chemicals from the blood and use them to synthesize other molecules. The process of using chemicals to obtain energy is called **catabolism.** The process of synthesizing new compounds is called **anabolism.** The two processes together are called metabolism. Although the catabolic and anabolic chemical reactions in the body are very complex, it is possible to measure the sum of all those reactions, the **metabolic rate.**

The sum of all the chemical reactions in the body results in heat production, which can be measured by placing a person in a sealed chamber kept at a known temperature by cooling it with water, as depicted in the figure on p. 138. (The total heat produced per unit time is a measure of the metabolic rate.) If the temperature of the entering and exiting water is measured as well as the volume of water flowing around the chamber, the heat conducted away from the chamber can be measured. That heat results from the metabolic processes of the subject in the (continued on next page)

HIGHLIGHT 5.3 (Continued)

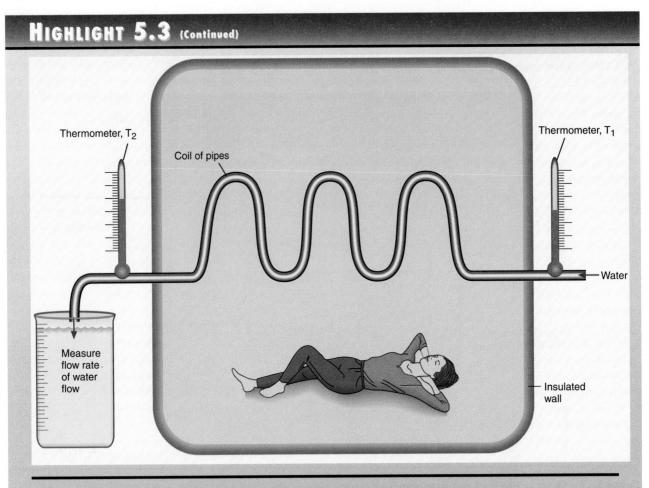

Diagram of the method used to measure basal metabolic rate. Calories produced per hour = $[T_2 - T_1] \times$ liters of water collected per hour.

chamber, and is measured in kilocalories (kcal). A kilocalorie is the amount of heat it takes to increase the temperature of 1 liter of water 1 degree centigrade.

The metabolic rate measured at rest is called the **basal metabolic rate (BMR);** it depends on many factors, including sex, weight, muscle mass, and circulating levels of hormones.

Although measuring heat production is an excellent method of determining the BMR, it is quite cumbersome and cannot be used routinely. A simpler method is to measure oxygen use by the resting subject. From many laboratory studies, we know that on average the consumption of 1 liter of oxygen corresponds to 4.8 kcal. The number of kcal produced from food each day is quite variable and depends greatly on physical activity. Should the energy (kcal) expended exceed the intake of food that is able to produce that number of kcal, the individual will lose weight. Should the kcal in the food intake exceed the kcal expended each day, the individual will gain weight. For most people, a good balance is struck between energy loss (activity) and

energy gain (food intake). Table 1 presents some common activities and the number of kcal needed to perform them.

Table 1	Aproximate kcal expended by a 70 kg person doing the following activities.
Activity	**kcal/hour**
Rest	65
Normal daily activity	100
Normal walking	200
Rapid walking	350
Swimming	700
Chopping wood	750

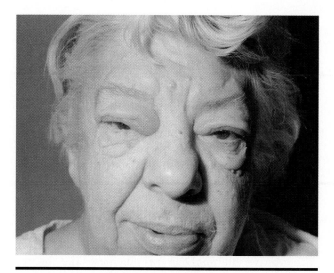

Figure 5.18 Photograph of an adult with a hypoactive thyroid gland, exhibiting characteristic facial changes associated with hypothyroidism.

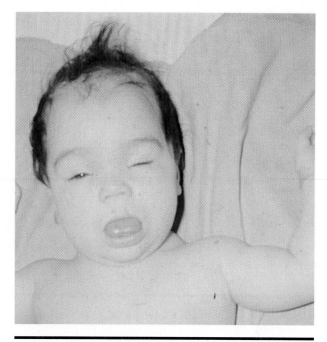

Figure 5.19 Photograph of a child with cretinism.

protrude. Figure 5.20 is a photograph of a person exhibiting typical features of hyperthyroidism. Note the characteristic protrusion of the eyes (exophthalmos).

Treatment for this condition may consist of surgical removal of the thyroid gland, followed by replacement therapy. Fortunately, thyroxine is not destroyed by the gastrointestinal tract and may be absorbed intact, allowing oral administration of thyroxine. Generally, a single thyroxine-containing pill each day is sufficient to restore the metabolic rate to normal.

Calcitonin

Researchers discovered that another hormone is produced by the thyroid glands; if injected into dogs, it lowers the calcium level in plasma. The hormone is called **calcitonin.** Further experiments showed that calcitonin is secreted by small clusters of cells near the follicular cells of the thyroid, called **parafollicular cells.** Any increase in the concentration of Ca^{++} in plasma stimulates the release of calcitonin, which acts on the kidney and bone. The kidney increases its excretion of calcium and bone dissolution is reduced; less Ca^{++} is released from bone into the plasma. The combined effect is a reduction in the circulating level of Ca^{++}. Although this appears to be an ideal regulatory system for maintaining plasma Ca^{++} constant, the significance of calcitonin in Ca^{++} homeostasis is not yet clear. People whose thyroids have been removed, who do not secrete calcitonin, have normal levels of Ca^{++}, with no apparant disruprion of Ca^{++} homeostasis. Thus, the role of calcitonin needs further investigation.

Parathyroid Gland

Embedded in the thyroid gland are four small pieces of tissue, the parathyroid glands (Figure 5.21). Although closely associated with the thyroid gland, they have an entirely separate function. The **parathyroid glands** secrete a peptide commonly called **parathyroid hormone (PTH),** also known as **parathormone.** PTH regulates the calcium balance and its

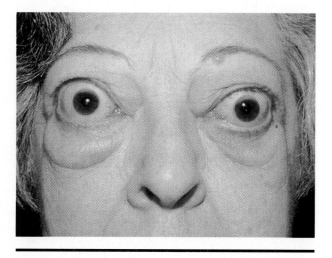

Figure 5.20 Photograph of an adult with long-lasting hyperthyroidism. The bulging eyes are a common characteristic of this hormonal imbalance.

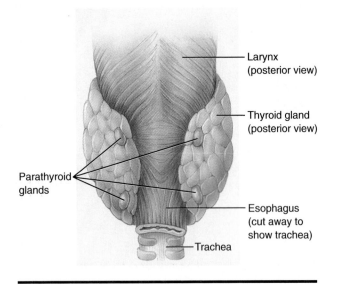

Larynx
(posterior view)

Thyroid gland
(posterior view)

Parathyroid
glands

Esophagus
(cut away to
show trachea)

Trachea

Figure 5.21 Anatomical relation of the parathyroid glands to the thyroid gland. Although the parathyroid glands are small and embedded in the thyroid gland, they are separate endocrine glands.

rate of secretion is controlled by the concentration of Ca^{++} in plasma. Figure 5.22 is a schematic diagram showing the negative feedback control system for PTH. A decrease in the calcium concentration of plasma causes the inhibitory influence on the glands to be reduced; this stimulates secretion. The increase in the PTH concentration of plasma causes a series of actions that increase the Ca^{++} concentration of plasma. PTH stimulates the kidney to retain Ca^{++} and so reduces urinary loss. It stimulates bone to increase dissolution of calcium from bone into the extracellular fluid and plasma, directly raising the calcium concentration of plasma. In addition to its direct action on bone, PTH stimulates the activation of a kidney enzyme that converts inactive vitamin D into its active form. Active vitamin D, in turn, increases the absorption of calcium from the intestines. All of these actions help to increase the concentration of calcium in blood plasma. The negative feedback arm is calcium itself. As the calcium concentration of plasma increases, the stimulus for parathyroid hormone secretion is removed.

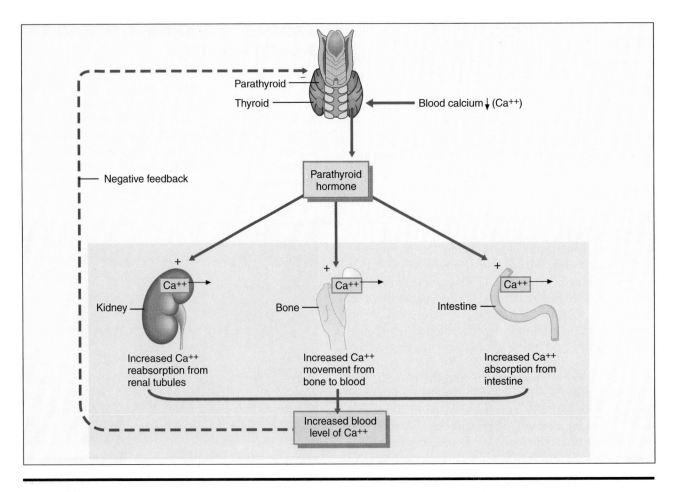

Parathyroid

Thyroid

Blood calcium ↓ (Ca^{++})

Negative feedback

Parathyroid
hormone

+
Ca^{++}

+
Ca^{++}

+
Ca^{++}

Kidney

Bone

Intestine

Increased Ca^{++}
reabsorption from
renal tubules

Increased Ca^{++}
movement from
bone to blood

Increased Ca^{++}
absorption from
intestine

Increased blood
level of Ca^{++}

Figure 5.22 Control of calcium levels in blood by parathyroid hormone.

The scheme outlined in Figure 5.22 indicates that PTH plays a pivotal role in the maintenance of calcium homeostasis. The parathyroid glands are able to respond to the calcium concentration of plasma and regulate PTH secretion so that plasma calcium levels remain constant. Because calcium balance is so important for normal neuromuscular activity (Chapter 6), lack of a functional parathyroid is incompatible with life. Complete lack of PTH results in death in a few days due to spasm of the respiratory muscles. A relative deficiency of PTH leads to increased irritability of voluntary muscles as well as increased nervous irritability.

Other Hormones

The hormone systems discussed in this chapter are some of the major hormones that affect body function. However, many other hormones exist, and some are discussed in different chapters. You will see that the gastrointestinal system secretes hormones that play important roles in the digestive processes. The kidney and the heart are also endocrine glands, and we discuss their hormones in later chapters.

Summary

The role of hormones is to help maintain body homeostasis. In contrast to the nervous system, the endocrine system acts slowly and is long-lasting. Hormone effects may last for hours, whereas neural influences react within fractions of seconds.

Endocrine glands are ductless glands that secrete their hormones into the interstitial space from which they enter the blood. Blood then carries the hormones to all parts of the body. The endocrine hormones are of three chemical types: peptide hormones, which are composed of chains of amino acids, amines that are derivatives of tyrosine, and steroid hormones, which are derived from cholesterol. Each hormone is specific for a particular organ or groups of organs, called targets. In order for a hormone to exert its effect, it must attach to a receptor molecule on the surface of the target cell or enter the cell and react within that cell. If it attaches to a membrane receptor, it causes another molecule or series of molecules to be liberated in the cell. In this case the hormone is called the first messenger and the second chemical is called the second messenger.

Some hormones, such as the steroid and thyroid hormones, enter the cell directly and react with cellular or nuclear receptors, which in turn regulate protein synthesis by DNA. All hormones have their secretory rate regulated by mechanisms that complete a negative feedback system. In general, the chemical regulated by a hormone acts to suppress hormone secretion. An example of this is seen with the hormone insulin, secreted by the pancreas. High blood levels of glucose increase insulin secretion, which causes glucose entry into cells, thus reducing blood glucose concentrations. As the blood glucose concentration falls, the stimulus for insulin secretion is reduced, so insulin secretion is reduced at the same time.

Major hormones discussed were insulin, thyroxine, steroid hormones, and some of the pituitary hormones. The lack of any of the hormones leads to a set of symptoms specific for each hormone. Insulin deficiency results in diabetes and lack of thyroxine results in goiter and a condition known as hypothyroidism. These deficiency diseases can be corrected by administering the appropriate hormone.

Some conditions lead to an oversupply of hormones, which results in a pathological set of symptoms. In the case of the thyroid, cancer of that gland may lead to oversecretion of thyroxine. The individual will experience increased metabolism and a series of symptoms such as nervousness, hyperactivity, and irritability. For maintaining good health it is important that the endocrine glands secrete their hormones in just the right amounts. The role of negative feedback in hormone regulation is most important.

Questions

Conceptual and Factual Review

1. Both the endocrine system and the nervous system act to regulate and maintain many bodily processes. In what ways do the two systems differ in that function? (p. 112)
2. What is the difference between an endocrine gland and an exocrine gland? (p. 112)
3. What is meant by the term *second messenger? First messenger?* (p. 113)
4. Although hormones are secreted into the blood, they act only on their target cells, not all the cells of the body. Explain why this is so. (p. 113)
5. How do the three different types of hormones enter cells? (pp. 113–115)
6. What are the three general categories of hormones? (p. 114)
7. Because the secretory rate of hormones is highly variable, what controls their secretion? (pp. 117–119)
8. List and identify the major functions of the hormones of the following endocrine glands: adrenal, anterior and posterior pituitary glands, pancreas, thyroid, parathyroid. (p. 119)

9. For each of the hormones, list the negative feedback arm. Explain how each works to maintain the homeostatic balance of the system it affects. (pp. 124–136)
10. For each of the hormones, name the major stimulus for their secretion. (pp. 124–136)
11. What are the clinical symptoms of a deficiency of the hormones secreted by the endocrine glands listed in Question 8? What are the symptoms of excessive secretion by these endocrine glands? (pp. 124–136)
12. Where are the endocrine glands named in Question 8 located in the body? (p. 124, p. 129, p. 131, p. 139)
13. Define a tropic hormone. (p. 126)
14. What is the relationship between the adrenal medulla and the sympathetic nervous system? (p. 127)
15. What is the relation of the hypothalamus of the brain to the pituitary gland? (pp. 129–131)

Applying What You Know

1. Insulin is the hormone lacking in people with diabetes mellitus. It now can be made in very pure form. The treatment consists of injecting insulin every day, even several times a day. Why is it that insulin cannot be taken orally?
2. Diabetics who inject themselves with insulin are advised to carry a candy bar with them. Why do you think this is necessary?
3. In the early days of insulin research, it was often found that injection of a pancreatic extract actually caused blood glucose levels to increase. After some time, the blood glucose decreased. Why do you think this happened? (It does not happen today.)
4. Popular belief maintains that if one eats too much sugar he or she will develop diabetes mellitus. Is this a correct belief? Why?
5. A mother brings her young son to a physician. She is worried about his strange behavior. He is listless, has a poor appetite, and eats a great deal

of salt. In fact she actually has seen him pour salt into his hand and eat it. Because she heard that salt is bad for a person, she is concerned about his health. What do you think may be a possible problem with her son?
6. Cancer of the thyroid gland may necessitate its surgical removal. However, this is difficult surgery, as the surgeon must spare the parathyroid glands, which are imbedded within the thyroid glands. Following surgery, the patients are given a high dose of radioactive iodide intravenously. Why do you think this is done?
7. A person with Addison's syndrome may exhibit low blood pressure. Why? That person may also be more susceptible to the ill effects of stress. Why?
8. Up to the early twentieth century, goiter was an endemic disease in many parts of the world. Now it is a relatively rare condition. What accounts for its rarity?

Chapter 6
Skeletal and Muscular Systems

Objectives

By the time you complete this chapter you should be able to

◆ List the general functions of the skeletal system.

◆ Understand the structure of bone.

◆ Explain how bone develops and how it is repaired and replaced.

◆ List the three types of muscle.

◆ Describe the general anatomy of skeletal muscle.

◆ Describe the sliding filament theory of muscle contraction.

◆ Explain how the contraction of skeletal muscle is regulated by nerves.

◆ Appreciate what factors contribute to the amount of force a skeletal muscle can produce.

◆ List the sources of energy available to skeletal muscles and describe their relative importance.

◆ Describe the three types of skeletal muscle.

◆ Explain the anatomical and functional attributes of cardiac muscle.

◆ Define *tetanus* and indicate why it can occur in skeletal muscle but not in cardiac muscle.

◆ Describe in general terms where smooth muscle is located and how its properties differ from those of skeletal and cardiac muscle.

Introduction

The first multicellular organisms most certainly lived in the primordial sea. Theirs was perhaps an easy life, but it was not without its shortcomings. For one thing, because they had no apparent means of locomotion they relied exclusively on the currents to move them from place to place and to provide them with food. The evolution of movement is part of the story of the development of a free and independent life. Indeed, the power of locomotion is one of the many characteristics that distinguishes animals from plants.

Because these first organisms were soft-bodied, the fossil record does not provide us with much information concerning their structure. However, it is generally accepted that the evolutionary process did confer on some of these creatures the ability to move from place to place. With movement came the ability to obtain food when it was not readily available nearby. Movement also afforded them the ability to avoid being eaten and to seek an environment conducive to their continued existence. The first structures devoted to movement probably were either hair-like structures (cilia) still found in some single-celled organisms or tail-like structures (flagella) similar to those of human sperm. These structures contain some of the structural elements found in the muscle tissue of humans. Muscle tissue is specialized for movement (and an excellent source of protein for carnivores).

At some point in evolution, animals began to move from the sea onto land. Here they faced additional obstacles, one of which was the force of gravity. How were the soft tissues to be supported against this force, and how was movement to be accomplished? In general, land animals evolved one of three types of support structures:

- The first type of support structure, typified by the earthworm, is not a rigid structure at all; the earthworm uses muscles to compress fluid-filled cavities, creating a form of rigidity somewhat similar to a water-filled balloon. This type of support system has limitations with regard to size and speed of locomotion.
- A second type of support structure is the **exoskeleton** (jointed external skeleton) common among insects. An exoskeleton is composed of a nonliving layer of several substances (chitin and protein) that combines both flexibility and strength, somewhat analogous to fiberglass. A limitation of an exoskeleton is that it cannot be made larger once it is deposited, so it must be shed as the animal grows.
- The third type of support structure is the **endoskeleton,** or internal skeleton, typified by humans. The human skeleton provides a strong, rigid framework to which the soft tissues are attached, thereby supporting the soft tissues against gravity. It also protects certain soft tissues; for example, the skull protects the brain and the ribs surround and protect the heart and lungs. In addition, it forms a framework for attachment of muscles so that movement can be accomplished.

The Skeletal System

The human skeleton (Figure 6.1) is composed of just over 200 hard elements (bones) and various connective tissues. Anatomists customarily subdivide the skeleton into two parts: the **axial skeleton,** consisting of the parts that form a central axis, and the **appendicular skeleton,** or the portion of the skeleton that forms the shoulders, the hips, and the four limbs, or appendages.

The Skeleton

In general, the axial skeleton protects some of the soft tissues, including the heart and the nervous system, and supports the soft tissues and the appendicular skeleton. The **vertebral column** forms the backbone of the axial skeleton. It is composed of twenty-six individual bones, called **vertebrae,** each named according to its location (Figure 6.2). The vertebral column serves either directly or indirectly as the attachment for the other bones of the skeleton and also forms a protective barrier around the spinal cord. In fact, the real danger of broken vertebrae is not in the problems associated with the degree of bony structural damage but in the risk of damage to the spinal cord or the spinal nerves. This is why one is cautioned against moving anyone who is suspected of having a broken neck until a professional assessment is made.

Between the vertebrae are the **intervertebral discs.** A disc, shaped somewhat like a hockey puck, is composed of a soft inner core and a stiff outer fibrous ring that is slightly compressible. The discs serve as shock absorbers and allow for limited motion of the spinal column while creating a strong union between the vertebrae.

Figure 6.3 shows the relationship between the discs, the vertebrae, and the spinal cord and the effects of normal compression and damage to a disc caused

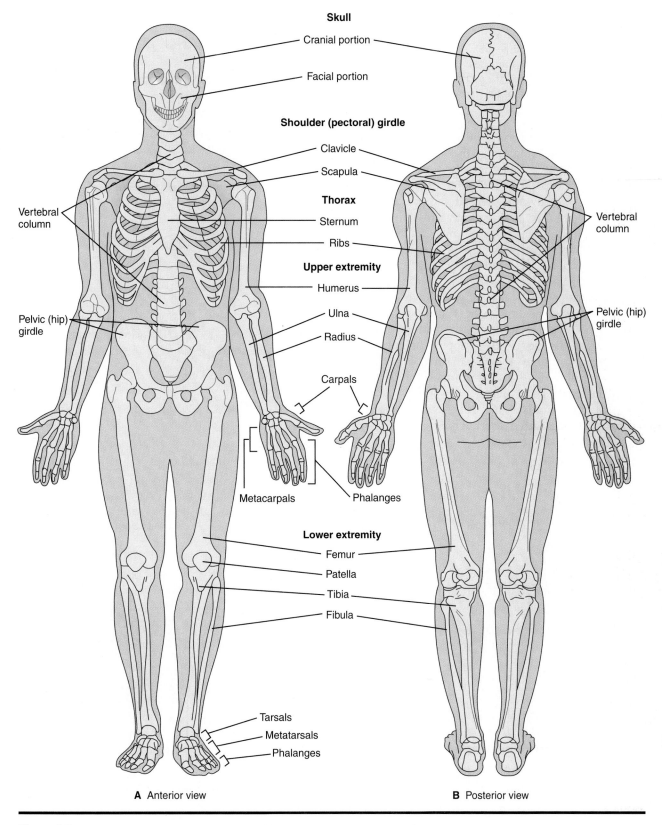

Figure 6.1 The human skeletal system. Most of the major bones are named. (A) Anterior view. (B) Posterior view.

Figure 6.2 The human vertebral column, right side view. All but the last two vertebrae are named by number and location: cervical (C_1–C_7), thoracic (T_1–T_{12}), and lumbar (L_1–L_5). The remaining two, the sacrum and coccyx, each consist of several bones that have fused into one.

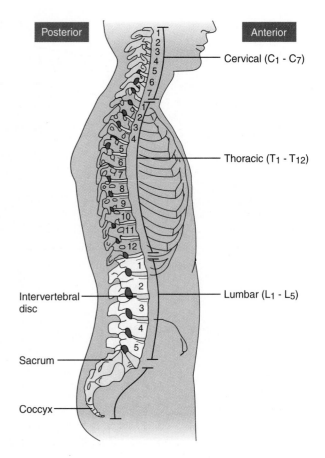

by excessive compression. With excessive compression, the stiff outer ring of the disc may actually rupture (Figure 6.3(C)) and the soft center may then herniate (protrude) into the outer ring. The major problem associated with a herniated disc (sometimes called a slipped disc) is that it may press on a spinal nerve. If that happens in the lumbar region (lower part of the back), intense back pain can occur. Severe compression of a spinal nerve may cause pain in the portion of the body supplied by the nerve, usually the leg on the same side of the body, and may also produce muscle weakness of the leg. When this happens, it may be necessary to remove the herniated portion of the damaged disc surgically. It should be noted that the vertebrae are joined together in such a way that regardless of how one twists and turns, the vertebral bones themselves are unable to compress a spinal nerve. In other words, spinal nerve compression can be caused by a herniated disc, but not by a slight twisting of the bones of the spine.

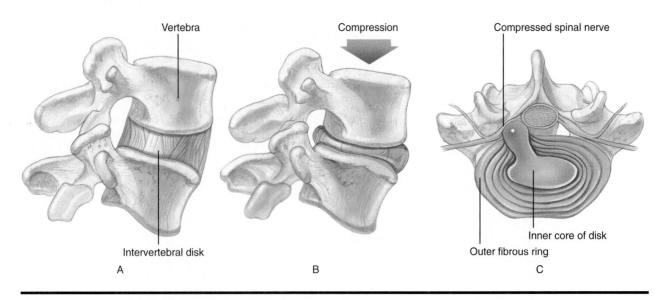

Figure 6.3 Relationship between two vertebrae and an intervertebral disc. (A) Normal position. (B) Effect of compression. (C) Rupture of the outer portion of a disc with herniation of the soft inner core, seen from above. Note that the herniated inner core of the disc presses on a spinal nerve.

Twelve pairs of **ribs** are attached posteriorly (at the back of the body) to the thoracic vertebrae. The first seven pairs are also attached anteriorly (at the front of the body) directly to the sternum, or breastbone, by means of **cartilage.** The next three pairs are attached anteriorly only by means of their attachment to the seventh pair, and the final two pairs are not attached at the front of the body at all (See Figure 6.1). The ribs surround and protect the heart and lungs and serve as the attachment for some of the muscles used in breathing.

The **skull** is composed of the cranium and the facial bones. The cranium encloses and protects the brain, and although at first glance it appears to be one unit, it is actually composed of eight separate bones. In the adult they are tightly joined, but in the newborn they are not yet completely fused, permitting the skull of the fetus to be compressed slightly during normal birth and allowing for expansion during normal growth and development. You may have felt the soft spot on the top of a baby's head where the bones of the cranium are not yet joined. The only movable portion of the skull in the adult is the mandible, or jawbone, used to chew food.

The appendicular skeleton is specialized not for protection, but for movement and support against gravity. It consists of the **pectoral** (shoulder) **girdle,** the **pelvic** (hip) **girdle,** and the bones of the four extremities (arms and legs). The two girdles attach to the axial skeleton and in turn anchor the four extremities.

Joints and Attachments

Based on the skeletal system shown in Figure 6.1, you might think that some of the bones seem to barely touch each other, if they touch at all. How exactly is the skeletal system held together? How does it provide the strength and rigidity needed for support and still allow a wide range of motion? The answer lies in the ways in which bones are attached to each other and to muscle, and in the type of joint (or articulation) that exists between the bones.

A **joint** is a point of union between bones. Joints are classified either functionally, according to how much movement they permit, or structurally, depending on whether there is a space between the bones and on the type of tissue that binds the bones together. We will ignore most of the technical names of joint types in favor of concentrating on the principles that govern joint strength and movement. A few joints allow essentially no movement, such as most of the joints of the skull. Although you should be aware that such joints exist, we will concentrate on joints designed for movement.

Most of the joints designed for motion are called **synovial joints** because they have a small, fluid-filled space between the bone surfaces called a **synovial cavity.** A generalized diagram of a synovial joint is shown in Figure 6.4. The watery fluid in the synovial cavity acts as a lubricant, just as oil acts as a lubricant between moving metal parts in a car. The amount of fluid is exaggerated in Figure 6.4 for emphasis; in fact it may be no more than a thin lubricating layer. The

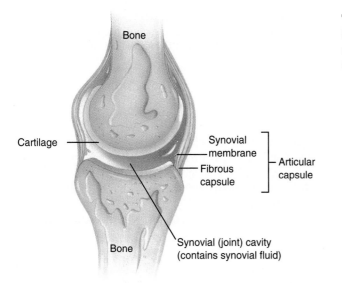

Figure 6.4 A synovial joint. A thin layer of fluid in the synovial cavity acts as a lubricant between the bone surfaces.

synovial fluid is secreted by cells that line the synovial cavity. The adjoining surfaces of the bones are covered with flexible, somewhat elastic cartilage and the entire joint is surrounded by a capsule of connective tissue known as the articular capsule, which keeps the fluid in the joint and binds the two bones together.

Although the capsule surrounding the joint maintains the integrity of the synovial cavity and offers some support, other connections also contribute to the strength and flexibility of the joint. **Ligaments** are connective tissues that join bones to other bones and **tendons** are connective tissues that attach muscles to bones. Ligaments and tendons are composed primarily of a few living cells (called fibroblasts) and a large quantity of extracellular **collagen,** a type of protein, which is produced by these cells. The collagen is arranged in parallel fibers, making ligaments and tendons very tough yet moderately flexible. Figure 6.5 shows a photomicrograph of a tendon.

The primary function of ligaments is to strengthen and stabilize the joint. Typically, there are a number of ligaments in any joint, each of which provides support in a different plane or dimension. An example of a joint with its various ligaments is the knee joint, depicted in Figure 6.6. Twisting injuries to the knee may tear some of these ligaments. Tendons and muscles also help to stabilize joints, which is why part of the therapy for torn ligaments of the knee is to try to strengthen the surrounding muscles through careful repetitive exercise.

Movement

Bones are made to move in relationship to each other around a common point (the joint) by the action of

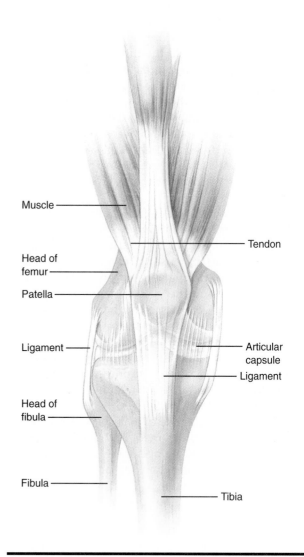

Figure 6.6 The knee joint (anterior view), showing some of the ligaments that attach bones to each other, supporting the knee joint. This view also shows several tendons that attach muscles to bones. Most of the muscles have been cut away so that the ligaments can be seen clearly.

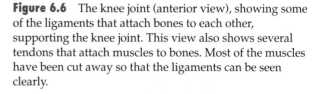

Figure 6.5 A photomicrograph (400×) of a section of a tendon.

muscles. Most muscles that affect skeletal movement are elongated, with an attachment at each end to bones (Figure 6.7(A)). The end of a muscle that is attached to the more stationary part of the skeleton is called its **origin.** The end of a muscle that is attached to the part of the skeleton that moves is called its **insertion.** Muscles can have more than one origin or insertion, as Figure 6.7(A) demonstrates.

Contraction of the biceps muscle (Figure 6.7(B)) shortens the distance between the radius bone (forearm) and the scapula (shoulder), decreasing the angle between the bones of the upper arm and forearm. This type of motion is called **flexion.** Conversely, contraction

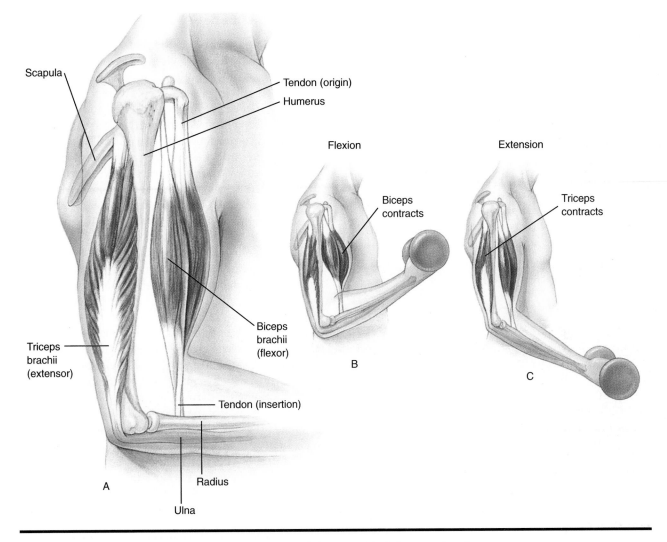

Figure 6.7 How muscles produce skeletal movement. (A) A simplified diagram showing the two major muscles that produce movement of the forearm, and their attachments to bone by tendons. Although it appears that one origin of the triceps brachii attaches the muscle directly to the humerus bone, there are actually microscopic tendinous connections that cannot be seen with the naked eye. (B) Flexion (of the forearm relative to the upper arm) is produced by contraction of the biceps brachii muscle. (C) Extension is produced by contraction of the triceps brachii muscle.

HIGHLIGHT 6.1

Joint Stabilization by Tendons, Ligaments, and Muscles

The patella (kneecap bone) is attached to the tibia (lower leg bone) by a ligament and to the muscles of the thigh by a tendon. In order to appreciate the role ligaments, tendons, and muscles play in stabilizing and strengthening a joint, try this simple experiment. While sitting in a low chair or lying down, straighten (extend) your leg out in front of you with your heel resting on the floor and relax your leg muscles. Try to move the kneecap from side to side gently with your hand; notice how easily it can be moved out of position. Then, without changing position, tense the muscle of your thigh and again try to move your kneecap. It remains held in position because the muscle contraction has put tension on the tendon and the ligament, which cross the surface of the kneecap, holding the kneecap firmly in place. If you move your hand to just below the kneecap, you can feel the tightening of the patellar ligament as you alternately tense and relax the thigh muscle.

of the triceps brachii muscle shortens the distance between the proximal end of the ulna bone (forearm) and the scapula, leading to **extension,** or an increase in the angle between the upper arm and forearm (Figure 6.7(C)). These are just two examples of the many types of skeletal motions; the student is referred to more advanced texts of anatomy for a more complete treatment of this subject and for the names of all the muscles in the body. The important point is that active movement of the skeleton is produced only by shortening of muscles. Put another way, muscles actively pull elements of the skeleton toward each other; they have absolutely no capacity to push them apart.

Bones

You already know some of the functions of bones because bones are the structural elements of the skeleton. Functions already discussed include structural support of soft tissues, protection of some organs, and sites of attachment of muscles so that movement can be produced. But there are several more functions of bones not generally thought of when we think of bones only as hard structural elements. These include storage of energy in the form of fat, storage of minerals, and production of blood cells. You should be aware of these latter three functions, but they will not be described in great detail.

Structure of Bone

Bone is composed primarily of extracellular nonliving material in which a few living cells are embedded. Although most of the bone material is nonliving, it is considered a living tissue. A cut-away view of a typical bone is shown in Figure 6.8. The middle portion of the bone is composed primarily of very hard, dense bone called **compact bone** with a central hollow space, called the **medullary cavity.** Compact bone in an adult is composed primarily of crystals of minerals (primarily calcium phosphate and calcium carbonate), which have formed around a framework of collagen fibers. Minerals compose about two-thirds of the weight of the bone, and most of the remainder is collagen. The collagen framework of bone is secreted by specialized bone-forming cells, which ultimately become surrounded by hardened crystals and thus are permanently embedded in the bone. However, even compact bone is not completely solid; there are blood vessels at regular intervals to nourish the bone cells and carry away waste. The medullary hollow cavity at the center of most adult long bones is filled with yellow marrow, which is composed primarily of cells that store energy in the form of fat.

The two ends of the bone consist of **spongy bone,** so named because the hard elements are interspersed with many small hollow spaces, giving it a spongy appearance (it does not *feel* spongy, however). Spongy bone is composed of a network of thin plates of bone with numerous interconnecting spaces between them. The red marrow in the spaces of spongy bone is re-

Figure 6.8 Photograph of a bone. Part of the bone has been removed to show the medullary cavity and the difference between spongy and compact bone.

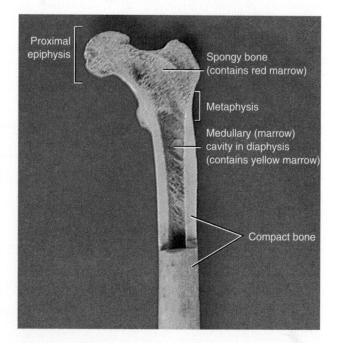

Compact Bone

Have you ever noticed that construction of concrete structures always involves the formation of a network of steel reinforcing rods embedded in the concrete? The interconnecting steel network gives the concrete flexibility; without it, the concrete would be brittle and prone to crumbling. The same holds true for bone; the collagen fibers form the framework and provide flexi-bility, and the mineral deposits that fill in the area around and between the fibers provide rigidity and strength. If you can imagine that the rocks in concrete are actually living cells that produce both the steel matrix and the cement that fills in the spaces, you will have a pretty good idea of the structure of compact bone.

---

sponsible for the producing blood cells, so marrow consists primarily of red blood cells and white blood cells in various stages of development. The surface of the bone not covered by cartilage is covered by a thin fibrous layer, known as the **periosteum,** which contains blood vessels, nerves and the cells responsible for the formation of new bone.

Bone as a Dynamic Tissue

Most mechanical devices wear out. Even a well-made automobile finally reaches the point where it must be replaced. How is it, then, that the human body can function for 70, 80, or 100 years? The answer is that most tissues have the capacity to replace whole cells, cell parts, or extracellular material as they wear out, sometimes at phenomenal rates. This is certainly true of bone. Even in adults who are not gaining in overall bone mass, approximately 5–10% of the bone mass is replaced every year. In some bones, particularly those that are exposed to mechanical stress, the turnover is even faster. Some portions of the thighbone may be completely replaced several times a year! Furthermore, bone undergoes adaptive changes in response to exercise, aging, and damage.

Normal Growth and Replacement Bone cells go through three stages of differentiation. During the first stage the cells are capable of producing only cartilage, during the next stage they produce cartilage *and* contribute to the deposition and removal of calcium, and finally the cells perform only calcium deposition and removal. The cells are named according to the stage of their life-cycle. During fetal development, **chondroblasts** (cartilage-forming cells) produce a cartilaginous framework, or model, that will ultimately be replaced by hardened bone tissue. Thus, in the earliest stages of fetal development the bones are no more than cartilaginous models of the hardened bones of adults (Figure 6.9).

Once the initial cartilage model has been formed, some of the chondroblasts, particularly those near the outer surfaces of the developing bone, differentiate into **osteoblasts** (bone-forming cells). Like chondroblasts, osteoblasts can secrete new cartilaginous material for the framework of bone, but they are also responsible for depositing the calcium minerals that give bone its rigidity and strength. Osteoblasts are involved in forming both spongy and compact bone. With time, the osteoblasts become embedded in the calcified bone tissue and lose the ability to form new cartilage. However, these mature bone cells, called **osteocytes,** continue to direct the processes of calcium deposition and removal that occur throughout life.

During growth and development, the activity of chondroblasts and the osteoblasts that develop from them is concentrated on the outer surface, particularly at the ends of the bone. Thus, bones tend to get longer as a child develops primarily because new bone is added to the ends. Also during growth and development, the chondroblasts that were present at the center of the developing bone die and the cartilage is dissolved away, forming the hollow medullary cavity. Both the formation of hardened bone at the outer surfaces and the formation of the hollow medullary cavity begin during fetal life, but these processes are by no means complete by birth.

There is a fourth type of cell in bone, the **osteoclasts** (from the Greek words for *bone + to break*). Osteoclasts do not differentiate from any of the other bone cells; rather, they arise from a type of white blood cell. These cells remove existing hardened bone tissue by burrowing through the hardened bone, dissolving the calcium mineral deposits and digesting the cartilage matrix in their path. They are often followed by osteoblasts, so that new bone is laid down where old bone is lost. Consequently, bone is in a constant state of replacement even though its overall mass may not change appreciably. The rate of growth, replacement, and repair of bone, therefore, depends on the balance tween and around the broken ends of the bone. With time, dead fragments of the original bone are resorbed by osteoclasts, and the temporary union becomes dense and hard as calcium deposits fill in the space. The entire process can take weeks to months depending on which bone is involved and on the age of the person. In general, the repair process slows with age.

Sometimes a fracture may not heal as rapidly as might be desired or may fail to heal at all. Recent research suggests that in these cases the rate of healing may be increased significantly by applying small electric currents in the vicinity of the fracture. Exactly how applying an electric current stimulates the healing process is unclear; one hypothesis is that it simulates the effect of compression stress on osteoblast activity.

Exercise Bones tend to get bigger when exposed to compression stress during weight-bearing exercise; indeed, the bones of trained athletes may be noticeably thicker and heavier than those of a nonathlete. For a long time scientists knew that the rate of bone formation increased in response to exercise, but not what caused it. However, it is now known that mechanical compression stress causes bone to produce tiny electrical currents that stimulate osteoblasts. Thus new bone tissue tends to be deposited in exactly the right place to counteract the stress.

Repetitive compression stress can actually modify the shape of bones slightly (Figure 6.10). This phenomenon is most noticeable in the case of a broken bone that has initially healed at a slight angle, although it also occurs under normal conditions throughout life. The greater compressive stress on the inner angle of a bent bone causes increased osteoblastic activity in that region, with the subsequent addition of more new bone to the bone's surface at the inner angle of the bend. At the outer angle of the bent bone, where compressive forces are less, osteoblastic activity is reduced. Because osteoclast activity remains fairly constant, there is net resorption of bone at the outer angle. Over time, then, the bent bone straightens as new bone is added to the inner angle and old bone is removed from the outer angle. This is actually a homeostatic mechanism, albeit an exceedingly slow one, that creates a similarity of compressive forces throughout all regions of the bone.

Aging Bone mass increases until about age thirty in women and thirty-five in men, after which it begins to decline. The rate of loss of bone mass in women is accelerated after menopause. Women may ultimately lose more than a third of their young adult bone mass over their lifetime, whereas men generally lose somewhat less. In general, this slow loss of bone is of little consequence in our daily lives except for an increased risk of fractures. However, severe decreases in bone mass may lead to **osteoporosis.**

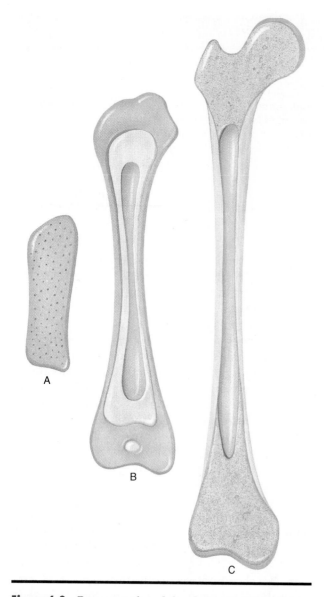

Figure 6.9 Bone growth and development. (A) Early in fetal development, a template of cartilage is laid down by chondroblasts. (B) During growth of the bone, chondroblasts continue to lay down new cartilage at the developing ends of the bone and osteoblasts add calcium minerals to the existing cartilaginous matrix. Some of the cartilage at the center of the bone is dissolved, forming the medulla. (C) An adult bone.

of the activity of osteoblasts and osteoclasts. The balance depends on several factors, including the availability of calcium and phosphorus, sufficient amounts of several vitamins, regulation by several hormones, and even the degree of mechanical compression stress.

Bone Repair When a bone is broken, or fractured, osteoblasts and osteoclasts migrate quickly to the area and begin the repair process. First, the osteoblasts form a temporary union of new collagen be-

HIGHLIGHT 6.3

Osteoporosis

Osteoporosis is a disease caused by excessive bone loss over time, leading to brittle, easily broken bones. It is particularly prevalent in women who have gone through menopause. The clinical signs and symptoms of osteoporosis include a tendency toward fractures of the vertebral column, the hips, and the forearm just above the wrists, a hunched posture, and difficulty walking. Osteoporosis is a major health problem, accounting for approximately 300,000 hip fractures every year.

Progressive bone loss occurs whenever a slight imbalance develops between the rate of bone formation and the rate of bone resorption. Over decades, even a very slight imbalance can result in a significant reduction in bone density and bone mass. For example, if the rate of bone loss were only 0.5% per year, one would lose just under 15% of one's bone mass over thirty years. A loss of this magnitude is less than the normal rate of loss with aging, and would not be sufficient to produce osteoporosis. However, if the rate of bone loss were to rise to 1% per year, one would lose 25–30% of one's bone mass over the same time period. At this point bones would begin to fracture easily, and the outward signs of osteoporosis would become apparent. This example illustrates why it is so important that physiological regulatory systems stay in nearly exact balance throughout life.

Exactly how an imbalance in bone formation and

A woman with osteoporosis, showing the characteristic hunched posture.

resorption comes about is not clear. Until recently, the prevailing hypothesis was that the bone-forming activity of osteoblasts declines with age, and that this effect is due to the decline in the concentrations of the sex steroids testosterone (in men) and estrogen (in women). Since estrogen declines substantially after menopause, this would fit with the finding that osteoporosis is more common in women than in men. However, very recent studies report that the rates of bone resorption and formation both actually increase after menopause, particularly in those women who lose the most bone mass. Thus, osteoporosis is not due to a decline in bone formation alone, but rather to an as of yet unexplained increase in the imbalance between resorption and formation.

A second contributing factor to the normal decline in bone mass in all people is the availability of calcium. Calcium is required for the formation of new bone. Part of the aging process is a decline in the availability of calcium, in part because calcium absorption by the intestinal tract declines with age. Thus it is important that one maintain an adequate calcium intake not just during childhood, when one is growing rapidly, but also throughout adulthood.

The best way to prevent the onset of osteoporosis in normal individuals is to maintain an adequate calcium intake combined with a consistent, moderate exercise regimen throughout life. Physicians may prescribe estrogen replacement therapy for some women after menopause, and some recommend calcium supplements to women in their twenties and thirties in the hope of preventing the later development of osteoporosis. Studies have shown that estrogen replacement therapy and calcium supplementation reduce the rate of bone loss in women after menopause. However, the evidence is less convincing that calcium supplementation alone in healthy women before menopause or in men will protect against osteoporosis later in life, as long as their normal supply of calcium is adequate. In all groups, consistent moderate exercise seems to be of benefit.

Muscles

As described at the beginning of this chapter, the ability to produce movement is one of the requirements for a free and independent life. Bones and joints provide the structural framework, but it is our muscles that generate the force required for movement.

Types of Muscle

It is customary to classify muscle tissue into three types based on some functional and anatomical differences. The muscle type with which most people are familiar is **skeletal muscle.** Skeletal muscle is usually attached to bones (via tendons) and is called voluntary muscle because we generally have control over its action.

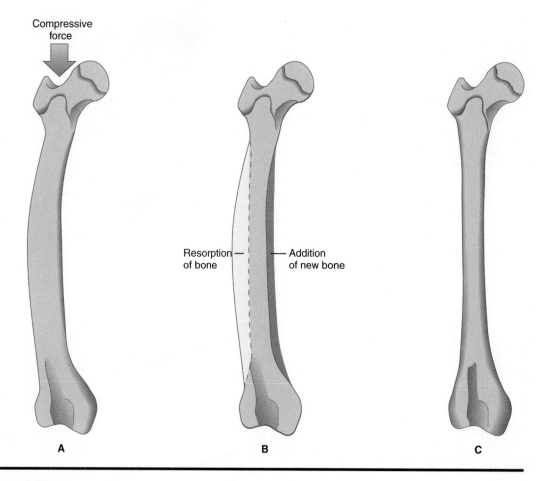

Figure 6.10 Bone remodeling and repair. (A) Compression stress on a slightly bent bone produces greater force on the inside angle of the bone, stimulating osteoblasts. (B) As a result, more new bone is added to the inside angle and more old bone is resorbed from the outside angle. (C) With time, the bone straightens.

Cardiac muscle is found only in the heart. Cardiac muscle cells are shaped somewhat differently from skeletal muscle cells and have specialized electrical connections between cells that are not present in skeletal muscle. Skeletal and cardiac muscle share several common features: both have a striated appearance when examined under the microscope (and thus both are called striated muscle) and they share a similar mechanism of contraction. However, the way in which contraction is initiated differs.

The third muscle type, **smooth muscle,** lacks the striated appearance of skeletal and cardiac muscle. Smooth muscle is found around blood vessels and around many of the hollow organ systems, including the bladder, the gastrointestinal tract, and the uterus. Cardiac and smooth muscle are called involuntary muscle because we have virtually no conscious control

over their activation. Photomicrographs of the three muscle types are shown in Figure 6.11.

Functions

Muscles perform several important functions related to their ability to generate force. The most obvious function to most of us is the generation of movement. Movement can be either voluntary, as in the case of conscious skeletal muscle movement, or involuntary, as in the pumping action of the heart or the contraction of the smooth muscle around a blood vessel. A second important function, at least of skeletal muscle, is the *prevention* of motion. A good example is the maintenance of posture. The importance of prevention of motion becomes clear when you consider what happens when a person loses control of his or her skeletal muscles during fainting. Main-

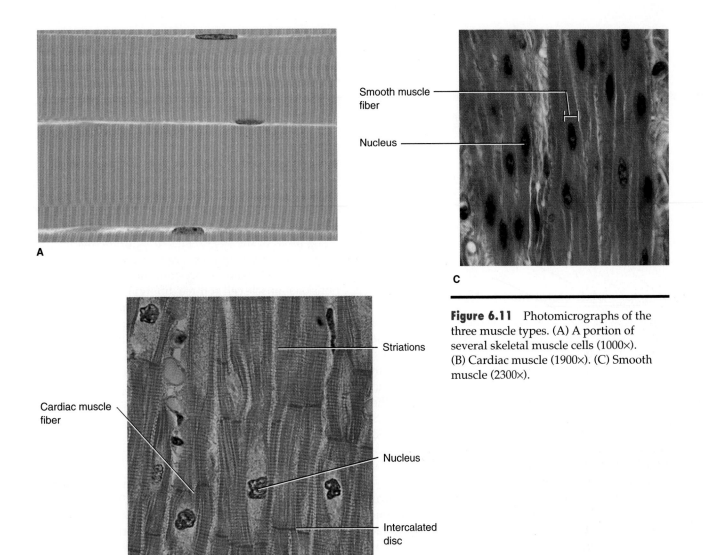

A

B

C

Figure 6.11   Photomicrographs of the three muscle types. (A) A portion of several skeletal muscle cells (1000×). (B) Cardiac muscle (1900×). (C) Smooth muscle (2300×).

taining posture is primarily a reflex action; we generally don't have to think about it except under special circumstances, such as when we are asked to stand at attention or when we are on a ship in rough seas.

A third function of muscle, again primarily of skeletal muscle, is generating heat. You may not think of this function as important under most circumstances. However, our body temperature must be kept fairly constant at about 98–99°F even in the face of fluctuating external temperatures, and heat is constantly being lost through the skin. Muscle contraction is responsible for over three-quarters of all the heat generated by the body. Under extremely cold conditions, skeletal muscle contraction can occur specifically to generate more heat without producing much movement; this is known as shivering.

Mechanism of Action

All muscles, regardless of type, have only one basic action: they contract (shorten in length) when activated. The many different, complex actions of the human body, from threading a needle to lifting heavy weights, are possible because of the relationships between bones and skeletal muscles. The pumping action of the heart by cardiac muscle, the constriction of blood vessels by smooth muscle, and the movement of food through the digestive tract (also by smooth muscle) are all accomplished by the same basic mechanism: contraction. Cardiac and smooth muscle, however, do not pull against bones; these muscles accomplish their action by virtue of their particular spiral or circular arrangements around a hollow chamber.

Each of the three muscle types will be examined in the following sections.

Skeletal Muscle

The hundreds of individual skeletal muscles in our bodies make up approximately 40% of our body mass. They range in size from the tiny muscles that move our eyes to the bulging thigh muscles of a weight-lifter. In order to understand how they function we must first understand their structural design.

Structure of Skeletal Muscle Tissue

A whole skeletal muscle, such as the biceps muscle of the upper arm, is covered by a connective tissue sheath. The sheath is continuous with the tendons, which anchor the muscle to bone. When viewed in cross-section (Figure 6.12(A)), the muscle appears to be arranged in bundles. Connective tissue forms the structure that holds the bundles together and anchors the bundles to the tendons. Each bundle is composed of many muscle cells (a muscle cell is also called a **muscle fiber**). In Figure 6.12(A), a muscle fiber is depicted as extending from the cut surface.

A magnified view of a single muscle cell is shown in Figure 6.12(B). The cell is surrounded by a cell membrane, called the sarcolemma, and a coating of collagen fibers. Muscle cells do not look like the typical cell depicted in Chapter 1. They are very elongated and appear to be filled with hundreds to several thousand long, thin **myofibrils.** Note that the muscle cell is also atypical in that each cell contains more than one nucleus. These nuclei are located just beneath the cell membrane because nearly all of the space in the muscle cell is occupied by myofibrils. Numerous mitochondria are squeezed into the small amount of space between the myofibrils, but they are too small to be seen at the magnification depicted in Figure 6.12(B).

An expanded view of a myofibril (Figure 6.12(C)) shows that it has a striated, or banded, appearance that is repeated at regular intervals. A single segment, from one thin dark band (called the **Z-line**) to the next is called a **sarcomere.** Sarcomeres are the repeating units in muscle cells that contain the contractile elements.

Figure 6.12(D) shows an expanded longitudinal section (along the long axis) of a myofibril and reveals the arrangement of the contractile elements in a sarcomere. A sarcomere consists of **thick filaments,** composed of a protein called **myosin, thin filaments,** composed of a protein called **actin,** and two additional proteins called **troponin** and **tropomyosin;** these latter two are so closely associated with each other that they are sometimes referred to as simply troponin–tropomyosin. Troponin and tropomyosin are involved in the regulation of contraction, as we will see in a moment.

The thick and thin filaments overlap each other for at least a portion of the sarcomere. Contraction of a muscle occurs when the thick and thin filaments slide relative to each other, causing shortening of the sarcomere. A single myofibril of a muscle cell in your biceps muscle may be composed of over 100,000 sarcomeres arranged end-to-end. As a consequence of the shortening of all of the sarcomeres in all of its myofibrils, the muscle cell becomes shorter, or contracts.

Figure 6.13 shows electron micrographs of portions of a muscle cell at three different magnifications. Figure 6.13(A) shows a section of several muscle cells. At this magnification, the striated appearance of the muscle cell is visible as a repeating pattern in the cell. At a higher magnification (Figure 6.13(B)), the individual myofibrils and the sarcomeres are clearly distinguishable. A cross-section through a myofibril (Figure 6.13(C)) shows the regular hexagonal arrangement of thick and thin filaments.

Mechanism of Contraction

Let us now examine the sarcomere more closely, because an understanding of how a sarcomere shortens is key to understanding how the whole muscle cell contracts. In a nutshell, sarcomeres shorten because the individual thick and thin filaments slide relative to each other; this is known as the sliding filament mechanism.

How does this sliding occur? The thick filaments are composed of many molecules of myosin, each of which is shaped somewhat like a pair of golf clubs with intertwining shafts. The shafts form the main part of the thick filament and the heads, called **cross-bridges,** stick out to the side. When the signal for contraction is given, the cross-bridges attach to actin of the thin filaments and pull the thin filaments toward the middle of the sarcomere. As you can see in the diagram of Figure 6.14(A), this has the effect of shortening the sarcomere without either the thick or the thin filaments getting any shorter. The limitation to the degree of shortening is that eventually the thick filaments touch both ends of the sarcomere at the Z-lines; at this point no further shortening is possible.

Figure 6.14(B) is a schematic view of a single thick filament and four thin filaments showing how the cross-bridges produce movement. Because a single cross-bridge can move only a short distance, the cross-bridges must attach, produce movement, and detach thousands of times in order to produce maximal

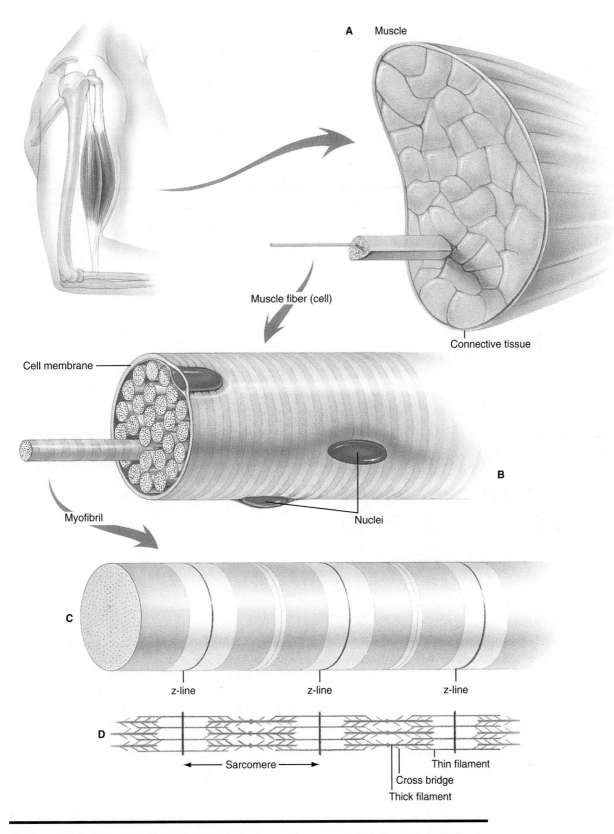

Figure 6.12 The organization of skeletal muscle. (A) The relationship of muscle to tendon and bone. (B) Closeup of a portion of a single muscle cell. (C) Expanded view of a portion of a myofibril. (D) Representation of a sarcomere showing the relationship between thick and thin filaments.

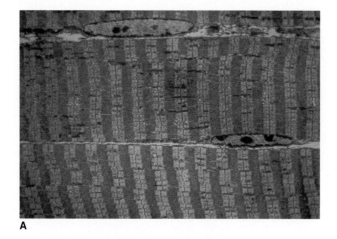

A

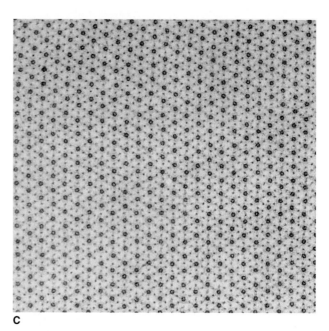

C

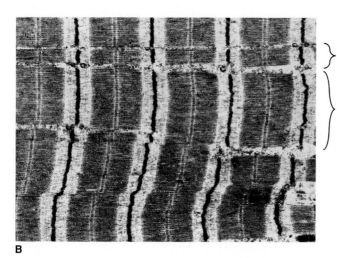

B

Myofibrils

Figure 6.13 Electron micrographs of myofibrils of a muscle cell. (A) Low-power view (2400×) showing many myofibrils in the longitudinal sections of three muscle cells. The oval structures are nuclei. (B) A longitudinal section of several myofibrils at 148,000×. (C) A cross-section through a myofibril in the region of the sarcomere that has overlapping thick and thin filaments.

shortening of the sarcomere. Thus not all cross-bridges are attached at once; indeed, while some are attached and actively producing movement, others are in the process of detaching and still others are returning to their original positions to attach again.

We have described how the sarcomere shortens, but what initiates the contraction, and how does contraction actually occur? In muscle, an activation signal is required; that signal is an increase in calcium concentration in the cytosol. Muscle cells have a complex system of membrane-bound storage sacs, called the **sarcoplasmic reticulum.** The sarcoplasmic reticulum is actually a modified endoplasmic reticulum; its function is to collect and store calcium. When calcium concentration in the cytosol is low, the muscle is in a relaxed state because the myosin cross-bridges are pre-

vented from interacting with actin of the thin filaments by the physical presence of troponin–tropomyosin. When an action potential occurs in the muscle cell, the sarcoplasmic reticulum releases calcium into the cytosol; calcium binds to troponin, causing troponin–tropomyosin to move away from actin sufficiently that the myosin cross-bridges can interact with actin, and contraction occurs. These relationships are shown in Figure 6.15 on p. 160.

The magnitude and duration of the contraction are directly related to the concentration of calcium in the cytosol. Contraction ends (and relaxation begins) when the calcium is transported out of the cytosol back into the sarcoplasmic reticulum, and troponin–tropomyosin again inhibits the attachment of the myosin heads to actin.

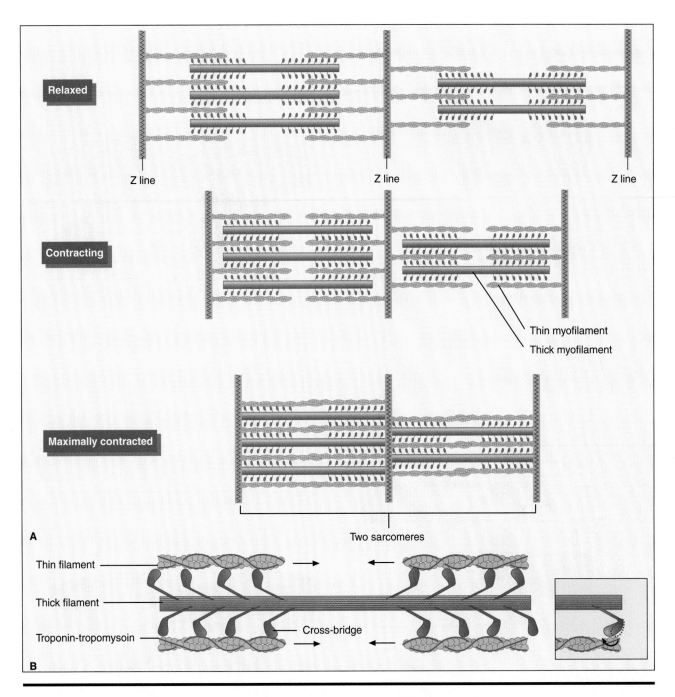

Figure 6.14 Sliding filament mechanism of contraction. (A) The relationship between the thick and thin filaments in the relaxed state and during contraction. (B) Closeup depiction of the relationship between the cross-bridges of the thick filament and the thin filaments.

Regulation of Skeletal Muscle Contraction

Recall that muscle cells, like neurons, are excitable; thus, they can generate action potentials. An action potential in a muscle cell is the stimulus for the release of calcium and the initiation of contraction. The muscle action potential, in turn, is caused by an action potential in the neuron that innervates it. The process whereby an action potential in a neuron produces an action potential in a muscle cell is called **neuromuscular transmission.**

In neuromuscular transmission, information is transmitted across a synapse just as it is between two neurons. A neuron-to-muscle synapse is called a **neuromuscular junction.** As is the usual case for a neuron, an action potential in the neuron causes the release of a neurotransmitter substance into the

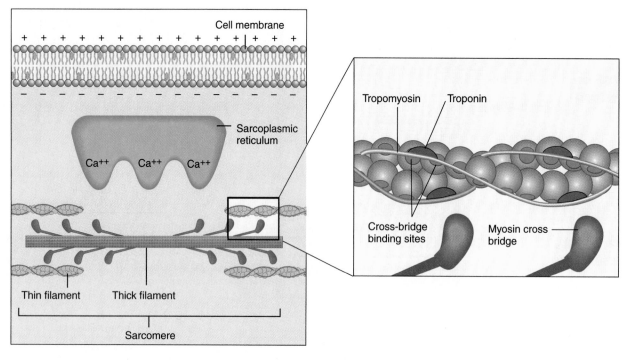

A

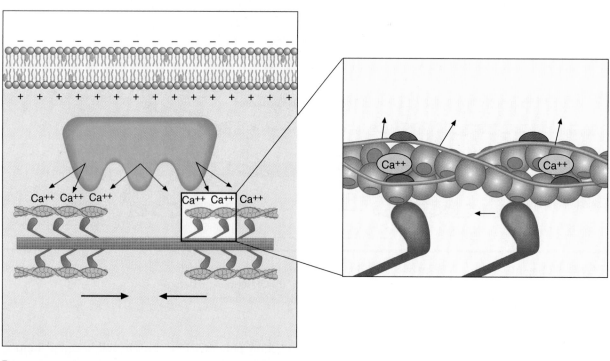

B

Figure 6.15 Excitation–contraction coupling. (A) Relaxed state; Ca⁺⁺ is stored in the sarcoplasmic reticulum, and thus the myosin cross-bridges cannot bind to actin molecules of the thin filaments because of interference by troponin–tropomyosin. (B) An action potential (excitation) causes C⁺⁺ to be released into the cytosol; Ca⁺⁺ binds to troponin–tropomyosin and causes it to shift position suffiently that the myosin cross-bridges can bind to actin, and contraction occurs.

synaptic cleft. The neurotransmitter at a neuro-muscular junction in skeletal muscle is always acetyl-choline (ACh). ACh diffuses across the synaptic cleft and binds to receptors on the muscle cell membrane, which in turn causes ion channels to open, depolarizing the muscle cell membrane in the region of the neuromuscular junction. Although the ion channels allow the passage of many small ions, it is primarily the inward movement of Na$^+$ that causes the membrane depolarization.

The local depolarization produced by ACh is a graded potential that is dependent on the amount of ACh released by the neuron. However, in addition to the ACh-gated channels concentrated at the neuromuscular junction, the entire muscle cell membrane (like that of a neuron axon) also has many *voltage-gated* channels. Therefore, the graded potential at the neuromuscular junction triggers an all-or-none action potential, which spreads across the entire cell mem-

brane of the muscle cell in all directions and to the sarcoplasmic reticulum, from which it causes the release of calcium. Like action potentials in neurons, muscle cell action potentials have a uniform amplitude and speed of conduction.

Although transmitting information from a neuron to a skeletal muscle cell is quite similar to transmitting information from one neuron to the next, there are some very important differences, as depicted in Figure 6.16. First, although a single neuron may activate several muscle cells (more on this later), *each individual muscle cell synapses with, and thus is stimulated by, only one neuron.* A good analogy for neuromuscular transmission is a military chain of command; one sergeant commands many privates, but each private is under the command of only one sergeant.

Second, *an action potential in the neuron always produces an action potential in the muscle cell* (a good soldier always does what he/she is told). This is different from

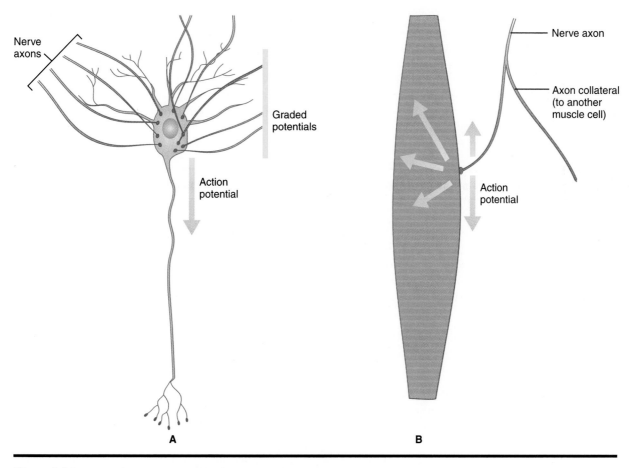

Figure 6.16 Neural transmission of information. (A) Transmission from neurons to another neuron. Multiple inputs from many neurons produce graded potentials, which become an action potential only at the axon hillock. An action potential travels only in one direction. (B) Neuromuscular transmission. A single neuron always produces an action potential in the muscle cell. The action potential travels outward in all directions from its point of origin.

information transmission between neurons, where the activity of many presynaptic nerve terminals may be required to bring the postsynaptic neuron to the threshold potential for an action potential. One-to-one transmission of an action potential from the neuron to the muscle cell is ensured because so much ACh is released by a single action potential that the threshold potential of the muscle cell is always exceeded.

Third, *muscle cell action potential transmission is outward from the neuromuscular junction in all directions*, so that the entire muscle cell membrane is activated as quickly as possible. A muscle cell does not have a specialized voltage-sensitive region of the cell membrane that could be considered analogous to the axon hillock of a neuron. Instead, the entire cell membrane is voltage-sensitive. This rapid spread of the action potential ensures that contraction can occur throughout the muscle cell as quickly and uniformly as possible.

These special attributes of neuromuscular transmission ensure that rapid, reliable contraction of a muscle cell occurs every time the neuron controlling that particular muscle cell is activated. There is no integration, no information processing, no "thought" required; the muscle cell contracts when the neuron is activated. The *decision* to activate a muscle cell is determined at a higher level (i.e., by whether the nerve cell to that muscle cell is activated), not from within the muscle itself.

The sequence of events leading to contraction of skeletal muscle, starting at the point of an action potential in a motor neuron, is summarized in Table 6.1.

Skeletal Muscle Mechanics

Now that we have described how a muscle cell contracts and how contraction is controlled by a neural stimulus, we are going to back up and look at the whole muscle again. How can the amount of force generated by a muscle be varied if a single action potential in a nerve always results in a single action potential in a muscle cell? For example, how is the biceps muscle controlled so that we can pick up a pin or lift a thirty-pound weight with the same muscle? The answer is that we can vary the force developed by a single muscle cell, at least to a limited degree, and more importantly we can vary the number of muscle cells that are activated.

Force Development by a Muscle Cell How can force be altered if neuromuscular transmission is always one-to-one and if the muscle action potential always has the same amplitude and speed of conduction? The answer lies in the fact that *the duration of muscle cell contraction is much longer than the duration of the action potential*. An action potential in a muscle cell membrane lasts only 1–2 msec, but the subsequent contraction is delayed in onset until well after the ac-

Table 6.1   A summary of the excitation–contraction process in skeletal muscle.

Step	Event	Description
1.	Nerve activation	An action potential is initiated in the motor neuron that innervates a skeletal muscle cell.
2.	Release of neurotransmitter	Upon arrival at the axon terminal, the action potential causes acetylcholine to be released into the neuromuscular junction.
3.	Initiation of an excitatory graded potential in the muscle cell membrane	Acetylcholine-gated ion channels in the muscle cell membrane open, allowing ions (primarily sodium) to diffuse, thereby producing a graded depolarization of the muscle cell membrane.
4.	Production of an action potential in the muscle cell membrane	The graded potential produced by acetylcholine causes voltage-gated ion channels to open. An action potential is initiated that is transmitted to the entire muscle cell membrane.
5.	Release of calcium	The action potential causes calcium to be released from the sarcoplasmic reticulum. Cytosolic calcium concentration rises.
6.	Initiation of contraction	The presence of calcium allows myosin cross-bridges to interact with actin.
7.	Contraction	In the presence of ATP and calcium, myosin cross-bridges attach to actin, bend, and then detach. This cycle is repeated as long as calcium and ATP are present. As a result, the myosin filaments slide past the actin filaments, and sarcomeres shorten.
8.	Relaxation	Calcium is transported out of the cytoplasm back into the sarcoplasmic reticulum. In the presence of ATP, myosin disengages from actin.

MILESTONE 6.1

How Muscles Contract

Compared to our knowledge in some of the older scientific fields such as astronomy, our knowledge of biology seems relatively recent. For example, the calendars of some ancient civilizations were extremely accurate even by today's standards, and by the sixteenth century sailors were navigating around the world by the stars. Yet the writings of seventeenth century physiologists indicate that they did not understand how muscles contract.

One of the foremost muscle physiologists of the time was Dr. William Croone. In a treatise published in 1664 and presented to the Royal Society of London that same year, Dr. Croone hypothesized that there was an "animal spirit" that came from nerve fibers, and that this animal spirit, or liquor, was released from nerve fibers by nerve impulses. Today we know that he was right; the animal spirit is the neurotransmitter acetylcholine. But the rest of the mechanism of contraction he had wrong:

> If this be so, it will also be highly probable that from the mixture of this liquor or spirit with the spirits of blood there occurs continuously a great agitation of all the spirituous particles which are present in the vital juice of the whole muscle, as when spirit of wine is mixed with the spirit of human blood . . . and no one is such a novice in Chemistry as not to know how great a commotion and agitation of the particles is accustomed to occur from different liquors mixed with each other.

He added that as a result of this chemical reaction, the muscle "swell'd like a Bladder blown up," causing it to get fatter but shorter. Apparently he was a rather provocative writer; his explanation for the rapid time course of muscle contraction and relaxation was that "the blood flows in and again flows out rapidly, as may be illustrated from that sudden erection of the penis from venereal thoughts and its flaccidity after the emission of the semen."

It seems that Dr. Croone despaired of ever actually being able to test his hypothesis, for he later wrote;

> These and several other Particulars I did endeavor to make out at large in those Lectures; yet only in the way of an *Hypothesis*, not as if I did presume to believe I had found out the true secret of *Animal Motion*, when I am most persuaded, no Man ever did or will be able to explicate either this or any other *Phenomenon* in Nature's true way and method: But because I reckon such Speculations among the best Entertainments of our Mind.

Although Dr. Croone was one of the top scientists of his day, it is easy to understand his pessimism. His work preceded the availability of even the most rudimentary light microscopes, and thus he was unaware of the existence of myofibrils, the arrangement of sarcomeres in the myofibril, and the presence of thick and thin filaments of protein in the sarcomeres. Photographs such as Figure 6.13 were not available until the middle of this century, following the development of the electron microscope. Only then were scientists able to understand the sliding filament mechanism of muscle contraction as we know it today.

---

tion potential itself is complete, and lasts much longer (50–100 msec). The delay is caused by the time it takes for the calcium to be released and for the muscle fiber to shorten by means of the sliding filament process. The prolongation of the contractile event is caused both by the time it takes to remove the calcium from the intracellular space and for the filaments to return to their original relaxed position. The force developed by a muscle cell in response to a single action potential is known as a **twitch.**

If a second action potential occurs while the twitch response to the first action potential is still ongoing, a second release of calcium occurs before the calcium released by the first action potential has been completely removed. As a result, the intracellular calcium concentration is higher than would occur in response to only one action potential, and it remains elevated for longer. Consequently, the muscle contraction is also larger and more prolonged because the force of the contraction is directly related to the intracellular calcium concentration. The cumulative effect of action potentials on force generation by a muscle cell is another example of the concept of summation.

If action potentials occur in such rapid succession that the muscle cell does not have any chance to relax between them, intracellular calcium concentration reaches a high and constant level, causing a sustained contraction of maximal strength known as **tetanus** (in this context, tetanus is a normal physiological event not to be confused with the disease called tetanus, or "lockjaw"). Tetanus is usually three to four times larger than the twitch response to a single action potential. Thus *the force generated by a single muscle cell can be regulated by the frequency of stimulation,* at least within this three- to fourfold range. These principles are illustrated in Figure 6.17. Twitch, summation, and tetanus apply to the whole muscle as well; indeed, most studies of these phenomena in the laboratory use whole muscles because they are easier to obtain and use. Summation in whole muscle includes one additional component: the stretching of tendons that are in series with the muscle fibers (analogous to stretching a rubber band).

Muscle Force and Motor Unit Recruitment

Contraction of skeletal muscle is controlled entirely by action potentials from **motor neurons** (so called because their activation causes movement). The axon of a motor neuron branches to form many axon terminals, each of which terminates on a single muscle cell. The result is that a single neuron activates more than one muscle cell at once. A motor neuron and the muscle cells it controls are called a **motor unit** (Figure 6.18). For simplicity, only two motor units are depicted in Figure 6.18, although there are many more than that supplying a whole muscle. All of the muscle cells supplied by a motor neuron contract simultaneously when an action potential occurs in the motor neuron.

The muscle cells of a single motor unit are scattered throughout the muscle so that their simultaneous contraction produces a smooth but rather weak contraction of the whole muscle. When we lift a pin, we may recruit only a few of the motor units, and the rate at which action potentials occur in these units may be relatively low. In contrast, when we attempt to bench-press a 100-pound weight we may need to recruit all of the motor units and to have action potentials occurring at such a rate that tetanus is achieved for most motor units.

The total force generated by a muscle depends on both the number of motor units recruited and the rate at which action potentials are generated in those units. When it is necessary to sustain a forceful muscle contraction for some time, such as during arm-wrestling, various motor units may begin to be activated asynchronously so that some are active while others are not. This on-and-off activation pattern for each motor unit helps to prevent muscle fatigue because the muscle cells of each motor unit are allowed an occasional moment of brief rest, yet it permits a sustained contraction to be maintained by the entire muscle.

In muscle mechanics, the opposite end of the spectrum from brute strength is precise control. Precision of muscular movement is determined primarily by the number of muscle cells per motor unit. Fewer muscle cells per motor unit means that the muscle is controlled by a relatively high number of nerve fibers. A relatively high number of nerve fibers permits more precise control because there can be greater variation in the number of motor units that are active at one time, and fewer muscle cells per nerve cell means that each motor unit has a smaller contribution to the overall response. A single motor unit in a muscle of the eye, one designed for fine control, may consist of only ten

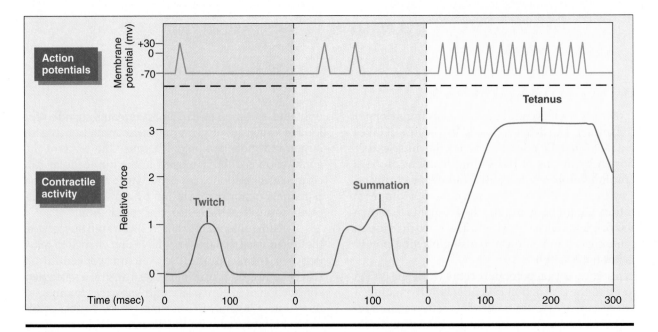

Figure 6.17 Force generation by a single muscle cell in response to action potentials generated in its cell membrane.

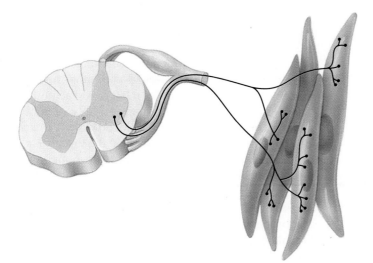

Figure 6.18 Motor units. Two motor neurons are shown, each innervating different muscle cells.

muscle cells. In contrast, the thigh muscles, capable of considerable force but lacking precise control, may have over a thousand muscle cells per motor unit. As a general principle, there is a trade-off between the generation of force and precise control that is determined by the number of muscle cells per motor unit.

Relationship Between Muscle Length and Force Research scientists are able to determine how much force a muscle can generate by removing a muscle from an animal, attaching it to a force-measuring device, and stimulating it electrically. Such experiments have proven that the force that a muscle can generate depends on the starting length of the muscle, or how much it is stretched before stimulation (Figure 6.19). This relationship between length and force can be explained by the degree of overlap of the thick and thin filaments. Maximum force is developed when there is optimal overlap between the thin filaments and the cross-bridges of the thick filaments (a). In most muscles of the body, this occurs at about the midpoint of the range of motion of a muscle. As a muscle is

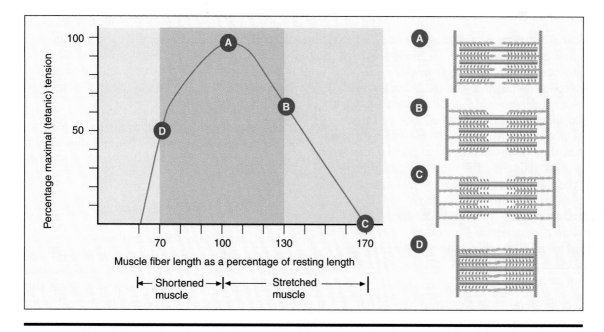

Figure 6.19 Length–tension relationship of skeletal muscle. The brown area represents the normal range of a skeletal muscle in the body. (A) Normal resting length. (B), (C) Increasing the muscle length reduces the degree of myosin–actin overlap. As a result, less tension can be generated. (D) As the muscle is shortened, less tension can be generated because of interference of overlapping actin filaments with the myosin–actin interactions.

stretched beyond its maximum force-generating length, force development falls off because fewer and fewer cross-bridges of the thick filament can interact with the thin filaments (b and c). When a muscle is shortened, force development again falls off as overlap of thin filaments hinders the attachment of the cross-bridges (d). Force generation may reach nearly zero at the point that the ends of the thick filaments are pressed against the ends of the sarcomere.

In our bodies, the attachments of muscle to bone prevent the kinds of extreme ranges of muscle length that are possible in a laboratory setting. As depicted in Figure 6.19, muscles are kept within a range of lengths that ensures reasonably efficient contractile force generation. Nevertheless, the force that can be generated at the ends of the physiological range of muscle lengths is only about 60% of the maximum force that can be generated at the optimum length. You intuitively take advantage of this length–force relationship when you lift a heavy object with your forearm slightly bent in relation to your upper arm, rather than with your forearm fully extended.

Although the optimum length of the muscle depicted in Figure 6.19 is near the midpoint of its physiological range of motion, not all muscles are the same in this regard. Different muscles vary somewhat with respect to where the maximum tension-generating ability is in relation to resting length. In general, however, muscles tend to be designed to capitalize on the relationship between length and tension, depending on the job each must perform.

Relationship Between Speed and Force You have probably noticed that you can move heavy loads only slowly, whereas you can move light loads rapidly. This is because speed (velocity) at which a muscle can shorten is inversely proportional to the force that must be generated to shorten the muscle. The basis for this phenomenon is that the force generated by a muscle is determined by the number of cross-bridges that are attached to the thin filaments: The more cross-bridge attachments, the more force that can be generated. However, it takes more time to attach, move, and detach all these cross-bridges, so the speed of contraction decreases. Imagine, for example, how many steps per mile (analogous to cross-bridge attachments per sarcomere) a racehorse makes compared to a draft horse pulling a heavy load.

Isotonic Versus Isometric Contractions Contractions of a muscle are sometimes classified as either **isotonic** (*same + tone*) or **isometric** (*same + length*). When you lift a light weight, it is presumed that in order to move the weight you are exerting a constant force (or tone) just slightly larger than the weight itself. In this case, the contraction is said to be isotonic; a constant force is applied and the muscle changes in

length. In contrast, when you attempt to move an object that is so heavy that it does not move, or when you "make a muscle" by voluntarily contracting opposing muscles, the muscle contraction is said to be isometric because the ends of the muscle do not move and the bones do not change position relative to each other.

Neither of these definitions is entirely accurate, however. In an isometric contraction, the muscle itself shortens somewhat as it pulls against the slightly elastic (stretchable) tendons attaching it to bone, even though the bones may not move. Furthermore, a true isotonic contraction is probably achieved only in the laboratory. Under normal conditions we must first get the object moving at the beginning of the motion and then allow it to come to a stop at the end. Therefore, more force is applied at the beginning of a contraction involving movement, and less at the end.

Nevertheless, the term *isotonic* is commonly used to describe muscle contractions in which movement occurs, and *isometric* is used to refer to contractions in which movement does not take place. If they are used only in this sense, the terms are useful in classifying types of muscle contractions. For example, in the muscles of the thigh, isometric contractions would be important for standing still, whereas isotonic contractions predominate during bicycling.

Muscle Energetics

The contractile activity of muscle tissue can vary over a very wide range, from completely inactive to a sustained contraction of great force. This great range of activity is made possible by the presence of stored energy in muscle cells, and by the special biochemical pathways in muscle cells for the use of that energy.

Metabolism As in most cells, the primary direct source of energy in muscle cells is the breakdown of ATP. ATP releases energy when it is split into adenosine diphosphate (ADP) and an inorganic phosphate ion (P_i):

$$ATP \rightarrow ADP + P_i + energy$$

The released energy is available to do work in the cell. This is the second time we have mentioned ATP; the first was in Chapter 2, in which we mentioned that ATP is also used as a source of energy for the active transport of molecules across cell walls. You will learn more about ATP in Chapter 14.

In muscle cells, ATP has three roles related to muscle contraction (Figure 6.20). First, it provides the energy for binding myosin to actin and for movement of the cross-bridges. In the resting state, ATP is already bound to myosin; myosin is energized, ready to begin the contractile response as soon as calcium binds to

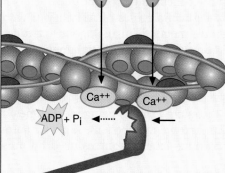

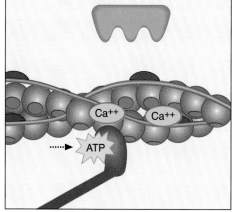

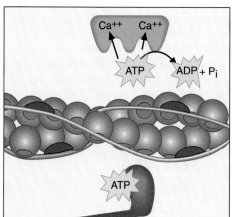

troponin–tropomyosin and makes the actin binding site accessible. During contraction this ATP loses a phosphate group, becomes ADP, and liberates enough energy to cause the myosin head to attach to actin and pull on the actin filaments.

The second role of ATP is in the process of relaxation, for myosin cannot detach from actin unless there is an ATP bound to it. After the contraction is complete, then, the original ATP that was used up is replaced by a fresh ATP from the cell's stores, myosin dissociates from actin, and relaxation occurs.

The requirement that ATP be bound to myosin before myosin can detach from actin is the explanation for **rigor mortis,** the stiffening of the muscles of the body that occurs about four hours after death. After death, calcium leaks slowly into the cell, causing muscle contraction. But the cell's stores of ATP are quickly depleted after death. Without ATP to replace those used up during contraction, myosin cannot dissociate from actin and thus the myosin cross-bridges remain "locked" to the actin in one position.

The third, less direct, role of ATP is in removal of calcium from the cytosol. In order for relaxation to occur, the calcium must be actively transported back into the sarcoplasmic reticulum, a process that requires the expenditure of energy.

The actual amount of ATP stored in muscle cells is so small that it can be completely depleted in less than 10 seconds of maximal muscle activity. If you think of energy as money, ATP would represent the cash we carry with us: It is readily available, but hardly sufficient to pay all of our monthly bills! However, the ATP can be replaced by several mechanisms, each with a different time course and different capacity for ATP production.

The first and most rapid source for ATP replenishment is **phosphocreatine,** another high-energy molecule that can be used to produce more ATP as the ATP is used up:

$$\text{Phosphocreatine} + \text{ADP} \rightarrow \text{ATP} + \text{creatine}$$

If we again think of energy as money, phosphocreatine would represent the money available from your bank account via an automatic teller machine, readily available to replenish your cash on short notice. There is roughly three to five times more phosphocreatine than

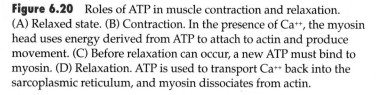

Figure 6.20 Roles of ATP in muscle contraction and relaxation. (A) Relaxed state. (B) Contraction. In the presence of Ca^{++}, the myosin head uses energy derived from ATP to attach to actin and produce movement. (C) Before relaxation can occur, a new ATP must bind to myosin. (D) Relaxation. ATP is used to transport Ca^{++} back into the sarcoplasmic reticulum, and myosin dissociates from actin.

ATP in a muscle cell, but even the phosphocreatine can be depleted fairly rapidly. During a 7-kilometer race, for example, the original ATP and all of the phosphocreatine stores are used up in about 30–40 seconds.

The next available source of new ATP comes ultimately from glycogen, the stored form of the sugar glucose (see Chapter 2). Glucose derived from glycogen stores can be metabolized by a process called **glycolysis.** The extent to which glycolysis is called into play depends largely on the amount of stored glycogen in the cell. This is why athletes, particularly those who engage in exercise less than several minutes in duration, may find it beneficial to eat a diet high in carbohydrates. Glycolysis does not require oxygen, so it is called anaerobic metabolism; its primary benefit is that it is relatively fast. However, it is inefficient, yielding only two ATP molecules per glucose molecule, and it results in the production of pyruvic acid, which in the absence of oxygen is converted to lactic acid. Ultimately the lactic acid must be metabolized by aerobic pathways, so the production of lactic acid places an individual in "oxygen debt." Lactic acid is thought to contribute to acute fatigue and to the sensation of burning pain in muscles during the first minute or so of very heavy exercise.

The production of ATP by glycolysis is initiated shortly after exercise begins, but the amounts of glycogen stored in the muscle cell generally cannot support a maximal muscle effort for more than about 2–5 minutes. The production of ATP by glycolysis can be compared to withdrawal of cash from your long-term savings account; with some difficulty it can be done, but it lowers your reserves, and you would want to replenish it as soon as possible.

For an activity that is sustained for more than 2–5 minutes, muscles must rely on **aerobic** (oxidative) **metabolism.** Aerobic metabolism uses a variety of chemical energy sources, including glucose obtained from the blood, the pyruvic acid generated from glycolysis, and even derivatives of fat (fatty acids) and protein (amino acids). You do not need to know the exact biochemical pathways at this point, but you should know that aerobic metabolism requires oxygen in order to create ATP from the energy sources; it is the predominant form of long-term energy production in muscle, and the pathways have an unlimited capacity for ATP production over the long run. In the presence of oxygen, the complete metabolism of one glucose molecule yields thirty-eight molecules of ATP. If you were to consider aerobic metabolism analogous to a full-time job, it would be one that pays enough money for your long-term needs, including replenishment of your depleted savings account. Within 5–10 minutes of beginning a sustained exercise such as running, essentially all of the energy that is being used comes from an increase in aerobic metabolism. Table 6.2 summarizes these various energy sources used by muscle.

Types of Skeletal Muscle Fibers So far we have described skeletal muscle as if all skeletal muscle cells were alike. Actually, there are at least two types of skeletal muscle cells in most animals. Humans have

---

HIGHLIGHT 6.4

Muscle Soreness and Acute Muscle Pain

All of us are familiar with the feeling of soreness that occurs 24–48 hours after participating in an exercise we have not been doing regularly. This type of soreness appears to be a consequence of damage to some of the sarcomeres in a muscle fiber, generally those near the center of the fiber. It is thought that the damage occurs because in the absence of regular use, some sarcomeres are not capable of contracting as well they should, and they become stretched to the point that the thick and thin filaments no longer overlap; at this point, they become physically disrupted and can no longer shorten. As these damaged sarcomeres disintegrate over several days and are degraded and removed in the repair process, they release a chemical substance (probably bradykinin) that causes the feeling of soreness. With time, the damaged sarcomeres are removed completely and new sarcomeres take their place. Once the muscles have become accustomed to the exercise, damage (and the accompanying soreness) no longer occurs.

The delayed soreness 1–2 days following an unfamiliar exercise is different from the more acute burning pain that weight-lifters feel during a sustained maximum effort for more than about 20 seconds. The latter type of pain is probably due to increased acidity in the muscle caused by lactic acid, produced during periods of increased muscle work when the oxygen supply is insufficient. Pain caused by the presence of lactic acid goes away within seconds of cessation of the exercise because lactic acid is quickly removed once the tissue is resupplied with sufficient oxygen.

Table 6.2   Energy sources for muscle.

Energy Source	Quantity	Time Course of Use	Comments
Stored ATP	Stored in only small quantities.	About 10 seconds	ATP is the only direct energy source; it must be replenished by the other energy sources.
Phosphocreatine	Three to five times the amount of stored ATP.	About 30 seconds.	Converted quickly to ATP.
Glycogen	Variable. Some muscle types store large quantities of glycogen.	Primarily used during heavy exercise within the first 3-5 minutes.	Glucose (derived from stored glycogen) can be metabolized to ATP without oxygen, but it yields only two ATP molecules per glucose molecule.
Aerobic metabolism	Not a stored form of energy; oxygen and nutrients are constantly supplied by the blood.	Always present; increases dramatically within several minutes of onset of exercise, when blood flow and respiration increase.	High yield; complete metabolism of one glucose molecule yields thirty-eight ATP molecules.

three, one of which is somewhat intermediate between the two types found in many other animals. The three types differ in terms of the speed at which they use ATP (and thus the speed at which they can contract) and the energy sources they use to provide the ATP:

> Type I: slow-oxidative
> Type IIa: fast-oxidative
> Type IIb: fast-glycolytic

Type I cells are called slow-oxidative cells because they have the slowest rate of ATP use and hence they contract rather slowly, and they have a high capacity for producing ATP by aerobic (oxidative) metabolic pathways. These cells can make ATP by aerobic metabolism as fast as they can use it. They store almost no energy in the form of glycogen, but they are richly supplied by blood vessels. Thus, they receive an ample supply of the oxygen and nutrients necessary for continuous ATP production in the blood that perfuses them. They have numerous mitochondria in which the oxidative metabolic processes that produce ATP are carried out. Type I cells also store some oxygen for immediate use when the level of activity increases suddenly. This oxygen is stored by iron-containing molecules in the cell called **myoglobin.** These cells have very high endurance as long as oxygen is supplied and a source of energy is available. However, they are not capable of work that requires a great deal of power.

Type IIa cells are a less common intermediate type of cell, between Type I and Type IIb. They are fast-contracting, and they are listed as oxidative because they have a moderate aerobic capacity. However, they

use ATP faster than they can make it by aerobic pathways, so they also use stored glycogen and anaerobic metabolism to some extent. They are capable of moderate power and they tend to become fatigued at an intermediate rate.

The fast-glycolytic Type IIb muscle cells contract very rapidly, thereby using ATP up much faster than they can produce it by aerobic pathways, and they have very limited aerobic capacity. They use glucose derived from stored glycogen (without needing oxygen) as their primary source of immediate energy. Type IIb cells are generally larger than Type I cells, and thus they are capable of great power. They cannot sustain the power for long, however, because the glycogen stores soon become depleted.

Most muscles in humans are composed of a mixture of the various types of muscle cells; the ratio depends on the use for which the muscle is intended. Muscles used for maintaining posture generally are made predominantly of Type I fibers, as maintaining posture is an activity that must be sustained for extended periods of time. On the other hand, arm muscles are made predominantly of Type II cells, as these muscles are used for forceful movements that are generally not sustained. For any given muscle, the ratio of Type I to Type II cells can vary between individuals. World-class marathon runners generally have a higher than average percentage of Type I fibers in their legs because running a marathon is an activity that is sustained for a long period of time and thus depends almost exclusively on aerobic metabolism. In contrast, sprinters rely almost exclusively on Type II fibers, as sprinting is an activity that depends on great power for very short periods of time.

Muscle Performance

Several factors influence muscle performance. The degree and type of exercise training, for example, influence the strength and endurance of a muscle. In addition, muscle performance tends to decline over time during sustained heavy exercise, a phenomenon known as **fatigue.**

Exercise Training The speed of a muscle cell contraction remains fairly constant despite the degree of exercise training. This is because the rate of ATP use, which determines contractile speed, is a function of the intrinsic biochemistry of each cell. In other words, in any one individual the ratio of Type I to Type II muscle cells in a muscle is fixed, and cannot be changed as a result of exercise training. However, exercise training *does* have an effect on the oxidative capacity of muscle.

Endurance training such as jogging is likely to produce at least a modest increase in the oxidative capacity of all cell types. This is accompanied by an increase in the number of blood capillaries supplying the muscle, an increase in the number of mitochondria per cell, and an increase in oxidative metabolic enzymes. Indeed, endurance training can produce a slow transformation of Type IIb cells (with low oxidative capacity) into Type IIa cells (with greater oxidative capacity), in addition to improving the oxidative capacity of the oxidative Type I cells already present. Muscle cell diameter and overall muscle mass and strength also increase, but only slightly; most of the change is in oxidative capacity, leading to improved endurance.

In contrast, weight training produces an overall increase in muscle strength and mass, but has very little effect on oxidative capacity. The increase in muscle mass and strength is due primarily to an increase in diameter of the Type II cells, which in turn is due to increased production of thick and thin filaments in these cells.

Fatigue There are several reasons why muscle performance tends to decline over time during sustained exercise. In muscular fatigue, the muscle simply does not respond to stimulation with the same level of performance. The availability of energy plays a key role, for once glycogen stores are depleted, only oxidative pathways are available for producing ATP. In addition, continuous muscle activity causes the loss of potassium from the cells and alters the hydrogen ion concentration, both of which may interfere with the contractile process. Neuromuscular fatigue may occur when the motor neurons to the muscle are stimulated at very high rates. Under these circumstances, the motor neurons may become sufficiently depleted of acetylcholine that motor neuron action potentials do not always produce a muscle cell action potential.

Finally, there may be psychological fatigue, in which the central nervous system fails to continue to activate the motor neurons at a high rate even though the muscles themselves are still capable of greater performance. Certainly the discomfort associated with high levels of exercise and boredom associated with repetitive tasks contribute to psychological fatigue, but exactly how they do so remains a mystery.

Cardiac Muscle

The muscle tissue of the heart is specialized for pumping blood. Individual cardiac muscle cells contribute to the overall function of the heart by one action only: contraction. The contraction of millions of cardiac muscle cells together produces the pumping action of the heart. We will now explore the anatomical and functional attributes of cardiac cells, and how they interact with each other.

Anatomy

Cardiac muscle cells (also called myocardial cells) are not nearly as elongated as skeletal muscle cells (Figure 6.21; review also Figure 6.11(B)). Each cell has only one nucleus. Like skeletal muscle cells, cardiac muscle cells contain thick and thin filaments of myosin and actin that are arranged in parallel within sarcomeres.

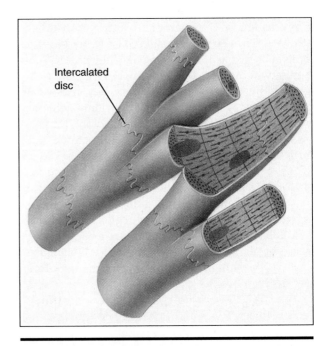

Intercalated disc

Figure 6.21 Cardiac muscle cells.

Thus cardiac muscle, like skeletal muscle, is also called striated muscle. The cells are somewhat rectangular, although they may branch occasionally. The ends of the cell do not taper to a point, as they do in skeletal and smooth muscle cells. The cell membrane and associated structures at the ends of the cell exhibit many foldings of the cell membrane that greatly increase the surface area and form a close attachment to an adjacent cell. Together, the cell membranes at the ends of two adjoining myocardial cells and associated structures are called an **intercalated disc.** Intercalated discs join cardiac cells together both physically and electrically.

A schematic representation of an intercalated disc is shown in Figure 6.22. The two cell membranes are joined together by **desmosomes,** or areas of physical attachment, and by gap junctions (first introduced in Chapter 3) that span the two membranes and allow direct passage of small molecules and ions from one cell to the next. Gap junctions permit an action potential in one cell to be transmitted directly to the next.

Function

The functional attributes of cardiac muscle cells make them ideally suited to carry out the pumping action of the heart. First, some cardiac muscle cells are capable of initiating action potentials spontaneously, even in the complete absence of neural input. Cardiac cells that can do this are called pacemaker cells, because they set the pace for the beating of the other cardiac muscle cells. Because of these pacemaker cells, a heart

will continue to beat rhythmically for a time even when it is removed from the body. This is quite different from skeletal muscle cells, which contract *only* when the cell receives an action potential from the one neuron that innervates it. Nerves do *modulate* the heart rate as well as the force of contraction of the heart, as you will learn in Chapter 8, but they are not essential for the basic functioning of the heart.

The second functional attribute of cardiac muscle cells, also related to electrical activity, is that electrical activity can be transmitted directly from one cardiac muscle cell to the next via gap junctions. The result is that an action potential in one myocardial cell ultimately can produce action potentials in all the cells of the heart. In terms of electrical activity, then, the heart behaves as if it were a **syncytium,** i.e., as if it were one large, multinucleated cell without individual cell membranes. The transmission of action potentials between cells (and thus the synchronization of contraction) is essential to the pumping action of the heart because the heart has no bones against which to pull. You will learn more about the pumping action of the heart in Chapter 8.

A third important attribute has to do with the relative timing of excitation and contraction. The action potential in a cardiac muscle cell lasts much longer than that of a skeletal muscle cell because there is a prolonged plateau phase of depolarization. The prolonged plateau of the depolarization is primarily due to the opening of calcium channels at this time, allowing the inward diffusion of calcium. Typical action potentials and contractile responses for a cardiac muscle cell are shown in Figure 6.23. The entire action potential in a cardiac muscle cell lasts about 250 msec, during which time a second action potential cannot be generated. The contractile response, lasting about 300 msec, is nearly complete even before the action potential is over. As a result of this prolonged action potential during which the cell is refractory to another action potential, tetanus cannot be produced in cardiac muscle. The inability to produce tetanus in cardiac muscle is an important adaptation, because if tetanus were produced (if a contraction were sustained, without a corresponding period of relaxation) it would mean the end of the heart's rhythmic pumping activity.

In terms of cell metabolism, under normal conditions, cardiac muscle cells rely entirely on aerobic metabolism to produce their energy supply (ATP). The heart beats throughout life without ever stopping to rest for more than a second or so. Any energy supply that could be rapidly depleted, such as phosphocreatine or even glycogen, would be of little use to the heart except as a very temporary or emergency measure. Because the raw materials for oxidative metabolism (oxygen and various energy substrates) are delivered to the heart via the blood, it is important that

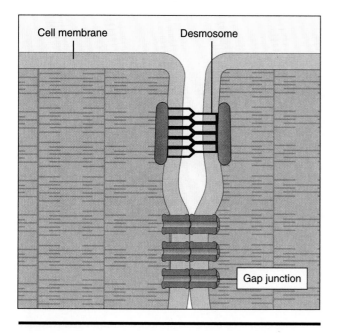

Figure 6.22 An intercalated disc joining two cardiac muscle cells.

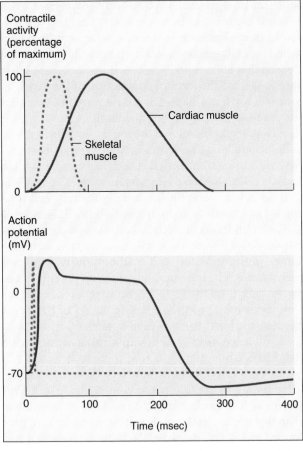

Figure 6.23 A single action potential and muscle twitch in a cardiac muscle cell. For comparison, the dotted lines show the responses of a skeletal muscle cell. Note the prolonged plateau phase of a cardiac muscle cell action potential.

the blood supply to the heart be maintained at all times. Compromise of the blood supply to the heart can result in damage and even death of cardiac muscle cells; this condition is known as **myocardial infarction,** or heart attack.

Smooth Muscle

Any discussion of muscle types invariably discusses smooth muscle cells last. Yet these tireless little cells are constantly working behind the scenes in a variety of organs. Smooth muscle in the walls of blood vessels controls blood vessel diameter, thus it plays a pivotal role in controlling blood pressure. Layers of smooth muscle surround the esophagus, the stomach, and the intestines, producing movement of food through the

digestive system. Sheets of smooth muscle surround the uterus and the urinary bladder, affecting processes as dramatic as childbirth and as mundane as controlling the bladder's capacity.

Anatomy

Smooth muscle cells are much smaller than skeletal muscle cells, and they are generally spindle-shaped, tapering at each end (Figure 6.24). The cells are generally joined by gap junctions so that electrical activity passes from cell to cell as it does in cardiac muscle. Some, but not all, of the cells receive input from nerves.

Smooth muscle lacks the striated appearance of skeletal and cardiac muscle (hence the name *smooth muscle*). Although smooth muscle cells contain thick and thin filaments of myosin and actin, the filaments are not arranged in the typical pattern of a sarcomere and are not visible under the microscope. An electron micrograph of smooth muscle cells reveals very little detail except for the presence of dark spots, called dense bodies, at regular intervals (Figure 6.25). The dense bodies, consisting of protein attached to the cell membrane, are the points of attachment for bundles of actin and myosin filaments that criss-cross the cell in a lattice pattern (Figure 6.26).

Function

As is the case for both skeletal and cardiac muscle, contraction of smooth muscle occurs when calcium enters the cell cytoplasm, causing myosin to interact

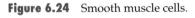

Figure 6.24 Smooth muscle cells.

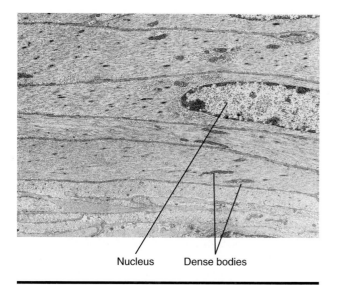

Figure 6.25 Electron micrograph (39,000×) of several smooth muscle cells, showing the dense bodies and lack of striations.

with actin and the filaments to slide past each other. Thus, the sliding filament theory accounts for the contractile action of all three muscle cell types, even though the arrangement of the filaments is slightly different in smooth muscle than in the other two muscle types. In the case of smooth muscle, contraction of bundles of actin and myosin filaments causes the dense bodies that anchor the bundles to be pulled closer together, which has the overall effect of shortening the cell (Figure 6.27).

The rate of contraction of smooth muscle is relatively slow; it may take several seconds for smooth muscle to contract. In contrast, recall that the contractile response of a skeletal muscle cell to a single action potential is complete within one-tenth of a second. In part, the slow rate of smooth muscle contraction is a result of the slower rate at which calcium starts the process. Smooth muscle has only a poorly developed sarcoplasmic reticulum. Thus, most of the calcium that is present during contraction must enter the cell from *outside* the cell, via calcium channels in the cell membrane that are opened by the action potential. In addition, the rate at which smooth muscle cells can split

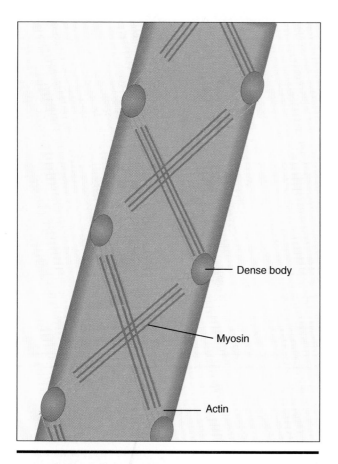

Figure 6.26 The arrangement of actin and myosin in a smooth muscle cell.

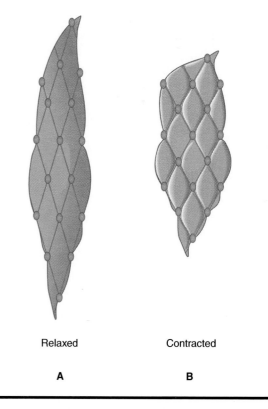

Relaxed   Contracted

A   B

Figure 6.27 Schematic diagram of the (A) relaxed and (B) contracted state of a smooth muscle cell. Contraction of the actin and myosin bundles pulls the dense bodies toward each other, producing a shortening of the cell. Contraction may cause dimpling of the cell membrane, to which the dense bodies are attached.

ATP, and hence the rate at which energy for contraction is produced, is slower than it is in skeletal muscle.

Smooth muscle actually has two types of electrical activity (Figure 6.28). **Slow-wave potentials,** lasting several seconds, occur at roughly 10-second intervals. Like the generation of action potentials in cardiac muscle, the generation of slow-wave potentials by smooth muscles is an inherent property of certain smooth muscle cells that generate this potential spontaneously. Slow-wave potentials are caused by cyclical changes in the rate at which these cells transport sodium ions out of the cell. Superimposed on these slow waves, whenever a threshold voltage is passed, are groups of action potentials lasting approximately 300 msec each. This is a slow action potential, compared to those in skeletal muscle (1–2 msec) or cardiac muscle (200 msec).

The transmission of electrical activity between smooth muscle cells is similar in many respects to the transmission between cardiac cells. Electrical activity in one smooth muscle cell is transmitted to nearby cells via gap junctions so that all of the cells are activated simultaneously. Also, like the heart, the electrical and contractile activity of smooth muscle can be modified by both stimulatory and inhibitory nerves. Nerves produce a slight change in the membrane potential; they do not generate action potentials directly. Stimulatory nerves produce a slight depolarization, making it more likely that threshold will be reached by the slow waves, thereby generating more action potentials. Inhibitory nerves, on the other hand, cause a slight hyperpolarization, and thus fewer or no action potentials are generated. The effects of stimulatory and inhibitory neural activities on smooth muscle

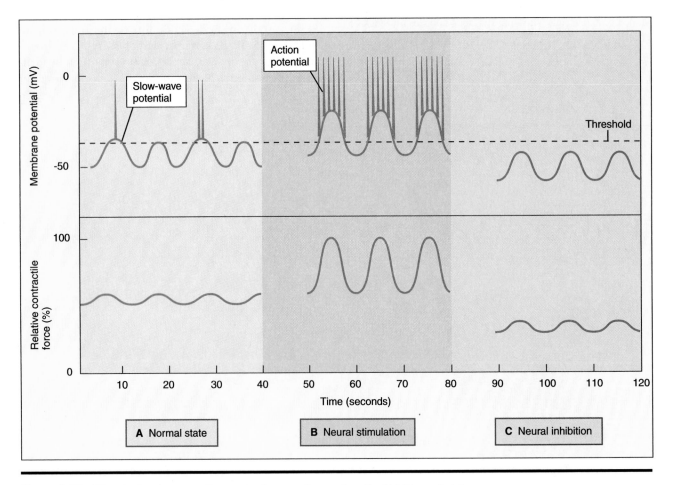

Figure 6.28 Electrical and contractile activity in smooth muscle cells. (A) Normal state. (B) Neural stimulation. (C) Neural inhibition.

membrane potential and muscle tension are shown in Figure 6.28(B) and (C). Note that the particular smooth muscle represented in the figure always maintains some muscle tension, or tone, even when no action potentials are being generated.

It would be a mistake to assume that because smooth muscle is not anatomically arranged into sarcomeres and is so slow, it is somehow an inferior or evolutionarily primitive type of muscle. On the contrary, smooth muscle cells are particularly well-adapted for the tasks they perform. Recall that the ability of muscle to generate force is inversely related to the speed of contraction. Precisely because the speed of contraction *is* so slow, smooth muscle cells are very energy-efficient, enabling them to maintain tension over prolonged periods of time at very low energy cost. This is important because unlike skeletal or cardiac muscle cells, *smooth muscle cells may never fully relax*. Most (but not all) smooth muscle cells maintain at least some contraction at all times, altering the *degree* of contraction (slightly more or less force) as required.

This is not meant to imply that whole skeletal muscle cannot maintain a constant state of contraction. Indeed it can; we know, for example, that skeletal muscles controlling posture can sustain a submaximal contraction for prolonged periods of time. But we must be clear on the distinction between the whole muscle and the individual cells of which it is comprised. In skeletal muscle, a sustained contraction is maintained by the cyclic contraction and relaxation of individual skeletal muscle cells, each of which cannot maintain a consistent contraction without running out of energy. In contrast, each smooth muscle cell is capable of continuous contraction.

Another important functional attribute of smooth muscle is its tremendous ability to maintain tension over a very wide range of muscle cell lengths. Recall that skeletal muscle loses most of its ability to generate contractile force when it is stretched to beyond about 170% of its resting length. Smooth muscle, on the other hand, can still generate contractile force even when it is stretched to several times its normal length (Figure 6.29). This capacity for generating force even when stretched is why smooth muscle is so ideally suited for the uterus, which expands many-fold during pregnancy, and for the urinary bladder, which may fill and empty several times a day.

Table 6.3 summarizes some of the attributes of the three muscle types.

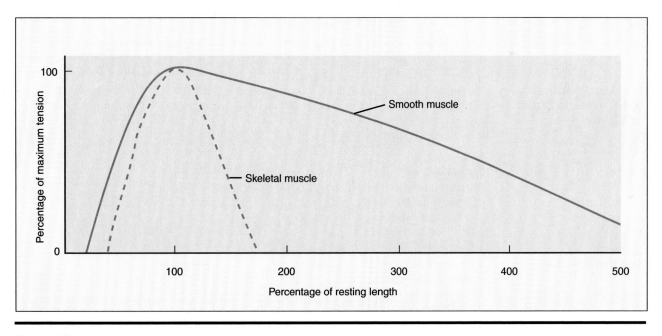

Figure 6.29 A representation of the length–tension relationship for smooth muscle. The dashed line represents the same relationship for skeletal muscle, adapted from Figure 6.19.

Table 6.3   **Characteristics of skeletal, cardiac, and smooth muscle.**

Characteristics	Skeletal	Cardiac	Smooth
		Type of Muscle	
Location	Attached to bones (skeleton).	Heart only.	Forms part of the structure of blood vessels. Surrounds many hollow organs.
Function	Movement of the body.	Pumping of blood.	Constriction of blood vessels. Movement of contents in hollow organs.
Anatomical description	Very large, cylindrical, multinucleate cells arranged in parallel bundles.	Quadrangular cells with occasional branching points. Joined to other cells by intercalated disks.	Small, spindle-shaped cells with the long axis generally oriented in the same direction.
Striated	Yes.	Yes.	No.
Presence of gap junctions	No.	Yes.	Yes.
Initiation of action potentials	Only by a neuron.	Spontaneous (pacemaker cells) or from another cardiac muscle cell.	Spontaneous, whenever slow-wave potentials exceed the threshold.
Role of nerve stimulation	Required for initiation of a twitch contraction. Summation and tetanus are possible.	Stimulatory and inhibitory nerves modulate the heart rate and force of contraction, but are not required for cardiac muscle to function.	Stimulatory and inhibitory nerves can modulate the degree of tension developed, but are not required.
Duration of electrical activity	Short-duration action potentials (1–2 msec).	Long-duration action potentials (200 msec).	Very long slow waves at roughly 10-second intervals, with occasional superimposed long-duration action potentials (300 msec).
Energy source for generation of ATP	Phosphocreatine, stored glycogen, aerobic metabolism.	Aerobic metabolism.	Aerobic metabolism.
Energy efficiency	Low.	Moderate.	High.
Likelihood of fatigue	Low to very high, depending on the source of energy and the work load. Under extreme conditions, may fatigue within seconds.	Low as long as blood supply is adequate.	Very low.
Rate of muscle shortening	Fast compared to other muscle types. Type II fibers are faster than Type I fibers.	Moderate.	Very slow.
Duration of contraction	As short as 100 msec for a single twitch. Tetanus may be prolonged.	Short; about 300 msec. Summation and tetanus not possible.	Very long. May be sustained indefinitely.

Summary

The skeletal system supports and protects the organ systems. Perhaps most importantly, however, it provides a framework against which the skeletal muscles can exert force, thus producing movement. The skeleton is subdivided into the axial skeleton (forming the central axis) and the appendicular skeleton (forming the appendages). Bones are the hard elements that make up the skeletal system. They are held together at joints by connective tissues called ligaments. Skeletal muscle is attached to bones by tendons. Bone is actually a specialized living tissue that is constantly undergoing replacement throughout life and can undergo substantial changes in response to growth, exercise, and damage repair.

Muscles are specialized tissues that generate force as they contract, or shorten in length. They are also important for the generation of heat. There are three different types of muscle: skeletal, cardiac, and smooth.

Skeletal muscle cells are long and cylindrical. Each muscle cell contains numerous fibrils arranged in parallel. The fibrils have a striated appearance because each is composed of numerous repeating structures, the sarcomeres, containing thick and thin filaments of protein. Contraction of the muscle cell occurs when these filaments are caused to slide relative to each other, shortening each sarcomere. Skeletal muscle cells are activated (stimulated to contract) only by nerves; otherwise they are in a relaxed state. The amount of force a skeletal muscle can generate depends on a number of factors, including the rate at which action potentials occur in the nerve, the number of nerves that are active, the length of the muscle, and the speed at which the muscle shortens.

All muscles use energy in the form of ATP in order to produce contraction. Muscles can fatigue if they run out of energy. How fast skeletal muscle uses energy (related to the speed of contraction) and what metabolic pathway they use to produce their ATP are the two criteria used to classify skeletal muscle cells into three different types.

Cardiac muscle is found only in the heart. Cardiac muscle cells are striated muscle. Important attributes of cardiac muscle cells are that some are capable of generating action potentials entirely on their own, nerves modify but do not completely control the electrical activity, action potentials can pass directly from one cardiac muscle cell to the next, the action potentials are prolonged and thus tetanus of the contractile response cannot occur, and cardiac cells rely solely on oxidative metabolism as their source of energy.

Smooth muscle lacks the regular pattern of striations found in skeletal and cardiac muscle, so it is called smooth. It is found around blood vessels and around hollow organs such as the uterus and bladder. Important attributes of smooth muscle are that it is much slower but more energetically efficient, it can maintain constant tension, it can generate its own electrical activity, which can be modified by nerves, and it can generate tension over a wide range of muscle lengths.

Questions

Conceptual and Factual Review

1. What is an intervertebral disc, and what is its function? (p. 144)
2. How are movable joints kept lubricated? (p. 147)
3. What is the difference between a ligament and a tendon? (p. 148)
4. What are the two main constituents of compact bone, and what cells produce them? (p. 150, p. 151)
5. How might prolonged athletic training affect bones, and why? (p. 152)
6. What are the three types of muscle? How do they differ? (pp. 153, 154)
7. What are three functions of muscle? (pp. 154, 155)
8. What fraction of your total body weight is skeletal muscle? (p. 156)
9. What is a sarcomere? (p. 156)
10. How does shortening of a sarcomere occur? (p. 156)

11. What initiates the contractile process in a muscle cell? (p. 158)
12. What factors determine the amount of force generated by a skeletal muscle such as the biceps? (pp. 162–166)
13. What is the direct energy source for muscle contraction? What are the ways in which it is produced? (p. 166)
14. In terms of contractile speed and energy metabolism, what are the three different types of skeletal muscle fibers? (pp. 168, 169)
15. What is the function of the intercalated discs between adjoining cardiac muscle cells? (p. 171)
16. How is an action potential in one cardiac muscle cell transmitted to an adjoining cell? (p. 171)
17. Where is smooth muscle found, and why is it called smooth? (p. 172)
18. What causes the slow-wave potentials commonly observed in smooth muscle? (p. 174)

Applying What You Know

1. Why do you suppose it is generally accepted medical practice to get bedridden patients up and walking as soon as medically feasible?
2. Why is it that when the spinal cord is severed in the neck, the skeletal muscles become completely flaccid (relaxed), whereas the heart and smooth muscles continue to function?
3. How would you expect the training regimen for a marathon runner to be different from that of a sprinter? Explain.

4. Why do you suppose that over the long term, the heart cannot rely on glycolysis as an effective source of ATP? What would you expect to happen when blood flow to the heart is substantially reduced, and why?
5. Why do heart rate and respiratory rate remain elevated for several minutes following heavy exercise?

Chapter 7
Sensory Systems

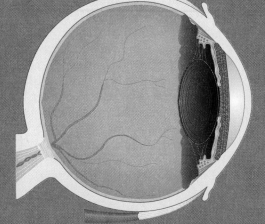

Objectives

By the time you complete this chapter you should be able to

◆ Understand the terms *receptor* and *transducer,* and be able to explain how sensory information from outside the body is transduced into action potentials.

◆ List the types of sensory signals the body receives.

◆ Describe the five general types of receptors used to receive these sensory signals.

◆ Understand how the central nervous system determines sensory modality and signal magnitude.

◆ Explain what is meant by *receptor adaptation.*

◆ List the types of sensory receptors present in skin, joints, and muscles.

◆ Compare and contrast the receptor/transducer mechanisms for taste and smell.

◆ Explain how sound waves are transduced by the ear.

◆ Compare and contrast the mechanisms for determining position and movement of the head.

◆ Describe how light is received and transduced by the eyes.

◆ Explain how light is focused on the retina and how the iris regulates the amount of light received.

◆ Appreciate what determines visual acuity and sensitivity to light.

◆ Explain how color vision occurs.

Introduction

The single-celled organisms floating in the primordial sea several billion years ago were without the special sensory organs you will read about in this chapter. With limited means of locomotion, they depended entirely on the conditions of the nearby external environment for everything from food to the proper temperature. Finding a mate was unnecessary because sexual reproduction had not yet developed. If a predatory organism approached, these organisms were likely to be eaten. If the temperature changed dramatically or nutrients became scarce, these organisms died. The ability to survive under a wide variety of conditions depends, in part, on the ability to make judgments about the external environment, and then to take action based on those judgments. Throughout the course of evolution, survival value has been conferred on organisms that can recognize and avoid danger, locate food and water when they are scarce, and find a potential mate.

Humans have developed many strategies for receiving and processing information about the world around us. We will focus on the four special senses that help us make judgments about this information: vision, hearing, taste, and smell. We will also discuss several other important sensory systems, including receptors in the skin that allow us to sense light touch, pressure, pain, heat and cold; receptors in the muscles and joints that signal the position of our bones and muscles (even when our eyes are closed); and the vestibular apparatus of the inner ear, responsible for sensing the direction of movement and position of the head.

In this book we concentrate on the physiology of humans, but the special sensory systems of humans are not unique, or even the most sensitive in the animal kingdom. Predatory birds have far better vision, the average dog has a better sense of smell, the cat has better balance, and deer have more sensitive hearing than humans. Because we can know the external world only as we experience it, the world must seem like a very different place to these animals than it does to us.

A common feature of all of the sensory systems discussed in this chapter is that they all are designed to provide information about the *external* world or to determine the position of our body in it. In addition, we are all equipped with other sensory systems that are designed to monitor and control the *internal* environment, such as the systems regulating internal body temperature, blood pressure, and the rate of breathing. These systems are discussed in other chapters.

General Principles of Sensory Systems

Sensory systems have several essential features in common, regardless of the type of sensory signal they receive. First, all systems have a mechanism for receiving the signal from the external environment and converting it into action potentials. These action potentials are sent to the central nervous system for processing. Second, the central nervous system must have a way of interpreting those action potentials correctly, not only in terms of the *magnitude* of the signal, but also in terms of the *type* and *location* of sensory signal that originated the action potentials in the first place. For example, the central nervous system must be able to distinguish action potentials generated by taste from those generated by sound waves. Third, the central nervous system must have a way to focus on selected inputs when necessary, so that we are not overwhelmed or confused by multiple inputs arriving at once.

Reception and Transduction of Sensory Information

The general problem faced by an organism is how to get sensory information that originates from outside the body to the inside in a form that can be used by the central nervous system to make decisions. More specifically, the organism must have a way to receive the input (reception) and a way to convert it into action potentials (transduction).

The first step in this process is the receiving of sensory information by a receptor. When you first learned of receptors in Chapter 2, it was in the context of receptor proteins in the cell membrane. Indeed, *receptor* is a general term meaning *that which receives*. In the present context, a receptor is a specialized cell (or region of a cell) that receives sensory input. This is the first step in the two-step process of **signal reception–transduction.** Some receptors are actually modified neurons that can complete the entire signal reception–transduction process themselves. Others are receptor cells that receive the sensory input and begin the transduction process, but they must synapse with a neuron before an action potential can be generated.

Receptor cells or regions are often classified on the basis of the type of signal or sensory modality received. **Chemoreceptors** in the nose and mouth respond to chemicals, **photoreceptors** in the eyes respond to light, and **thermoreceptors** in many parts of the body respond to heat or cold. **Mechanoreceptors** respond to deformation of their cell membrane. Mechanoreceptors include the receptors for touch, pressure, and vibration in the skin, stretch receptors in joints and muscles that signal position and tension, hair cells in the ear responsible for hearing, and the cells in the ear responsible for our sense of balance. Last, but certainly not least, are the receptors for pain, called **nociceptors.** Table 7.1 summarizes some of the attributes of the various receptor types that will be described in this chapter.

Once a receptor receives a sensory signal, it must be converted into action potentials for transmission to the central nervous system. In order to do this, a **transducer** is required. In the most general sense, a transducer is any device that converts (or transduces) energy input in one form into energy output in another form. In living systems, transducers convert all of the various sensory inputs such as chemicals (taste and smell), sound (hearing), light (vision), and mechanical distortion (touch) into just one form of output: action potentials.

The mechanisms whereby signals are received and transduced into action potentials are surprisingly similar for most of the different receptor cell types (Figure 7.1(A)). Generally, when the receptor is activated by the stimulus, small ion channels open in the receptor cell membrane. Usually the ion channels are nonselective in that the permeability to *all* small ions is increased, but the predominant effect is an inward movement of sodium ions. This leads to a local, graded depolarization of the membrane potential. The local graded depolarization is called a **receptor potential** because it occurs in a receptor cell or region. It is essentially the same as the graded potential described in Chapter 3, which occurs in the postsynaptic neuron when neurotransmitter is released into the synaptic cleft. Like the potential in the postsynaptic neuron, it is a graded potential in that its amplitude and duration vary with the intensity and sometimes the rate of application of the stimulus.

In a receptor cell that cannot itself generate an action potential, the graded receptor potential causes the graded release of a chemical transmitter, which in turn produces a graded potential and ultimately action potentials in a nearby neuron. The connection between a receptor cell and a neuron can properly be called a synapse (introduced in Chapter 3) because the information is chemically transmitted. In essence, the receptor cell substitutes for the presynaptic neuron. Note that in the process of receiving sensory information and passing it to the neuron, the original sensory modality has been converted (transduced) into action potentials.

Some receptors are just special terminal regions of modified nerve cells. In this case the same cell can both generate the receptor potential and subsequently produce the action potentials (Figure 7.1(B)). The receptors

for the sense of smell are a typical example; they are neurons that respond to a specific chemical substance (odorant) in much the same fashion that a postsynaptic neuron responds to its chemical neurotransmitter.

Interpretation of Sensory Input

As Figure 7.1 implies, the central nervous system receives sensory input from the outside world only in the form of action potentials. If this is the case, how can the central nervous system possibly distinguish between very different external sensory stimuli? How can it distinguish taste from vision, or pain from sound? The answer is twofold. First, each receptor responds best to only one type of stimulus (called its adequate stimulus because only that stimulus is adequate to activate it). Second, the receptors and the neurons to which they synapse are "wired" very specifically to the central nervous system, so that action potentials going to specific regions of the central nervous system originate only from certain types of sensory receptors.

How can a receptor respond to one type of sensory stimulus and not to another? The key seems to lie in the particular structure of each receptor cell and of the proteins that serve as the channels for small ions. As you will soon learn, the ion channels in photoreceptor cells respond best to light, but not to mechanical distention or movement, ion channels in the hair cells of the ear respond to the mechanical movement caused by sound waves, channels in temperature receptors respond only to temperature and not to pressure on the skin, and so on.

The second factor, the specificity of how the neural input is wired to the central nervous system, is also important in determining how the brain uses sensory input. As you learned in Chapter 4, the central nervous system receives sensory input via twelve pairs of cranial nerves. Information from the eye to the brain travels only in the optic nerve, and most of the information goes directly to the portion of the cortex responsible for vision. Any action potentials arriving in the visual cortex, then, are interpreted as light. In contrast, information from the ear travels in the vestibulocochlear nerve and goes primarily to areas responsible for hearing, and so on.

Although under normal conditions each receptor responds only to its adequate stimulus, under extreme conditions other stimuli may be able to activate the receptor. For example, you may "see" flashes of light if you are hit in the eye. In this case, the physical deformation of the photoreceptor cell membrane by externally applied pressure is so great that it may induce a change in membrane permeability, even though physical pressure is not the photoreceptor's adequate

Table 7.1 **Types of sensory receptors that provide information about the external environment and/or our body position.**

Receptor Type	Location	Sensed Variable	Cell Type	Stimulus	Comments
Chemoreceptors	Mouth, but especially on the tongue	Taste	Receptor cells	Chemicals	Four different receptor cells respond to sweet, salty, sour and bitter.
	Nasal passages	Smell (olfaction)	Modified neurons	Chemicals	Humans have thousands of different olfactory neurons, each highly specific for a particular odorant molecule.
Photoreceptors	The retina of the eye	Vision	Receptor cells	Light	The receptor cells for light are the rods and cones. Cones account for color vision; rods are highly sensitive to dim light but do not permit us the ability to sense color.
Thermoreceptors	The skin (and various other tissues and organs)	External (and internal) temperature	Modified neurons	Heat, cold	Different receptors respond to heat and cold.
Mechanoreceptors	The skin	Pressure, light touch, and vibration	Modified neurons	Mechanical (deformation of free or encapsulated endings)	These are usually rapidly adapting receptors. Receptors for heavier pressures generally are located more deeply than receptors for light touch.
	Muscles	Muscle length	Modified neurons	Mechanical (stretch)	Neuron receptor endings make contact with or wrap around modified muscle cells in a structure called the muscle spindle.
	Tendons	Muscle tension	Modified neurons	Mechanical (stretch)	Neuron receptor endings are located in a structure called the Golgi tendon organ.
	Joints	Joint position	Modified neurons	Mechanical (stretch)	Provide precise information about the position of joints, and hence of our limbs.
	The ear	Sound	Receptor cells	Mechanical (bending of hairs)	Receptor cells are called hair cells. Different cells respond to different frequencies based on their location in the cochlea.
	The semicircular canals of the vestibular apparatus	Angular acceleration of the head	Receptor cells	Mechanical (bending of hairs)	Respond to changes in angular acceleration to detect changes in the position of the head.

(continued)

Table 7.1 Types of sensory receptors that provide information about the external environment and/or our body position. *(cont'd.)*

Receptor Type	Location	Sensed Variable	Cell Type	Stimulus	Comments
	The utricle and the saccule of the vestibular apparatus	Static position of the head; also linear acceleration	Receptor cells	Mechanical (bending of hairs)	Movement of crystals called otoliths cause the hairs to bend.
Nocireceptors	The skin (and various other tissues and organs)	Pain	Modified neurons	Extremes of temperature and mechanical deformation; also locally released chemicals	Some adapt, but most are slow to adapt or do not adapt at all.

stimulus. This will lead to action potentials in the neurons to which the photoreceptors synapse. The central nervous system then interprets these action potentials as light because they arrive at the visual cortex via the optic nerve. This example illustrates how the central nervous system's interpretation of sensory input is based on the *source* of the action potentials and where they go in the brain, not on the shape or frequency of the action potentials themselves.

Stimulus Intensity Coding

Our senses have the ability to perceive a wide range of stimulus intensities, even though the brain receives action potentials that are all of equal magnitude. The perception of stimulus intensity is encoded in several ways. First, the frequency of action potentials in the afferent neuron depends on the magnitude of the graded potential, at least once threshold has been reached (Figure 7.2). A more intense stimulus produces a greater graded potential and more action potentials in the neuron. Second, more intense stimuli are likely to activate a greater number of the total available population of receptor cells; therefore, more nerve cells will be sending action potentials to the brain. Stimulus intensity is thus encoded by both the frequency of action potentials in each neuron and the number of neurons that are active.

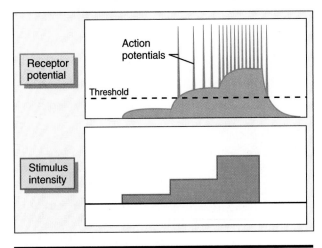

Figure 7.2 Stimulus intensity coding. Once a threshold is passed, the number of action potentials in a neuron is directly related to the magnitude of the graded receptor potential. This is an example of a receptor cell that is a modified neuron because the same cell generates both a receptor potential and action potentials.

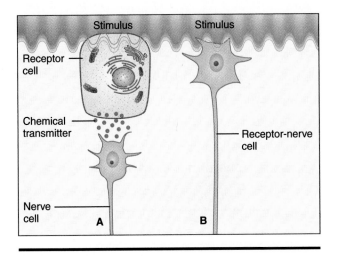

Figure 7.1 How a sensory stimulus is transduced into action potentials. (A) A typical receptor cell that synapses with a nerve cell. (B) A modified neuron that also functions as a receptor cell.

Receptor Adaptation

Try stroking a single hair on your arm lightly, and notice how easily you can feel it. What if every hair on your body were that sensitive all of the time? You can imagine that the brain might be faced with sensory overload if it received continuous high levels of input from every possible sensory receptor in the body! Fortunately, stimulus intensity is not the sole determinant of the firing rate of action potentials in an afferent neuron. Many receptors adapt quickly, so that a continual stimulus is *not* accompanied by a sustained high frequency of action potentials in the afferent neuron. **Receptor adaptation** describes the decline in the sensory neuron's receptor potential during a constant stimulus (Figure 7.3). Rapidly adapting receptors are especially well-suited for signalling a *change* in the stimulus intensity. Most skin receptors for light touch are rapidly adapting, which explains why you are not always conscious of the touch of your clothes, and why a light stroking motion is more easily felt than constant light pressure applied in one location. Similarly, visual receptors adapt to changes in light intensity. Visual adaptation exemplifies a second useful function of adaptation besides prevention of sensory overload; it permits a much wider range of sensory intensities to be sensed accurately.

Receptor adaptation is not a universal phenomenon; for certain receptors, most notably those for pain, the *failure* to adapt has survival value. Nonadapting pain receptors have a receptor potential that is always directly related to the stimulus intensity. The persistence of pain while a painful stimulus is applied evokes a continued attempt to withdraw from the stimulus.

Brain Selectivity of Perception

Whereas receptor adaptation limits the sensory input *to* the brain, the brain itself still has the ability to process selectively the many different sensory inputs it receives. Consider, for example, how often you have been accused of not paying attention to what someone is saying. Your receptors for hearing (the hair cells in the ear) are responding to the sound waves, and the afferent neurons are sending action potentials to the brain in the usual manner, but your brain is de-emphasizing the information, concentrating on something else instead. You can quickly bring the hearing input to the forefront of your consciousness if you choose. This type of processing is useful for preventing sensory overload; it allows you to shift your awareness from one type of input to another, almost at will. Stop for a moment and listen to the various sounds around you; were you thinking of them a moment ago when you were concentrating on this book? Clearly, our knowledge of the world around us is determined by our limited abilities to perceive it. The

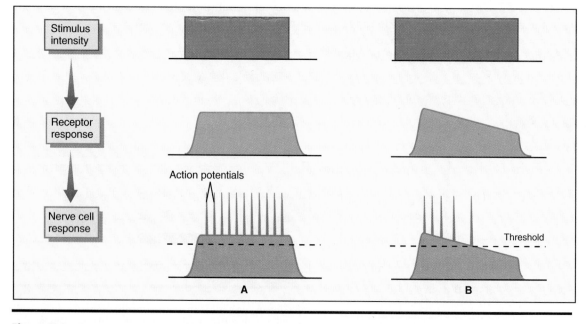

Figure 7.3 Receptor adaptation. (A) A nonadapting receptor, in which the receptor potential is maintained as long as the stimulus is present. (B) An adapting receptor, in which the receptor potential declines over time despite the continued presence of the stimulus. These are examples of receptor cells that are not neurons. Rather, they synapse with neurons, producing graded potentials and action potentials in the neurons.

brain can even suppress the perception of pain at times, an adaption that has survival value in that we can focus on escaping from a dangerous environment rather than on the pain we may be in while making our escape.

Sensory Receptors in Skin

The skin and the sensory receptors within it are critical to how we interact with our environment. Thermoreceptors for heat and cold allow us to select a comfortable environment, pain receptors help us avoid injury, and receptors for light pressure and touch allow us to explore the fine details of the world around us. A common feature of all of these receptors is that they are either free or modified axonal endings of sensory neurons. Figure 7.4 illustrates the various sensory nerve endings of the skin.

Thermoreceptors

Thermoreceptors are the least understood of the sensory receptors, for we do not yet know exactly how the sensations of warmth and cold are transduced into action potentials. The current consensus hypothesis is that the temperature in the vicinity of the free dendritic endings of the sensory neuron affects the opening and closing of specialized gated channel proteins (recall that the tertiary structure of proteins can be temperature-sensitive). This could lead to an inward movement of sodium and a depolarizing receptor potential. For technical reasons it has not yet been possible to test this hypothesis.

The neurons that detect cold are not the same ones that detect warmth. The ability to detect cold with a separate population of neurons is especially puzzling because there is really no such thing as cold, only the relative lack of heat. Nevertheless, these neurons allow us to perceive the sensation we call cold.

Thermoreceptors adapt rapidly to a sustained change in temperature; that is, they are phasic

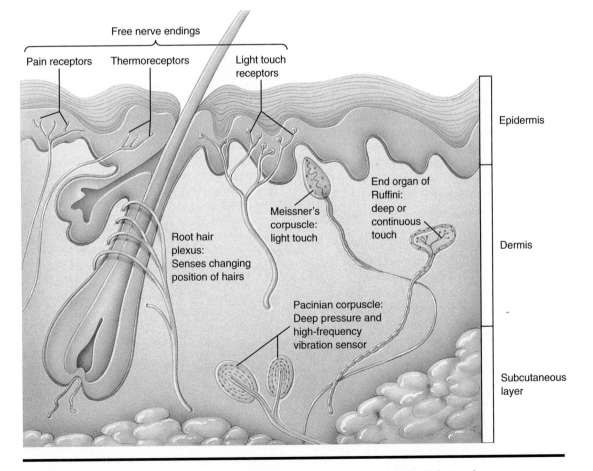

Figure 7.4 Receptors of the skin. Thermoreceptors, some of the receptors for light touch, and the receptors for pain are the free endings of nerve cells. Other receptors for touch and for pressure and vibration are the encapsulated endings of nerve cells.

receptors. This allows us to both assess a change in temperature accurately and adjust the sensory input so that it becomes more bearable. An example of this is stepping into a hot shower; at first the water seems quite hot, but we soon become accustomed to it.

Pain Receptors

For human survival, there are probably no more important receptors than those for pain. Like thermoreceptors, pain receptors in the skin are the free endings of specialized neurons. There are two types of pain sensations. Fast pain, also called sharp or acute pain, begins within a tenth of a second of a stimulus being applied. Receptors for fast pain are generally located near the surface of the body, such as in the skin. The receptors for fast pain respond to excessive mechanical or thermal stimuli. The prick of a needle and the burning sensation of a hot flame are examples of fast pain receptor activation. In contrast, slow pain begins seconds or even minutes after the sensory stimulus is applied. Some slow pain receptors are located in the skin, but most are located more deeply in the body, such as in muscles or abdominal organs. Slow pain receptors can also respond to mechanical and thermal stimuli, but most are chemical receptors that respond to chemicals (bradykinin, histamine, acids, potassium ions, and others) released by damaged cells.

Unlike thermoreceptors, pain receptors generally do not adapt. The tonic (nonadapting) nature of pain receptors has obvious survival value in alerting you to the continued presence of a harmful situation from which you may need to withdraw. However, the nonadapting nature of pain receptors also means that pain can sometimes persist long after the initial cause of the harm or injury has been removed, as anyone who has suffered a burn knows.

Pain receptors are not just about avoiding harm from the external world; they also signal danger from within. Examples are the pain of excessive stomach acidity and distention (signalling damage to the stomach lining or excessive stretch of the stomach), the pain associated with the passage of a kidney stone through the ureter to the bladder, and the pain of muscle injury.

Receptors for Light Touch and Pressure

You may think of light touch and pressure as different degrees of intensity of the same stimulus, the stimulus being a physical deformation. In general, receptors located near the surface of the skin sense light touch, whereas receptors located more deeply sense stronger pressures. Some of the receptors for touch and pres-

sure are free endings of sensory neurons, but others are encapsulated in an associated structure (see Figure 7.4). There are several different types of encapsulated structures: Meissner's corpuscle, organ of Ruffini, and **Pacinian corpuscle.** Except for the Pacinian corpuscle, exactly how each one functions is unknown.

Free nerve endings of certain sensory neurons sense light touch. Some sensory neurons form a network or tangle of free nerve endings around the root of each hair. The hair root network is responsible for the exquisite sensitivity of a single hair to light touch or vibration. Pacinian corpuscles are located in the deeper layers of the skin and thus are responsive to pressures greater than light touch. In a Pacinian corpuscle, the sensory ending of the neuron is encapsulated in multiple layers of connective tissue. When pressure is applied, the initial response is a receptor potential proportional to the stimulus. If the stimulus is maintained, the layers of connective tissue surrounding the nerve ending slip past each other, dissipating the energy of the original stimulus (Figure 7.5). Sensory receptors for touch and pressure are rapidly adapting, making them particularly useful for sensing a *change* in a stimulus. The ability to respond to vibration can now be understood. A vibration is simply a stimulus that changes direction (moves back and forth) rapidly.

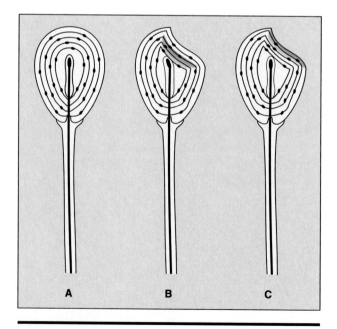

Figure 7.5 How Pacinian corpuscles adapt. (A) Unstimulated state. (B) Initial application of pressure. (C) Adaptation of the receptor despite the maintenance of pressure.

Sensors for Body Position and Movement

Though not strictly for sensing the external world, receptors that sense body position and movement, called **proprioceptors,** are essential for us to produce coordinated and useful movements. Proprioceptors in joints, tendons and muscles provide information about the position of the joints (joint receptors), muscle length (muscle spindles), and the amount of tension being exerted on tendons by muscle contraction (Golgi tendon organs).

Joint Receptors

Joint receptors tell us the position of our limbs even though we are not consciously aware of it. This allows a basketball player to focus his or her attention on the basket, even as joint receptors provide the brain with precise information about the exact angle of each joint throughout the shooting motion. Joint receptors are mechanoreceptors located in and around joints that respond to mechanical deformation with action potentials. They may take a number of different forms, ranging from free endings to encapsulated structures.

Muscle Spindles

Muscle spindles are specialized bundles of modified muscle cells and nerve cells that sense involuntary changes in muscle length. That is, they respond primarily to changes in muscle length that are brought about by forces in the external environment, not by voluntary contractions of muscle.

A muscle spindle tapers at each end and attaches to regular muscle fibers within a whole muscle (Figure 7.6). The bulk of the spindle is made up of modified muscle cells; these muscle cells have contractile regions only at each end. The contractile regions are innervated by motor neurons called **gamma motor neurons.** The central regions of these modified muscle cells have lost the ability to contract, and can be thought of as an elastic rubber band. Sensory nerve endings innervate the central, elastic portion. When the whole muscle is stretched, the muscle spindle also stretches (like a rubber band) because it is embedded in the muscle. This stretches the sensory nerve endings, thereby producing receptor potentials in the nerve endings and ultimately action potentials in the nerve.

Two types of sensory neurons make contact with the central region of the muscle spindle. One type has endings that wrap around the muscle spindle cells in a spiral fashion and the other type makes branching contacts. Both types of sensory neurons respond to stretch of the central portion of the muscle spindle with an increased frequency of action potentials. The spiral neurons adapt very rapidly, providing information primarily about how quickly the stretch occurs, or rate of change of the stretch. The branching neurons adapt less quickly, providing information about both the rate of change and the magnitude of the change during stretch.

How the muscle spindles prevent *involuntary* changes of muscle length (passive stretch) that are imposed by the external world around us, while at the same time permitting smooth and coordinated *voluntary* changes in muscle length, is best demonstrated by an example. Imagine that you are standing in a cafeteria line and that you wish to pick up a cafeteria tray from a low table and place it on a high shelf. This motion requires voluntary contraction of skeletal muscles, an event that is initiated by the motor neurons to the muscle fibers (these motor neurons are called **alpha motor neurons**). However, the gamma motor neurons to the contractile ends of the muscle spindle are also activated during voluntary contractions. Thus, the muscle spindle shortens in parallel with the entire muscle and normal tension is maintained on the center, sensory portion of the spindle. As a result, the sensory fibers innervating the spindle do not change their firing rate, and the muscle spindle does not respond to the voluntary change in muscle length at all.

However, as a consequence of the parallel shortening of both the whole muscle and the muscle spindle during voluntary movements, the spindle remains poised to respond to involuntary changes in muscle length. For example, imagine that a friend sets a heavy textbook on your tray while you are holding it in the air but looking in another direction. The weight of the book causes an involuntary stretching of your muscles, simultaneously stretching the muscle spindle. As a consequence, the sensory nerves to the muscle spindle increase their firing rate, and a reflex is very quickly set in motion to increase motor neuron activity to that muscle, increasing the force of contraction and returning the stretched muscle to its prestretched, voluntary length. In other words, you can quickly adapt to the increased load without dropping the tray or overreacting by throwing the entire tray over your shoulder.

The bottom line is that the muscle spindles tend to prevent involuntary changes in muscle length while ignoring all voluntary changes in muscle length.

Golgi Tendon Organs

Golgi tendon organs are our sensors of tension (a measure of force) within muscles. The structure of a Golgi tendon organ is not complicated; it is basically an encapsulated nerve ending within the tendons that join

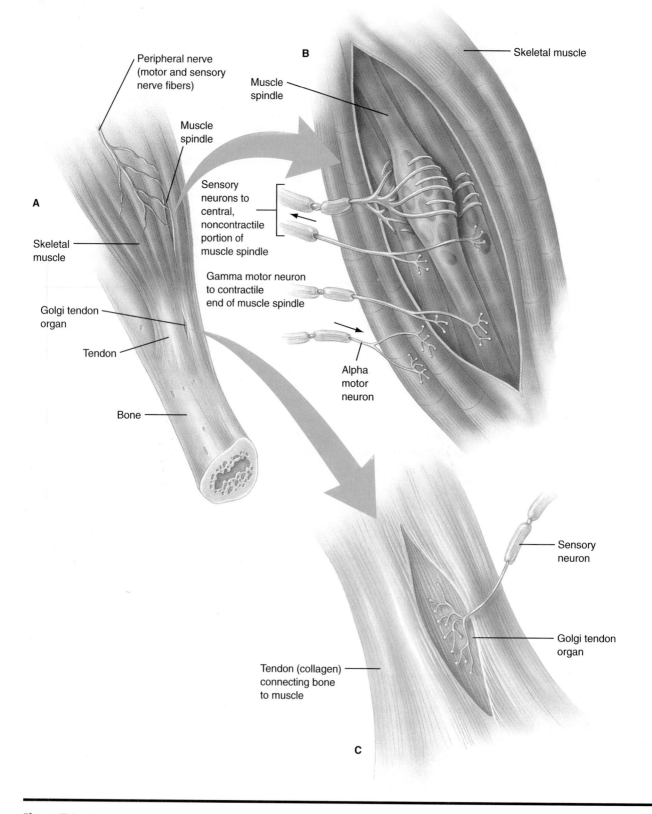

Figure 7.6 Sensors for muscle stretch and muscle tension. (A) The general location of muscle spindles and Golgi tendon organs. (B) The muscle spindle. The muscle spindle is completely surrounded by contractile skeletal muscle fibers. Two types of sensory neurons innervate the central portion: rapidly adapting neurons with spiral endings and slowly adapting neurons with branching endings. Motor neurons innervate the contractile ends of the muscle spindle cells (gamma motor neurons) and the skeletal muscle cells (alpha motor neurons). (C) The Golgi tendon organ. The Golgi tendon organ is innervated only by a slowly adapting sensory neuron.

muscle to bone (Figure 7.6(C)). Golgi tendon organs are good sensors of muscle tension primarily because of their *location*, not because their structure is unique. Recall that tendons are composed of connective tissue only; they do not have any contractile elements. Although they are not very elastic (think of a tendon as a stiff rubber band), they do stretch slightly whenever tension on them is increased. However, because the tendon is in series with the muscle rather than in parallel (joined end-to-end rather than side-by-side), any increase in tension in the muscle stretches the Golgi tendon organ, stretching the sensory nerve endings that innervate it. Thus, the degree of tension in a muscle is encoded in the number of action potentials coming from sensory neurons of Golgi tendon organs.

Golgi tendon organs provide the central nervous system with information about the degree of tension in muscles. Unlike the muscle spindles, however, the Golgi tendon organs make no distinction between tension generated voluntarily and tension generated involuntarily. Beyond simply providing information to the central nervous system about tension, however, Golgi tendon organs may be particularly important in preventing muscle injury. For example, stimulation of Golgi tendon organs by excessive muscle tension produces a reflex inhibition of alpha motor neuron activity to the same muscle. This lessens the degree of active muscle contraction that can be developed. At very high muscle tensions, this reflex can cause the muscle to relax suddenly. In other words, the Golgi tendon organs initiate a reflex that encourages an overloaded muscle to give up before serious injury occurs. For example, you *cannot* land on the ground in a full upright position when you jump to the ground from a height of about eight feet or more. This is because the Golgi tendon organs sense excessive tension, and the action potentials that go to the central nervous system initiate a reflex that forces your thigh muscles to relax.

The Chemical Senses

The ability to taste is important to our ability to identify food and to avoid potentially dangerous foods. Our ability to smell is also important in this regard; just try holding your nose as you eat! Children learn to use this maneuver rather quickly when asked to eat foods they consider unpalatable. Of course, taste can be used to identify only food actually placed in the mouth. In contrast, the ability to sense chemicals in the air (olfaction) can be very useful for learning about the more distant external environment. Olfaction is used to identify food, to sense danger in the environment (such as the smell of smoke), and to obtain information about members of our own species. Many animals release identifying chemicals called **pheromones** that serve as sexual attractants. Although humans also release pheromones, their effect on our behavior is not entirely clear. However, we often notice a change in our emotional attitude when a person is wearing perfume or cologne, which contain chemicals intended specifically as attractants.

The chemical senses have a common general mechanism, relying on the binding of chemicals to specific receptors on the surface of a receptor cell. Our sense of taste is limited to four taste qualities; sweet, salty, sour, and bitter. In contrast, our sense of smell can distinguish between a very wide range of different chemicals.

Taste

Taste receptors are located primarily on the tongue (Figure 7.7(A)), although there are a few scattered about in other locations in the mouth. As mentioned previously, taste receptor cells are sensitive only to the taste qualities of sweet, salty, sour (acidic foods such as lemons are sensed as sour), and bitter. The subtlety of taste variations is due to the combinations of these qualities, modified by our sense of smell. The receptors for each of the four taste qualities tend to be concentrated at different locations on the tongue; the receptors for sweet are located on the tip, sour on the sides, salty on the sides *and* tip, and bitter at the back. Very few receptors are located on the center of the tongue. As a simple demonstration, dip a cotton-tipped applicator into sweet or salty solutions and then touch it to different areas of your tongue. You should be able to appreciate the specific localization of these two types of receptors. To gain the maximum taste sensation, particularly from liquids, we move the liquid around in our mouths before swallowing. On the other hand, if we are thirsty or do not want to taste the liquid, we swallow repetitively, passing the liquid primarily over the center of the tongue. It is probably no accident that the receptors for a bitter taste are located at the center of the back of the tongue. A bitter taste often is a sign of an unwholesome or even poisonous substance, and the receptors for a bitter taste are the hardest to bypass.

The receptors for taste are located in **taste buds,** clusters of about fifty cells arranged in a bud-like structure, much like cloves of garlic. Taste buds are clustered at the base of the **papille,** which are small circular or mushroom-shaped elevations on the surface of the tongue that give the tongue its rough feel and appearance (Figure 7.7(B)).

A single taste bud is shown in Figure 7.7(C). Some of the cells in each taste bud are **taste receptor cells,** whereas others are **supporting cells** not involved in sensing taste. The tip of each receptor cell terminates in a single hair-like **microvillus** that extends toward the surface of the tongue, and the base of the receptor cell makes contact with (synapses with) the dendrites of a neuron. When the appropriate chemicals come in contact with receptors on the cell membrane of the microvillus, ion channels open, producing a depolarizing receptor potential in the receptor cell. The receptor potential causes the release of a chemical transmitter from the receptor cell at the synapse with the neuron, which in turn sends action potentials to the brain. Figure 7.8 is a photomicrograph of a taste bud.

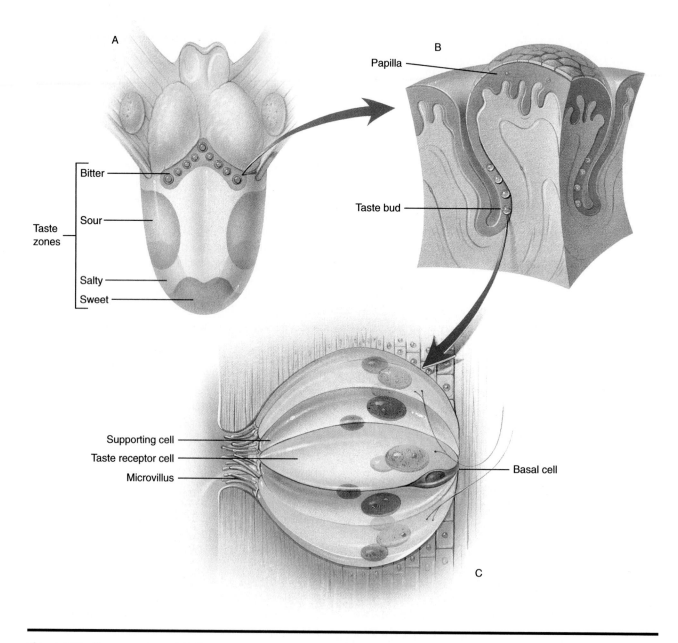

Figure 7.7 Taste receptors. (A) The location of the receptors for bitter, sour, salty, and sweet tastes on the tongue. (B) Closer view of a papilla on the surface of the tongue, showing the location of the taste buds. (C) A taste bud.

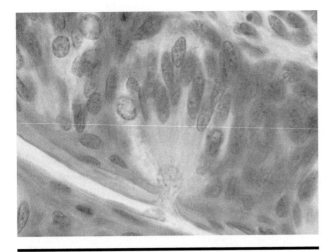

Figure 7.8 A photomicrograph (×450) of a taste bud.

The taste bud cells are actually exposed to a fairly harsh environment that includes foods of very hot and cold temperatures and strong chemicals such as very acidic foods. Consequently, the taste bud cells become damaged and must be replaced frequently. At the base of the taste bud are cells, called **basal cells,** that divide and grow into new taste receptor cells at a rate that results in the nearly complete replacement of taste bud cells every two weeks.

Smell

Our sense of smell originates from receptors located in the upper part of the nasal passages (Figure 7.9). The surface layer of the nasal passages, called the nasal epithelium, contains three cell types: olfactory receptor cells, supporting cells, and basal cells. **Olfactory receptor cells** are modified neurons with hair-like receptor endings called **olfactory hairs.** The hairs are located right at the surface of the olfactory epithelium, embedded in a layer of mucus. Olfactory receptor cells

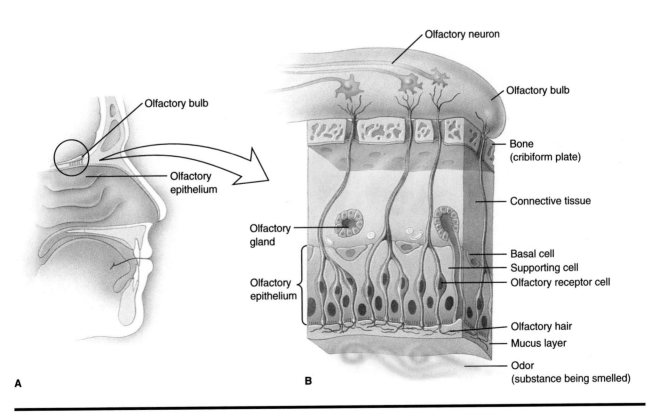

Figure 7.9 Olfactory receptors. (A) Sagittal section through the skull, showing the location of the olfactory epithelium. (B) Enlarged view showing the olfactory receptors and their synapses with neurons of the olfactory nerve. Note the olfactory glands that produce the layer of mucus covering the surface of the epithelium.

are the only neurons that are directly exposed to the external environment, so the layer of mucus is particularly important for protecting the hairs from drying out by exposure to air. Odor molecules (odorants) in the air dissolve in the mucus, eventually making contact with specific protein receptor molecules that are part of the cell membrane of the hairs. When an odorant molecule makes contact with a receptor molecule, ion channels open, producing a receptor potential. Once a threshold membrane depolarization is reached at the axon hillock of the olfactory receptor cell, action potentials are generated.

The axons of several olfactory receptor cells travel in close proximity to each other through the connective tissue and the thin bone of the nasal passageway, to synapse on other neurons that form the **olfactory nerve.** The neurons of the olfactory nerve carry information (in the form of action potentials) to the brain.

Between and surrounding the olfactory receptor cells are supporting cells that, like those in the taste buds, lend structural support to the receptor endings of olfactory cells. Near the base of the epithelial layer are basal cells that, again like those in the taste buds, differentiate into new receptor cells. The turnover rate of olfactory receptor cells is relatively high; they die and are replaced approximately every two months.

Just beneath the olfactory epithelium is a connective tissue layer that contains the **olfactory glands.** The olfactory glands produce mucus, which is deposited on the surface of the epithelium via ducts that pass through the epithelium.

Humans are thought to have receptors for nearly a thousand different kinds of odorants. Some of these receptors are exquisitely sensitive; the odorant we identify as garlic, for example, can be detected at a concentration of less than one odorant molecule in every 10 billion molecules of the gases in air. But in addition to having a specific receptor for its detection, an odorant must be volatile in air (to reach the nose) and able to dissolve into the watery mucus (to reach the receptors) before it can be detected. There are many substances that do not fill at least one of these three requirements, including natural gas (methane). Because natural gas is deadly, an odorant is added to it to warn us of its presence.

Stimulation of an olfactory receptor cell ends either when the odorant molecule becomes detached from the receptor or when the odorant molecule is destroyed. Recently, it has been discovered that at least part of the destruction of odorants is carried out by enzymes that are present in the mucus layer. You already know that enzymes are catalysts for the biochemical reactions *within* a cell; here they act *outside* a cell as a "molecular cleanup crew." At this point, it is not clear which cells produce and secrete these enzymes.

The Ears: Hearing

The ears are organs specialized primarily for hearing, the sensation associated with sound. Closely associated with the ear is another small organ, the vestibular apparatus. Anatomically, the vestibular apparatus is part of the inner ear. However, the vestibular apparatus is not involved in hearing; its role is in the maintenance of equilibrium. We will concentrate on the role of the ears in hearing in this section, but first we must understand the nature of sound.

What Is Sound?

Sound is waves of compressed air traveling in all directions from their source, like the ripples caused by a stone tossed into a pool. Sound is characterized by its intensity and frequency. The **intensity** of sound, or its loudness, is determined by the *amplitude* of the wave and is described in units of decibels (db). The intensity of a sound diminishes with distance from the source. The tone (or pitch) of a sound is determined by its **frequency,** the number of sound wave cycles that pass a given point per second. Frequency is generally reported in units of hertz (Hz), the designation for cycles per second. High tones are those with high frequencies. Examples of sound waves of different frequencies and intensities are shown in Figure 7.10.

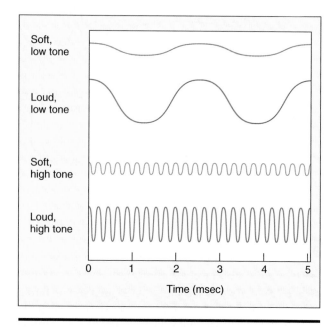

Figure 7.10 Properties of sound waves. Intensity is determined by sound wave amplitude. Pitch (or tone) is determined by sound wave frequency.

The rate at which sound travels through air is constant, regardless of its intensity or frequency. Sound travels through air at about 1100 feet per second. Knowing this, you can judge the distance of an approaching thunderstorm: Every 5-second delay between the lightning and the thunder means that the storm is about a mile away.

Hearing

The ear is the organ that transduces sound waves into action potentials in nerve cells. In brief, sound waves reaching the ear are transmitted through a membrane, three small bones, and a second membrane to a fluid-filled chamber of the inner ear. Fluid movement in the fluid-filled chamber causes mechanical movement of the hairs of specialized receptor cells, causing receptor potentials. Receptor potentials in the receptor cells in turn produce action potentials in nerve cells with which they make contact.

The general structures involved in hearing are shown in Figure 7.11(A). We can categorize the structures as belonging to the outer ear, the middle ear, or the inner ear. The outer ear consists of three parts: the fleshy external structure we know as our ears, an air-filled canal, and the **tympanic membrane,** or eardrum. Sound waves are funneled into the air-filled canal, where they strike the tympanic membrane, causing it to vibrate.

The middle ear is an air-filled chamber bridged by three small bones, named the **malleus** (hammer), the **incus** (anvil), and the **stapes** (stirrup). The middle ear is not in direct contact with the air in the outer ear because of the complete seal provided by the tympanic membrane. The middle ear is kept at the same air pressure as the external air pressure by way of the **eustachian tube,** a narrow tube connecting the inner ear to the upper portion of the throat (nasopharynx). The eustachian tube normally is closed off by the muscles along its sides, but it opens briefly when we yawn or swallow so that the air pressure in the middle ear can be equalized with that of the outer ear.

Vibration of the tympanic membrane by sound causes vibrations in the malleus, incus, and stapes, thereby transmitting sound across the middle ear. In turn, vibrations of the stapes produce vibrations in another membrane (called the oval window) that is part of the inner ear. Thus the three bones of the middle ear bridge the gap from outer to inner ear. More importantly, the combination of the tympanic membrane and the three bones concentrates the force collected from the large surface area of the tympanic membrane onto the small surface area of the oval window. This is necessary because it takes more force to produce waves of pressure in the fluid-filled inner ear than could be generated directly by sound waves of air.

The ability to concentrate the force of sound waves leads to a potential problem, and that is the risk of damage to the inner ear by excessively loud sounds. However, the structure of the middle ear tends to protect us from such damage. First, although the three bones of the middle ear make contact with each other, they are not bonded rigidly together. Each bone must cause the next one to vibrate in an orderly fashion in order for sound to be faithfully transmitted to the inner ear. Very loud sounds have so much energy that they cause the bones to buckle or shift slightly relative to each other rather than vibrate, and thus some of the energy is lost before it strikes the inner ear. As a result, very high-intensity sounds are not concentrated as much as lower-intensity sounds.

There is a second way in which the structure of the middle ear protects us from loud noise. The malleus and the stapes are attached to bones of the skull by two muscles (see Figure 7.11(B)). These muscles contract in response to high-intensity sounds. As a result, high-intensity sounds are attenuated because the contracted muscles partially prevent the bones from vibrating against each other. This reduces the force applied to the oval window, in effect turning down very loud sounds before they can reach the inner ear and cause damage. Because the latency period for contraction of these muscles is about 15 msec, this mechanism is not effective at protecting us against sounds that reach a high intensity very rapidly (such as a gunshot at close range).

Even with these protective mechanisms, repeated exposure to loud sounds may damage the cells responsible for hearing. At risk are members of rock bands and people exposed to loud machinery on the job without protective headgear. The next time you are in your local airport, notice the protective headgear worn by the ground crew personnel.

Although the muscles of the middle ear protect against loud sounds, that may not be their only function. Recent evidence suggests that these muscles may contract reflexively just a fraction of a second before voluntary speech and with gross body movements such as walking. The effect would be to attenuate the hearing responses to self-initiated sounds, such as our own words or the rustle of our own clothes as we walk. Although this may be an advantage because we generally are not as interested in such sounds, it would also make it harder to hear well while we are moving about or talking. This is why it may be advantageous to stand completely still when trying to listen intently to a weak sound from another source.

The inner ear consists of two structures: the **cochlea,** an organ of hearing, and the **vestibular apparatus,** associated with equilibrium (the subject of the next section). The cochlea is a bony, coiled, fluid-

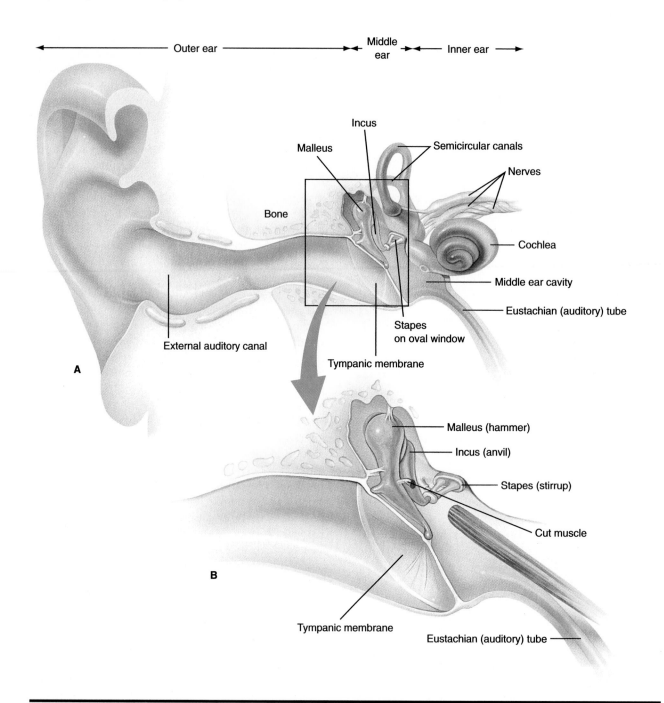

Figure 7.11 Structures involved in hearing and equilibrium. (A) General diagram showing the relationships between the structures of the outer, middle, and inner ear. (B) A closer view of the middle ear, showing the muscles that attach the malleus and the stapes to the bones of the skull.

filled structure shaped like a snail. Its structure can be revealed only in cross-sectional slices, but if you could uncoil it, the cochlea would appear as the schematic drawing in Figure 7.12(A). It appears to be composed of three fluid-filled chambers, but a closer look shows that top and bottom chambers are actually one continuous chamber that completely surrounds the center one. In cross-section (Figure 7.12(B)), these appear as three chambers. The lower membrane of the center chamber is called the **basilar membrane.** The receptor

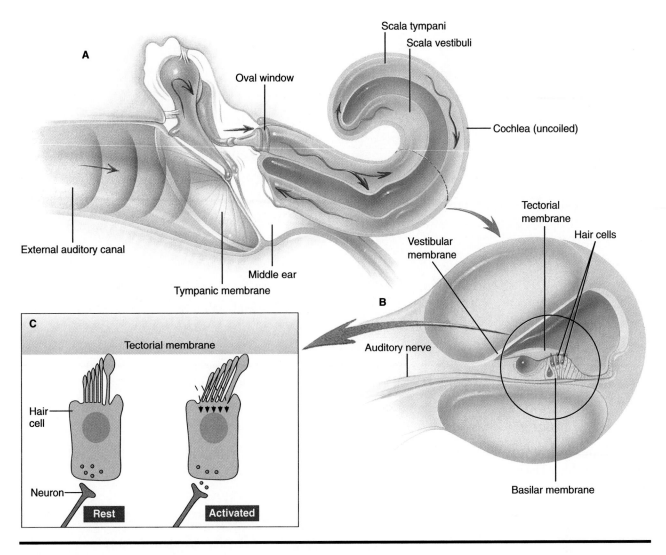

Figure 7.12 Structures and receptor mechanisms for the transduction of sound. (A) An uncoiled view of the cochlea and its relationship to the outer and middle ear. The upper and lower fluid-filled chambers are actually one continuous chamber that surrounds and encloses the center chamber. (B) A cross-section through the cochlea showing the location of the hair cells on the basilar membrane. (C) How a hair cell is activated.

cells for hearing, called **hair cells,** are located just above the basilar membrane. When the stapes vibrates, it causes the oval window to vibrate, setting up waves of fluid pressure in the upper chamber of the cochlea that are the same frequencies as the sound waves. These waves are transmitted to the lower chamber, with the result that the basilar membrane between them vibrates. The vibrations of the basilar membrane force the hairs of the hair cells to bend against the **tectorial membrane,** and this movement of the hairs produces a depolarizing receptor potential in the hair cells (Figure 7.12(C)). The intensity (loudness) of the sound is encoded by the magnitude of the movement of the hairs of a hair cell and presumably

also by the number of hair cells activated. As usual for sensory transduction, the receptor potential in a hair cell ultimately produces action potentials in a neuron that carries the message to the brain. Photomicrographs of normal hair cells, as well as hair cells that have been damaged by loud noise, are shown in Figure 7.13.

The human ear is most sensitive to frequencies of sound ranging from about 800 to 6000 Hz, although we can actually detect a much broader range (20 to 20,000 Hz). As indicated in Figure 7.14, higher frequencies (higher tones) tend to cause greater movement of the basilar membrane nearer the oval window, whereas lower frequencies (lower tones) tend to cause

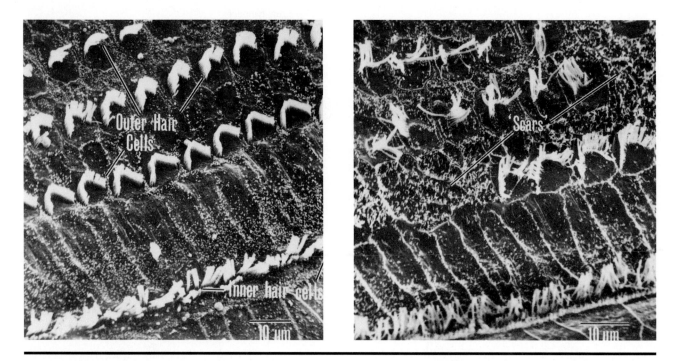

Figure 7.13 Hair cells of the cochlea. (A) Normal healthy hair cells. (B) Hair cells that have been damaged by loud sounds.

greater movement of the basilar membrane at the far end of the cochlea. The brain's perception of tone is determined by which hair cells are activated; that is, where they are located along the cochlea and thus which neurons are activated. The location-specific coding for tone also explains why one can be deaf to certain frequencies of sound and not to others. Damage to hair cells or neurons at the beginning of the cochlea, for instance, would result in selective loss of hearing of the higher frequencies, but might still leave intact the ability to detect the lower frequencies.

Beyond being able to detect the loudness and the tone of a sound, we also have the ability to determine the direction from which a sound is coming. Sound arrives at the ear nearer to the source of the sound just a fraction of a second before it arrives at the opposite ear. Furthermore, the intensity of sound is lower at the opposite ear because the head acts as a shield. The auditory regions of the brain judge the location of the sound by comparing the timing of arrival and the number of action potentials arriving from the two ears. Thus, the ability to determine location is a function of the brain, not of the receptors themselves.

Many humans suffer from partial or complete deafness. The two general types of deafness and some of their causes are listed in Table 7.2. The hearing deficits caused by some, but not all, forms of deafness

are correctable with a hearing aid. In general, hearing aids amplify incoming sound waves.

The Vestibular Apparatus: Equilibrium

The vestibular apparatus consists of three semicircular canals that detect changes in the direction of movement (rotational velocity) of the head, plus two chambers that detect both the static position and changes in linear velocity of the head. The ability to detect the position and motion of the head are important for our sense of equilibrium and coordination of movements.

Detecting Movement of the Head

The **semicircular canals,** which detect head movement, are hollow, fluid-filled spaces surrounded by bone. They are continuous with the two chambers that detect position and linear velocity of the head (the **utricle** and the **saccule**), and with the cochlea (Figure 7.15). Each of the three semicircular canals is oriented on a different plane so that we can detect movement

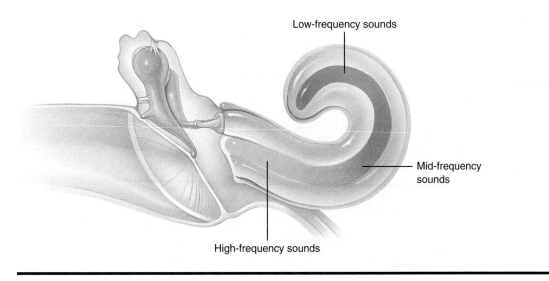

Figure 7.14   The location-specific transduction of sound by the cochlea. The brain is thus able to interpret tone by determining which neurons are activated.

in any direction. A widened section at the base of each canal, called the **ampulla,** contains a group of hair cells. These hair cells are different from those in the cochlea in that each cell has several hair-like structures and one larger extension called a **kinocilium,** which is attached to the hairs. The hairs and the

kinocilium are embedded in a gelatinous material composed of glycoproteins, called the **cupula,** that attaches to the ampulla. Bending the hairs toward the kinocilium causes a depolarizing receptor potential and increases the number of action potentials in the connecting neuron, whereas bending the hairs in

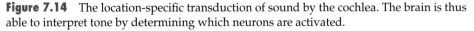

Table 7.2	**Types of deafness.**	
Type	**Conductive Deafness**	**Nerve Deafness**
General definition	Damage to the conduction system of the outer and middle ear, which transmits sound to the oval window of the cochlea.	Damage to the cochlea, the auditory nerve, or auditory pathways in the central nervous system.
Specific causes	• Excessive earwax. • Rupture of the tympanic membrane. • Thickening of the tympanic membrane (common in aging). • Adhesions of the bones of the middle ear. • Fluid in the middle ear. • Middle-ear infections.	• Damage to hair cells by loud noise or loss of hair cells as a consequence of aging (most common). • Damage to the auditory nerve (rare). • Damage to the auditory cortex (very rare).
Benefitted by hearing aids?	Often yes. The most common hearing aids amplify sound to overcome a partial deficit in the conduction system.	No. Hearing aids cannot overcome defective auditory nerve transmission or brain damage. Cochlear implants may be effective in cases of damage to the cochlea.

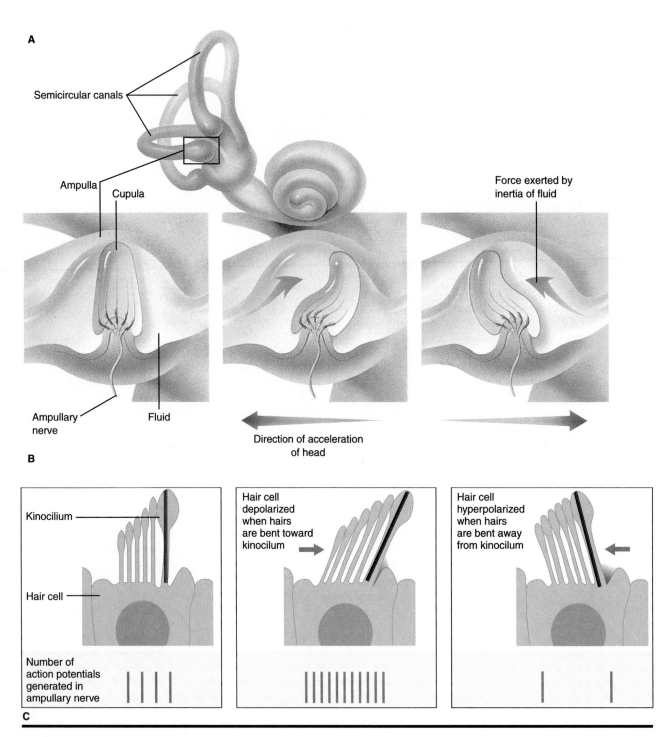

Figure 7.15 Detection of movement of the head by the semicircular canals. (A) A general schematic showing the location of one of the three ampulla in a semicircular canal. (B) An enlarged view of an ampulla showing the location of the hair cells and the cupula at rest, and during acceleration of the head. (C) How movement of the head bends the hair cells. The hair cells of the vestibular apparatus can transduce not only movement, but also direction of movement.

the opposite direction hyperpolarizes the membrane, reducing the number of action potentials. Thus, these receptors can determine not only movement, but also the direction of movement.

To understand how the semicircular canals detect changes in position, you must understand what causes the hairs and the kinocilium of the hair cells to bend. When the head changes position, the fluid in the

HIGHLIGHT 7.1

A Model of Fluid Movement in the Semicircular Canals

The principle of fluid movement in the semicircular canals relative to the movement of the head can be demonstrated with a bucket of water. Fill a bucket half-full of water and place a floating object in the center of it. Grasp the bucket by the handle and quickly rotate it a half-turn or so in a clockwise direction. Notice that the floating object tends not to begin moving as quickly as the bucket, especially if the movement of the bucket is rapid. Thus it *appears* as if the water and the object (though relatively stationary at first) are turning in a direction opposite to the bucket itself. Stop the bucket. When you stop the bucket, the opposite happens; the water and the object are now in motion and they continue to spin in a clockwise direction for a short time after the bucket stops. This is precisely how fluid moves in the semicircular canals.

The delay in change of movement of fluid in the semicircular canals (because of inertia) explains why the brain can sometimes be tricked by signals coming from the vestibular apparatus. To demonstrate this, close your eyes and have someone spin you rapidly while you are seated on a chair that can rotate. Then have him or her stop you quickly, and at the same time open your eyes. For a brief second it may appear as if your field of vision is turning in the opposite direction from the direction in which you were spinning. This is because once the fluid in the semicircular canals is set in motion, it remains in motion for a very brief period even when your head stops moving.

semicircular canals moves in a direction *opposite* to the direction of movement of the head. The reason for this is that the fluid tends not to move along with the semicircular canal. The force exerted by the fluid movement relative to the semicircular canal bends the cupula, and thus the hairs of the hair cells. Figure 7.16 is a photomicrograph of the kinocilium and hairs of several hair cells.

Detecting Position of the Head

The sensors for static head position are hair cells located in the utricle and saccule (Figure 7.17). They rely on a simple design that uses gravity to cause the hairs to bend, producing a receptor potential. The hairs are embedded in a thick gelatinous material in which are also embedded numerous crystals of calcium carbonate, called **otoliths** (*oto = ear, lith = stone*).

The otoliths are heavier than the gelatinous material. Thus, whenever the head is tilted, the otoliths tend to cause the gel to sag, thereby bending the hairs of the receptor cells (Figure 7.18 (A) and (B)). The hairs will *stay* bent as long as the head remains tilted, and thus the utricle and saccule can continue to accurately reflect static head position as long as the head stays in that position. This is an important difference between these hair cells and those of the semicircular canals and the cochlea; the latter two types return to their original positions once the movement is finished or the sound waves have passed.

Although we think of these receptors as sensors of static head position, they do respond to linear acceleration and deceleration as well (Figure 7.18(C)). When you are in a car that is accelerating rapidly, you

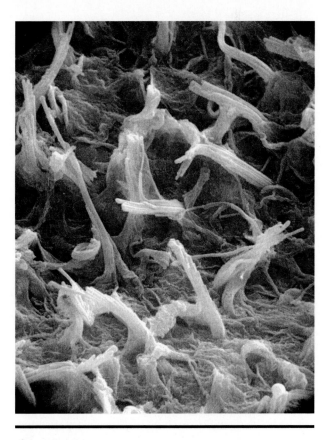

Figure 7.16 A photomicrograph of hair cells of the vestibular apparatus (5800×).

feel that you are being pushed backward in your seat because you are being pushed by the force of acceleration acting against your own inertia. Likewise, the otoliths are pushed backward as long as you are

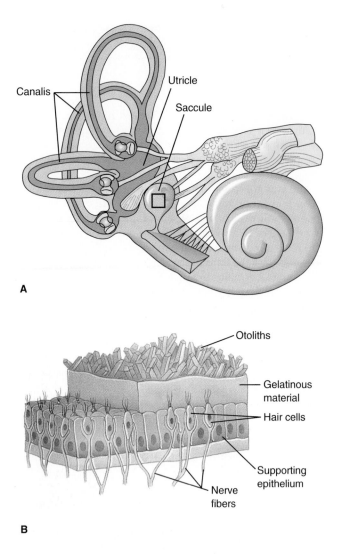

A

B

Utricle

Canalis

Saccule

Otoliths

Gelatinous material

Hair cells

Supporting epithelium

Nerve fibers

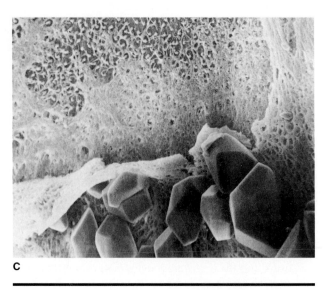

C

Figure 7.17 Sensors for head position. (A) The location of the utricle and saccule in the ear. (B) Enlarged view of a section of the saccule showing the relationships between the otoliths, the gel layer, and the hair cells. (C) A photomicrograph of the otoliths.

accelerating; only when your velocity becomes constant do they return to their original position. Thus, although the otiliths can detect acceleration and deceleration, they do not respond to linear velocity (a constant rate of speed in one direction).

The Eyes: Vision

Vision is the most far-reaching of our sensory systems. Our eyes can detect light originating from stars, the nearest of which is more than four light years away. Light, a form of electromagnetic energy, is detected and transduced into action potentials by special photoreceptor cells in the eye. But the ability to detect and interpret light is somewhat more complicated than for some of our other senses because we must first collect and focus the light on the receptor cells, a function that requires a rather specialized organ, the eye.

What we call light is a portion of a wide spectrum of electromagnetic radiation (Figure 7.19). Electromagnetic radiation is made up of packets of energy called photons that are traveling in a wave-like fashion. The wavelength of light, and of all electromagnetic radiation, is the distance from the beginning of one wave to the beginning of the next. Visible light is only a very small portion of the electromagnetic energy spectrum, which includes the very long wavelengths of radio and television waves at one end of the spectrum and the very short-wavelength X rays and cosmic rays at the other end. Imagine if our eyes could see all of these wavelengths as well!

MILESTONE 7.1

Motion Sickness

A sailor who sailed around the world alone once remarked, "If it were not for seasickness, all the world would be sailors." Seasickness is just one form of motion sickness. Motion sickness is not a disease; rather, it is a normal response of the body to abnormal conditions of motion and visual experience. Nearly everyone is susceptible to varying degrees. The primary symptoms of motion sickness are nausea and vomiting, often accompanied by cold sweats, pallor, hyperventilation, and headache.

Motion sickness has been well-documented since humans first traveled in boats (indeed, the Greek word *nausia* means *seasickness* and the Greek word for a ship is *naus*). One of the first references to motion sickness comes from the Greek physician Hippocrates, who wrote in the late fourth century B.C., "sailing on the sea proves that motion disorders the body." Today we know that motion sickness can be brought about by nearly any artificial mode of transportation, including travel in outer space. The common denominator of the types of motion that cause motion sickness seems to be that the motion must be involuntary (not caused by one's own action).

The history of our understanding of motion sickness is an example of how our knowledge evolves over time. Until the late nineteenth century it was thought that motion sickness was caused by excessive motion of the stomach and its contents. Treatment of the symptoms ranged from wearing a specially designed seasickness belt to the administration of all kinds of medicines and special diets, none of which had much effect. The best advice was that of a writer who suggested in 1891 that "as a rule, medicine of all kinds should be eschewed by those who do not wish to aggravate what is already hard to bear."

A breakthrough in our understanding of motion sickness came with the recognition of the function of the vestibular apparatus. A British physician, Dr. J.A. Irwin, wrote in 1881, "our bodies are endowed with what may be termed a *supplementary special sense*, quite independent of, but at the same time in the closest alliance with, our other special senses, the function of which is to 'determine the position of the head in space.' . . . Beyond doubt its principal seat is in the semicircular canals of the internal ear, which may for practical purposes be regarded as 'the organ of equilibration.'" Subsequent research led to the hypothesis that motion sickness was due to over-stimulation of the vestibular apparatus, which in turn sent neural signals to the areas of the brain that produced the symptoms of nausea and vomiting.

However, research since World War II has shown that vestibular overstimulation alone does not fully account for motion sickness. We now know that inputs from the visual system and from muscle and joint receptors can also contribute to motion sickness. According to the current sensory conflict hypothesis, motion sickness results when the central nervous system receives conflicting and confusing signals from the vestibular apparatus, the eyes, and the various proprioceptors. The symptoms are thought to be the result of the inability of the central nervous system to integrate conflicting signals into an understandable whole. The adaptive value of motion sickness is not clear, but one theory is that it may incapacitate or immobilize an afflicted person when there is a danger of falling or being injured by the environment.

The sensory conflict theory explains why one does not get motion sickness from voluntary motions such as running (voluntary motions, because they are initiated by the brain, are understood by the brain). It also explains the phenomenon of adaptation. The symptoms of motion sickness go away over time even though the motions that caused it continue, because over time the central nervous system comes to accept these motions as normal.

The first effective medical treatment for motion sickness came about as the result of a chance observation. In 1949 a woman who was susceptible to motion sickness was given dimenhydrinate (now known as Dramamine) for a skin condition. She reported to her physician that she seemed to be immune to motion sickness while taking the drug. Subsequent careful scientific verification with larger test groups proved this to be a general phenomenon. Today, Dramamine and a second drug, Scopolamine, are the two most effective drugs available for motion sickness. However, both are best given *before* the symptoms occur as neither is very effective once nausea or vomiting have developed. Scopolamine is the most effective when the exposure to motion will be fairly severe, and it can be administered by a transdermal skin patch. However, because Scopolamine blocks the onset of the symptoms, it also blocks the adaptation to motion sickness. Thus it must be taken throughout the period of exposure.

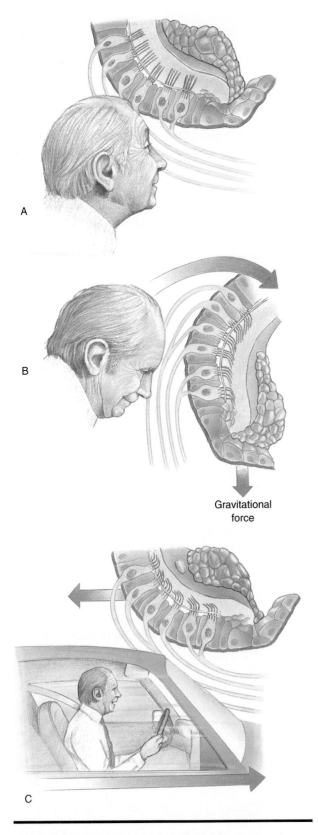

Figure 7.18 How the utricle and saccule sense static position and linear acceleration. (A) Normal position at rest. (B) Activation by a change in static position. (C) Activation by linear acceleration.

Gravitational force

Structure of the Eyes

The eyes receive and transduce light energy. However, their structure reflects the fact that they must do more than just receive and transduce light. First, because we cannot see in all directions at once, our eyes must be able to move so that we can bring our attention to bear on a small portion of the visual field that surrounds us. Thus, our eyes are controlled by muscles that allow them to rotate in their sockets. Second, because light energy varies greatly in intensity, we must have a way to increase or decrease the amount of light that reaches the very light-sensitive photoreceptor cells. This is the function of the **iris,** a muscular structure that determines how much light enters the eye. Third, we must have a way to focus the incoming light, directing it to the photoreceptor cells located at the back of the eye. This is a shared function of the **cornea** and the **lens.** Finally, we must transduce the light energy into action potentials. This final step is the function of the **retina,** the layer at the back of the eye that contains the photoreceptor cells and other nerve cells that send action potentials to the brain.

The general structure of an eye is shown in Figure 7.20. A tough outer coat known as the white of the eye (the sclera) covers most of the eye except in the very front, where it is continuous with the transparent cornea. Light passing through the clear cornea passes through a fluid-filled chamber and then through the **pupil,** the circular hole at the center of the iris. The light then passes through the lens, which focuses the light on the back layer of the eye, the retina. Light reaching the back of the eye activates photoreceptor cells in the retina, which in turn activate other nerve cells. Some information processing occurs in these other nerve cells, but ultimately the information is transmitted from the eye to the brain via the **optic nerve.**

Providing the Proper Amount of Light: The Iris

The iris is a thin, pigmented smooth muscle structure that gives our eyes their color. The iris also determines how much light enters the eye. This function is essential to our ability to detect a wide range of light intensities, because the photoreceptor cells are so sensitive to light that they could be overstimulated by high intensities. The iris consists of two sets of smooth muscles. Radial muscles are aligned along the radius of the eye (like spokes of a wheel), and circular muscles are arranged in a circular fashion around the pupil (Figure 7.21). When the radial muscles contract in response to stimulation of the sympathetic nerves which innervate them, they enlarge the pupil, admitting more light to the eye. In contrast, contraction of the circular muscles (in response to stimulation of the para-

Figure 7.19 The electromagnetic spectrum. Visible light is only a small portion of the entire spectrum.

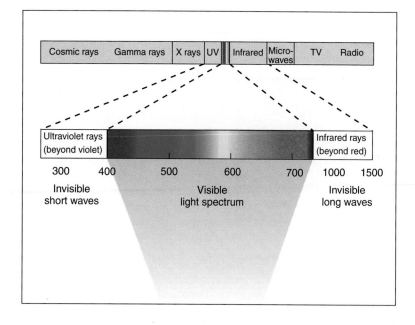

sympathetic nerves) reduces the size of the pupil, reducing the amount of light entering the eye in much the same way that a shutter reduces the aperture of a camera. Note that the opposing sets of muscles are under the control of the opposing branches of the autonomic nervous system. In effect, both enlargement *and* reduction of the pupillary size are under active neural control.

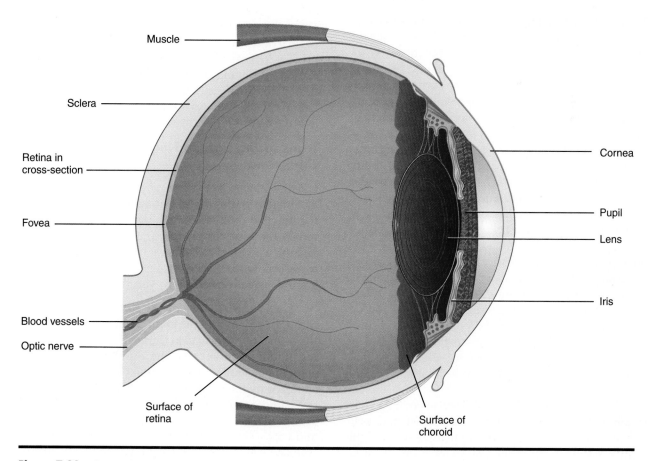

Figure 7.20 General structure of an eye.

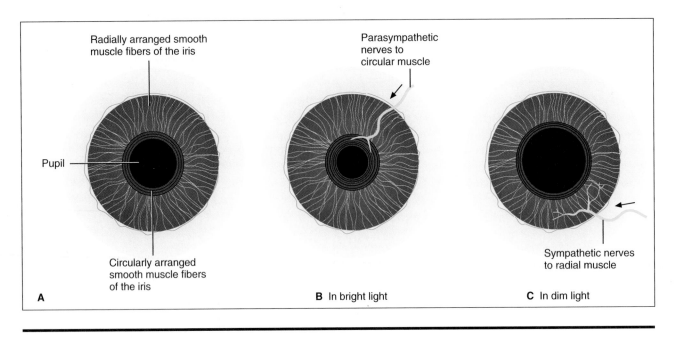

Figure 7.21 The muscles of the iris. (A) The normal condition. (B) In bright light, contracting the circular muscle decreases the size of the pupil and admits less light. (C) In dim light, contracting the radial muscle increases the size of the pupil and admits more light.

Focusing the Light: The Cornea and the Lens

The curved cornea and lens together bend the incoming rays of light so that they can be focused on the photoreceptor cells of the retina. The cornea is responsible for most of the bending of the incoming light. However, the curvature of the cornea is not adjustable; the *regulation* of the degree of bending of incoming light is accomplished solely by adjustment of the curvature of the lens.

The lens of the human eye is composed of transparent fibers that give it elasticity, and thus the ability to change shape. It is attached to a ring of circular muscle (called the ciliary muscle) by connective tissue fibers (suspensory ligaments). Like the circular muscles of the iris, the ciliary muscle contracts to reduce the inner radius of the muscle, reducing the tension exerted by the connective tissue on the lens. This allows the lens to return to its more natural curved shape, bringing nearer objects into focus. In contrast, relaxation of the ring of muscle increases the tension on the lens, stretching and flattening it and bringing more distant objects into focus (Figure 7.22). The ability to change the curvature of the lens so that either near or distant objects can be focused is called **accommodation.**

The most common defect of vision in young people is **myopia,** or nearsightedness, in which distant objects remain out of focus. The cause of myopia is that light from a distant source is brought to focus on a spot in front of the retina, rather than on it, because the eyeball is too long from front to back. As people get older, many develop **presbyopia,** the loss of the ability to focus on objects very near to us. Those who develop presbyopia begin to notice that they must hold objects further away in order to focus on them. The primary cause of presbyopia is a declining elasticity of the lens with age. Thus despite maximal contraction of the circular muscle, the lens may be less able to return to its more curved shape in order to bring near objects into focus.

Transducing Light to Neural Activity: The Retina

Like the other sensory systems, the eyes transduce their adequate stimulus (light) into action potentials. This is the function of the retina. The unique structure of the retina, the types of photoreceptor cells in the retina, and the way these cells interact with neurons allow us to see in color, to perceive accurate images of the world around us, and to adapt to varying light intensities.

General Structure of the Retina Light passing through the lens and the main chamber of the eye hits the retina, the structure at the back of the eye consisting of four layers of cells (Figure 7.23). The outermost layer is a layer of pigmented cells that absorb light not captured by the photoreceptor cells themselves; these cells do not play a role in vision. Next is the layer of photoreceptor cells called **rods** and **cones.** The rods and cones synapse with the third layer, the **bipolar cells,** which pass the information on to the fourth and innermost layer, the **ganglion cells.** Because light en-

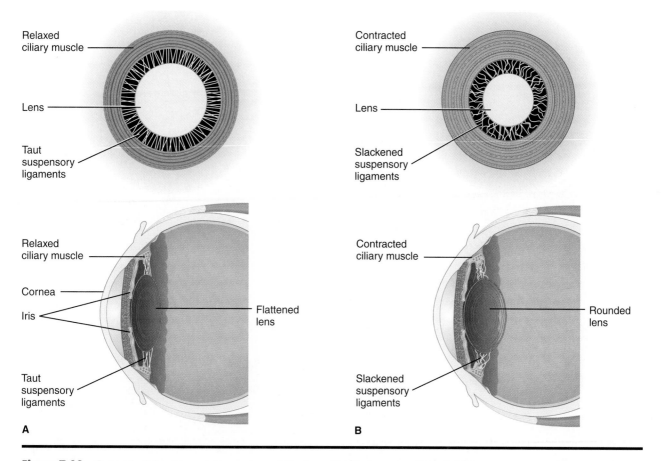

Figure 7.22 Control of the degree of curvature of the lens. (A) To focus on distant objects, relaxation of the circularly arranged ciliary muscle pulls on the ligaments attached to the lens, stretching and flattening the lens. (B) To focus on nearer objects, contraction of the ciliary muscle reduces the tension on the lens, allowing it to return to its natural, more rounded shape.

tering the eye reaches the innermost layer of the retina first, it must pass through the ganglion and bipolar cell layers, and in fact through the cells themselves, to reach the rods and cones. The seemingly backward arrangement of the three innermost layers comes about because of how the retina is formed during embryologic development; it does not significantly hinder the ability of the photoreceptor cells to receive light. Once activated by light, the rods and cones send information back to the bipolar and ganglion cells. The axons of the ganglion cells ultimately carry information to the brain in the form of action potentials. The axons of all of the ganglion cells converge at one site to become the optic nerve, which passes back through the retina in a region called the **optic disc.** Blood vessels supplying the retina also enter the eye through the optic disc (Figure 7.24). Because the blood vessels and the optic nerve pass through the retinal layer, there can be no photoreceptor cells in this region and thus there is a blind spot in our vision from each eye.

Transducing Light into Action Potentials
Light is transduced into graded potentials by the rods and cones. The structures of the rods and cones are shown in Figure 7.25 on p. 209. One end of the cell consists of a series of flattened discs arranged to form either a rod-like or cone-like structure, as the names imply. The flattened discs contain **photopigments,** molecules that change shape when they absorb energy in the form of light. In addition (and this cannot be seen on any microscopic examination), rods and cones have a rather unique membrane potential. Their membrane potential at rest (in the dark) is considerably less negative (closer to zero) than most other cells. This is because they have many sodium channels that are open at rest, causing a constant leak of sodium into the cell. Furthermore, the less-negative membrane potential is accompanied by the constant release of a neurotransmitter from the synaptic endings of the rods and cones. With this in mind, let us now see what happens when light enters the eye:

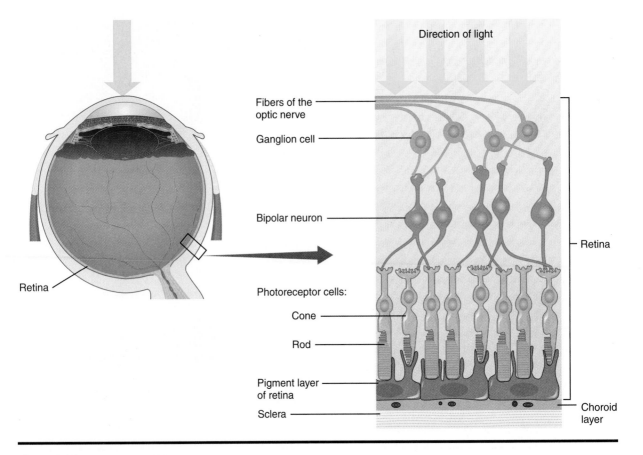

Direction of light

Fibers of the optic nerve

Ganglion cell

Bipolar neuron

Retina

Photoreceptor cells:

Cone

Rod

Pigment layer of retina

Sclera

Retina

Choroid layer

Figure 7.23 Cell layers of the retina.

1. Light falling on the discs of a rod or cone is absorbed by the photopigment molecules. The absorption of energy causes the photopigment molecule first to undergo a change in shape and then to break apart into two separate molecules.

2. Through a series of chemical steps, the change in shape of the photopigment molecules closes the cell membrane sodium channels that are normally open in the dark. This causes a graded hyperpolarization of the cell membrane. The degree of hyperpolarization is a function of the intensity of the light stimulus.

3. Graded hyperpolarization of the cell membrane causes a graded decrease in the normal rate of release of neurotransmitter substance from the synaptic endings of the rods or cones.

4. The transmitter released by the rods or cones is inhibitory to the bipolar cell. Thus, when light activates the rods or cones, less inhibitory transmitter is released and the bipolar cells are activated (depolarized) in a graded fashion.

5. Graded depolarization of bipolar cells causes them to release a stimulatory neurotransmitter, producing a graded, depolarizing potential in the ganglion cell.

6. The graded potential in the ganglion cell produces action potentials. Note that the ganglion cell is the first in the sequence to transmit information over long distances (to the brain) via action potentials rather than via local, graded potentials.

In essence, rods and cones function in reverse of most other receptor cells or nerve cells: They are relatively depolarized and have their highest rate of neurotransmitter release *at rest* rather than when they are activated by their adequate stimulus. Nevertheless, light ultimately causes action potentials in the optic nerve because it inhibits the release of an inhibitory neurotransmitter. In other words, a double-negative produces a positive result.

Visual Acuity and Sensitivity The rods and cones are quite different with respect to their importance in providing the brain with a highly detailed and accurate image (acuity) and their ability to respond to very dim light (sensitivity). Rods are more important in very dim light, whereas the cones are used primarily in bright light. However, the cones provide for greater acuity. These differences are caused not only by differences in the sensitivities of the rods and cones themselves (the rods are about 300 times as

HIGHLIGHT 7.2

Focusing: How the Eye Is Different From A Camera

Light from a distant source arrives at a camera or the eye as nearly parallel rays. These rays must be focused on a common point to obtain a clear image. This can be done with a lens, a curved structure that bends the rays of light as they pass through it. In a camera and in the eye, a lens bends the parallel rays so that they all fall on a common point, called the focal point, and the image is in focus. The focal point in a camera is where the film is placed. The focal point in the eye is the retina.

In contrast to light from a distant source, light from a nearer source (say, less than 20 feet) is still radiating out from its source when it arrives at the camera or the eye. The same lens that focused the more parallel rays from the distant source would now have the rays from the nearer source out of focus if the focal distance (the distance from the lens and the focal point) were the same.

Here is where the camera and the eye differ. Because the camera uses a glass lens with a fixed curvature, focus can be reestablished only by adjusting the distance between the lens and the focal point.

In contrast, the structure of the eye is such that the distance between the lens and the retina of the eye must remain constant, but the lens itself is sufficiently flexible that the curvature of the lens can be altered by muscles. Thus the eye maintains focus by adjusting the curvature of the lens rather than by adjusting the distance between a fixed lens and the focal point.

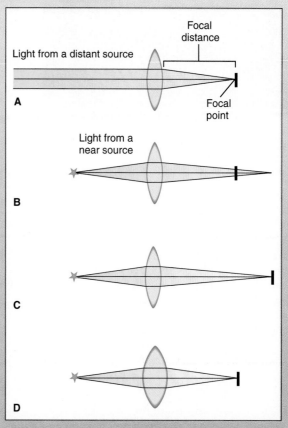

(A) How a lens focuses light from a distant source. (B) What happens when light from a near source strikes a lens positioned to focus light from a distant source. (C) A camera adjusts focus by adjusting the distance between the lens and the film. (D) An eye adjusts focus by adjusting the curvature of the lens.

sensitive as cones), but also in how they are connected to bipolar cells and ganglion cells.

The human eye may have 120 million rods, 6 million cones, and only about 1 million ganglion cells. The rods and cones are not distributed evenly throughout the retina (Figure 7.26). At the center of the retina is a region called the **macula** that contains primarily cones. The **fovea** is a small region at the very center of the macula that contains only cones. In the fovea, the neural connections do not converge; one cone connects to one bipolar cell, which in turn connects to one ganglion cell. In addition, at the very center of the fovea the bipolar cells and ganglion cells are pulled to the side so that light falls directly on the cones. In contrast, the more peripheral regions of the retina contain primarily rods. Many rods may converge on each bipolar cell, and many bipolar cells may converge on a single ganglion cell. Convergence of the rods is why the areas of the retina farthest away from the fovea, where most of the rods are located, are more sensitive to weak light. Recall that an action potential occurs in a neuron (in this case a ganglion cell) only when a graded potential is produced that is of sufficient magnitude to pass a threshold. In the eye, the summation of the activity of many very-light-sensitive rods (and thus many bipolar cells) brings a ganglion cell to threshold much more easily than stimulation by a single less-light-sensitive cone.

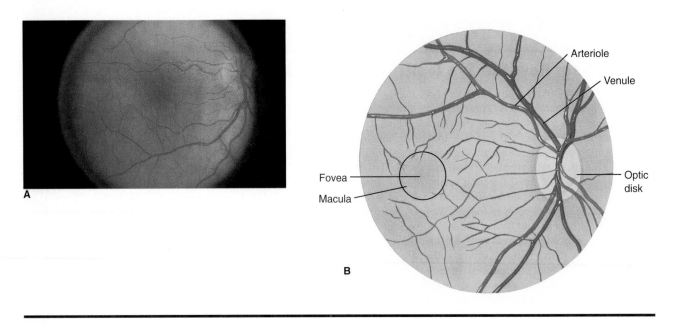

Figure 7.24   The retina. (A) A photograph of how the retina appears through an ophthalmoscope. Blood vessels are visible on the surface. (B) A schematic representation of the same view. The optic disc is the region where the axons of the ganglion cells converge and pass back through the retina, forming the optic nerve. Blood vessels converge at this point as well. The fovea, a region made exclusively of cones, is at the center of the less-distinct macula.

The difference in the "wiring" of rods and cones to the ganglion cells also explains why cones provide more detailed information about the image (have a higher acuity), provided that the light is sufficiently strong to activate the less-light-sensitive cones. When a ganglion cell connected to a cone sends an action potential to the brain, the brain can determine precisely from which cone the information came because the neural connection is direct. In contrast, an action potential in a ganglion cell connected to many rods may have come from strong activation of a single rod or from weak activation of many rods.

When we look directly at an object, we are focusing the image on the fovea to take advantage of the fact that the acuity of the image is highest there. As a demonstration, focus on a single object and, while still focusing on it, try to describe the detail, or even the exact color, of another object off to the side of your point of focus. Notice how inaccurate your image is. This is because the peripheral regions of the retina contain mostly rods. A second demonstration will reveal the greater sensitivity of rods to dim light. Go outside on a clear night, focus on a single spot in the sky, and try to pick out a dim star that can be seen in the periphery of your vision. Now focus directly on it; notice that it gets weaker or disappears entirely.

Color Vision   Humans see in color, but not all animals do. Humans are able to perceive color because we have three kinds of cones, each with a different photopigment. The different structures of the three photopigments causes each to absorb light energy of a different wavelength (color). Figure 7.27 shows the absorption curves for the three types of cones (blue, green, and red). Note that the names of the cones do not necessarily correspond to the color of light that activates them best. Although the blue cones respond best to blue light, the red cones actually respond best to orange light. Nevertheless, they are called red cones because they are the only ones that are activated at all by red light. Similarly, the green cones are the best of the three at responding to green, even though they are most easily activated by yellow light. Our ability to distinguish all of the other various colors is due to the way the brain interprets the *ratios* of action potentials coming from the various ganglion cells connected to the three different types of cones (blue, green, and red). The yellow color indicated by the dashed line in

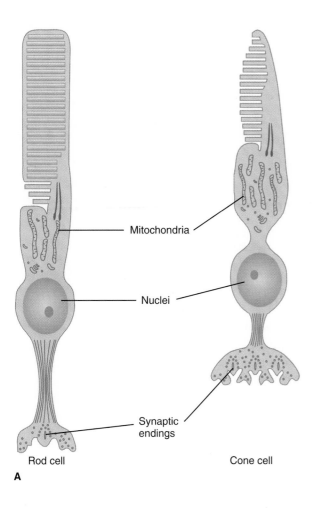

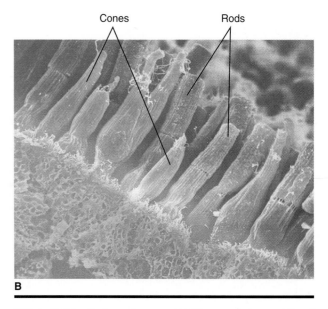

Figure 7.25 Rods and cones. (A) The general structure of rods and cones. (B) An enhanced scanning electron micrograph of rods and cones.

Figure 7.27 is not absorbed at all by the blue cones but is absorbed equally by the pigments of the green and red cones. The brain then "interprets" the resultant ratio of action potentials from the ganglion cells connected to the three cones (0:84:82) as yellow. The specific green color shown in the figure is represented by a ratio of 37:66:30, blue as 97:0:0, red as 0:0:50, and so on for the myriad colors we can distinguish. The perception of white light is caused by activation of all three types of cones by light of many different wavelengths; there is no specific wavelength corresponding to white light.

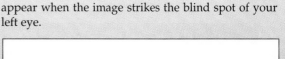

Demonstration of the Blind Spot

You can reveal the presence of the blind spot by doing the following exercise. With your left eye closed, hold the image at the right about 4 inches directly in front of your face while focusing on the left-hand image (the square) with your right eye. Now move the book slowly away from your face until the circle disappears. The circle disappears because at this distance from your eye, the image of the circle is striking the blind spot of the right eye. You can do the same demonstration with your left eye by focusing on the circle with your right eye closed. In this case the square will disappear when the image strikes the blind spot of your left eye.

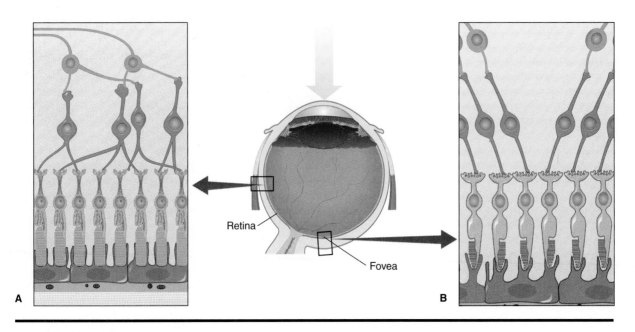

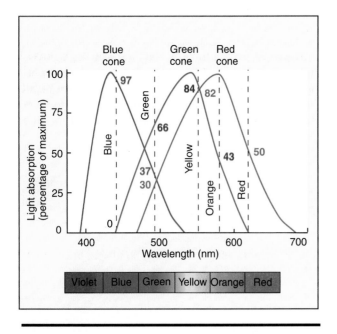

Figure 7.26 Differences in the neural connections between rods and cones. (A) In the periphery, where the photoreceptor cells are nearly all rods, there is marked convergence of neural input such that many rods converge on a single bipolar cell and many bipolar cells converge on a single ganglion cell. This is the region of highest sensitivity to light. (B) In the fovea, where there are only cones, there is no convergence of neural information. In addition, the bipolar cells and ganglion cells are pulled to the side so that the light can fall directly on the cones. This is the region of highest visual acuity.

Figure 7.27 The absorption curves of the photopigments of the three types of cones.

The single photopigment in rods, called **rhodopsin,** also has a certain wavelength of light to which it responds best. The peak of light absorption sensitivity for rhodopsin is about 500 nm, a green color (Fig-

ure 7.28). Recalling that the rods are roughly 300 times more sensitive than are the cones, and thus only the rods are activated by dim light, one might expect to see a pure green color in dim light. However, dim lights of wavelengths shorter or longer than 500 nm still can have only one effect, and that is the generation of action potentials in the ganglion cells connected to the rods. As Figure 7.28 illustrates, in dim light the brain has no way of distinguishing between lights of 450 nanometers (yellow) and 550 nanometers (blue); both provide the brain with exactly the same number of action potentials from the rods, and neither is sufficiently strong to activate the less-sensitive cones. In other words, without the ability to interpret *combinations* of action potentials originating from different receptor cells, the brain is limited to interpreting any action potential as just like any other. In this case they are all interpreted simply as white light. To have color vision, then, we must have at least two different pigments with roughly the same sensitivity to light. Of course, if we *did* have only two cone pigments instead of three, we would see colors but not the range of colors we are used to seeing.

Table 7.3 summarizes the differences between rods and cones we have described above.

Adaptation An important feature of our vision is the ability to see light over a very wide range of light intensities. In part, this is brought about by adjusting

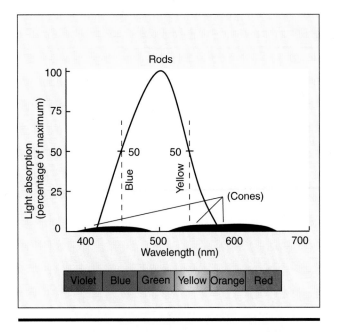

Figure 7.28 The absorption curve of rhodopsin, the photopigment of rods. Rhodopsin is nearly 300 times more sensitive to light than are the photopigments of the cones, so the maximum absorption for the rods occurs in light so dim that it would hardly activate the cones at all. The absorption curves for the photopigments of the cones are exaggerated slightly so that they can be seen on this scale.

the amount of light entering the eye by changing the size of the pupil, as you have already learned. In addition, the rods and cones themselves can adapt to light of very different intensities.

Adaptation makes sense when you understand that the transduction of light energy "uses up" the intact photopigments, breaking them down into two other molecules. These molecules can be resynthesized into photopigments again, but the synthesis takes some time. Thus, when you have been out in the bright light, the very-light-sensitive rods have already had some of their rhodopsin used up. When you then enter a dimly lit room, not enough rhodopsin is available to respond in the usual fashion. With time (about 5 to 15 minutes), the rhodopsin is resynthesized from the two breakdown products and once again you can see in the dim light. Conversely, when you go outside into bright light from a dimly lit room, the light seems very bright because you have the maximal amount of intact photopigments available in both the rods and cones. Over time, sufficient photopigment is broken down, particularly from the more-sensitive rods, so that less photopigment is available to respond to the bright light. In other words, photoreceptor adaptation takes place because of the natural tendency for an inverse relationship between the intensity of the light and the amount of photopigment available to receive it.

An additional factor in the ability to adapt to dim light is the availability of vitamin A. Vitamin A is readily converted to one of the molecules used to form intact photopigments. This is why it is said that carrots, high in vitamin A, are good for the eyes. Severe vitamin A deficiency can lead to night blindness, or the inability to see in very dim light. Although vitamin A affects the ability to synthesize the photopigments of the cones as well as the rods, there is generally enough pigment in the cones to respond to very bright light.

Table 7.3	A comparison of rods and cones.	
	Rods	**Cones**
Number per eye	120 million.	6 million.
Location	Primarily in the peripheral regions of the retina.	Primarily in the center of the retina (the fovea).
Neural connections	Convergence of many rods onto a single neuron (bipolar cell).	Little convergence. Some neurons (bipolar cells) receive input from only a single cone.
Sensitivity	High.	Low.
Acuity	Low.	High.
Primary function	Vision in dim light.	Vision in strong light.
Visual perception	Shades of grey.	Color vision.
Number of different types	One. All rods have the same photopigment, called rhodopsin.	Three. Each of the three has a different photopigment, so each is maximally responsive to a different wavelength of light.

HIGHLIGHT 7.4

Color Blindness

Color blindness is a general term for the inability to distinguish certain colors. Most forms of color blindness might better be described as color defectiveness or color deficiency because they are brought about by a deficient number of one or more of the three types of cones; such people can see most colors, but certain colors are hard to distinguish from each other. Less common is a type of color blindness in which the person completely lacks one of the three types of cones. Such people also see many colors, but not the full range seen by most people. The inability to see any color at all is quite rare in humans, for it occurs only when two of the three types of cones are completely missing.

Many types of partial color blindness are inherited traits. The most common inherited form of color blindness is red–green color blindness, brought about by an inherited deficiency or lack of either the red or the green cones. Such people have difficulty distinguishing shades of color in the green to red range, and thus they are unable to distinguish green from red or orange from yellow (review Figure 7.27).

The inherited form of red–green color blindness is fairly common in men but relatively rare in women. Nearly one in twelve males has difficulty distinguishing red from green, whereas fewer than one in 100 women are affected. The explanation for the gender differences is that the genes that encode for each type of cone are located on the X chromosome. Females have two X chromosomes (XX) whereas males have only one

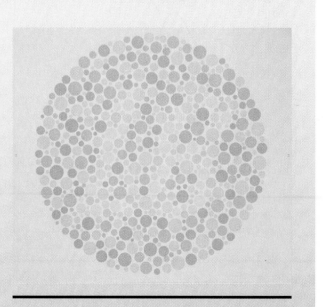

(XY). Because one normal gene is all that is necessary to produce normal cones, females are color-blind only if they inherit defective X chromosomes from *both* of their parents, whereas males are color blind if they inherit the defective X chromosome from their mothers.

Color blindness is easily tested by means of colored charts such as the one above. Many red–green color-blind or color-deficient people cannot tell that there is a number in this chart.

Summary

The sensory systems described in this chapter provide us with detailed information about the external world. In general, each sensory system transduces (or converts) a particular type of sensory input into neural signals, the only kind of information that can be processed by the brain. In the skin, thermoreceptors respond to heat or cold and a variety of mechanoreceptors respond to light touch, pressure, vibration, and pain. Other mechanoreceptors in muscles, joints, and tendons monitor the position of our limbs and the tension in our muscles. Chemoreceptors located on our tongue and in our nasal passages determine our senses of taste and smell, respectively. Our ears and eyes are specialized for transducing sound (hearing) and light (vision). Associated with the inner ear is the vestibular apparatus, a structure designed specifically to determine the position and motion of the head.

Although each of these sensory systems is uniquely suited to transduce only one type of stimulus, the first step in the signal transduction process is remarkably similar for many of the sensory systems. In general, the receptor cell or nerve ending is specialized so that it responds best to one type of stimulus, which alters the permeability of the receptor cell membrane to certain small ions. The change in membrane permeability in turn produces an alteration of the membrane potential (the electrical charge across the cell membrane) in the vicinity of the stimulus. From this point, mechanisms differ depending on whether the receptor region is part of a modified neuron capable of an action potential, or whether it is part of a specialized receptor cell incapable of generating an action potential. In the latter case, the graded potential produced by the receptor cell causes the release of a chemical transmitter at a synapse with a neuron, and an action potential is thus generated in the neuron and sent to the brain.

Thus, all sensory information, regardless of its type, ultimately is encoded as action potentials. Stimulus intensity is encoded in the number of action potentials generated and by the number of neurons active at any one time. However, stimulus intensity is not always transmitted to the brain accurately because many receptor types adapt to a continuously applied stimulus, responding less over time or even ceasing to respond at all. The brain, too, can selectively process information, concentrating on one sensory input at the expense of another. In addition, the sensitivities of most sensory systems can be adjusted so that they can respond to a wide range of stimulus intensities. The ability to adapt to a continuously applied stimulus, to adjust the sensitivity to sensory input, and to concentrate on one sensory input at the expense of another confers survival value by preventing sensory overload while allowing fine discrimination of sensory detail.

Questions

Conceptual and Factual Review

1. What are the four types of receptor cells, based on the type of signal they receive? (p. 180)
2. What is a transducer? (p. 181)
3. In general, what happens at the level of the cell membrane when a sensory receptor is stimulated? (p. 181)
4. How is the brain able to determine stimulus type? Stimulus intensity? (p. 181, p. 183)
5. What is meant by receptor adaptation? (p. 184)
6. What types of receptors are present in the skin? Which of these do not adapt? (p. 185, p. 186)
7. What is the function of sensory receptors in joints? (p. 187)
8. What is a muscle spindle, and what is its function? (p. 187)
9. What are the Golgi tendon organs designed to monitor? (p. 187)
10. What are the four taste qualities to which our taste buds respond? (p. 189)
11. Taste bud cells are replaced rather frequently. Why? (p. 191)
12. What is the function of the eustachian tube? (p. 193)
13. What is the function of the muscles attached to the bones of the middle ear? (p. 193)
14. What is the name of the cells in the cochlea that transduce sound waves into receptor potentials? Where in the cochlea are they located? (p. 194, p. 195)
15. How do we determine from which direction a sound is coming? (p. 196)
16. Which structures sense movement of the head? (p. 196)
17. What is the mechanism for determining the static position of the head? (p. 199)
18. What is the function of the iris of the eye? The lens? (p. 202, p. 204)
19. How is the degree of curvature of the lens adjusted? (p. 204)
20. What is presbyopia, and what causes it? (p. 204)
21. Which of the cells of the retina are specialized for the transduction of light energy? (p. 205)
22. Why do we have a blind spot in the vision from each eye? (p. 205)
23. What is a photopigment? (p. 205)
24. Of the two types of photoreceptor cells, which are responsible for color vision? Which are the most sensitive in dim light? (pp. 206–208)
25. Why is our visual acuity best at the very center of our vision (when we look directly at an object), whereas the sensitivity to dim light is greater at the periphery of our vision? (p. 208)
26. Why is vitamin A important to the ability to adapt to very dim light? (p. 211)

Applying What You Know

1. With her eyes closed, would an astronaut in a weightless environment be able to touch her nose with her index finger? Would she be able to detect movement or static position of the head with her eyes closed? Explain.

2. How would you explain the finding that dogs are completely color-blind (they see only in black and white)?

3. Why might you expect the repeated crashing of cymbals to cause more severe loss of hearing than would an equally loud but more sustained source of sound, such as a trumpet?

4. Why do you suppose that you are not normally aware of the blind spot in each eye?

Chapter 8

Cardiovascular System: Heart

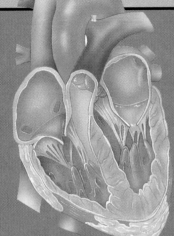

Objectives

By the time you finish this chapter you should be able to

◆ Describe the anatomy of the heart and the circulatory system, and be able to describe the function of each component.

◆ Understand that the movement of blood depends on the heart acting as an effective pump.

◆ Understand the factors that determine both systolic and diastolic pressures.

◆ Describe how systolic and diastolic blood pressures are measured in humans.

◆ Know what causes the heart sounds.

◆ Understand how the heart rate is controlled.

◆ Understand the relationships that exist between the heart rate and force with which the heart pumps blood, the resistance against which it pumps, and the flow through the circulatory system.

Outline

Introduction

During the seventeenth century, the theories of the third-century physician Galen were still accepted as fact. He thought that blood was like the tides of the sea, ebbing and flowing in the arteries, first away from the heart, then toward the heart. This movement carried with it the "humors," good and bad, that determined one's well-being. Should disease strike, physicians would bleed the patient, as this was thought to drain off the corrupted humors and restore health. Conclusions regarding function were arrived at by reasoning from the old beliefs rather than from any experimental evidence. Not until the experimental approach, or scientific method, came into being could any significant change in thinking occur.

William Harvey, a seventeenth-century physician, was the first to realize that Galen's views regarding the role of the heart could not be correct. Using a variety of animals, such as frogs, cats, and dogs, Harvey carefully studied the anatomy of the heart, always asking, "What is its function?" He tried to make out how the heart beat, but when he looked at the beating hearts of mammals, he was unable to discern anything of their motion. He said they beat in a "twinkling." However, when he studied the hearts of frogs their beat was so slow that he noticed that the **atria,** the small, thin, muscular upper chambers of the heart, beat a moment before the single **ventricle** of the frog heart. The ventricle is the lower, thick muscular chamber of the heart. (Notice in Figure 8.1 that the mammalian heart has two ventricles.) This fact suggested to him that blood flowed from atrium to ventricle with each beat of the heart.

Harvey was the first to show that the heart contains valves forcing one-way movement of fluid, a finding incompatible with an ebb and flow and suggestive of a circular motion of blood. (This will be seen in more detail later.) Nevertheless, some physicians argued that the heart or the lungs kept supplying new blood to the system. However, Harvey measured the volume of the heart chambers and counted the number of times that the heart contracted in one hour. He calculated that if the heart emptied only half of its estimated volume with each beat, the entire quantity of blood contained in the body would be pumped in a matter of minutes. These findings forced him to propose that blood moved in a circular pattern in the body. We now know that Harvey was right; in one minute the heart pumps the entire blood volume, which is approximately five liters, or slightly more than five quarts.

The simple but elegant set of experiments Harvey performed changed the history of medical science. Blood was now shown to circulate rather than ebb and flow. Harvey showed how the left side of the heart pumps blood into the **aorta,** the large, thick-walled vessel that arises from the heart. The aorta branches into smaller and smaller arteries until blood reaches the smallest vessels, the **capillaries.** It is across the thin, porous, capillary walls that the transfer of nutrients and waste products occurs. Capillaries merge into successively larger, thin-walled vessels known as **veins,** which return the blood to the right side of the heart. When the right ventricle pumps blood to the lungs, circulation is completed. After flowing through the capillaries of the lungs, the blood returns to the left side of the heart. The cardiovascular system is outlined in Figure 8.1. Although the pulmonary system is less extensive than the systemic system, the same amount of blood is pumped by the left and right sides of the heart.

However, Harvey found no evidence of the presence of vessels so small as to escape unaided vision. Yet the presence of small vessels was necessary to link the arteries to the veins. Harvey simply referred to the "porosity" of tissue in order to make his theory correct. It was not until anatomist Antonie van Leeuwenhoek made the first simple high-power microscope with good resolution (Figure 8.2) that capillaries were discovered. Another anatomist, Marcello Malpighi, was the first to view them. He studied the lungs of a living frog in 1661, and was able to show that small vessels connect the arteries with the veins. Others soon followed up these observations, demonstrating that capillaries exist in all tissues. Although the circular motion of blood was established in the seventeenth century, this discovery did not immediately eradicate the ancient ideas regarding evil and good humors. That giant step in thinking had to wait two more centuries.

Although a circular motion of blood was established, it was also recognized that the rate of blood flow through different parts of the body must change as the requirements for energy sources change with activity. An active muscle requires more oxygen and energy substrates than an inactive muscle. Because it is the blood that supplies muscle with these molecules, the blood supply must vary as the demand for energy changes. But how does the circulatory system keep the pumping rate of blood at the appropriate level so as to meet changing requirements? And how can the needs of an individual organ be met when its requirements are altered? These are the questions that confronted modern researchers.

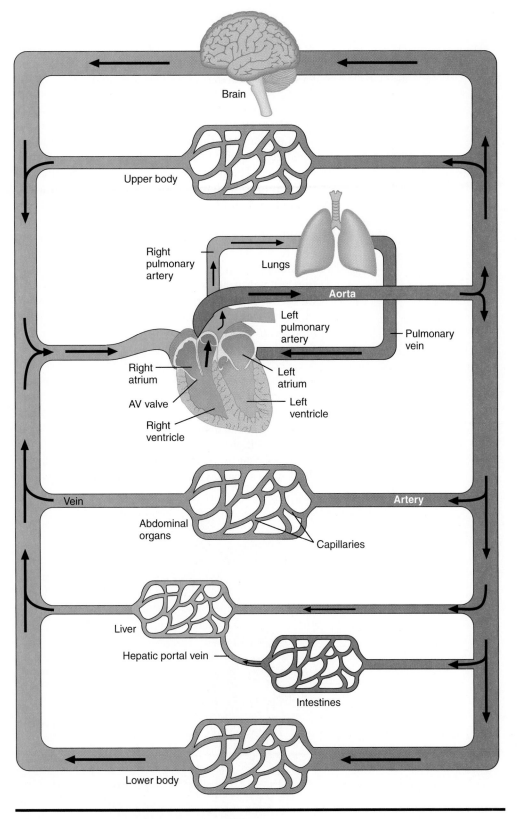

Figure 8.1 Schematic representation of the cardiovascular system. Note that there are two separate circuits: the pulmonary circuit supplying the lungs and the systemic circuit supplying the remainder of the body.

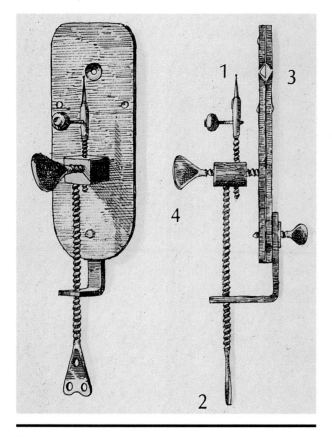

Figure 8.2 Photograph of one of Antonie van Leeuwenhoek's microscopes, used to study blood cells and spermatozoa.

The Heart as a Pump

Ever since Harvey described the circulation of the blood, we have known that the heart acts as a pump, forcing blood through all the vessels in the body. In order to understand just how this is accomplished, we must understand the anatomy of the heart and its associated vessels. Figure 8.3 illustrates the internal structures. The heart is a hollow, muscular organ divided into four chambers: the left and right atria and the left and right ventricles. Separating the left and right sides of the heart is the septum, effectively partitioning the heart into two pumps, as diagrammed in Figure 8.3. The muscular wall of the left ventricle is considerably thicker than that of the right ventricle, as the left ventricle pumps blood into the aorta and thence to all parts of the body and at a higher pressure than the right ventricle. The right ventricle pumps blood into the pulmonary artery, which then flows through the shorter pulmonary circuit.

Perhaps the most significant finding is that there are four one-way valves in the heart. Two are between the atria and ventricles and two are located at the junction of the right and left ventricles and the pulmonary artery and aorta, respectively. If the excised heart of an animal is filled with any fluid, and if the left ventricle contracts, that fluid is forced out through the aorta. No fluid flows backward into the atria. The valves

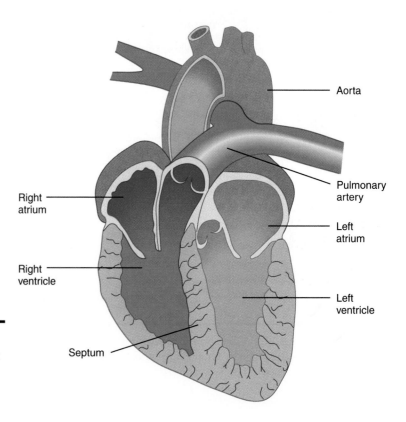

Figure 8.3 Anatomical cross-section of a human heart. Note the location of the four heart valves and the difference in thickness between the left ventricular muscle and the right ventricular muscle.

MILESTONE 8.1

Demonstration of Venous Valves

During the course of his research, Harvey described other properties of the cardiovascular system. He demonstrated that the veins in the arms and legs also contain valves. Using the simple technique demonstrated in the figure below, he showed that blood in the veins can move only in the direction of the heart; a reverse flow is not possible. The experiment, as pictured in the figure, is as follows: (1) Place a loose tourniquet around the upper arm, applying sufficient pressure to stop the venous flow of blood but not that of arterial blood. (Arterial pressure is higher than venous pressure.) The veins will bulge as they fill with blood that cannot continue the journey to the heart. Along the veins you will notice a series of small lumps or nodules,

each of which marks the location of a valve. (2) Have a second person place a finger over one vein between two valves, at point H, compressing it so that no blood from the lower part of the arm flows beneath the finger. (3) With another finger, squeeze the vein as with a tube of toothpaste, so the blood is forced away from the finger at point H and beyond the valve at point O. The vein will remain empty as long as the lower finger is in place. Note that blood flows back only until it reaches the upper valve at point O. Back flow is prevented beyond that point. If the lower finger is removed blood flows from below, filling the veins completely up to the tourniquet.

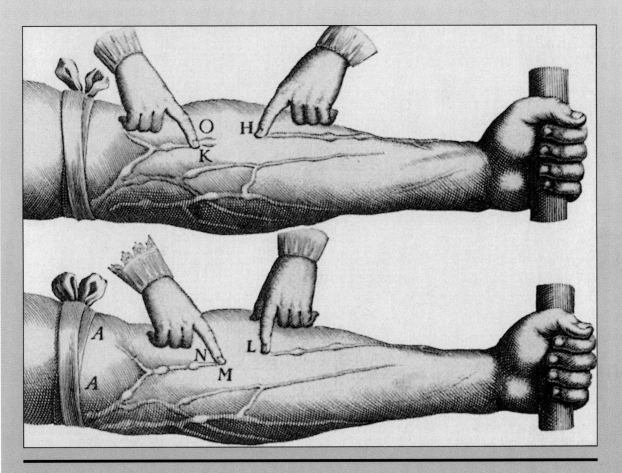

Harvey's experiment demonstrating the presence of valves in the veins. Valves are located at points H and O.

responsible for this unidirectional flow are remarkably well-made. During the course of an average human lifetime, the heart will contract about 3 billion times, yet those valves rarely wear out or leak.

Figure 8.4 illustrates in detail the interior anatomy of the heart. The leaflets of the valves are connected to muscular tissue known as the **papillary muscles** by tendons called the **chordae tendinae.** As the ventricle contracts, so do the papillary muscles, preventing the valves from being forced into the atrium and leaking. However, there are some disease states in which leaky valves do occur and cause significant difficulty. You will see later the effect this might have on one's health.

Although the atria contract and force blood into the ventricles, atrial contraction is not responsible for the major filling of the ventricles. As blood continually flows into the atria from the veins, it also passes directly into the ventricles. During the period of ventricular relaxation or **diastole,** the ventricle continues to fill. At the beginning of atrial contraction, the atria empty their contents into the ventricles, filling them

more. Immediately after the atrial contraction, the ventricles contract, in **ventricular systole,** forcing blood into the pulmonary artery and aorta. Blood is prevented from flowing back into the atria during ventricular systole by the presence of the valves between the atria and ventricles, the left and right **atrioventricular** (AV) valves. Following ventricular systole, as the ventricles relax, blood is prevented from flowing back into the heart by the aortic and pulmonary valves, or **semilunar valves.** This arrangement of the four valves ensures that blood can move in only one direction: from the right heart via the **pulmonary artery** to the lungs, to the left heart via the **pulmonary veins,** and then to the aorta. Although the arrangement of valves prevents back flow, the ventricles do not empty their entire contents into the aorta or pulmonary artery by the end of ventricular contraction. There is always a residual volume left in the ventricles, called the **end diastolic volume.**

The total volume of blood pumped into the aorta each minute in the average person at rest is approxi-

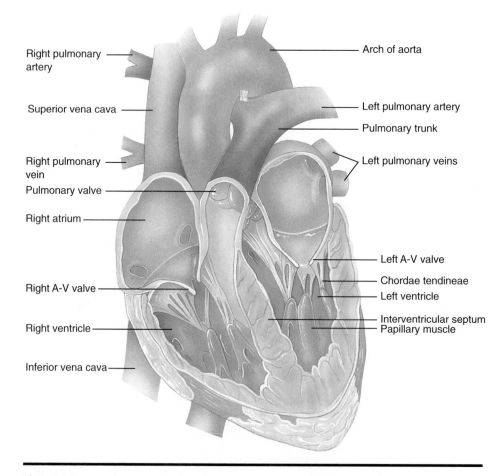

Figure 8.4 Detailed internal anatomy of the heart. Note the location of the chordae tendinae and the papillary muscles. As the ventricle contracts, they prevent the AV valves from inverting into the atria. The right and left ventricles pump blood into the pulmonary artery and aorta, respectively.

mately five liters. This is called **cardiac output.** Consider the two variables that determine the cardiac output: The volume of blood pumped by the heart per minute must be the product of the volume pumped with each beat (**stroke volume**) times the number of beats per minute (**heart rate**). In mathematical terms, this is stated as

Cardiac output = Stroke volume × Heart rate

The cardiac output is determined by the heart rate, the force of contraction, and the **total peripheral resistance** against which the heart pumps. The total peripheral resistance is the absolute resistance to blood flow from the left ventricle to the right atrium. These factors, in turn, determine the mean blood pressure.

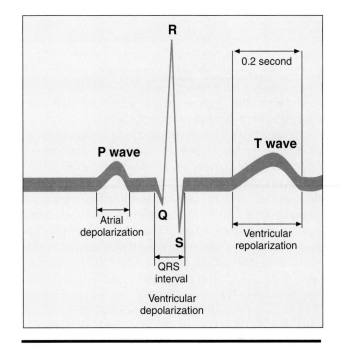

Figure 8.5   The electrical events (recorded on an EKG) during a single cardiac cycle.

Cardiac Cycle

The sequence of events that occur from one beat of the heart to the next is called the **cardiac cycle.** As the term *cardiac cycle* implies, the events are repetitive. Thus, one may begin to describe it at any stage in the cycle. For simplicity, it is easiest to begin at the start of atrial contraction and follow the cycle to the beginning of the next atrial contraction. A record of the cardiac cycle may be divided into mechanical and electrical events.

To monitor the workings of the heart, small hollow-bore tubes called **catheters** are placed into the left atrium, ventricle, and aorta, and are connected to a pressure-measuring device. At the same time, one can also record the heart sounds and the electrical activity of the heart by connecting electrical leads from both arms and one leg to an **electrocardiograph** machine, abbreviated **ECG** or **EKG.** (The reason for the use of *EKG* is that the *K* comes from the German spelling of *electrocardiograph,* as the first recordings were made by the Nobel Prize winner Wilhelm Einthoven.) The record of electrical activity is called an **electrocardiogram.**

Electrical Activity

The tracing in Figure 8.5 shows the electrical activity recorded during a cardiac cycle. Just before the end of the diastolic period, when the heart is filled with blood, one can record a small rise followed by a decline in the electrical activity. This is called a **P-wave** and it precedes contraction of the atria by a very brief period, reflecting a wave of depolarization spreading relatively slowly over both atria. The presence of a

clearly formed wave of electrical activity indicates that the depolarization of the muscle must originate in a single spot and then spread in some fairly uniform manner over the atria. If the depolarization were initiated in a random manner, or at different places along the atria simultaneously, one would not see a smooth wave form.

The question then is, "Where does the depolarization originate?" By carefully probing the heart muscle with electrodes, one finds that a small area in the upper part of the right atrium depolarizes first. Because heart muscle is a syncytium, that is, each fiber is connected to at least one more fiber by tight or communicating junctions (see Chapter 6), a single point of depolarization initiates a spreading wave over the entire atrium. The depolarization spreads just as it does along a single striated muscle fiber. Because all the cardiac fibers are interconnected, the entire muscle is activated by a single focus of depolarization.

A very small, well-delineated region in the wall of the right atrium near the entrance of the veins is responsible for initiating the electrical activity of the cardiac cycle. This specialized area is known as the **sinoatrial node,** or **SA node.** The tissue spontaneously goes through a cycle of slowly rising membrane potential from about –60 mV to about –40 mV. At –40 mV the threshold for an action potential is reached and the cells then show a rapidly rising potential to near +20 mV. At this point, the cells begin repolarization to −60 mV and repeat the cycle. This excitable area has no single, stable membrane potential as do most cells, but rather drifts toward the threshold potential that sets off an action potential. Thus, the SA node acts as a **pacemaker,** for its constant drift in potential starts the entire sequence of events, and it is responsible for a rhythmically beating heart.

Soon after atrial contraction, the ventricles contract, forcing blood into the aorta and pulmonary artery. In order to expel blood into the aorta efficiently, the ventricular muscle must contract as a unit rather than with a wave of contraction spreading over it, as happens in the atria. At the junction of the atria and ventricles is a node similar in some respects to the SA node, known as the **atrioventricular node,** or AV node, a system of conducting fibers that serves to spread the electrical activity very rapidly over the entire ventricle. The specialized group of conducting fibers leaving the AV node is the **bundle of His,** which runs down either side of the inner wall of the heart (the septum) that separates the two ventricles. The left

and right branches of the bundle of His, in turn, branch into many smaller fibers, called the **Purkinje system.** The specialized conducting system of the heart is shown in Figure 8.6.

The Purkinje system distributes the electrical depolarization simultaneously throughout the left and right ventricles, ensuring that both ventricles beat with the same rhythm and that all the ventricular muscle fibers contract almost simultaneously. This rapidly spreading wave of depolarization is seen in Figure 8.5 as the **QRS-wave.** It is both shorter in duration and of greater magnitude than the P-wave. Last seen on the EKG is the **T-wave.** This relatively small, smooth wave represents repolarization of the ventricular muscle. When the T-wave is completed, the ventricles are ready for another depolarization.

The sequence of events described above implies that the rhythm of the heart is determined by the cyclical nature of the membrane potential of the SA node, and that the heart rate is a constant. That clearly conflicts with everyday experience. The heart rate may

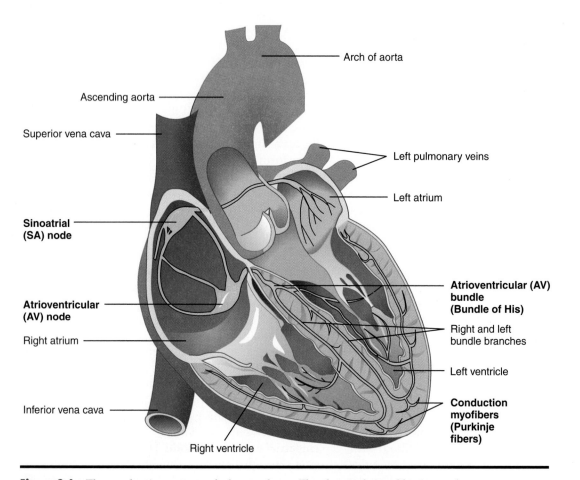

Figure 8.6 The conduction system of a human heart. The electrical signal begins at the AV node, spreading over the atria through the atrial muscle. When the impulse meets the AV node it is spread rapidly by the bundle of His and the Purkinje system through the entire ventricle.

double or even triple during certain conditions. In fact, one of the signs of adequate exercise to improve one's fitness is an increase in heart rate during the exercise period. (An interesting side note is that with training, the resting heart rate is reduced. Well-trained athletes may have an enlarged, muscular heart, just as weightlifters have enlarged muscles, with a heart rate of 50–60 beats per minute, compared to 70+ beats per minute for sedentary people.) How then, does the heart rate change if it is driven by the cyclical changes in the membrane potential of the SA node?

The electrical activity of the SA node is illustrated in Figure 8.7. As stated above, the membrane potential steadily moves toward the threshold value, which sets off the sequence of events resulting in contraction. The diagram illustrates how the rate of action potentials is changed if the drift of the membrane potential of the SA node first slows, and then increases. The slower the drift, the longer it takes for the next action potential to occur. The heart rate, then, is set by the time it takes for the threshold potential to be reached after recovery from the preceding action potential. The more rapidly the membrane potential of the SA node rises to the threshold potential, the more rapid is the heart rate. Conversely, if the rate of change in the membrane potential is reduced, the heart rate is also reduced. This leaves us with the question, "What mechanisms control the heart rate?"

Control of Heart Rate

During early investigations of factors controlling the heart rate, researchers recognized that nerves to the heart play an important role. They anesthetized animals and stimulated the two major nerves supplying the heart: the sympathetic nerve and the parasympathetic nerve, or **vagus nerve.** (Refer to Chapter 4.) If the sympathetic nerve was artificially stimulated the heart rate increased. If the vagus nerve was stimulated the heart rate decreased. In addition, it was shown that the greater the frequency of neural stimulation, the greater was the effect on heart rate. As in so many other systems of the body, two opposing forces exist. In this case, one serves to speed the heart and one serves to slow it down. The heart rate at any moment is determined by the balance between the two nerves. In experiments with anesthetized animals, if the vagus nerve was severed, the heart rate increased. If the sympathetic nerve was severed, the heart rate decreased. Thus, both nerves are active simultaneously, and the balance between them determines the heart rate. However, in a person at rest, the parasympathetic activity is relatively greater than the sympathetic activity. During the day, as circumstances change, so do the relative activities of the parasympathetic and sympathetic nerves. This situation is similar to driving an automobile with one foot on the gas pedal and the other on the brake. Although that

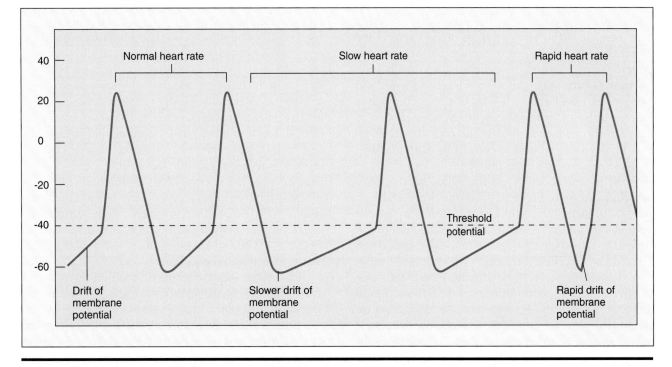

Figure 8.7 Changes in the drift of the membrane potential of the SA node cause the heart rate to change; the steeper the drift, the more rapid is the heart rate.

would be bad for the car, wearing out the brake linings, it would allow for more careful and rapid adjustments in speed.

From what is known of the pacemaker (SA node), it is reasonable to assume that the nerves in some way exert an effect on the rate at which the membrane potential moves toward the threshold and initiates an action potential. If electrodes are placed on the SA node to monitor the change in membrane potential before and during sympathetic or vagal stimulation, it is found that stimulation of the vagus nerve reduces the rate at which the membrane potential of the SA node drifts toward the threshold potential, causing the heart rate to decrease. Conversely, stimulation of the sympathetic nerve increases the rate of drift toward the threshold value and increases the heart rate. Neural impulses act on the membrane of the excitable tissue.

Although many researchers thought that nerves might release some active chemical that in turn would excite or inhibit the rate of change in the membrane potential, they could find no evidence for such a chemical until Otto Loewi performed certain experiments in 1921. He anesthetized two frogs and removed their hearts, arranging them so that fluid bathing one heart was carried to and bathed the other. Following a control period, the vagus nerve to the first heart was stimulated electrically. As expected, the heart rate slowed in response to the stimulation. Shortly after slowing of the first heart, the second heart also slowed, even though its vagus nerve was not stimulated. This sequence of events could be explained only by liberation of a chemical at the nerve endings of the vagus of the first heart, altering the rate of change of the membrane potential. This chemical then diffused into the bathing medium, which was carried to the second heart and exerted its effect there. It was the first conclusive demonstration that nerves release a chemical that in turn acts on cell membranes. In this case, the chemical was finally isolated and shown to be acetylcholine.

As a result of this experiment, it was found that an increase in neural activity of the vagus nerve causes acetylcholine to be liberated from the nerve endings. Acetylcholine diffuses the short distance to the SA node and acts on the membrane of those cells to decrease the rate at which the membrane potential drifts toward the threshold potential. The decrease in that rate causes the heart to slow. The decrease in the slope of the membrane potential is a result of the action of acetylcholine on the K^+ channels. Acetylcholine increases the permeability of those membrane channels to K, allowing K^+ to leak out of the cells more rapidly. The result of an increase in the outward K^+ movement is a hyperpolarization and reduction in the drift of the potential toward threshold.

A similar sequence of events can be shown to occur with stimulation of the sympathetic nerve. However, the chemical released from their endings in humans is norepinephrine. Norepinephrine, in contrast to acetylcholine, causes the drift of the membrane potential to accelerate and so acts to increase the rate of firing of the SA node. The exact mechanism of action of norepinephrine on the SA node is not known. However, norepinephrine is thought to increase the membrane permeability to Na^+ and Ca^{++}, which increases the slope of the membrane potential toward threshold. Under normal conditions, the heart rate depends on the rate of release of both norepinephrine and acetylcholine.

As noted, the heart receives innervation from the two components of the autonomic nervous system. Under normal conditions both sets of nerves are continuously active. This relationship is governed by a group of neurons in the medulla, in which a small area serves as a governor of cardiac activity. It receives nervous input from different parts of the body and from other areas of the brain. This cardiovascular center then integrates the neural inputs and, in turn, signals the heart via the autonomic nervous system.

The cardiovascular center may also react to psychological stimuli. You have probably experienced a racing heart while watching an exciting movie or being startled. This reaction is a response to your thoughts. The cortex of the brain signals the cardiovascular center, which in turn causes an increase in the sympathetic stimulation to the heart and a decrease in parasympathetic stimulation. The heart rate then increases. This has an adaptive value, as it prepares you for a situation in which you might have to expend energy rapidly in order to escape a threatening situation.

Although the heart receives input from both the sympathetic and parasympathetic nervous systems, it is able to beat rhythmically without those inputs. It is this ability that allows surgeons to transplant a heart from a donor to a recipient. Some people suffer such severe damage to their heart that their life can be sustained only by replacing the failing heart with a healthy one. The normal heart is obtained from people who have died because of some trauma not associated with heart disease, such as an automobile accident, and have a flat line EEG, as discussed in Chapter 4. The transplanted heart will function normally, assuming no rejection.

The major reflex systems controlling the rate and strength of the heart beats are depicted in Figure 8.8. The cardiovascular center receives input from special receptors, called baroreceptors, that are located in the carotid artery and the aorta and are able to sense blood pressure. Afferent nerves from the carotid and aortic baroreceptors continually relay blood pressure information to the cardiovascular center. The cardio-

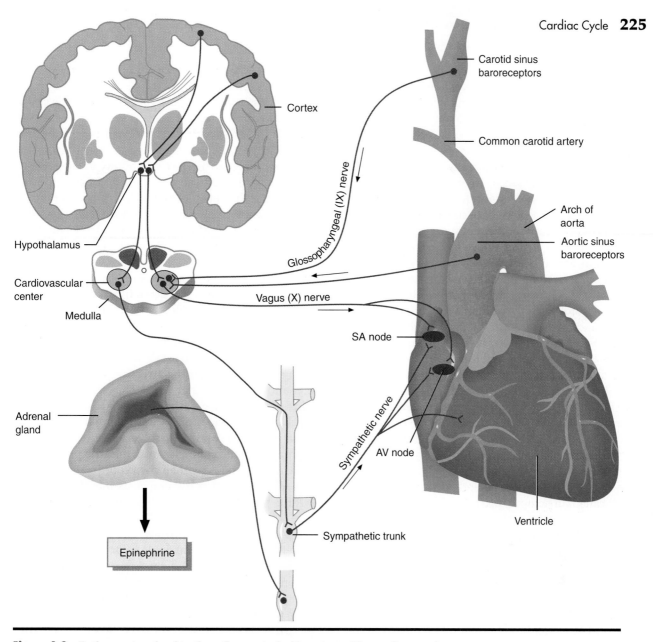

Figure 8.8 Pathways involved in the reflex control of heart rate. The cardiovascular center in the medulla receives input from baroreceptors in the carotid sinus and the aortic sinus, as well as information from the cortex. The efferent pathways from the cardiovascular center run via the vagus nerve and the sympathetic nerves to the heart.

vascular center acts as an integrating center, also receiving stimuli from the cortex, integrating them with other stimuli, and sends out the appropriate messages to the heart. If blood pressure falls the heart is signaled to increase its rate. Conversely, if blood pressure is increased, the signal is to reduce the heart rate.

Arrythmias

Thus far, it seems that the heart rate is perfectly regulated and that little can go wrong, yet you know that is not the case. Often people experience irregularities in their heart rhythm, sometimes of life-threatening

severity. The generic term used to denote irregularities of the heart rhythm is **arrhythmia.** What are these arrhythmias, and what causes them?

Irregularities of Heart Rate Probably at one time or another, most people have experienced a peculiar feeling in the chest that feels as if your heart stopped or skipped a beat. The description is apt. On occasion, in healthy people, the regular heart rate is interrupted. One or two or even more extra beats may be inserted between normal ones, and these beats are called **extrasystoles.** Sometimes it is a sequence of very irregular heart beats that lasts for minutes, or even hours, and can be felt as a peculiar feeling in the chest.

Although the causes may vary, one general type of extrasystole is most common. Recall that the regularity of cardiac rhythm is dependent on the SA node firing at equal intervals as the membrane potential drifts at some specified rate. However, on occasion, other parts of the atrium may become electrically excited; that is, a small piece of cardiac tissue spontaneously depolarizes and may have its membrane potential reach the threshold potential before the SA node does. If this happens, that piece of tissue acts as the source for atrial depolarization. It initiates a spreading action potential over the atrium, which leads to a premature contraction of the heart. We say it is premature because it occurs during the interval between two regular beats. This extra systole cannot be felt. However, the extra systole leads to a slightly longer interval between it and the next naturally occurring beat. This in turn allows more blood to enter the atrium and ventricle, as more time is allowed for filling. The increased filling stretches the heart more than normal. An increase in the stretch or length of the heart muscle fibers causes them to develop more force than normal on contraction. Within wide limits, the force of contraction increases with the fiber length just before contraction begins. This more forceful beat may be felt as a thump in the chest. Because the beat of the extrasystole originates at a spot other than the SA node, that spot is called an **ectopic focus.**

Although occasional extrasystoles generally are inconsequential, a continued irregularity of the heart rate is a more disturbing phenomenon. On occasion that irregularity may last for days, or weeks, in otherwise healthy people. As stated above, stimulation of the sympathetic nerves causes the release of norepinephrine, which has the ability to alter the rate of change of the membrane potential of the SA node. However, norepinephrine may be released from sympathetic nerve endings in many other parts of the body. In addition, the adrenal gland may release large quantities of the related hormone, epinephrine (adrenaline). Should a person suffer undue stress for any reason, the sympathetic system and the adrenal medulla may be activated and the concentrations of epinephrine and norepinephrine in plasma increase, sometimes dramatically. In some people, this may lead to a very rapid and irregular heart beat, as a result the person may experience many episodes of cardiac arrhythmia each day.

Cardiac arrhythmias of atrial origin, although annoying, do not generally compromise the ability of the heart to pump blood. The atria are thin-walled chambers of the heart and their pumping action contributes very little to the total amount of blood that returns to the ventricles. Thus, even though atrial pumping may be compromised somewhat, this has no serious consequences for the amount of blood that the ventricles may pump.

If stress is the initiating cause, relief of the stressful situation is often sufficient to reestablish normal rhythm. If not, and if the condition is very bothersome, drugs may be used. One class of drugs is **beta blockers.** These drugs have the ability to block the effect of norepinephrine on its receptors on the cell membrane, the **beta receptors.** However, other factors such as excessive smoking or drinking large quantities of coffee or other caffeine-containing drinks may also initiate an irregular train of atrial contractions.

Atrial Fibrillation A somewhat more troublesome arrhythmia is called **atrial fibrillation.** With this condition, no single well-defined location in the atria, such as the SA node, rhythmically stimulates the heart to beat. Instead, for reasons not well-understood, the wave of depolarization arises in multiple sites in the atria and sweeps over the atria in an almost continuous and rapid irregular path. This results in a ventricle that beats irregularly as the atrial depolarizations reach the AV node rapidly and irregularly. Although the atria in this condition are very inefficient pumps, separation of the ventricles from the atria prevents the ventricles from losing their own coordinated contractile activity. The ventricular beats may be irregular, but the ventricles still act as efficient pumps.

Ventricular Fibrillation The most serious cardiac arrhythmia is **ventricular fibrillation.** For reasons not well-understood, the ventricular muscle may become very excitable, with impulses arising from different parts of the ventricular muscle itself rather than from the electrical conducting system. Because the depolarization waves that flow over the ventricle in this condition are not coordinated and occur in a circular manner, different segments of the ventricle contract at different times. If you were to look at a ventricle during fibrillation, it would appear almost as a mass of quivering jelly rather than as a rhythmically contracting muscle. Because the contractions are not coordinated, the ventricular muscle is incapable of exerting any appreciable force on the blood within it. The flow of blood out of the heart slows to a trickle, blood pressure falls to very low levels, and the person loses consciousness rapidly. If the normal beat is not restored quickly, death ensues.

It would seem that a person who experiences an episode of ventricular fibrillation has no chance of survival. However, it is possible to survive for a few minutes with no blood flow to the brain, the organ most easily affected by loss of circulation. Resuscitation occurs in two steps. First, it is necessary to reestablish the pumping action of the heart sufficiently to raise blood pressure. Next, it is necessary to stop the fibrillation and reestablish the normal rhythm.

The first goal may be met by rhythmically pressing on the chest so that the pressure is transferred to the heart (**cardiopulmonary resuscitation** or CPR). If this is done properly, the ventricles may be forced to act as a pump, even though the muscle is still fibrillating.

As soon as possible the second step must be taken to stop the fibrillation. Figure 8.9(A) illustrates fibrillating muscle showing noncoordinated waves of depolarization sweeping in a circular motion over the heart. To stop the fibrillation, one must understand the sequence of electrical events of membrane depolarization. If the entire ventricle can be made to depolarize simultaneously, the circular wave of depolarization will stop and a normal sequence may begin. A simple and effective way to depolarize the entire ventricle simultaneously is to apply a strong electric current across the entire heart in a single burst. If this is done, the entire heart muscle will depolarize and, most importantly, repolarize simultaneously. If the initiating cause of the fibrillation is no longer active, the SA node

may recover and reestablish its normal rhythm. This procedure requires the use of two large flat electrodes. They are placed on both sides of the chest so that a strong brief current, often sufficient to make the patient's other muscles contract, is sent between the electrodes to depolarize the entire heart muscle, as shown in Figure 8.9(B). The heart muscle then repolarizes and may respond to the next normally occurring impulse from the conducting system with a normal beat. If the entire procedure is accomplished quickly, the patient may recover with no permanent damage. This method of cardiac resuscitation was made possible only because careful experiments led to a good understanding of the basics of membrane potentials and of the characteristics of fibrillation.

Cocaine

As described in Chapter 4, cocaine has two major actions. It is a local anesthetic, blocking conduction of impulses along membranes, and it prevents the reuptake of catecholamine. (In the brain, cocaine interferes

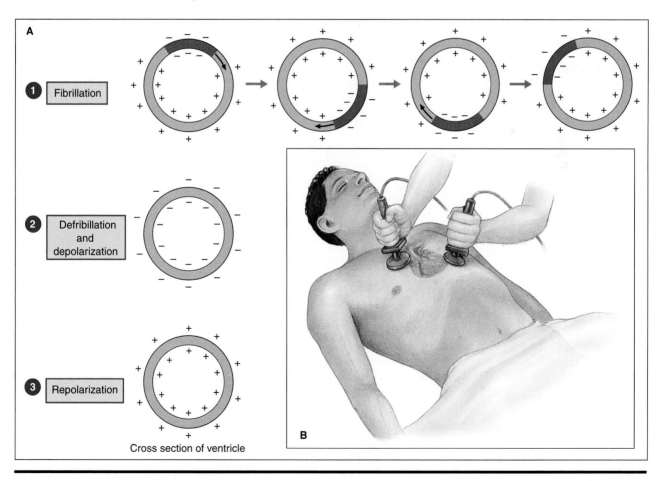

Figure 8.9 (A) 1. An illustration of the electrical potentials in a fibrillating heart; 2. Application of a defibrillating current depolarizes the entire heart; 3. Repolarization of the heart. (B) Defibrillator being applied to the chest of a person whose heart has stopped.

HIGHLIGHT 8.1

Cardiac Action Potentials

If atrial fibrillation results in a very rapid series of electrical stimuli reaching the ventricle, why does the ventricle not undergo a sustained contraction as would striated muscle? Remember that if a striated muscle fiber is stimulated with increasing frequency, it maintains a sustained contraction. That is what happens when you flex a muscle. If this were to occur with cardiac muscle, the heart could no longer pump blood, as the heart muscle would remain in the contracted state. Yet it continues to pump blood. Why?

The answer lies in the form of the action potential of cardiac muscle, which prevents a sustained contraction of the muscle. If electrodes are placed on a single fiber, the action potential recorded is as shown in the figure. Note that the membrane shows a sustained depolarization. The membrane remains depolarized for the entire time of the contractile response. Thus, it is not possible to initiate a second depolarization during the contractile phase. The membrane is completely refractory during that period. This is in contrast to striated muscle, which has a very rapid depolarization and repolarization sequence. This unique characteristic of cardiac muscle serves to prevent a lethal situation. Rapid firing of the atria cannot cause a sustained ventricular contraction. Such a contraction would prevent pumping of blood and quickly result in death.

Comparison of an action potential of a voluntary striated muscle with that of a cardiac muscle. Cardiac muscle has a refractory period 50–60 times longer than that of the voluntary muscle.

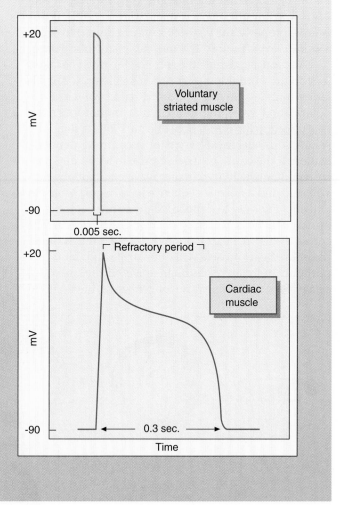

with the reuptake of dopamine, which is primarily responsible for its psychological effects.) Both of these effects can wreak havoc with the heart. High doses of cocaine cause increased stimulation of the heart as norepinephrine levels at the cardiac sympathetic nerve terminals increase. This results in an increased heart rate. At the same time, the anesthetic action of cocaine disturbs the normal conduction of impulses sweeping over the heart and reduces contractility. The combination of these effects may be lethal.

The rapid heartbeat with disturbed conduction and contractility may lead to severe arrhythmias and, in some cases, to sudden death from stoppage of the heart. Cocaine is a particularly dangerous drug for those who have cardiac problems. The combination of alcohol and cocaine is even more dangerous than either one alone. For reasons not well-understood, al-

cohol increases the lethal effect of cocaine, allowing lower doses of cocaine to cause fatal arrhythmias.

Generation of Arterial Blood Pressure by the Heart

Very shortly after the beginning of the ventricular contraction, the pressure in the ventricle begins to rise very sharply (1) (see Figure 8.10). At the same time, the first heart sound is heard (2), signaling closure of the AV valves (3). Ventricular pressure rises to the level of the aortic pressure in about 0.04 seconds. As soon as the pressure in the ventricle exceeds that in the aorta, the aortic valve opens (4) and blood is ejected into the

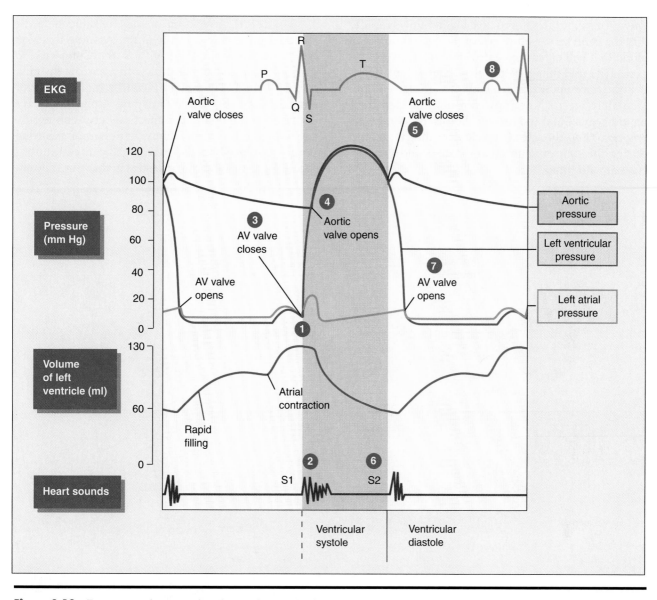

Figure 8.10 Pressure, volume, and auditory changes in the heart during one cardiac cycle. The simultaneous EKG is reproduced to show its relationship to pressure changes. The pressure and volume values are those that occur in the left side of the heart. Changes in the right heart are similar, but the pressure values are lower.

aorta. The maximum pressure obtained in the aorta is known as **systolic pressure.**

The ventricle continues to contract for another 0.25 seconds or so, and then begins to relax. As the ventricle relaxes, the intraventricular pressure drops rapidly. When it falls below aortic pressure, the aortic valve closes (as well as the pulmonary valve) (5) and the second heart sound is heard (6). (Although closure of the heart valves causes a sound, the turbulent flow of blood resulting from the closure of the valves also contributes to the heart sounds.) Following closure of the aortic valve, ventricular pressure continues to decline rapidly until pressure in the ventricle becomes

lower than that in the atrium and the AV valve opens (7), allowing the ventricle to refill. In contrast to ventricular pressure, aortic pressure falls relatively slowly after closure of the aortic valve as the aorta slowly recoils, propelling blood forward into the tissues. The lowest pressure achieved in the aorta is known as the **diastolic pressure.** About 0.4 seconds later, a new P-wave is initiated and the cycle begins again (8).

The second tracing from the bottom of Figure 8.10 shows the changes in ventricular volume during the cardiac cycle. During ventricular systole, blood continues to flow into the atria, expanding them. As soon as the ventricles cease contracting and relax, the AV valves open and the stored atrial blood flows rapidly into the ventricles. About 70% of the entire volume of blood that will enter the ventricle does so during this period of rapid filling, before atrial contraction. It is as if a trap door opened in the base of a storage bin. Immediately on opening, flow through the trap door would be maximal and then decline as the storage bin empties. Thus, the rate of ventricular filling is greatest during the start of ventricular diastole, and then decreases with time.

Figure 8.11 presents an analogy for understanding the pressure changes in the aorta. Assume that you have a rubber bulb filled with water. Connected to that bulb is a soft rubber tube with a narrow opening at the far end. The opening can be made larger or smaller by any device you imagine. A one-way valve is placed at the point of connection between the rubber bulb and the tube. Finally, a pressure-measuring device is placed in the tube so that a continuous recording is made of the pressure in the tube. The sit-

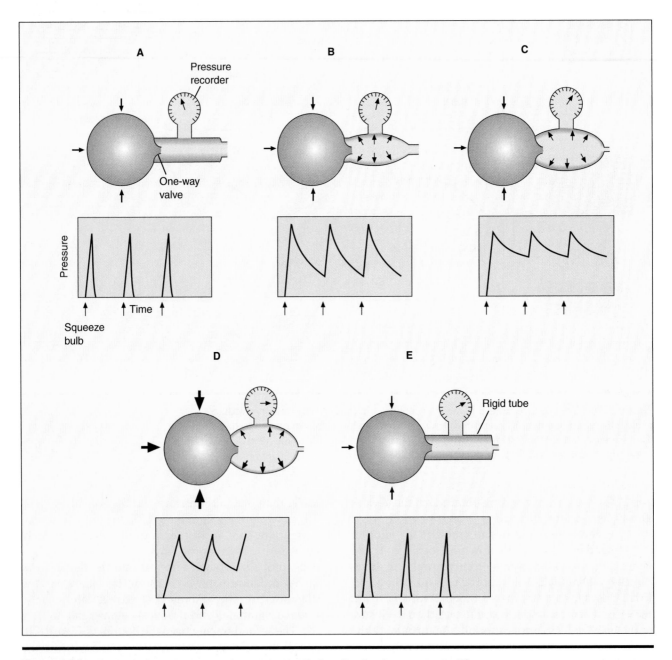

Figure 8.11 Pressure changes in a simple system simulating the circulatory system. The thickness of the arrows pointing to the rubber bulb represents the force applied to the bulb.

uation is analogous to the heart connected to the aorta with a one-way valve in between. The constriction at the end of the rubber tube is comparable to the resistance of the arterioles at the end of the arteries.

If the bulb is squeezed rapidly, the pressure in the bulb rises above that in the tube, and water is forced into the tube. As soon as the bulb is released, the pressure falls. If the resistance at the end of the tube is small, i.e., the end has a large opening, the record of pressure during the "cardiac cycle" is as shown in Figure 8.11(A). Pressure rises rapidly as the bulb is compressed and falls rapidly when the bulb is released. During compression, water leaves the tubing almost as fast as it enters from the bulb. Successive compressions result in successive pressure fluctuations, as shown.

But now constrict the end of the tube somewhat so that there is appreciable resistance to flow. When the bulb is rapidly compressed, the ejected water cannot flow out of the end of the tube as fast as it enters from the bulb. The rubber tubing expands as the pressure increases and the excess water is stored in the tube. Now when the bulb is released, the tubing recoils and continues to eject water out of the end. No water flows back into the bulb because the one-way valve between the two compartments prevents such movement. In Figure 8.11(B) the pressure recorded under these conditions is shown. Note that the rate of pressure drop in the tubing is slower than that which occurs in the first example. If the bulb is compressed soon enough after the first compression, pressure does not fall to zero in the tubing. The bulb is compressed before this can take place. This situation is similar to what happens with each beat of the heart. More blood is pumped into the aorta during systole than can run off through the arteries, arterioles, and capillaries during that same time. The elastic aorta swells during ventricular systole as the excess blood is forced into it. During the period of ventricular diastole, when the aortic valve closes, the aorta recoils and propels the stored blood toward the arterioles. Now we can consider what happens in a few circumstances that mimic physiological changes.

If the diameter of the end of the rubber tubing is constricted still more, raising the resistance again, the outflow is impeded even more and more blood is stored in the tubing during compression of the bulb (Figure 8.11(C)). The slope of the pressure decrease is shallower than in Figure 8.11(B) and the low point of pressure in Figure 8.11(C) is higher than in Figure 8.11(B). In cardiovascular terms, the diastolic pressure increases. From this it should be clear that alterations in the peripheral resistance, the sum total resistance to flow between the left ventricle and right atrium, has a large effect on the diastolic pressure.

The more forcefully the pump contracts, the higher is the systolic pressure (Figure 8.11(D)). If one considers the simple system diagrammed in Figure 8.11(A)–8.11(D), one can see that the force of the pump has a greater effect on systolic pressure than on diastolic pressure. Conversely, the peripheral resistance has a greater effect on the diastolic pressure than on the systolic pressure.

The rubber bulb example can be modified as shown in Figure 8.11(E), where the stiffness of the tubing is changed. This is analogous to the aging process, which often results in stiffening of the arteries, a condition known as **arteriosclerosis.** If the tubing is made rigid, the systolic pressure increases a bit as the tubing cannot expand to accommodate all the fluid squeezed out of the bulb. The only way outflow can equal inflow is for the pressure in the tube to rise sufficiently high to drive the water through the narrow opening. However, as soon as the bulb is released, the pressure falls toward zero rapidly because the stiffened walls do not expand and recoil as easily as an elastic tube does. The difference between the highest and lowest pressures obtained increases. In terms of the cardiovascular system, the **pulse pressure,** which is defined as the difference between the systolic and diastolic pressures, increases.

The simulation just described approximates the circulatory system and the factors that determine both the diastolic and systolic pressures, as well as the **mean blood pressure.** The period between beats, or the diastolic period, lasts longer than the period of contraction. Thus, the mean pressure is not the simple average of systolic and diastolic pressures. Rather, mean blood pressure is approximately equal to diastolic pressure plus one-third of the difference between systolic and diastolic pressures. In healthy humans, systolic pressure is about 120 mm Hg and diastolic pressure about 80 mm Hg (120/80 mm Hg). Therefore,

$$\text{Mean blood pressure} = 80 \text{ mm Hg} + \tfrac{1}{3}(120 \text{ mm Hg} - 80 \text{ mm Hg})$$
$$= 80 \text{ mm Hg} + \tfrac{1}{3}(40 \text{ mm Hg})$$
$$= 93 \text{ mm Hg}$$

Heart Sounds

The sounds generated by the heart reveal some of its workings. In Figure 8.10, the lowest tracing is a visual recording of the heart sounds heard through a stethoscope placed on the chest. The first sound occurs as the atrioventricular valve closes. The second heart sound signals the closure of the aortic valve. These sounds are often described as *lub-dub.* If one of the valves does not close completely because of some anatomical abnormality, the normal lub-dub sound may be supple-

HIGHLIGHT 8.2

Measuring Blood Pressure

It is a rare person who has not had his or her blood pressure taken with the use of a **sphygmomanometer.** The way this works is as follows: A cuff is fastened around the arm and is inflated with air. The pressure in the cuff is increased so as to compress the major artery in the arm, the brachial artery. While the physician or nurse listens with a stethoscope placed over the brachial artery, the pressure is slowly lowered. The figure on the right illustrates the sequence of events allowing that determination.

When the cuff pressure is higher than the systolic pressure, the brachial artery is collapsed. No blood flows through it, and no sound is heard. As the cuff pressure is reduced below systolic pressure but remains higher than diastolic pressure, the artery alternately opens and closes, causing an interrupted flow of blood through the artery. This is heard as a series of sounds coincident with the heart beat. The pressure at which the first sound is heard is taken as the systolic pressure. As the cuff pressure is lowered to just below diastolic pressure, the artery remains open during the entire cardiac cycle and no sounds are heard over the brachial artery. Thus, the pressure at which the sounds disappear is taken as the diastolic pressure.

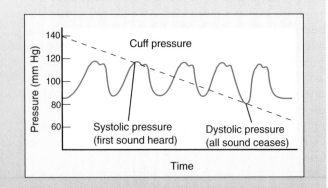

The genesis of sounds during the measurement of blood pressure using a sphygmomanometer. The first heart sound is heard as the cuff pressure falls just below systolic pressure. Silence occurs when the cuff pressure falls below diastolic pressure.

mented by an additional one, which is known as a **murmur.** A murmur is produced by an abnormal flow of blood being forced backward through the incompletely closed valve. In the case of an incompletely closed aortic valve, the murmur is heard between the second (dub) heart sound and the first (lub) heart sound. On the other hand, if the opening of the aortic valve is narrowed for any reason, but the valve can close during diastole, a murmur is heard as blood is forced out of the ventricle under high pressure through the narrow opening. In this case the murmur is heard between the first and second heart sounds. In addition, the quality of the sound allows a clinician to gain further insight into the nature of the pathology. Thus, careful attention to heart sounds with a stethoscope may provide useful information in diagnosing some forms of heart defects.

High Blood Pressure

As you will learn in the next chapter, the body contains many special sensors of blood pressure that act as negative feedback systems to maintain blood pressure within normal limits. However, often for reasons not well-understood, those regulators fail and blood pressure may rise well above normal. By convention, any person with blood pressure greater than 140/90 mm Hg is considered to have high blood pressure (**hypertension**). If untreated, hypertensive people are at greater risk for heart attacks and strokes. The high pressure in the arterial system seems to be an initiating factor in producing vascular pathology in the heart and brain. Any medical treatment that acts to reduce blood pressure from hypertensive levels reduces the risk of heart attacks and strokes.

Summary

Blood flows in a circular path from the right atrium to the right ventricle to the lungs, and back to the heart. It enters the left atrium, moves to the left ventricle, and is pumped into the aorta. Blood moves through the arteries, arterioles, capillaries, venules, and veins back to the heart. Flow is regulated by the rate and force of contraction of the heart and by the total peripheral resistance. Small changes in the diameter of the arterioles create large changes in the resistance of the circuit. The rate and force of ventricular pumping are controlled both intrinsically and extrinsically. Intrinsic control resides in the SA node, which has its own rhythm of depolarization. Extrinsic forces such as sympathetic and parasympathetic nerves as well as circulating catecholamines act on the SA node and the heart muscle itself. Sympathetic nervous stimulation and circulating epinephrine and norepinephrine increase the rate and force of contraction. Parasympathetic stimulation does the reverse.

Blood pressure is normally set at about 120 mm Hg for systolic pressure and 80 mm Hg for diastolic pressure. The difference between systolic and diastolic pressures is the pulse pressure. Systolic pressures may be altered by changes in the stiffness of the aorta, resistance to flow in the circulatory system, and force of the heartbeat. High blood pressure is a serious risk factor for heart attacks and stroke. The risk may be reduced by altered behavior that causes a reduction in blood pressure. The systolic and diastolic pressures are determined by the force of contraction of the heart, the total peripheral resistance, and the distensibility of the arterial system.

Questions

Conceptual and Factual Review

1. Where are valves located in the heart? What is their function? (pp. 218–220)
2. Other than the heart, where in the circulatory system are valves also located? What is their function? (p. 219)
3. How did William Harvey use animals to demonstrate the circulation of blood? (p. 219)
4. What role does the SA node play in the excitation of the heart? (p. 221)
5. What factors affect cardiac output? (p. 221)
6. What events do the P-, QRS-, and T-waves of the EKG signal? (pp. 221, 222, 228, 229)
7. Which hormone acts to slow the heart? At which site on the heart does it act? What is its mechanism of action? (p. 223, p. 224)
8. What effect do norepinephrine and acetylcholine have on the drift of the SA node? How does this affect the heart rate? (p. 223, p. 224)
9. What factors determine systolic pressure? Diastolic pressure? (p. 230, p. 231)
10. What is the value for normal blood pressure? (p. 231)
11. What is the relationship between cardiac output, peripheral resistance, and blood pressure? (p. 231)

Applying What You Know

1. What would be the consequences of a valve that is very stiff? Such valves do not open easily or entirely. How might a physician go about diagnosing this pathological condition?
2. People with arteriosclerosis, or hardening of the arteries, have a large difference between their systolic and diastolic pressures. Why?

3. Some people have such severe heart disease that they require a heart transplant. The transplanted heart has no neural attachments, yet it functions quite well. How is this accomplished?

4. A trained athlete has a lower resting heart rate than a sedentary person, yet the resting cardiac output of both is essentially the same. What do you think accounts for this?

5. As a person ages, his or her systolic and diastolic blood pressure tend to change. What changes would you expect, and why do they change in the direction you describe?

6. How would the following affect systolic, diastolic, and mean blood pressures:
   a. Stiffening of the large arteries.
   b. An increase in the total peripheral resistance, for any reason.
   c. An increase in catecholamine levels in blood, which serve to increase both the rate and force of contraction. Assume no other changes.

Chapter 9

Cardiovascular System: Blood Vessels

Objectives

By the time you complete this chapter you should be able to

◆ Understand the multiple functions of the circulatory system.

◆ Understand the function of the different blood vessels and their relationship to one another.

◆ Appreciate the role of the cellular elements of blood and the dissolved solutes in plasma.

◆ Understand how blood pressure determines the fluid balance of the body.

◆ Explain the relationship between blood pressure and fluid movement into and out of capillaries.

◆ Describe the nervous and hormonal mechanisms that regulate blood pressure as well as the factors that cause alterations in blood flow to local areas.

◆ Understand the physiology of some of the more common pathologies of the cardiovascular system.

Introduction

In order for an animal cell to survive, it must have ready access to an environment that supplies all the necessities of life. In addition, that environment must be able to carry away the waste products of metabolism. Although unicellular animals and very small multicellular ones may satisfy their demands for nutrients by simple diffusion of molecules directly to and from the external environment, larger animals cannot. One of the simplest multicellular animals living in the sea is the sponge. Although it may be quite large, oxygen and carbon dioxide must still be carried to and from all of its cells. How does the sponge meet this challenge? Most sponges are largely hollow segments joined to form a single organism (see Figure 9.1). Water is drawn into the hollow body and expelled from it through pores, thus bringing sea water close to all cells. As water flows through the sponge, oxygen is extracted, food particles are engulfed, and carbon dioxide is given off. One might consider this a primitive circulatory system because it carries vital molecules to and from the cells.

As the size and complexity of animals increase, the circulatory system becomes more important. The cells of larger and more complex animals require a constant supply of many different molecules. For example, hormones produced in one part of the body must be delivered to distant cells; insulin produced in the pancreas must be carried to cells far from that organ. Other molecules must also supply distant parts of the body. For example, the liver constantly adjusts its production of a large variety of vital molecules such as glucose and amino acids that are transferred to the blood and are then taken up by the cells of remote organs.

Large animals accomplish this through an effective circulatory system. A series of pipes, or blood vessels, carry blood throughout the body so that every cell of the body is either in direct contact with a small blood vessel or is so close to one that it can receive all the molecules necessary for life, as well as ridding itself of the waste products of its own metabolism.

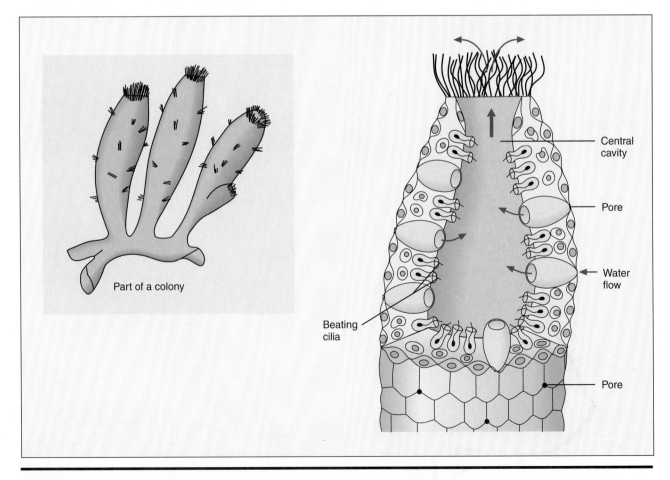

Figure 9.1 Simple circulatory system of a sponge. Water flows in through the pores on the side of the sponge and out through the top.

Blood Vessels

Although we generally speak of the heart as a single organ, it is actually two separate pumps joined together, as described in Chapter 8. Venous blood returning from all parts of the body is carried to the upper chamber of the right side of the heart, the right atrium. From there it moves into the larger, thicker-walled, lower chamber of the right heart, the right ventricle. When the right ventricle contracts, it forces blood into the lungs via the pulmonary artery. In the lungs, CO_2 is eliminated from the blood and O_2 is picked up. The oxygenated blood returns to the heart via the pulmonary veins, which carry the blood to the left atrium and then into the left ventricle. The left ventricle is the largest, most muscular chamber of the heart, and pumps blood into the aorta to supply the remainder of the body.

Arteries and Arterioles

The blood pumped by the left ventricle flows into a large, thick-walled vessel, the aorta. The aorta is the largest artery in the body. As it makes a 180° turn to travel down the trunk, it branches into many smaller arteries, which in turn branch into even smaller and more narrow arteries. The smallest of these arteries are called **arterioles.** Arterioles branch into capillaries.

Arteries and arterioles have a very important ring of smooth muscle located between the inner and outer vessel walls. This ring of smooth muscle can contract or relax, to decrease or increase the diameter of the vessel. Change in vessel diameter is the most important way in which blood flow to the tissues is regulated. Although both arteries and arterioles can change diameter, the arterioles are the primary regulators of blood flow. When tissues require an increase in blood flow, the smooth muscle of the arterioles feeding that tissue relaxes. This causes an increase in the diameter of the arteriole and blood flow to the tissue increases. This regulatory process is explained in greater detail later in this chapter.

Even though the heart pumps blood into the aorta intermittently (only when the left ventricle contracts), blood flows continuously through the vessels. How can this be? The smooth muscle of the aorta and arteries is responsible for this phenomenon. When the left ventricle contracts and blood flows into the aorta, the arteries act as a reservoir of blood that can recoil between beats of the heart. The aorta stretches to accommodate the large volume of blood ejected by the heart. In between beats, the aorta recoils and shrinks back to its resting diameter, propelling the blood toward the capillaries (Figure 9.2). This is an important characteristic of the arteries and accounts for the continuous flow of blood through the blood vessels, rather than an intermittent flow. Refer back to Figure 8.9 of the previous chapter.

Capillaries

Unlike arterioles, capillaries contain no smooth muscle. Their walls are made up of a single layer of flattened endothelial cells. These thin-walled vessels form

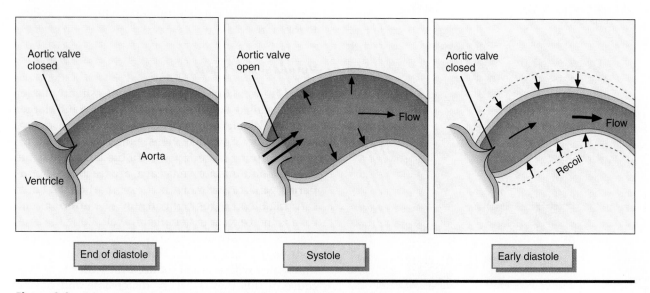

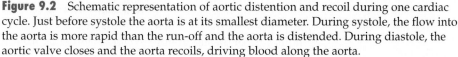

Figure 9.2 Schematic representation of aortic distention and recoil during one cardiac cycle. Just before systole the aorta is at its smallest diameter. During systole, the flow into the aorta is more rapid than the run-off and the aorta is distended. During diastole, the aortic valve closes and the aorta recoils, driving blood along the aorta.

an intricate network and provide the largest surface area between blood and tissues in the body. It is in the capillaries that the primary business of the cardiovascular system takes place: the transport of molecules between blood and tissue. Figure 9.3 is a photomicrograph of a capillary. You can see that it is a thin-walled vessel with numerous pores between the cells that make up its walls, which allow for passage of water and small molecules. Despite the fact that the capillaries are the smallest vessels, there are so many that their total cross-sectional area is about 1000 times that of the aorta.

Although the aorta is always filled with flowing blood, this is not true of the capillaries. Blood flow through them is controlled largely by the arterioles and **precapillary sphincters,** small rings of smooth muscle located at the junction of the arterioles and capillaries (see Figure 9.4). Their location allows them to act as regulators of flow into the capillaries. Contraction of the arterioles or sphincters reduces capillary flow, sometimes even to zero. Precapillary sphincters are analogous to a faucet that regulates flow through a pipe. This ability to change the distribution of capillary flow is responsible for the change in the color and temperature of skin seen in a variety of conditions. Blushing is the result of capillaries opening in the face, just as is the reddening and increase in temperature of a place on the skin that has been irritated. Conversely, blanching or cooling of the skin following

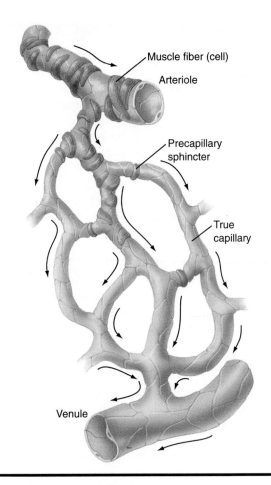

Figure 9.4 Drawing of a capillary network showing the relationship of the capillaries and arteriole leading to a venule.

exposure to cold is a reflection of a reduced blood flow to the skin.

Venules and Veins

Blood flows out of the capillaries into slightly larger vessels, called **venules.** Venules coalesce into larger venules. When venules become larger than microscopic, they are called veins. Small veins come together forming larger ones, until the largest two, the superior and inferior vena cava, are reached. The **superior vena cava** drains the upper part of the body, carrying blood to the right atrium. The **inferior vena cava** is the very large vein that travels along the back wall of the body and returns blood from the lower body to the right atrium. Veins, unlike arteries, have thin walls, with a much reduced muscular layer. Because of this they are very distensible and have a larger total cross-sectional area than arterial vessels. About 65% of the total blood volume resides in the venous side of the circulatory system. The remainder of

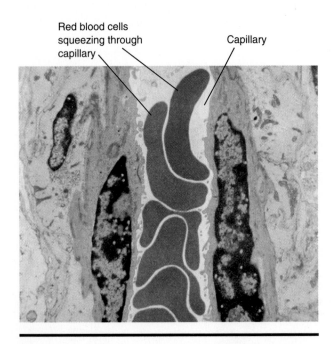

Figure 9.3 Photomicrograph showing red blood cells squeezing through capillaries.

the blood is contained in the arteries, arterioles, capillaries, and heart. Thus, the veins act not only as simple channels for returning blood to the heart, but also as reservoirs. Should blood be lost, as in hemorrhage, venous contraction reduces the volume of the venous reservoir and forces more blood to the heart, helping to maintain blood pressure near normal.

The valves in the larger veins have a function other than preventing back flow of blood. They also make many of the contracting muscles of the body act as ancillary pumps, making sure that the venous blood moves toward the heart. Figure 9.5 shows what happens to a vein interposed between two muscles during contraction of those muscles. As the muscles contract, they shorten and become more rounded. You see this every time someone shows off by flexing his or her biceps, as it shortens and bulges. Rounding out the muscles squeezes the vein between them, much as you might squeeze a toothpaste tube from the middle. The veins act similarly to that toothpaste tube. Venous valves in the lower part of the arm prevent movement of blood toward the fingers, much as the closed end of the toothpaste tube prevents its contents from coming out the wrong way. The upper valve, however, opens and allows blood in the compressed vein to flow toward the heart. In effect, every time you contract your muscles, you propel blood along the venous system toward the heart. It is analogous to the toothpaste coming out of the top of the tube when the tube is squeezed in the middle.

Blood

Before discussing the mechanical and hydraulic characteristics of the circulatory system, it is important to consider the nature of blood. When the early investigators first looked at the lungs of a frog through the newly discovered microscope, they saw movement in small channels; that is, they saw cells moving in single file through microscopic vessels. Thus, based on their observations they concluded that blood must be made up of at least two components: fluid and cells, which are carried in that fluid. These are broken down further into red blood cells, white blood cells, and the lightest fraction, a straw-colored liquid called **plasma** (Figure 9.6). It is possible to separate each component for analysis. About 55% to 60% of blood is liquid plasma, 40%–45% is red blood cells, and about 1% is white blood cells.

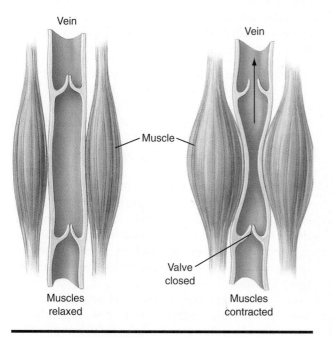

Figure 9.5 Diagram indicating how venous valves propel blood forward when muscles rhythmically contract.

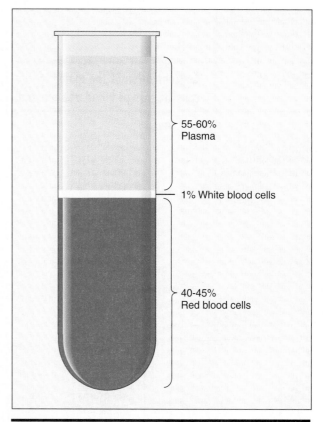

Figure 9.6 Following centrifugation, blood separates into three distinct layers: plasma on top, followed by white blood cells, and red blood cells at the bottom.

Plasma

Plasma not only serves as the medium in which the blood cells are carried to all parts of the body, but also contains the dissolved solutes required for cellular life. The major solutes are simple salts composed primarily of Na^+, K^+, Cl^-, and HCO_3^-. In addition, larger molecules such as glucose, amino acids, and hormones are present. The largest molecules are the proteins **albumin** and **globulin**. Table 9.1 presents a more complete listing of the constituents of plasma, with their normal or average values for humans. Although it is unlikely that any healthy individual exhibits exactly those values in his or her plasma, the deviation is small. (Remember that an *average* is made up of a range of values, none of which need be exactly equal to the average.) Despite very wide variations in lifestyle, food intake, and activity, the concentrations of solutes in plasma differ very little among healthy people, and remain constant from day to day. A person who is "addicted" to salted peanuts will have a concentration of Na^+ in plasma very similar to one who eschews salt. (This is explained further in Chapter 12.)

Two major groups of proteins can be isolated from plasma. Albumin is the group of smaller proteins found in plasma and it makes up about 5% of the plasma volume; that is, there are about 5 g of albumin per 100 ml of plasma. The movement of albumin across the capillary wall is severely restricted, and therefore contributes an effective osmotic pressure across the capillary wall. The smaller solutes, such as glucose and salts, pass freely across the capillary walls and so do not contribute to the effective osmotic pressure. (See Chapter 2.) The osmotic pressure contributed by albumin is an important factor that determines the balance between **interstitial fluid,** the extracellular fluid that bathes all cells, and plasma volume. However, albumin also performs other functions, such as carrying fatty acids and other substances dissolved in the plasma.

Globulins constitute the second group of plasma proteins. They are large protein molecules, representing about 2% of the plasma volume. These proteins also function as carrier molecules and **antibodies,** which are specific proteins involved in immune responses.

Cells

Microscopic examination reveals many different cell types in plasma, each with a different function. By far, the major cell type is the **red blood cell (RBC),** or **erythrocyte.** As the name implies, RBCs give blood its color. In addition to this abundant cell are the **white blood cells,** or **leukocytes.** The third and smallest cellular element is the **platelet** or **thrombocyte.**

Red Blood Cells The primary function of these cells is to transport O_2 and CO_2. They are the most abundant cells in blood, about $5,000,000/mm^3$, (approximately 150,000,000,000 per ounce of blood), and are shaped somewhat like a doughnut with a thin filled-in center (Figure 9.7). Red cells make up approximately 40–45% of blood. This percentage is called the **hematocrit,** i.e., the fraction of whole blood that is composed of red blood cells. Unlike the other cells in the body, red blood cells do not contain a nucleus. An RBC is in effect a small living bag containing within its cell membrane a homogeneous solution of its main constituent, **hemoglobin,** the oxygen-carrying molecule.

Chemical analysis reveals that hemoglobin is a complex molecule containing four subunits, or chains. Each chain is composed of a large polypeptide called **globin** and a red-colored, smaller unit called **heme,** which contains an atom of iron. The iron atom of this complex molecule can form a reversible bond with a molecule of oxygen. Thus, each hemoglobin molecule

Table 9.1	**Concentration of the major constituents of blood and plasma.**

Blood	
Red blood cells	5 million per mm³*
Hemoglobin	15 grams per 100 ml
White blood cells	8,000 per mm³
Neutrophils	5,000 per mm³
Eosinophils	200 per mm³
Basophils	40 per mm³
Lymphocytes	2,500 per mm³
Monocytes	400 per mm³
Platelets	250,000 per mm³

Plasma	
pH	7.4
Sodium	140 mM
Potassium	4.4 mM
Chloride	110 mM
Bicarbonate	27 mM
Phosphate	1 mM
Albumin	4.5 grams per 100 ml
Globulin	3.5 grams per 100 ml
Fibrinogen	0.5 grams per 100 ml

Note: The above values are average values. In any individual the concentration of any of the blood constituents may deviate from that reported above by 5%–10% and still be considered within the normal range.

*There are 1000 mm³/ml

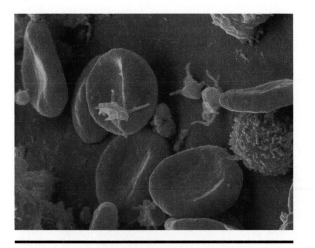

Figure 9.7 Photomicrograph (about 9000×) of red blood cells. Notice their doughnut shape.

can bind or carry up to four molecules of oxygen. Figure 9.8 presents a schematic representation of a single hemoglobin molecule. One molecule of oxygen may

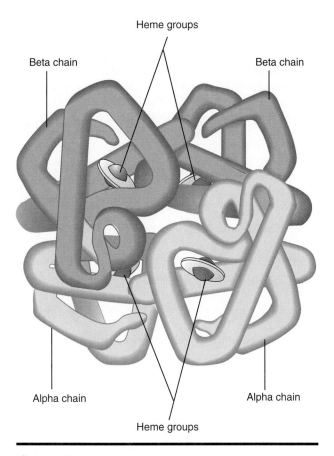

Heme groups

Beta chain

Beta chain

Alpha chain

Alpha chain

Heme groups

Figure 9.8 Model of the hemoglobin molecule. Note that an atom of iron is attached to each of the four heme groups.

be attached to each of the four heme groups as the molecule of hemoglobin passes through the lungs.

The normal concentration of hemoglobin in adult males is about 15 grams per 100 ml of blood. For adult females, the concentration is slightly less, about 14 grams per 100 ml of blood. A hemoglobin concentration of 15 grams per 100 ml blood is able to carry approximately 20 ml of O_2 per 100 ml of whole blood. Should the hemoglobin concentration fall to half normal, so does the oxygen-carrying capacity of blood.

When blood passes through the capillaries of the lungs, oxygen is added to hemoglobin and the blood turns bright red. When saturated with oxygen, hemoglobin is called **oxyhemoglobin.** In the reduced state, when oxygen is given up to the tissues, the color changes to a darker blue–red shade. This accounts for the bright red color of arterial blood and the deeper blue of venous blood. This difference in color can be seen by looking at the veins of the forearm in fair-skinned people. They have a distinctly bluish cast. The red color of oxygenated blood may be seen in the whites of the eyes, where the small vessels carrying oxygen-rich blood are easily visible.

Anemia The hematocrit in a healthy person is approximately 45%, but some people have much lower values. This condition is called **anemia.** Most anemias result from a reduced rate of production of red blood cells or hemoglobin. The average life of a red blood cell is about four months. This means that the bone marrow must replace all the red cells in the body in that same period. When this does not occur, anemia results. One cause of decreased red blood cell production is cancer of the bone marrow. When cancer is present it decreases the ability of the bone marrow to manufacture red blood cells.

A more common cause of anemia is iron deficiency. Because each hemoglobin molecule has four iron atoms, iron must be available for hemoglobin production. In the absence of an adequate supply of iron, the bone marrow is unable to produce sufficient hemoglobin to meet the requirements necessary for replacement of red cells. If there is a shortage of iron available to incorporate into the hemoglobin molecule, fewer cells or ones containing less hemoglobin will be made. Although the same number of cells may appear in the blood, they are smaller than normal and the amount of hemoglobin contained in each milliliter of blood is lower than normal. Both the hematocrit and the hemoglobin concentration of blood decrease, as does the oxygen-carrying capacity.

Jaundice When red blood cells are destroyed, the released hemoglobin molecules are metabolized by the liver. **Bilirubin** is one of the products of this metabolism. When there is too much bilirubin in the

HIGHLIGHT 9.1

Sickle-Cell Anemia

The precise configuration of the hemoglobin molecule is very important. Very slight alterations in the structure often lead to severe problems. For example, a change in a single amino acid in two of the four hemoglobin subunits is responsible for the disease **sickle-cell anemia,** common among people of African heritage. The alteration in the hemoglobin causes the red cell to assume a bent shape, so that from the side it looks like a sickle, as illustrated in this photomicrograph. When the cells change shape, it is difficult for them to move through capillaries. They may then form clumps in the smaller vessels, which can shut off blood flow entirely to a region of tissue, leading to severe pain and tissue damage.

In addition, sickle cells are more fragile than normal cells, and are easily damaged and destroyed. This may cause a reduction in the total number of red cells in blood, resulting in anemia. Red cell destruction might increase to a rate that cannot be counteracted by an increased rate of production. In sickle-cell anemia, the odd characteristics of the cell membrane cause a greater rate of destruction of red blood cells than normal. Most red blood cells are destroyed in the spleen, as its capillary network has the smallest diameters in the body. The rate of destruction is so great that the bone marrow cannot keep up with it by producing new cells, so the number of red cells in the blood decreases.

The disease in its homozygous state (both sickle-cell genes are present, one from each parent) is serious. However, in the heterozygous state (only one gene for sickle-cell anemia is inherited, a condition called the sickle-cell trait), the consequences are less severe, but do

Erythrocyte beginning to sickle · Crenated erythrocyte · Normal erythrocyte · Sickled erythrocyte

Photomicrograph (×9200) of a sickled red blood cell.

constitute a health problem. For many years it was unknown why a gene that confers so much danger to health should persist in a population. Evolutionary theory holds that a potentially lethal gene should slowly recede from a population as people who are homozygous for the gene generally die before they reach reproductive age, yet its incidence remains relatively high. This paradox was solved when it was recognized that somehow the sickle trait bestows partial resistance to malaria. Thus, a seemingly "bad" trait actually helps to allow survival of a population living in an area infested by the malaria-carrying mosquito.

blood, as results from excess red blood cell destruction, bile pigment is deposited in the skin, the whites of the eyes, and the mucous membranes of the mouth and nose, causing a yellow discoloration. This condition is called **jaundice.** An excess of bilirubin occurs when too much bilirubin is released into the blood or when bilirubin clearance from the liver is impaired. Syndromes such as sickle-cell anemia and hemolytic anemia cause an increase in red blood cell production with a resultant increase in bilirubin production and the development of jaundice.

Red Blood Cell Production If the number of red cells in blood falls for any reason, such as increased red cell destruction or bleeding, the bone marrow is stimulated to increase the rate of production of red blood cells. Although the process is elaborate, it was

recognized many years ago that the kidney is somehow involved. People with kidney failure often suffer from severe anemia without signs of jaundice. Because there was no reason to believe that they suffered from increased red cell destruction, it was reasonable to assume that inadequate production of red blood cells was the cause. Early experiments showed that if extracts of kidney tissue were injected into experimental animals, there was a rapid increase in the number of circulating red blood cells. As researchers refined experiments, they were able to isolate a hormone, called **erythropoietin,** that is secreted by the kidney and stimulates bone marrow production of red cells. Recently the gene for this hormone was isolated and is used to clone erythropoietin. Sufficient quantities are now available so that it can be given to people with kidney failure.

The Effect of Altitude on Blood Although it is not clear just how erythropoietin stimulates the bone marrow, we know that cellular **hypoxia** (inadequate oxygen) of the kidney leads to increased secretion of erythropoietin. If blood with lower-than-normal concentrations of hemoglobin or with a low concentration of oxygen is delivered to the kidney, increased levels of erythropoietin appear in the blood. Simple experiments made clear that it is the concentration of oxygen to which the kidney responds. Oxygen-poor blood was delivered to one kidney of an anesthetized animal while the other kidney was perfused with normal blood. The venous blood coming from each kidney was collected and analyzed for erythropoietin. Researchers found that the kidney perfused with oxygen-poor blood produced more erythropoietin than the other.

White Blood Cells The normal number of circulating white blood cells (also called leukocytes) is 5,000 to 10,000/mm³. Despite the relatively low number (remember that the normal number of circulating red blood cells is about 5,000,000/mm³), white blood cells are essential to survival. Microscopic examination of the blood reveals that there are a number of different types of white blood cells, which are discussed in detail in Chapter 10. These different types of cells act together to defend the body against disease. In a healthy person, the white blood cells consist of granulocytes (neutrophils, eosinophils, and basophils), lymphocytes, and monocytes. Neutrophils, the most common white blood cells, engulf and kill invading bacteria, a process known as **phagocytosis.** The next most common white blood cell is the lymphocyte, which is involved in the cellular and immune processes. Monocytes and eosinophils make up the remaining 5–10% of the circulating white blood cells. Monocytes perform a number of critical functions such as phagocytosis of certain types of invading microorganisms and the tissue debris of inflammation, and the clearance of aging red blood cells, platelets, denatured plasma proteins, and bloodborne microorganisms. This phenomenon may be observed following a cut or other injury to the skin. If the injury becomes infected, pus may form. **Pus** is a collection of dead white blood cells and tissue fluids. Eosinophils are active at sites of allergic reaction and sites of invasion by helminthic (worm) parasites that are too large to be ingested by phagocytic monocytes.

Platelets Platelets (also called thrombocytes) are the smallest and second-most numerous cellular elements found in blood. They are formed when large cells in the bone marrow called megakaryocytes shed their cytoplasm. The shed fragments of cytoplasm, lacking a nucleus, become platelets. They are responsible for the clotting that occurs when the endothelium

of any blood vessel is damaged. Platelets quickly adhere to the exposed subendothelium. In small vessels they pile up, forming an effective plug that acts as a physical barrier, preventing blood from leaking out of the damaged vessel (Figure 9.9). However, in more serious injuries, this alone cannot stop bleeding. Clot formation in injured vessels is the major means by which bleeding is halted. Exposure of the tissue layers below the thin layer of cells lining the blood vessels initiates a series of chemical reactions that end in the formation of a dense tangle of fibers that trap blood cells and plasma, as shown in Figure 9.10.

The series of chemical reactions that results in clot formation is complex, ending in the creation of a tangle of protein strands (**fibrin**) that form the basis of the clot. Figure 9.11 shows the series of reactions involved in clot formation. Following tissue damage, a cascade of **clotting factors** act in plasma in a sequential manner to form a substance that acts on **prothrombin** to form **thrombin**. Thrombin, in turn, is an enzyme that liberates fibrin from **fibrinogen**. Both prothrombin and fibrinogen are proteins that are always present in plasma.

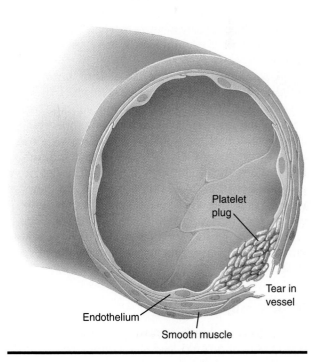

Figure 9.9 A drawing of a platelet plug. The tightly packed platelets prevent blood from flowing out through the broken wall of the small blood vessel.

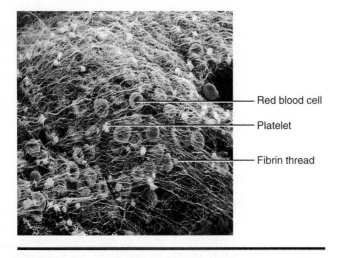

Red blood cell

Platelet

Fibrin thread

Figure 9.10 A photomicrograph (×1550) of a clot showing fibrin strands and red blood cells trapped in the clot.

As noted in Figure 9.11, many of the reactions require Ca^{++} in order to proceed normally. It should also be evident from the figure that the first series of reactions requires formation of the clotting factors. They are usually designated by Roman numerals and number about a dozen. Clotting factors are released by injured tissue, and they are always present in plasma. Vitamin K is required for the synthesis of six of those clotting factors, many of which are synthesized in the liver. Thus, liver damage or a deficiency of vitamin K may interfere with the clotting mechanisms. The best dietary sources of vitamin K are green leafy vegetables and beef liver.

A hereditary deficiency of a clotting factor is responsible for **hemophilia,** a disease in which the time it takes for clot formation is prolonged excessively. Because the genes controlling this factor are on the X chromosome and are recessive, heterozygous females do not have hemophilia. On the other hand, males who inherit the X chromosome with the defective gene are not protected by a normal one, so they develop hemophilia. These people lose large amounts of blood from relatively small cuts or bruises. Consequently, they require frequent transfusions of whole blood or blood constituents containing the normal clotting factors.

Blood Groups

Although early humans must have recognized that severe bleeding could lead to death, there was little that they could do except attempt to staunch the bleeding. Unfortunately, this was often difficult and too late. That blood loss could result in death led to attempts in the mid-seventeenth century to replace blood with blood from another source. Sir Christopher Wren, the noted English architect and astronomer, was among the first to devise a method to transfer blood from a donor to a recipient. He used a hollow quill feather of a bird attached to a bladder for the purpose. The first transfusions, however, were of dog's blood into a human. Although superficially it looked the same as human blood, it led to very serious consequences, often ending in death of the recipient.

It was not until 250 years later that the puzzle relating to the mortality following such transfusions began to be solved. Researchers noted that if the blood of two individuals was mixed in a small tube and placed under a microscope, they often observed clumping of red cells, as pictured in Figure 9.12 on p. 246. Note the absence of fibrin strands or platelets, as one sees in a clot. On occasion, however, no such clumping was seen. By carefully checking and cross-checking different combinations of red blood cells and plasma, researchers established that four different blood groups could be distinguished among people. They were named group A, B, AB, and O. (Dogs do not have the same blood groups.) This nomenclature was based on the finding that if group A cells were mixed with plasma from group B blood, clumping or **agglutination** of red cells occurred. The cells stuck together in bunches rather than floating freely in the plasma. If the reverse was done, group B cells placed in plasma from group A blood, clumping was also seen.

This indicated that red cells have at least two different proteins or **antigens** on their surface. Antigens are large complex molecules that stimulate the immune system. They are discussed in more detail in Chapter 10. Type A cells contain antigen A and B cells contain antigen B. Furthermore, the plasma from blood of both types has an antibody to the opposite antigen. That is, type A blood has the antibody to antigen B in its plasma. Type B blood carries the antibody to antigen A. In the two cases cited above, a mixture of cells with an antigen and plasma with the antibody against it causes the red blood cells to clump together or agglutinate. Based on these findings, it was found that nonhuman blood cannot be used for transfusions into humans. The antigens on the surface of animals' red cells are not compatible with human blood and cause massive agglutination as antibodies to those foreign antigens are formed.

Two other blood types were also noted. Red cells of AB blood contain both A and B antigens on the surface of the cell, but lack the antibodies to either in the plasma. Cells of type O blood have neither antigen on the surface but have the antibodies against both A and B antigens in the plasma. If AB cells are mixed with plasma of any other group, agglutination occurs be-

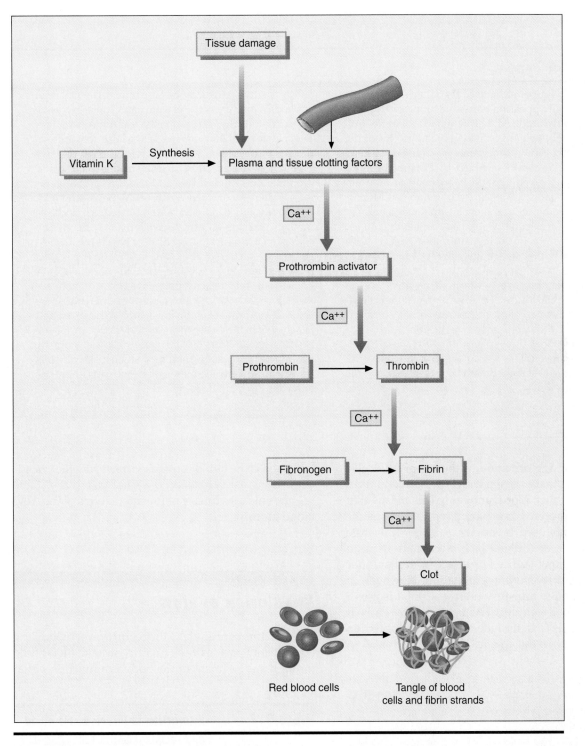

Figure 9.11 The sequence of reactions involved in clot formation. Note that Ca⁺⁺ is required at almost every step in the chain.

cause both antibodies are present on the cell surface and antigens to them are present in the plasma of A, B, and O blood. If red cells from O blood are mixed with the plasma of any other type, no agglutination occurs because type O red cells contain no antigens. How is it that type O blood, which has antibodies against both A and B antigens in its plasma, does not cause the recipient's cells to agglutinate when those antibodies are introduced into the circulation?

The answer is that donated plasma is so diluted by the recipient's own plasma that the final concentration of donor antibodies is too low in the recipient to cause

MILESTONE 9.1

Rh Blood Type

Many years ago, people noticed that some newborn babies, although appearing normal in all other respects, were severely jaundiced. This remained a puzzle until scientists who were investigating blood groups of the rhesus monkey found a new antigen–antibody system. In honor of the rhesus monkey, it was named the **Rh factor.** We now know that humans may or may not carry the Rh antigen on their red cells. If one is Rh positive, i.e., has the antigen, there are no Rh antibodies in the plasma. Rh negative people have neither the antigen nor the antibody. But if they are transfused with Rh positive blood, they will begin to produce antibodies to that antigen.

A problem may arise when an Rh negative woman conceives a child with an Rh positive man and the fetus is also Rh positive. Following a normal birth or an abortion, fetal red blood cells may enter the mother's bloodstream. In response, maternal antibodies to the Rh antigen are produced. This has no effect on the first child. However, because the mother is now sensitized to the Rh factor, a subsequent pregnancy in which she carries an Rh positive fetus may be hazardous. The maternal antibodies may cross the placenta and cause excessive destruction of the fetal red cells, leading to jaundice. The high level of bilirubin in the newborn constitutes a danger to the child. Moreover, with each successive pregnancy, the reaction to this mismatched blood grouping gets more severe and could result in neurological damage or even death of the newborn child.

Today, this condition can be treated. Rh negative mothers carrying an Rh positive fetus may be given a commercial product that binds to or covers the Rh antigen on the surface of the red cell. The mother is then protected from the exposure and does not make antibodies to it. When such a protective product is not administered, however, children may be born with severe jaundice resulting from the mismatch of maternal and fetal blood. In these cases doctors perform a transfusion, exchanging the entire blood volume of the newborn for fresh blood. Although there is a small risk with this transfusion, it protects the newborn from the toxic bilirubin.

---

any significant agglutination of the recipient's cells. Thus, it follows that a person with type AB blood lacks antibodies to either A or B antigens. This means that he or she can receive blood of any type as long as the volume of transfused blood is not large. If a large enough volume of A blood or B blood is transfused into an AB patient, sufficient plasma from the donor may be present to cause agglutination. However, in order to minimize transfusion errors, blood is cross-matched. A small sample of recipient plasma is mixed with donor cells to determine whether agglutination occurs. Only when the absence of agglutination is es-

tablished is the blood transfused, ensuring that an adverse transfusion reaction does not take place. Table 9.2 presents a summary of the interactions of the ABO blood groups.

Regulation of Flow

A few factors contribute to the regulation of blood flow through any part of the circulatory system. Blood flow through any segment of the body is a function of the

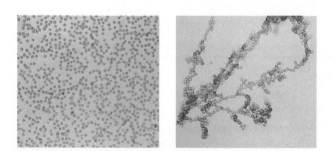

Figure 9.12 Photomicrographs (×450) of normal and agglutinated red blood cells in mismatched blood. Although the red blood cells are clumped together, no fibrin strands or platelets are present.

Table 9.2			Interaction between donor blood and recipient blood following a transfusion.			
Recipient's Blood Type	Antigen	Antibody	Donor's Blood Type			
			A	B	AB	O
A	A	B	-	+	+	-
B	B	A	+	-	+	-
AB	A and B	None	-	-	-	-
O	None	A and B	+	+	+	-

- = no agglutination; + = agglutination.

pressure in the arteries (P) and the resistance (R) of the vessels through which the blood is flowing. The resistance is directly proportional to the length of the vessel and inversely proportional to the diameter of the vessels. The higher the driving pressure, the greater the flow. Conversely, the greater the resistance to flow, the lower the flow. This is expressed by the equation

$$\text{Flow} = \frac{\text{Pressure}}{\text{Resistance}}$$

For the entire circulatory system, this is expressed as

$$\text{Cardiac output} = \frac{\text{Mean blood pressure}}{\text{Total peripheral resistance}}$$

Recall from Chapter 8 that cardiac output is the volume of blood pumped per minute. Total peripheral resistance (TPR) equals the sum of the resistances of the entire circulatory system excluding that of the pulmonary circuit. The average adult has a resting cardiac output about equal to his or her total blood volume of 5 liters per minute. Average blood pressure is about 95

mm Hg. The total resistance of parallel circuits may be calculated. The parallel circuits shown in Figure 8.1 are all the paths connecting the artery on the right hand side of the figure with the vein on the left side of the figure. The equation expressing the relationship is

$$\frac{1}{\text{Total peripheral resistance}} = \frac{1}{R_1} + \frac{1}{R_2} + \cdots \frac{1}{R_N}$$

This equation indicates that any increase in the number of resistance circuits, such as the opening of more capillaries, causes a decrease in the total peripheral resistance. This happens often during exercise, as the number of open circuits in the active muscles increases. During rest, not all muscle capillaries are open. As the closed capillaries open, they add additional parallel circuits to the muscle, reducing the total peripheral resistance. In order to get a feel for the meaning of the equation, try putting some arbitrary numbers in it. You will note that as the number of individual resistances increases, the total peripheral resistance decreases.

The relationship between pressure, flow, and resistance is illustrated in Figure 9.13. In Figure 9.13(A), a column of water supplies a single pipe with

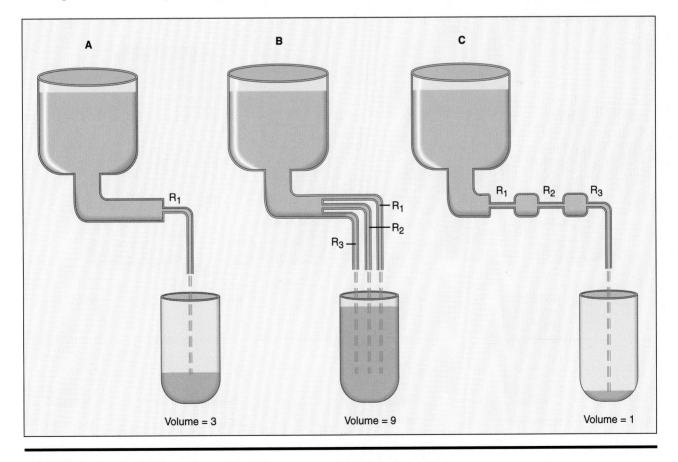

Figure 9.13   Relationship between pressure, flow, and resistances in series versus parallel channels.

a narrowing, representing a single resistance to flow, R_1. In Figure 9.13(B), the same reservoir supplies three parallel resistances (R_1, R_2, and R_3) that have the same diameter as the single resistance in Figure 9.13(A). The total resistance of the system in (B) is one-third that of (A), and the flow rate is three times greater. In Figure 9.13(C), the resistances are placed in series. In this case, the total resistance is the sum of the three resistances, and the flow through the pipe is one-third of that through (A).

What determines the resistance of each individual channel? By measuring the flow through a single channel at any given pressure, the resistance to flow can be calculated. As the radius of the channel increases, flow also increases as the resistance of the channel decreases. Experiments demonstrate that the resistance of an individual vessel is inversely proportional to the radius of the vessel, but the relationship is not a simple linear one; rather, the resistance of a vessel is inversely proportional to the fourth power of the radius.

$$\text{Resistance} = \frac{1}{\text{Radius}^4}$$

This relationship is quite important, as very small changes in the diameter of a vessel lead to relatively large changes in resistance. For example, if a small vessel constricts so that its diameter decreases by 50%, the resistance is multiplied by 16. By far, the major changes in resistance are controlled by the arterioles. Thus, small changes in arteriolar diameter effect large changes in the resistance to flow. For this reason, the arterioles are called the resistance vessels.

Figure 9.14 shows this relationship schematically. The driving pressure is the same for all three outlets. Outlet (A) has a radius twice that of outlet (B), so its resistance is only $\frac{1}{16}$ that of (B). Therefore, 16 times more water flows through outlet (A) than through outlet (B). Outlet (C) is constricted over a small area so that its diameter is reduced to one-half that of outlet (A) over the area of constriction. This is analogous to constriction of the arterioles of a single organ. The increased resistance in this small area significantly increases the total resistance of the vessel. However, there is a small oversimplification in this example. The resistance to flow through a tube is not only inversely proportional to the radius, but is directly proportional to the length of the tube. If we consider the arteries

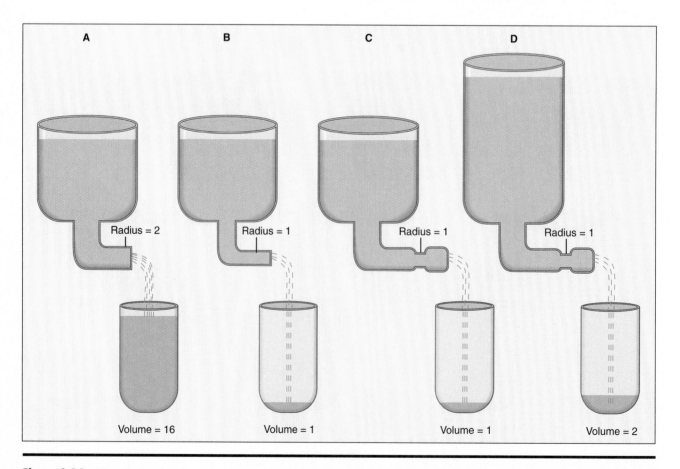

Figure 9.14 The relationship between flow, resistance, and pressure in rigid pipes.

and arterioles of the circulatory system, no significant change in their length can occur. Only their diameters can change. In panel (D) the resistance is the same as in (A) and (C), but the hydrostatic pressure is twice that of (A), (B), and (C). In this last case, the volume flow is double that of (B) and (C).

In order for the circulatory system to work effectively, it is essential that blood flow only in one direction. Should contraction of the ventricles cause blood to flow in the opposite direction, as would be the case if the heart valves worked improperly, the system would fail. In addition, the output of both the left and right sides of the heart must be perfectly matched. If the pumping rate of either ventricle falls below that of the other, blood will pile up behind the weaker ventricle like water behind a dam. This happens if the ventricular muscle is injured for any reason, and is known as **heart failure.**

The heart must also be able to adjust its pumping rate to meet changing demands for blood by the body; for example, during exercise blood flow to the exercising muscles must be increased. Finally, the contractile force of the heart must be sufficient to move blood to all parts of the body. However, the pressure must not become too high, as it could injure the blood vessels, even rupturing them.

Fluid Exchanges Across the Capillary Walls

The capillaries not only bring red blood cells in proximity to all tissues, but are also responsible for moving water and solutes into and out of tissues. As mentioned earlier, capillaries are very thin-walled tubes perforated by spaces between the endothelial cells that form the walls of each capillary. These spaces are sufficiently large that the small solutes of plasma and the water in which they are dissolved cross the capillary wall, much as a solution will cross filter paper.

Figure 9.15 depicts the osmotic and hydrostatic forces along a single capillary. Because the pressure of blood in the capillaries is, on average, higher than the hydrostatic pressure on the tissue side, a bulk flow of fluid moves in the direction of the tissue space. However, the protein molecules are sufficiently large that their movement is restricted. Thus, the capillary acts as a selectively permeable membrane. While restricting the movement of protein molecules, the pores in

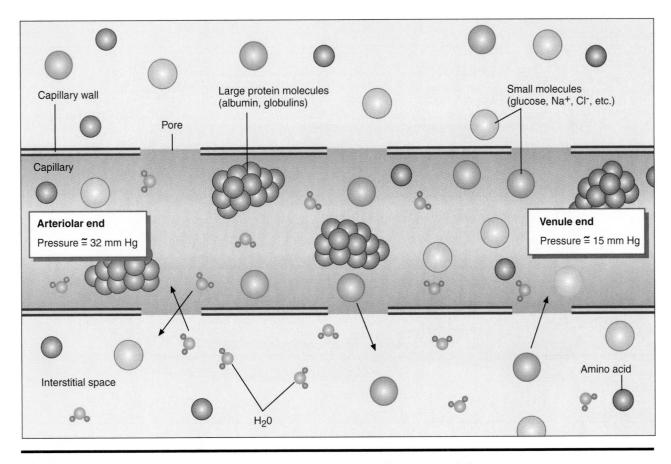

Figure 9.15 Schematic representation of a capillary showing the capillary pores and the movement of molecules through these pores.

the capillary allow smaller molecules such as water, electrolytes, and glucose to pass through them freely. What this means is that only some of the constituents of plasma are filtered across the capillary walls, in much the same way as freshly squeezed orange juice passes through a fine strainer while the larger pieces of pulp are caught by the meshwork and held back.

If a very small glass tube is placed in either end of a capillary, the hydrostatic pressure can be measured. At the arteriolar end, that pressure is about 32 mm Hg and at the venule end, the pressure falls to about 15 mm Hg. Because the hydrostatic pressure of the tissue space is very low (about 0 mm Hg), all along the capillary there is a force driving water and solutes out of the capillary into the tissue spaces. However, because the plasma proteins are very large molecules and cannot pass through the pores, they contribute an osmotic pressure tending to draw fluid back into the capillary. The force of that back pressure is about 25 mm Hg, and is called the **colloid osmotic pressure.** Figure 9.16 shows the consequences of this arrangement. At the arteriolar end of the capillary, the hydrostatic forces exceed those of the colloid osmotic pressure, resulting in the net filtration of water and its associated small

solutes out of the arteriolar end of the capillary. The fluid that filters out of the capillary is an ultrafiltrate of plasma, resembling plasma in every way except that the formed elements (red cells, white cells, platelets) and the plasma proteins are absent.

At the venule end of a capillary, the hydrostatic pressure falls to a value less than that of the colloid osmotic pressure. (See Highlight 9.2 for a discussion of colloid osmotic pressure.) Thus, fluid is drawn back into the capillary at that end. However, the relative area of outward filtration along the capillary is greater than that for absorption, so that on average, more water and small solute molecules leave a capillary than enter it. Clearly this action could not continue unless some other action were present, for tissues would continually swell and blood would become increasingly concentrated. However, capillary filtration and absorption, as outlined above, serve to stir the fluid contents of the body. Fluid is kept in constant motion from blood to interstitium and back to blood. This is important as it helps to promote the even distribution of metabolic energy sources as well as the end products of metabolism. In addition to filtration and absorption as outlined above, simple diffusion of small

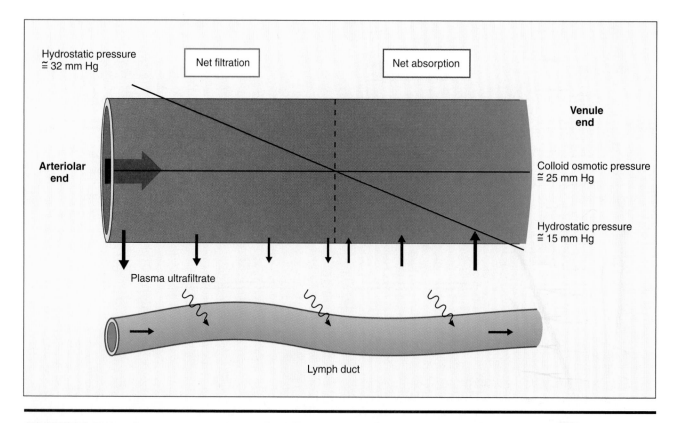

Figure 9.16 Illustration of the pressure gradients along a single capillary. The total area in which the net force for filtration exceeds that for reabsorption is greater than that for net reabsorption.

HIGHLIGHT 9.2

Fluid Movement and Colloid Osmotic Pressure

Fluid movement across capillaries is determined by two forces. Water and small dissolved molecules such as glucose and salts move out of the capillaries in response to the blood pressure within the capillaries. Opposing that force is the osmotic pressure contributed only by the large protein molecules within the capillaries. This pressure is called the colloid osmotic pressure.

Consider the situation pictured below. A selectively permeable membrane separates two solutions. The membrane has pores that allow the sugar molecules to pass freely across it, but are small enough that the larger molecules cannot cross. Each solution contains sugar molecules dissolved in water. No water movement takes place. Now protein is introduced into side 1. The protein molecules are so large they cannot pass through the pores in the membrane. Water moves into side 1 at a faster rate than water leaves (see Chapter 2) because the

concentration of water in side 1 is lower than in side 2. Because the proteins cannot cross the membrane, net movement of water and sugar continues from side 2 to side 1. As the water level rises in side 1, the rate of movement of water molecules in that compartment increases. When the increased hydrostatic pressure becomes sufficiently large, the water molecules on both sides begin to move at the same rate, as in (C).

If the difference in the height of the column of water between the two compartments is measured when net movement stops, we can determine how much pressure must be applied to the system to prevent water movement. In this case, the hydrostatic pressure difference is called the colloid osmotic pressure. Another way to measure this pressure is to apply a direct pressure to the solution side, sufficient to stop any net movement of water. This is diagrammed in (D).

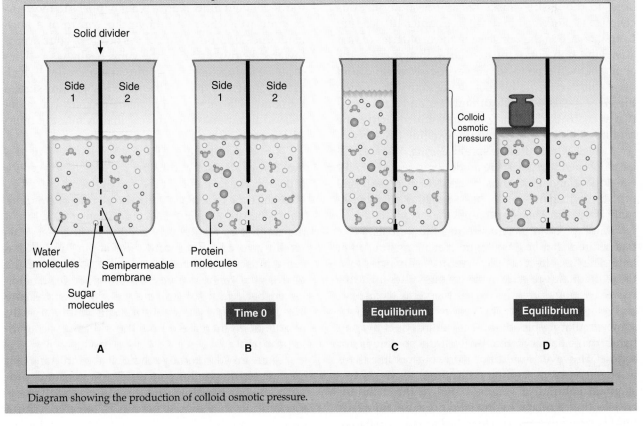

Diagram showing the production of colloid osmotic pressure.

molecules across the capillary occurs and accounts for the major exchange of those molecules. Diffusion is approximately 40 times as great as filtration. However, only filtration is responsible for **bulk flow** of fluid into the interstitial compartment. The combination of bulk flow and diffusion ensures that the interstitial compartment is well-mixed.

Lymph and Fluid Balance

Throughout all tissues of the body are other channels that take up the excess fluid that was filtered out of the capillaries and carry it back to the circulatory system. These channels are the **lymph vessels.** They are separate from the circulatory system, yet they form an intricate network of channels, known as **lymphatics,** that merge into larger and larger vessels as they approach the vena cava. At the level of the thoracic vena cava in the chest, the large lymph vessel coming from the lower extremities empties its contents back into the blood. In this way the excess fluid filtered out of the capillaries is returned to the blood. A similar network for the upper part of the body exists and empties its contents back into the blood. This system acts to maintain a constant fluid balance; the excess fluid that filters out of the capillaries is drained back into the blood by the lymph vessels.

Often you may experience its workings. A swelling such as a swollen ankle is evidence of a large imbalance between capillary filtration and absorption in the area of swelling. More fluid filters out of the capillaries than lymph vessels are able to return to the vena cava, and accumulates in the tissues. The accumulation is seen as the swollen ankle.

If you fall and bump your shin, swelling occurs a short time later at that spot. The mechanical damage done to the tissue causes release of some noxious chemicals (described in detail in Chapter 10), leading to an increase in the pore size in the capillaries of that area. Filtration out of the capillaries then exceeds the ability of the lymph to carry away the excess fluid, and one experiences swelling following injury. With time the injury heals, pore size diminishes, and fluid is carried away by the lymph. Some fluid returns across the capillary into the blood, and the swelling disappears. An athlete who suffers a sprain during a basketball game immediately has the ankle wrapped in a tight elastic bandage. The bandage serves to increase tissue pressure, which not only opposes swelling but may actually force fluid back into capillaries.

Another example illustrating this imbalance is that soldiers standing at attention for a long time often faint. The first effect of quiet standing is an increase in the hydrostatic pressure in the venous system. The increased venous pressure causes the veins to distend and retain more blood. This venous pooling removes some circulating blood and acts almost as if a small hemorrhage has occurred. By itself the pooling is not serious. However, in addition to venous pooling there is an increase in capillary filtration, as the semi-stagnant column of venous blood in the veins increases the capillary hydrostatic pressure. So much fluid leaves the capillaries that the blood volume may be severely depleted, causing a reduction in blood pressure. When blood pressure falls low enough, as in the case of the standing soldier, the brain can no longer be supplied with sufficient blood and the soldier faints. Fainting is an appropriate homeostatic, physiological response to the prolonged standing. As the body becomes horizontal, it allows for the movement

of venous blood and lymph back into the circulation. A simple remedy is available. Soldiers standing at attention for a long time are instructed to contract their leg and arm muscles periodically. This ensures that venous return is increased and that capillary pressure does not rise to levels sufficient to cause fainting.

The net filtration out of the capillary can be increased in several ways. The most direct way is an increase in the hydrostatic pressure at the arteriolar end of the capillary. Such an increase raises the pressure line along the capillary shown in Figure 9.16 and causes a greater length of the capillary to be devoted to filtration. This does happen, but not too often. Increased capillary pressure is more often due to a change in the pressure at the venous end. If the hydrostatic pressure at the venous end is increased, the average pressure in the capillary increases. This happens quite often. For example, if a limb is left relatively immobile in a vertical position for some time, the column of blood in the veins increases the venous pressure in the dependent limb. Notice what happens to the veins in your arm when you move the arm from a horizontal position down to the vertical. As your arm is lowered below the heart, the veins begin to swell, indicating an increase in pressure in those veins, which in turn increases the capillary hydrostatic pressure in the hands. The increase in capillary pressure causes increased filtration out of the capillaries into the interstitial space.

You experience this if you go to a movie and remove your shoes for comfort. Your legs dangle with little or no movement. At the end of the movie, when it comes time to replace your shoes, they seem tight. That is because the increased filtration of fluid out of the capillaries resulted in swelling of the feet. This swelling recedes soon after you begin to walk because muscular contraction forces blood toward the heart as the venous valves prevent blood from being forced backward toward the feet. Venous pressure falls, increasing capillary absorption. Simultaneously, lymph flow increases, and the shoes once again fit properly.

There are also some pathological conditions that may result in swelling, at times to enormous proportions. For example, the disease elephantiasis is a condition in which a person is infected with a parasitic worm. This parasite sometimes blocks the lymphatic drainage of a portion of the body. Because the blockage reduces the ability of the lymphatic system to carry away the fluid capillaries normally lose, that portion swells. In severe cases, a limb may swell to grotesque proportions, as shown in Figure 9.17. Any blockage of lymphatic drainage results in swelling.

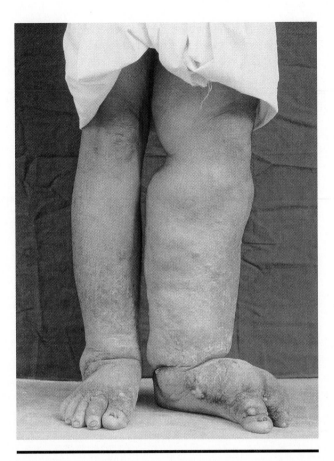

Figure 9.17 Photograph of a person with an advanced case of elephantiasis.

Control of Skin Blood Flow

You have all seen evidence of altered blood flow to parts of your skin. A most noticeable one is the increase in blood flow to the face, which we call blushing. Although the exact mechanism of blushing is not clear, there is some evidence that the central nervous system initiates the set of reflexes that causes blushing. In this case, the cerebral cortex is the initiator.

A more physiological local response is one that results from changes in environmental temperature. The skin is the major route of heat loss or gain, depending on the temperature of the surrounding air. For this reason, it is vital to regulate blood flow to the skin. The heat lost to the environment ultimately comes from the blood that perfuses the skin. Following exposure to cold, the skin of lightly pigmented people is seen to blanch, or become whiter in appearance. With continued exposure to the cold, the skin becomes redder. (Changes in skin color are most easily seen on the fore-

arm, where skin pigmentation is the least.) In the first instance, blood flow to the exposed area is reduced in response to cold. This is an adaptive mechanism, as heat loss through the skin may be great. In fact, in cold climates, sufficient heat may be lost through the skin to be life-threatening. Homeostatic mechanisms may be challenged. In the above case, the need to conserve heat by vasoconstriction may be counterbalanced by the need to supply sufficient blood to carry off waste products of cellular metabolism.

Reactive Hyperemia

Skin color changes to a more reddish hue following intense exercise of a muscle group or following occlusion of the blood supply to a limb. If a tourniquet is applied to an arm and kept in place for a few minutes, on its release, the skin color below the tourniquet reddens appreciably. The resultant vasodilation is due to a very localized reaction. During the period of occlusion, no blood flows to the skin or underlying muscle. Metabolism, however, continues even in the absence of an oxygen supply; this process is called **anaerobic metabolism.** Although metabolic processes may continue in the absence of sufficient oxygen, oxygen must be supplied later to make up for the period during which an **oxygen debt** accumulated. When the tourniquet is removed, the small vessels dilate, allowing the end products of metabolism to be carried away and more oxygen to be carried back to the tissue. Within a short time, the oxygen debt is repaid. (Oxygen debt is described in more detail in Chapter 14.) Thus, there appears to be a mechanism built into the circulatory system to help control local blood flow depending on the metabolic demands of the tissue.

White Reaction and Triple Response

If the skin is stroked lightly with a blunt instrument such as a dull pin, a whitish line forms along the track of the pin. The white line is most evident on the skin surfaces that are lightest in color, such as the forearm, and are due to direct mechanical stimulation of the underlying vessels, causing them to constrict.

If the pressure on the pin is increased, the track turns a reddish color instead. This **red line** is the result of dilation of the small vessels due to a stronger mechanical stimulation. A short time later, a reddening of the skin flares out, away from the track of the pin. This **flare** is followed by the **wheal,** a whitened area over the pin track, which is slightly raised from the surrounding skin. A strong mechanical stimulus injures the small blood vessels slightly, causing the release of histamine. Histamine has the ability to make capillaries more permeable, allowing some plasma proteins and fluid to escape from them. In effect, a very localized swelling or **edema** of the tissue occurs. This set of responses is known as the **triple response** and may be seen following a mosquito bite.

Control of Blood Pressure

As noted earlier, the average blood pressure of healthy people is approximately 120 mm Hg/80 mm Hg and is relatively constant. There are momentary changes in blood pressure, such as when a person changes from the lying position to the vertical one, but these do not constitute significant variations. When such a change does occur, the blood, which has pooled in the large veins in the upper part of the body, shifts to the lower portion of the body. As much as a half liter of blood may be transferred in this way. This has the effect of reducing the volume of blood returning to the heart. If venous return to the heart decreases, so must the cardiac output, as the heart cannot pump more blood

than it receives. However, blood pressure is restored to normal very quickly. How is this accomplished?

Baroreceptors

To learn how blood pressure is restored to normal, investigators in cardiovascular physiology worked out the controlling mechanisms with a series of simple experiments. The experiments were the outgrowth of observations that showed that each of the two major arteries in the neck, called the **carotid arteries,** contains a special area, called the **carotid sinus,** located at the point where the carotid artery divides into the internal and external carotid arteries. There is a slight swelling at that site, which is richly innervated. Investigators continuously monitored the blood pressure in the femoral artery of a leg of an anesthetized animal while the common carotid arteries below the carotid sinus were constricted. This procedure reduces the blood pressure at the level of the carotid sinuses. Figure 9.18 shows the results of such an experiment on an anesthetized dog, before and after cutting both vagus nerves. Almost immediately, systemic blood pressure increased and remained high, about 50 mm

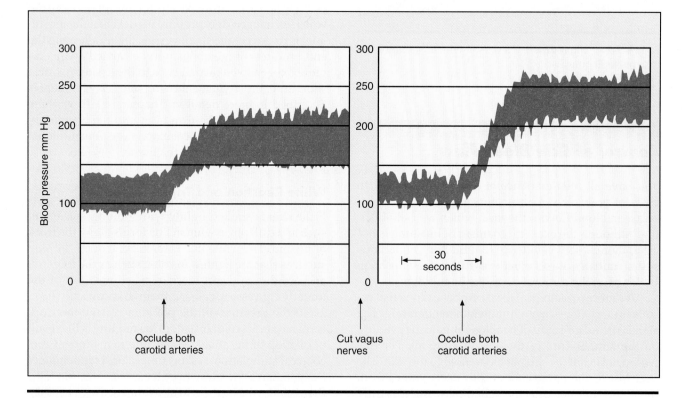

Figure 9.18 Change in blood pressure in an anesthetized animal after occlusion of the carotid arteries, before and after severing the vagus nerves. After the vagus nerves are severed, the blood pressure response to carotid occlusion is much greater than when the vagus nerves were intact.

Hg greater than control blood pressure. Following release of the carotid occlusion and section of the vagus nerves, the blood pressure was slightly elevated. However, on occluding both carotid arteries again, the increase in blood pressure was almost 100 mm Hg. The absence of the inhibitory effect of the vagus nerves on the heart allowed the blood pressure to rise much higher than in the presence of the vagus nerves. This experiment shows that the carotid sinuses are able to sense pressure and signal the central nervous system to respond appropriately. In addition, it illustrates the homeostatic control of blood pressure that involves both sympathetic and parasympathetic nerves.

Further investigation revealed that another site located at the beginning of the aortic arch has the same properties. Sensory fibers from that area have the effect of altering blood pressure in the same manner as the carotid sinus. Both these sites, called **baroreceptors,** are able to sense pressure and relay that information to the central nervous system. When blood pressure in the area of the carotid sinus or aortic arch falls, a decrease in action potentials ascends the nerves from those sites to a region of the medulla that controls blood pressure, called the **cardiovascular center.** That center then alters both the sympathetic and parasympathetic outflow to the heart, causing it to beat faster. In addition, generalized vasoconstriction is initiated, increasing the total peripheral resistance. The combination of increased heart rate and vasoconstriction serves to increase blood pressure back to the normal level, maintaining cardiac output and blood flow to the brain.

Occasionally, we are aware of this sequence of events. For example, if one stands up rapidly after sitting quietly for a while, that person might experience a slight feeling of dizziness. The dizziness is a result of a brief period of inadequate perfusion of the brain. This is even more evident following an illness during which a person is confined to bed for a few days. When he or she gets up for the first time, the dizziness may be sufficiently severe to require the person to sit down quickly. The extended time of disuse of the baroreceptor reflex leaves it a little sluggish. Instead of rapidly acting to reestablish blood pressure there is a short delay, during which time the brain is underperfused with blood, which causes the feeling of prolonged dizziness; on some occasions, this may cause the person to faint.

As stated above, one of the responses to a decrease in pressure at the carotid sinus and aortic arch receptors is a reflex vasoconstriction of the small arterioles in the body. Recall that the arterioles are responsible for the major resistance to flow. Should the total peripheral resistance rise, blood pressure also rises.

(Referring back to the analogy of the rubber bulb and tube in Figure 8.11 will be helpful in understanding this relation.)

Hemorrhage

The cardiovascular system contains many redundant feedback systems, all of which serve to maintain constant blood pressure. However, the most important mechanism is that of the baroreceptors, as described above. An excellent example of a cardiovascular homeostatic response is noted following loss of a moderate volume of blood. If one loses 500–1500 ml of blood, a series of events takes place to restore both the blood volume and blood pressure. First there is a decrease in blood volume, causing a significant decrease in venous return. The reduction in blood delivered to the heart causes a reduction in cardiac output. A reduction in cardiac output results, in turn, in a decrease in blood pressure. Once again, we see the relationship between flow, resistance, and pressure.

The series of immediate responses initiated following hemorrhage is illustrated in a flow diagram in Figure 9.19. The arterial baroreceptors sense the decrease in blood pressure and signal the cardiovascular center in the brain. In response to that signal, the sympathetic nervous system is activated while the parasympathetic system is inhibited, causing an increase in the rate and strength of the heart beat. Following the flow diagram, you see that the changes produced by the autonomic nervous system act in concert to restore blood pressure to near normal levels. Activation of the sympathetic nervous system causes the heart to increase its rate as well as the force of contraction. Cardiac output increases as a result of both these changes. Inhibition of the parasympathetic nervous system adds to the effect of sympathetic stimulation because inhibition of the heart rate is diminished, allowing the heart to beat faster. An increase in cardiac output increases blood pressure from the reduced value following hemorrhage.

Increased sympathetic discharge has two other effects. One is to cause generalized vasoconstriction at the arteriolar level, which results in an increase in total peripheral resistance. Because blood pressure is a function of total peripheral resistance, an increase in TPR has the effect of causing an increase in blood pressure.

Additionally, sympathetic discharge causes constriction of the large veins. This forces blood stored in the veins back to the heart, increasing venous return and thus increasing CO.

$$\text{Mean arterial blood pressure} = \text{CO} \times \text{TPR}$$

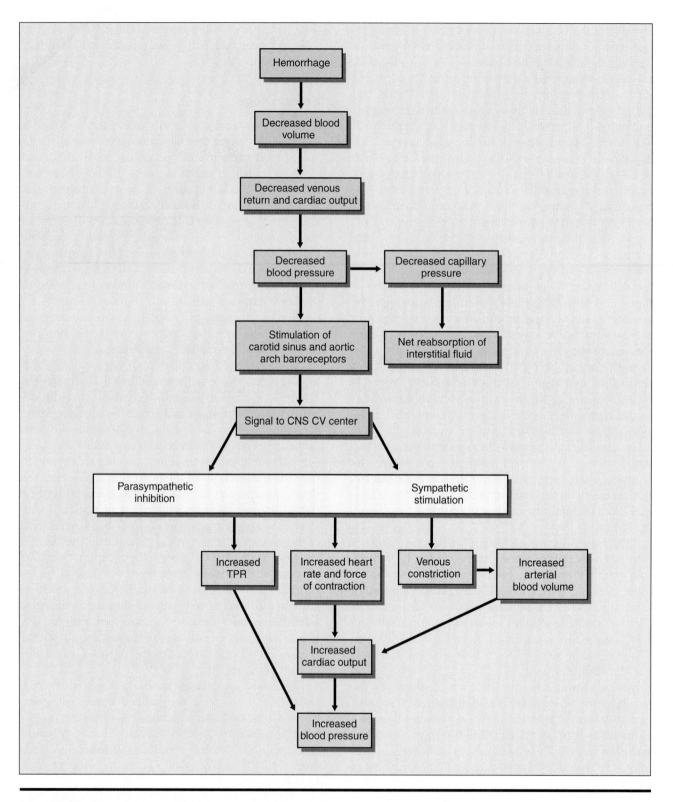

Figure 9.19 Flow diagram of the homeostatic cardiovascular responses to hemorrhage.

At the capillary level, other mechanisms are set into play to restore blood volume. As blood pressure falls with hemorrhage, capillary pressure also falls (refer back to Figure 9.16). If the hydrostatic pressure in a capillary is reduced for any reason, less fluid leaves the capillary and more fluid is reabsorbed from the interstitium than is filtered out of the capillary. The net effect is to increase the blood volume. Again, an increased blood volume serves to increase venous return and CO. We find evidence of this from hematocrit changes following hemorrhage. Shortly after blood loss, the hematocrit diminishes. This is the result of interstitial fluid being drawn into the capillaries, diluting the remaining blood in the system and lowering the fraction of red cells in that blood. However, secondary responses are delayed. Loss of blood triggers the feeling of thirst, often quite intense. Drinking water helps restore the blood volume. At the same time, the kidney is signaled to reduce urine output, helping to conserve body water. The bone marrow is stimulated and red cell production increased so that within a few days to a week or so, depending on the severity of hemorrhage, the red cells are replaced. All these events are sufficient to restore blood pressure to near normal values. However, if the hemorrhage is more severe, loss of consciousness occurs, and if blood pressure cannot be supported, **shock** results. Shock is a condition of circulatory failure. It is marked by pallor and clamminess of the skin, increased heart rate, and often unconsciousness. Unless a blood transfusion is initiated, shock may be irreversible.

Behavior Leading to Heart Problems

Factors other than physiological ones may alter blood flow and affect the heart. Many drugs cause local changes in vascular resistance; others may act on the heart to change its rate or even interfere with its normal function. Behavior may have a serious effect on the cardiovascular system.

Alcohol

The effects of alcohol on the cardiovascular system range from benign to deleterious. Often a common medicinal use by laypeople is to provide a feeling of warmth following exposure to cold. Although alcohol does provide this feeling, the warmth is misleading. Alcohol causes vasodilation of skin vessels. The increase in blood flow to the skin warms the skin, giving the person an overall feeling of warmth. However, this increase in blood flow to the skin actually means that more heat is lost to the cold air, as the skin becomes warmer due to the increase in blood flow. Thus, despite the feeling of warmth, the individual actually loses more heat than if alcohol had not been administered. For this reason, drinking alcohol in a cold climate may be unsafe.

Although the major acute effects of alcohol are usually associated with disturbances of the central nervous system, alcohol has many far-reaching effects. It is a drug that may lead to profound changes in metabolism. In fact, when high blood concentrations are attained during binge drinking, alcohol can produce atrial arrhythmias of the type discussed in Chapter 8. All arrhythmias are potentially dangerous and can be fatal.

Chronic alcohol abuse is a more serious problem than occasional binge drinking. Chronic alcohol abusers have a much higher mortality rate than the general population. Their alcohol abuse leads to a high incidence of hypertension and cerebrovascular accidents. If a cerebral vessel ruptures or a clot forms within it, or some other constriction reducing blood flow occurs, the person may become paralyzed in some area, depending on the site of the vessel. This condition is known as a **stroke.**

Smoking

Although smoking is often thought of as just a risk factor for cancer, it is also a risk factor for heart disease. Those who smoke on a regular basis have triple the death rate from heart attacks, particularly in the young. There are some clues as to why this is so. Smoking may increase damage to the interior of blood vessels, making them more susceptible to clot formation. Formation of clots in the coronary arteries may occlude blood flow to large areas of the heart muscle, causing it to die. In addition, the nicotine inhaled increases the susceptibility of the heart to arrhythmias. On top of this, the carbon monoxide released from the tobacco smoke reduces the ability of the blood to carry oxygen. Adding all these deleterious effects together means that smoking adds a significant risk of heart attack.

Special Situation: G Forces

Modern technology places many unusual demands on the body, one of which appeared with the advent of advanced aircraft. It was noted during World War II

MILESTONE 9.2

Diet and Heart Attacks

Heart disease is one of the most common causes of death in industrialized societies, and is the leading cause of death in the United States. Although there is reason to believe that some of the mortality from heart disease is associated with psychosocial factors such as aggressive competitiveness, strong striving for success, and anger, this has not been proven. However, other risk factors have been clearly associated with coronary artery disease. For example, diet appears to play a major role in the risk.

Cholesterol. The coronary arteries are prone to form **plaques** on the inside of the vessels, and these plaques narrow the diameter of the coronary arteries. In some people, the narrowing is so great that insufficient blood can be pumped through them to supply the heart muscle with sufficient oxygen. When this happens, the person is said to suffer a myocardial infarction, or heart attack. We now know that these plaques are composed of a fatty substance called **cholesterol.** It would seem at first that if one simply refrained from eating foods containing cholesterol, plaque formation would not be possible. However, the solution is not quite so simple.

The liver is able to make cholesterol from certain foods. Therefore, simply avoiding ingestion of cholesterol does not in itself afford protection. Long-term epidemiological studies, in which a large number of people are studied for many decades, have helped to show that genetics is a major contributor to cholesterol levels. However, diet is also a contributing factor. Diets high in saturated fats stimulate cholesterol production, raising its level in blood. These fats are most abundant in fatty red meats, butter, and some milk products. On the other hand, evidence exists that another type of fat, the polyunsaturated or monounsaturated ones such as olive oil and corn oil, do not promote cholesterol production in the same way.

Actually, total cholesterol is the sum of two different types: **low-density lipoprotein** (LDL), sometimes called "bad" cholesterol, and **high-density lipoprotein** (HDL), or "good" cholesterol. HDL is called "good" cholesterol because it transports cholesterol from cells to the liver for elimination and offers some protection against plaque formation in the coronary arteries. LDL, on the other hand, carries cholesterol to cells, where it may be deposited in the arteries, and so is implicated in the formation of plaques. For this reason, it is important to reduce ingestion of saturated fats, replacing them with mono- or polyunsaturated fats. When fat accounts for more than 40% of the calories in the diet, the risk of heart disease increases.

Salt. People with high blood pressure are often advised to refrain from using salt because a high salt intake may lead to water retention. If fluid is retained, the blood volume increases. An increase in blood volume may cause blood pressure to increase as the volume of the entire vascular system increases. Although this set of responses can be shown to occur in some susceptible people, it is not a universal phenomenon. The genetic disposition of individuals may determine their response to salt intake. In experiments with rats, researchers have shown that certain rats are salt sensitive; that is, their blood pressure is a direct function of salt intake. It seems that the conservative approach may be best for the general population. Not knowing our precise genetic makeup, it may be wise to go easy with the salt shaker.

Obesity. Excessive body fat, or obesity, is also associated with risk of heart attacks. Long-term studies have clearly shown that this is an independent risk factor that is not solely a reflection of a diet high in fat. One risk factor associated with obesity is the lack of physical activity. Active people are less likely to suffer heart attacks than those who are sedentary. Physical activity usually prevents obesity, and an active lifestyle in itself increases the HDL component of cholesterol, so there is a benefit to activity beyond a lack of obesity.

that pilots often fainted during the pull-out from a steep dive. This hazard was investigated using a linear deceleration sled. A person was placed in it so that gravitational (G) forces could be applied to the volunteer in a controlled situation, as shown in Figure 9.20. It quickly became evident that as the G forces increased, blood was shunted away from the upper part of the body. It was instead forced into the lower parts, expanding the large distensible veins, trapping a significant proportion of the total blood volume. This reduced venous return to the heart, causing cardiac output to fall. The brain was deprived of sufficient blood and the person fainted. In theory there is a simple cure for this: Design a flying suit such that when G forces exceed some minimum value, the lower part of the body is compressed by the suit. If the compression matches the G force, the veins will be prevented from stretching and venous pooling of blood will not occur. Although this is a phenomenon very few of us have to worry about, it does illustrate an important point: Excessive venous pooling reduces cardiac output and may have serious consequences.

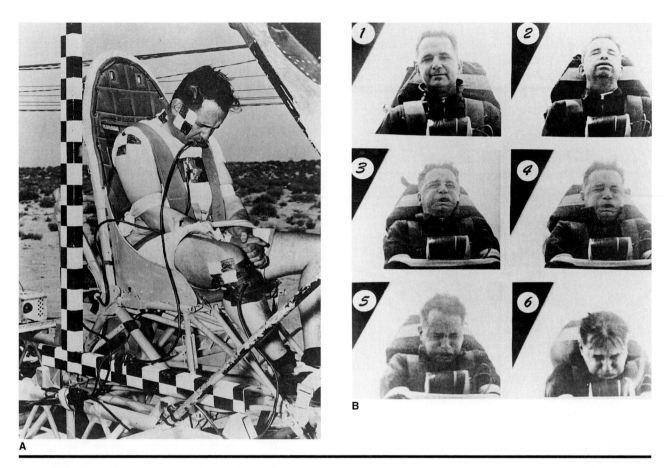

Figure 9.20 (A) Photograph of a volunteer in a human linear deceleration sled. (B) Sudden stop from supersonic speed.

Summary

The circulatory system serves as the transportation system for the body, carrying both nutrients and end products of metabolism to and from all the cells of the body. Included in the transportation scheme are oxygen and carbon dioxide, which are carried by the erythrocytes of blood. Other blood cells are the white cells and the platelets. White cells function mainly as a defense system against invading bacteria and virus particles. Platelets are responsible for normal clotting at the site of vascular damage.

Plasma contains the dissolved molecular elements of blood. All the salts (Na^+, K^+, Ca^{++}, Cl^-, etc.) are dissolved in the plasma and freely pass across the capillary walls. In this way, all cells are bathed in the same medium, which is necessary to sustain life. In addition to these small molecules are the larger protein molecules, albumin and globulin. The plasma proteins function as carrier molecules and antibodies. Additionally, the proteins determine, in large part, the net movement of fluid into and out of the capillaries.

At the local level, blood flow may be altered by neural and local factors. Local injury results in small areas of vasodilation and damage to the capillary epithelium. This allows fluid to escape the capillaries, causing local swelling.

The carotid and aortic baroreceptors sense blood pressure. When blood pressure falls, a signal is sent to the cardiovascular center in the brain. The response is to increase sympathetic output and decrease parasympathetic output, restoring pressure to normal levels. The elasticity of the arteries may also determine blood pressure. The stiffer the arteries, the higher the systolic blood pressure. High blood pressure is a serious risk factor for heart attacks and stroke. The risk may be reduced by behavior changes that cause a reduction in blood pressure. Low-fat diets and exercise aid in this respect. For some people, a diet low in salt is also advisable. Also, avoiding drugs that pose a health risk as well as avoiding smoking reduce the risk of cardiovascular disease.

Questions

Conceptual and Factual Review

1. What are the two circuits in the circulatory system? (p. 237)
2. Describe the components of blood. (pp. 239–240)
3. What is hemoglobin? What is its composition? Its function? (p. 241)
4. What is anemia? What is sickle-cell anemia? (p. 241, p. 242)
5. Where are red blood cells produced? (p. 242)
6. What hormone regulates red blood cell production? Where is that hormone produced? (p. 242)
7. What is the function of red blood cells? White blood cells? Platelets? (p. 240, p. 243)
8. What is the effect of altitude on blood? (p. 243)
9. Describe the clotting process and why and when it occurs. (p. 243, p. 244)
10. What is the difference between clotting and agglutination? (p. 243, p. 244)
11. What is hemophilia? (p. 244)
12. What are the major blood types? (pp. 244–246)
13. What forces determine fluid movement into and out of capillaries? (pp. 249–251)
14. What function do the lymphatics play in the homeostasis of fluid balance? (pp. 251–252)
15. Why does a swelling often appear at the site of injury? (pp. 253–254)
16. What is the function of the aortic and carotid sinus baroreceptors? (pp. 254–255)
17. If the sympathetic nervous system is stimulated, how is the heart rate affected? Force of contraction? Total peripheral resistance? (p. 256)
18. If the parasympathetic nerves are stimulated, what effect does that have on the heart rate and the force of contraction? (p. 256)

Applying What You Know

1. What are the advantages or disadvantages of holding the Olympic games in Mexico City, which is at an altitude of 7200 feet above sea level?
2. The largest animal to have lived was the dinosaur Mamenchisaurus. It was a long-necked creature stretching about 130 feet from head to tail, weighing many times more than present-day elephants. Its neck alone was about 25 feet long and its legs were about 10 feet long. What circulatory problems would this creature have had that humans do not have? How might its physiology have differed from ours to accommodate the unusual demands of its body shape and size? Think about the problem of moving its head from ground level to the vertical position, a distance of about 35 feet. Consider the capillary pressure differences in its legs, which were as much as 15 feet lower than its heart.
3. Following a bite by a mosquito, a small area on the skin becomes raised and whitish. Why?
4. If a person sprains an ankle, he or she is told to apply an ice pack to the painful area for some time. If it is still painful and swollen the following day, that person is advised to apply warm packs to the same area. What is the reason for the cold treatment followed by warm?
5. What would be the consequences of a leaky AV valve? A leaky aortic valve? How might they be diagnosed?
6. Children suffering from severe malnutrition, particularly from protein deficiency, often have large, swollen abdomens. Can you think of any reason for this? Remember that the protein concentration of plasma affects fluid movement into and out of the capillaries.
7. A person with type A blood needs an immediate transfusion. What blood type may the donor be? If the patient is type AB, what blood type could the donor be? If the patient is type O?
8. What are the consequences of a deficiency of white cells?

9. What changes in heart rate, skin temperature, color, and blood pressure would be expected in a person who sustains a moderate hemorrhage? What factors at the capillary level would come into play to restore blood volume?

10. A person inadvertently ingests a small amount of a foreign chemical. Following ingestion, the person complains of feeling ill, and you notice that her skin and the whites of her eyes have a yellowish cast. What might be the reason for these symptoms?

11. A man visits his doctor complaining of periods of very rapid heartbeat. After a thorough examination, nothing is found to be physically wrong with the patient. The physician notes that her patient wears a starched shirt with a tight collar and necktie. "Aha," says the physician. What is a probable cause of the patient's complaint?

Chapter 10
Defense Systems

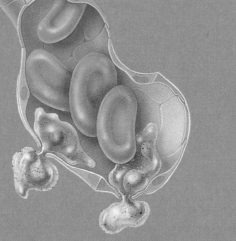

Objectives

By the time you complete this chapter you should be able to

◆ Understand that many bacteria, viruses, and parasites have the ability to harm us, yet we are able to ward off most of those attacks without medical help because our bodies have many complex defense systems.

◆ Understand that the first line of defense is to prevent foreign organisms from entering the body and the second line of defense attempts to destroy the organisms that breach the first line.

◆ Explain the difference between nonspecific responses and specific responses.

◆ Describe the cell types that play a role in protecting the body from invading organisms, as well as the site of their production and the signals that cause their release.

◆ Understand the limitations of the defense system and the problems that arise when it does not function properly, as in autoimmune disease.

◆ Describe the nature of hypersensitivity.

◆ Have a working knowledge of the immune system, which should allow you to recognize some of the problems of organ transplantation.

Introduction

All life is continuously exposed to potentially harmful invasion by foreign organisms or damage by toxic substances. Bacteria and viruses surround us, and if they enter the body, many can cause serious harm. Daily life puts us at risk of infection from a variety of substances, both living and nonliving, yet modern men and women may be expected to live into their seventies with relatively good health. The potential sources of danger are usually warded off by a host of defense mechanisms that act at the precise points of possible attack by harmful organisms, called **pathogens.** These organisms can be as small as virus particles or as large as parasitic worms. Considering the variety and universal nature of the potential threats to our health, our defense systems are truly a wonder of evolution.

Defense Systems

Some of the systems of the body that defend it against invasion by microscopic organisms and toxic substances are quite simple, and even obvious. Others are more complex, but still evident to the astute observer of human anatomy. The most complex defense is that against pathogenic organisms that are invisible to the unaided human eye, and are only now understood in any detail. Most of this chapter is devoted to this defense system, called the **immune system.** First, however, we discuss the first line of defense against pathogens.

Skin

The outer covering of your body, the skin, not only determines much of your shape by covering the underlying bone, muscle, and fat, but also serves as the first line of defense against invasion by harmful microorganisms. It is impermeable to most chemicals and protects you from a host of toxic agents. As long as the skin remains intact, neither bacteria nor viruses can penetrate it and gain access to the internal structures of your body. However, if your skin suffers a cut or an opening of any kind, such as a sore or an abrasion, microorganisms have access to the interior of your body.

The importance of a barrier to infection may be appreciated when one considers the rate of death from infections of a few centuries ago. In the mid-nineteenth century, the death rate of women in childbirth was as high as 20% in some hospitals. They died mainly from infections caused by the attending physician's dirty hands. Bacteria on the physician's hands were transferred to the woman during childbirth, as there was usually some break in the lining of the birth canal during delivery. As a result, the patient developed a generalized bacterial infection, called childbed fever, that often resulted in death. In 1861, Ignaz Semmelweiss published *The Cause, Concept and Prevention of Child-Bed Fever,* pointing out that organisms on the physician's hands were responsible for the disease. When it was finally realized that the attending physicians carried the disease organisms on their hands, they began to wash their hands thoroughly before attending to patients. The introduction of soap and an appreciation of cleanliness reduced the incidence of life-threatening infections to a small fraction of what it had been until that time.

In addition to the physical barrier the skin provides, it also contains cells called fixed **macrophages,** or Langerhans cells, that are capable of engulfing and destroying foreign particles, bacteria, or viruses. These cells add another protective barrier.

Cilia

Although your body is covered with the impermeable barrier of skin, other routes of entry to the interior exist. The mucous membranes of the respiratory tract are potential sites of entry for particulate matter. As you breathe, whatever is in the air is inhaled and may be drawn into the trachea and lungs. However, **cilia** line the respiratory airways and continually beat upward. If any particle touches the surface of the respiratory tract, the beating cilia move it (along with some mucous) toward the throat, where it is usually swallowed and may be destroyed by the acid contents of the stomach.

The cilia act as one-way brooms, sweeping out debris. If the particles are large enough to cause a slight irritation, reflex coughing helps expel the offending particle from the deeper respiratory passages. Sneezing is also a response to particulate intrusion into the respiratory tract, and usually clears the upper passageways. Although some bacteria or viruses may reach the lungs and cause disease, ciliary movement reduces your chance of exposure to harmful agents.

Stomach

Another source of entry into the interior of your body is through your mouth. Although saliva contains enzymes that can kill many bacteria, usually not all are destroyed. It is not possible to eat without swallowing millions of potentially harmful microorganisms, yet most of us remain healthy most of the time. Why is it that the organisms we swallow every day generally do not harm us? As you will learn in Chapter 13, the stomach contains a very strong solution of hydrochloric acid; this acid is so strong that it kills most bacteria and viruses that come in contact with it. There are times, however, when not all organisms are killed in the stomach, particularly if a large number are ingested. In such a case, some organisms may escape the sterilizing action of the stomach acids and cause illness. For this reason, people are warned not to eat food that has spoiled, or raw eggs, which may contain an overwhelming number of harmful bacteria. Even the best defenses may be overwhelmed.

Urinary Tract

A potential route of entry to the body is via the urinary tract. Pathogens may move from the environment into the urinary tract, the bladder, and the kidney. However, the constant flow of urine serves to wash out those pathogens. In addition, urine is usually acidic and does not provide a good environment for bacterial multiplication. The washing action and the acidity of urine greatly reduce the chance of contracting a bladder infection. Women are more prone to urinary tract infections because their urethras are shorter than those of men.

Immune System

If bacteria or viruses enter a person's internal tissues, disease may result, but this does not mean that every cut and scrape will lead to infection. In this section, we discuss how the immune system functions and the mechanisms in our body that have evolved either to ward off illness or to respond to illness.

Immunity As early as the fourth century B.C., it was recognized that certain diseases could be experienced only once. In 430 B.C., Thucydides noted during a plague year in Athens, "it was with those that recovered from the disease that the sick and dying found most compassion. These knew what it was from experience, and had now no fear for themselves, for the same man was never attacked twice, never at least fatally." Although Thucydides knew nothing of what we call immunity, he described it quite well. What accounts for immunity?

Inoculation A partial explanation for immunity was put forth in the eighteenth century. One of the scourges throughout the world was the disease smallpox. "The pox" swept through communities, killing 10–20% of the population within a few months. At times, smallpox killed most of the people in a single village or town.

We know that the greatest killer among Native Americans of the eighteenth and nineteenth centuries was disease, particularly diseases to which the population had never before been exposed. However, it was discovered that any person who contracted smallpox and survived would never contract it again. It was this small piece of information that led to the technique of **inoculation.** A small amount of the pus from a smallpox sore of an infected person was introduced into the skin of another. The recipient soon got a mild case of the disease, which then produced a lifelong resistance to the disease. Soon, however, this method was abandoned as the inherent dangers were recognized. People immunized in this manner sometimes contracted severe cases of smallpox, which sometimes led to their death.

In 1798, physician Edward Jenner electrified the world with his paper on cowpox. Cowpox was contracted from cows and was similar to smallpox but much less severe. He discovered that milkmaids who had daily contact with cows often contracted cowpox. These women were immune to smallpox. He then inoculated humans with fluid taken from cowpox sores. As a result, inoculated humans developed an immunity to smallpox. The inoculation caused the body to manufacture some substance capable of warding off a specific disease. The power of inoculation is evident in that today smallpox has been entirely eliminated from the world through worldwide inoculation efforts.

Nature of the Immune System

Throughout life we are all subject to physical insults that are potentially harmful. We cut ourselves, suffer skin abrasions, and become targets for a host of invading organisms, yet we have the ability to recover from each encounter. Our ability to recover is due to the two types of responses: a **specific immune response** and a **nonspecific immune response.** The former is immunity conferred against a single pathogen, such as the smallpox virus. The latter is the body's ability to ward off invasion by a new pathogen, as when many bacteria enter the body through cut skin. This system defends nonselectively against most foreign organisms.

Specific Immunity

Immunity bestowed by inoculation is not universal in nature. Whatever happens to the physiology of the body, it happens only with respect to a single disease, such as smallpox. This means that when a person is inoculated for one disease, the person develops immunity for only that disease. (The reason cowpox confers immunity against smallpox is that the cowpox virus is very similar to that of the smallpox virus. This is an unusual occurrence.)

It wasn't until 1890 that the first clue arose as to the nature of the mechanism of immunity. In that year, Emil von Behring and Shibasaburo Kitasato discovered that the active component bestowing immunity was contained in a group of large protein molecules found in blood called **gamma globulin,** or **immunoglobulin (Ig).** These proteins are capable of binding to foreign bodies such as bacteria and viruses, and are called antibodies because they are literally substances that oppose foreign bodies. As noted in Chapter 9, each antibody is specific for a single foreign substance, called an antigen.

When an antibody binds to an antigen, it may inhibit the antigen's ability to harm the individual, or the binding may stimulate the destruction of the antigen by circulating white blood cells, which engulf the antigen along with the bacteria or virus particle carrying the antigen. Consequently, an antibody may attack one virus, but have no effect on other viruses or bacteria. Another feature of the antibody is that its ability to confer immunity may last as long as a lifetime.

Nonspecific Immunity

Perhaps without recognizing it, you have witnessed the nonspecific defenses of your body in action. At some time in your life you must have suffered a cut or an abrasion that did not heal quickly, and became slightly infected. The infection was manifest by an accumulation of a thick, whitish liquid at the site of injury, a fluid called pus. If a drop of that fluid were placed under a microscope, you could see millions of white cells, or leukocytes, as well as cell debris. This concentration of white cells occurred in response to the injury, as the formation of pus is not a normal occurrence. We now know the events that lead to the accumulation of white cells at the point of injury.

Local Phagocytosis

When a foreign body enters your tissues through a break in the skin, chemical signals are given off because the chemical nature of the foreign body is different from that of the cells in the body. That difference in chemistry acts as a signal to attract certain white blood cells, or leukocytes, to the damaged area. The chemicals that attract white cells include those that are produced by the invading organisms, as well as specialized molecules released by the damaged tissues.

Some white cells are always nearby, and they are quickly drawn to the invading organisms. These cells are the tissue macrophages, or large white blood cells that engulf foreign particles and destroy them. If the tissue macrophages are unable to destroy the foreign particles quickly, other white cells are recruited from the blood. Chemical signals attract them to the area of invasion, where they aid in the destruction of the bacteria or viral particles. The response of the white cells to the chemical signals is called **chemotaxis.**

From what you learned about blood vessels in Chapter 9, it seems that it should not be possible for a white cell to move from blood to tissue. However, the white cells attracted to the site of invasion are able to squeeze through gaps between endothelial cells in the vessel walls, moving much like amoebas. Figure 10.1 shows the sequence of events as white blood cells move across the capillary wall and engulf a single bacterium, destroying it. In Figure 10.1(A), leukocytes are shown exiting a capillary by squeezing through small pores in the capillary membrane. Once a leukocyte enters the interstitial space and contacts a bacterium, it extends **pseudopods,** temporary protrusions which engulf the bacterium and form a vacuole that traps the bacterium. During this process, cellular lysosomes may contact the bacterium before engulfment is completed (see the left side of Figure 10.1(B)), in which case lysosomal enzymes are released into the area of the infected tissue and may act directly on other bacteria. If the bacterium is engulfed completely (see the right side of Figure 10.1(B)), lysosomes enter the vacuole while it is within the leukocyte and destroy the bacterium internally. This process of engulfment is called phagocytosis and cells that engulf the bacteria are called **phagocytes.**

General Phagocytosis

In addition to the localized phagocytic response, there is a general one. Phagocytes not only circulate in the blood but are also fixed in other specific tissues. Phagocytic cells are found in the liver, **lymph nodes** (special tissue located along the lymphatic vessels), and the spleen. As blood and lymph flow through these organs, the fixed phagocytes remove foreign particles. Normally a few circuits through the body are sufficient to remove most, if not all, foreign organisms. However, if the number of invading organisms is large enough, the fixed phagocytes may not be able to "sterilize" the blood and a general infection may ensue. Figure 10.2 shows the wide distribution of lymphatic tissue in the body. You may have become aware of this tissue if you suffered a general infection that caused swelling and attendant pain in one or more of your lymph nodes.

Many white cell types participate in the response to foreign particles, although not all the details of the differences among the involved cells are known. Figure 10.3 presents a summary of the leukocytes present in blood. All cellular elements of the immune system (and the blood) are produced in the bone marrow by a class of progenitor cells called hemopoietic **stem cells,** although it is not known just what prompts stem cells to make any particular cell.

Two general categories of leukocytes exist, as pictured in Figure 10.3: the **polymorphonuclear granulocytes,** containing a many-shaped nucleus and cytoplasmic granules. The other general category is the agranular monocytes, which have a simple nucleus and lack granules. Within the category of polymorphonuclear granulocytes, the **neutrophils** are the cells that are most actively engaged in phagocytosis. The **eosinophils** have some phagocytic action, but the **basophils** probably do not. These latter two cell types are involved in the inflammatory responses to be described later. Of the agranular leukocytes, the **monocytes** are phagocytic, whereas the **lymphocytes** are not phagocytic, but some produce antibodies, as will be discussed shortly. Figure 10.3 also indicates that two classes of leukocytes in blood make up the majority of blood cells: The neutrophils and lymphocytes account for about 93% of all leukocytes in blood.

Leukocyte Proliferation   Within an hour or two of invasion by a foreign pathogen, leukocytes gather at the site of invasion to begin the process of host-defense. The first leukocytes to arrive are the neutrophils, followed by the slower-moving monocytes. As little as six hours later, the blood count of neutrophils may increase fivefold. This increase in the white count of blood can be used clinically as an indication of an infection.

Complement   Phagocytes are capable of binding to and destroying invading bacteria as well as engulfing dead cells and debris in the area of infection. However, they do not attack normal cells of the host. The phagocytes recognize both foreign and abnormal tissue and ignore normal cells. The surface of target cells is different from that of normal cells, allowing phagocytes to locate and destroy them.

Although unaided phagocytes attack foreign cells, they do so slowly. However, the ability of phagocytes to attack foreign cells and abnormal ones is increased many times by a class of molecules called **complement,** containing about 20 different proteins, that normally circulates in the body fluids. Activation of one of the proteins sets off a cascade of reactions that results in the production of two types of complement. Some molecules are able to attach directly to a bacterium and kill it. Another type of complement protein binds nonspecifically to the foreign

Figure 10.1   (A) Leukocytes leave a capillary and engulf bacteria in response to chemical signals released during infection.

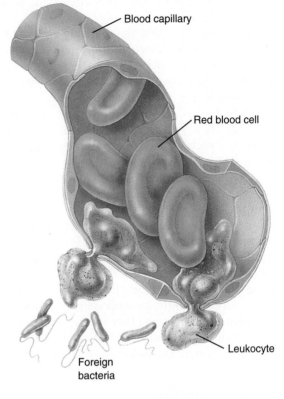

A

bacterium and to a specific receptor on the phagocyte, as shown in Figure 10.4. This couples the bacterium tightly to the phagocyte and allows for easier destruction of the bacterium.

Specific Immunity

The ability of leukocytes to bind to foreign bodies or produce antibodies that are able to bind to the invading organism is a very puzzling feature of the immune system. How is it that a leukocyte may bind to a specific cell it has never encountered before, yet not bind to any other cell? Binding requires the presence of a specific receptor molecule on the surface of the leukocyte for each specific invading antigen. As you learned in Chapter 2, the precise nature of a protein synthesized within a cell is dependent on genetic information stored in the genes of that cell. Because antibodies are specific for only one

(B) Diagram of a leukocyte destroying a bacterium. If the bacterium is engulfed entirely (right path), the internal enzymes of the leukocyte kill it. If it is only partially engulfed (left path), the leukocyte secretes enzymes, which kill it.

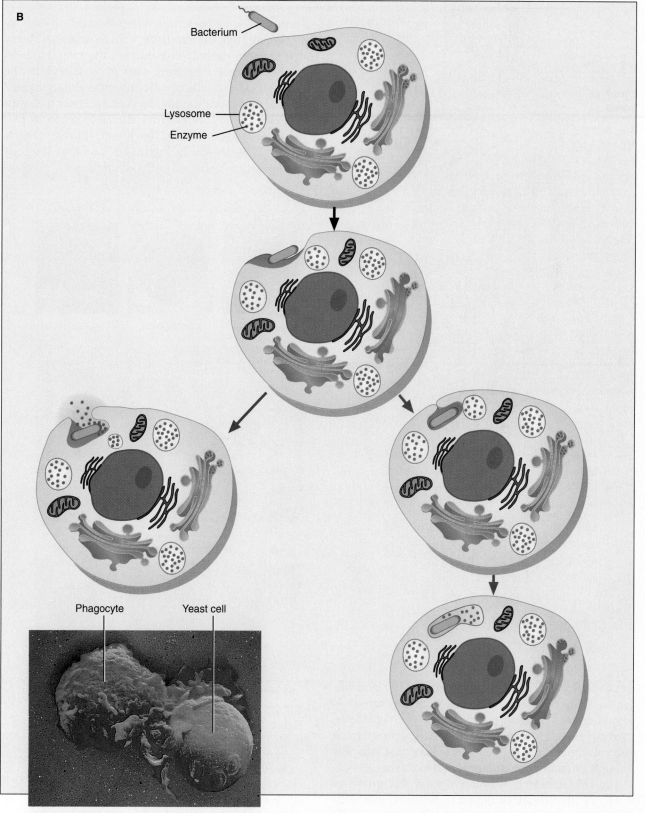

(C) Scanning electron micrograph of a macrophage engulfing a yeast cell (about ×7000).

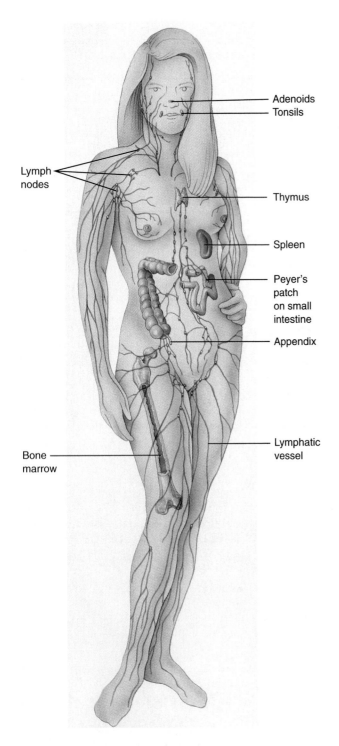

Adenoids
Tonsils

Lymph
nodes

Thymus

Spleen

Peyer's
patch
on small
intestine

Appendix

Lymphatic
vessel

Bone
marrow

Figure 10.2 Distribution of lymphatic tissue throughout the body.

surface. When a bacterium enters the body, it eventually encounters a lymphocyte with a surface antibody that matches the bacterial antigen, allowing it to bind to the antigen. The next step is to stimulate the lymphocyte to divide rapidly and produce large quantities of that specific antibody. We now know that antibodies are composed of four chains of proteins, two identical light chains and two identical heavy chains, arranged as shown in Figure 10.5, so that the four chains form a Y-shaped antibody molecule. The com-

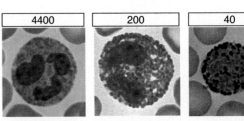

| 4400 | 200 | 40 |

NEUTROPHIL   EOSINOPHIL   BASOPHIL

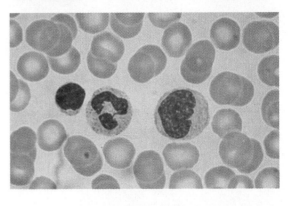

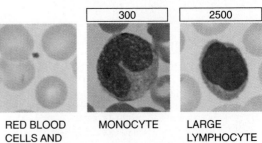

| | 300 | 2500 |

RED BLOOD
CELLS AND
PLATELET

MONOCYTE

LARGE
LYMPHOCYTE

Figure 10.3 Different types of leukocytes in the body. The two major divisions are the polymorphonuclear granulocytes (top row) and the agranular monocytes (the two cells on the right of the bottom row). The top row shows a neutrophil, eosinophil, and basophil. The bottom row shows the two agranulocytes; a monocyte and a large lymphocyte.

antigenic site on a foreign organism, how is it possible for so many different antibodies to be produced? It would require millions of genes, each one capable of signaling a cell to produce one specific antibody for each unique invading antigen. There just are not enough genes in your chromosomes to do the job.

In 1957, researchers began to solve that riddle as evidence from many laboratories indicated that each lymphocyte contains only one kind of antibody on its

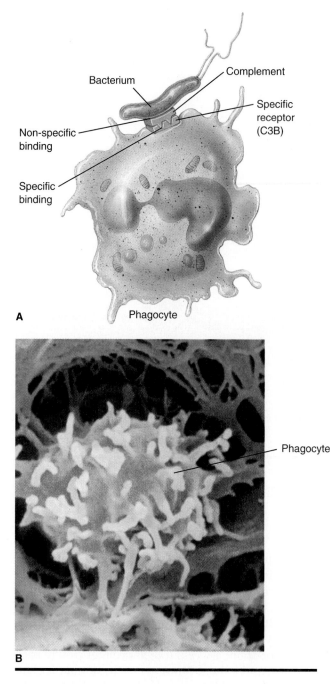

Bacterium
Complement
Non-specific binding
Specific receptor (C3B)
Specific binding
A
Phagocyte

Phagocyte

B

Figure 10.4 (A) Binding of a bacterium to a phagocyte with the aid of a complement molecule. The receptor on the cell surface binds to a molecule of complement, which in turn binds nonspecifically to an invading bacterium. (B) Photomicrograph (9000×) of a phagocyte.

bination of the light and heavy chains produces a specific site capable of recognizing one specific antigen. This means that literally millions of lymphocytes circulate, each one carrying a specific antibody. How can that be if, as stated above, there are insufficient genes to supply that many different protein molecules?

Research into this question unveiled an unusual finding. The genetic coding for the antibodies resides in genes within a gene. Four widely separated small genes, carrying different information in their DNA, randomly link with the others to form one larger gene coding for a single protein. Because of the number of these small genes and their ability to link together in a random manner, a million or more different proteins may be made. But in any single lymphocyte, only one type of antibody protein is assembled. Thus, each person may have a few million different kinds of lymphocytes circulating, each kind containing one specific antibody.

The variety of the antibodies is so large that the probability of a foreign antigen meeting an antibody that fits it is great. It is as if our lymphocytes make up a vast storehouse of receptors waiting for the right antigen to come along. When an antigen on the surface of a pathogen comes into contact with a matching antibody on the surface of a lymphocyte, binding occurs. This, in turn, is followed by a series of responses detailed below. At this point, it is necessary to understand only that there are literally millions of different antibodies incorporated into the membranes of the lymphocytes, with each lymphocyte containing only one specific antibody. Most of them remain functionless throughout a person's life because only a relatively small number will engage in an immune response.

Classes of Lymphocytes

Until the early 1960s, it was believed that only one type of lymphocyte existed. Then researchers recognized that there were actually two distinct types of lymphocytes, each beginning its life in bone marrow but maturing in a different organ. Each was also found to have a different function in the immune response. The lymphocyte maturing in the bone marrow is called a **B cell.** (The name *B cell* comes from the bursa of Fabricius in birds, where the B cell was first discovered.) The other type of lymphocyte is liberated into the circulation, but matures in the thymus gland beginning a few months before birth and continuing for some months following birth, and is called a **T cell.** The thymus gland, located in the upper part of the chest, is relatively large at birth but slowly atrophies until shortly after puberty, when it is essentially replaced by adipose tissue. Thus, if the thymus gland is removed a few months after birth, T cell immunity is not affected.

During adult life, the T cells are formed in lymphoid tissue throughout the body. There are three general types of T cells. One is a **killer T cell,** capable of destroying a target cell; another general type of T cell aids B cells to function and are called **helper T cells.** [Helper T cells respond to an antigen by proliferating

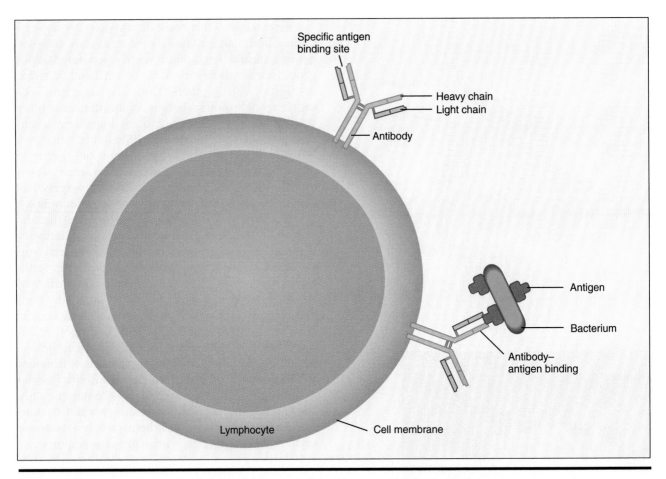

Specific antigen
binding site

Heavy chain
Light chain

Antibody

Antigen

Bacterium

Antibody–
antigen binding

Lymphocyte

Cell membrane

Figure 10.5 Binding of a bacterial antigen by a specific lymphocytic antibody. Lymphocytes containing a specific antibody to an antigen on the bacterial surface bind to the antigen and destroy the bacterium.

and secreting a class of compounds called **cytokines** that stimulate macrophage activation and antibody responses.] The third type is the **suppressor T cell.** Although less is known about the function of these cells, it is known that they suppress both the killer and helper T cells. It may be that through the mutual action of the helper and suppressor T cells, a negative feedback system acts to limit overreactivity of T cell production. The general scheme is shown in Figure 10.6.

B Cells When a B cell residing in lymphoid tissue is exposed to an antigen to which it contains an antibody, the B cell is activated. In addition, activated helper T cells also contribute to activation of the B cell. The B cell then enlarges and differentiates into a **plasma cell** that is carried by the lymph into the circulating blood. Plasma cells, in turn, release their antibodies into the blood and are the only cells that release antibodies into the circulation in response to an antigen. The circulating antibodies bind to and destroy the antigen-carrying pathogen.

If a foreign antigen is injected into an animal, the change in the antibody concentration in plasma may be measured. Following the first challenge with an antigen, the antibody concentration rises as shown in Figure 10.7. In a few days a measurable rise in the antibody concentration occurs and continues to increase for about a week. It then declines slowly back to near basal values. However, if the animal is challenged a second time, perhaps months later, the response is quite different. The increase in the antibody concentration may be more than a hundred times greater than that resulting from the first challenge, and the concentration will remain high for a much longer time. This simple experiment indicates that there must be at least two types of B cells. This is shown in schematic form in Figure 10.8.

Following the initial challenge with an antigen, the specific B cell containing the appropriate antibody proliferates, creating many copies of itself. Some of them actively secrete antibodies, and are called **effector cells.** Some of the newly formed B cells act as a pool of **mem-**

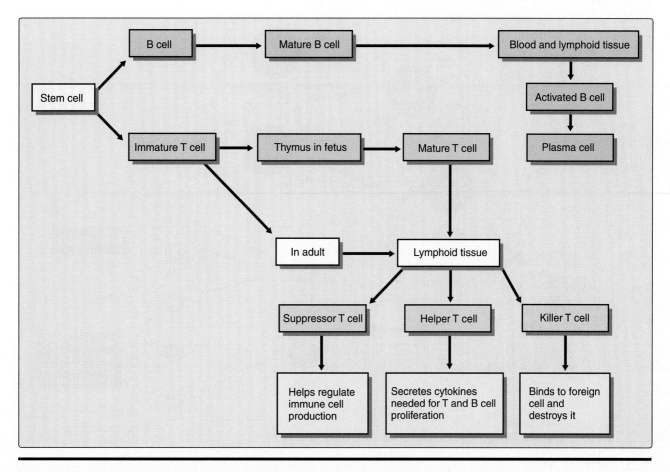

Figure 10.6 Scheme of T and B cell production from stem cells.

ory cells. They remain in the body for a long time after the first challenge, but do not secrete their antibodies into the circulation. However, if the same antigen is again injected into the animal, the memory cells rapidly proliferate, producing many more effector cells as well as more memory cells. This response confers complete or partial immunity on a person following a single exposure to some pathogens.

T Cells As discussed previously, two general types of T cells facilitate the immunity conferred by the antibodies of B cells: the killer T cell and the helper T cell. As their names imply, they either kill an invading pathogen or help in the destruction of that pathogen. However, unlike B cells, T cells cannot recognize an entire antigen. They recognize only antigenic fragments of the pathogen that become inserted in the membrane of an infected cell. When a virus enters a normal cell, some of its protein is fragmented. The fragments are brought to the surface of the infected cell by one of a class of molecules called the **major histocompatibility complex (MHC).** This exposes the antigenic fragment to the T cell, which may bind to it. The T cells then attack the infected cell, destroying it and inhibiting viral

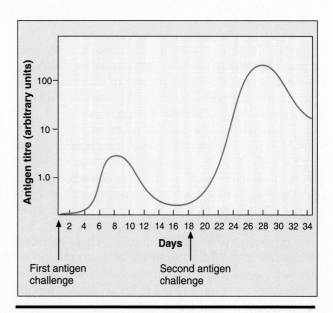

Figure 10.7 Graph of changes in antibody concentration in blood following the first and the second antigen challenge, with the ordinate as a logarithmic scale. The second exposure to the antigen results in a much larger response than the first exposure.

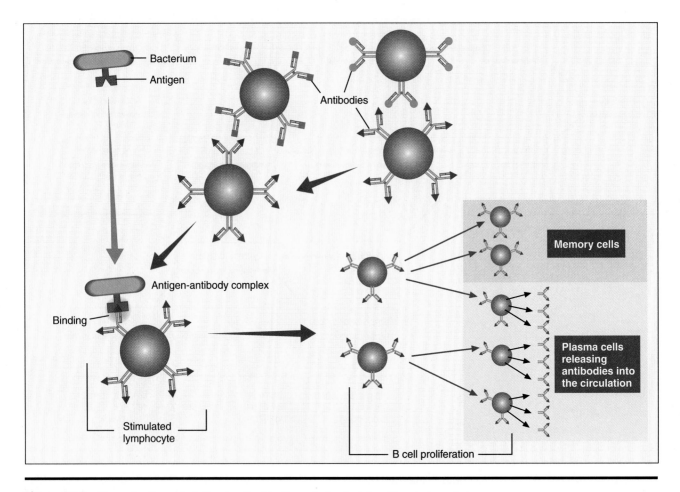

Figure 10.8 Cascade of events following the binding of a bacterial antigen to a surface antibody, causing proliferation of both memory cells and antibody-secreting cells.

replication. At the same time, recognition of the antigenic fragment by T cells triggers a set of responses described below. What is important to point out at this juncture is that these cells allow the immune system access to viruses that are buried deep within host cells and hidden from any circulating antibody.

If a killer T cell binds to an antigen, it destroys the cell invaded by the virus and kills the virus at the same time. If a T cell binds to the viral fragment, it secretes one of the cytokines. (At present, many cytokines have been identified, all of which are small protein molecules or peptides. For our purposes it is not important to identify each one.) These chemical messengers play important roles in immune responses, serving to communicate from cell to cell and modulate host responses to a challenge to the immune system. Cytokines activate other T cells, as shown in Figure 10.9, increasing the likelihood that the infected cells will be attacked by killer T cells. Additionally, cytokines stimulate B cells to enlarge and proliferate. Each activated B cell produces and releases a large number of antibodies. It is estimated that each acti-

vated B cell may liberate as many as ten million antibody molecules. Thus, invasion by a single type of pathogen stimulates a cascade of responses, all of which act to suppress the infection.

Leukemia

An uncontrolled production of white blood cells is called **leukemia** and is characterized by a large increase in the number of circulating leukocytes. This may be caused by uncontrolled, cancerous growth of leukocytes in lymph tissue or in the bone marrow. The leukocytes produced are usually abnormal and nonfunctional ones that cannot participate in the immune responses described above. People suffering from leukemia not only are open to infection due to the loss of functional white cells, but also have fewer red cells in circulation, as the production of red blood cells in the bone marrow is decreased. They are also severely debilitated as cancerous white cells lodge in lymphoid tissue, the liver, and other tissues. If not treated, the person may die of severe infections.

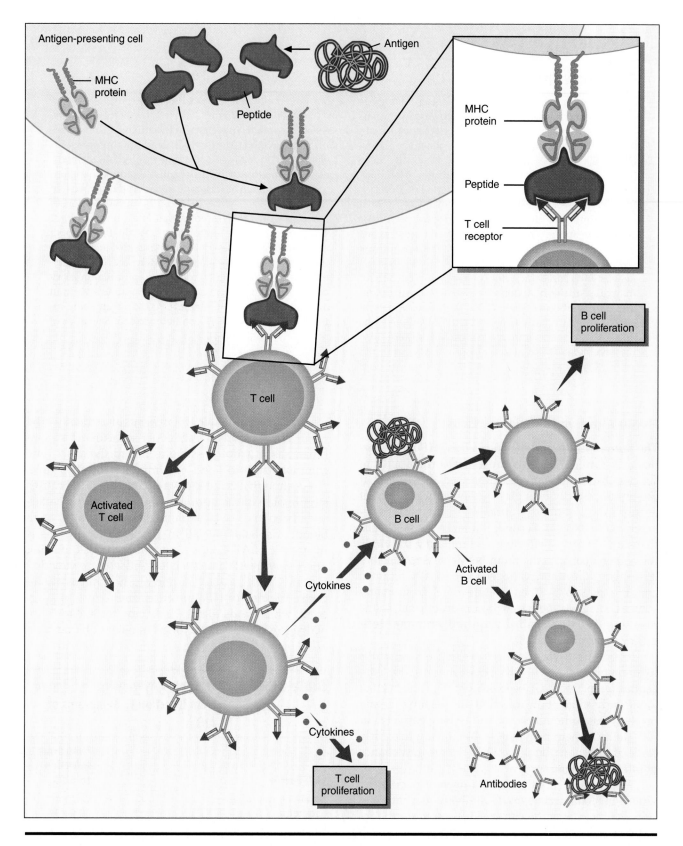

Figure 10.9 Binding of a T cell to an antigenic fragment, with the aid of an MHC protein, and the resulting cascade of events.

HIGHLIGHT 10.1

Autoimmunity

The immune system has the ability to discriminate between self and nonself. When invaded by foreign organisms it recognizes them as nonself, and appropriate responses are initiated to destroy or inactivate those organisms. But the immune system occasionally loses the ability to make that distinction and, as a result, fails to recognize normal cells. When that happens, T cells or B cells misdirect their antibodies to normal cells. Such an **autoimmune** disease may be organ-specific or non–organ specific, attacking individual target organs or even a few target organs. Although it is not entirely clear why this may happen, it appears that either genetic or environmental factors or a combination of both may cause the normal immune system to break down. The most common disease of the central nervous system is multiple sclerosis (MS), which is an autoimmune disease. The body's immune system fails to recognize the myelin sheath of neural tissue and attacks it, destroying many neural functions in the process.

Other autoimmune diseases, directed against the thyroid gland, the pancreas, or the adrenal glands, have been discovered. More widespread autoimmune disease may attack connective tissue throughout the body, as in rheumatoid arthritis, or cause a more general autoimmunity such as systemic lupus erythematosus. In the latter disease, widespread autoimmune responses result in a host of symptoms that affect many organs. The purpose here is not to enumerate the full range of autoimmune diseases, but to illustrate that at times the immune system acts inappropriately and causes disease. Treatment usually combines anti-inflammatory drugs with those that suppress the immune system in an effort to reduce the damage to the organ being attacked. Although this doesn't cure the disease, it does reduce the symptoms.

Human Immunodeficiency Virus

The **human immunodeficiency virus (HIV)** is a particularly virulent virus because it has the ability to survive the immune process described above. First, it can mutate rapidly, changing the antigenic portions of its surface, so antibody formation to the virus is severely handicapped. Second, it attacks the very cells that should be most effective in eliminating the virus: the helper T cells. As their number decreases, the ability to ward off other pathogens is critically reduced and the person is open to infections from many sources. The immune system becomes deficient, so the resulting disease is called **acquired immune deficiency syndrome (AIDS)**.

It takes from a few weeks to as long as six months following infection with HIV before the antibody concentration in blood can be measured. This means that in the early stage of the infection, the person can pass on the virus although he or she tests negatively for HIV. In fact, there is evidence that during this early stage of infection, the person has the greatest chance of infecting others.

After a few months, the number of circulating viruses in blood declines as the antibody concentration rises. At this stage, the chances of passing on the infection declines somewhat. During this period, which may last as long as ten or more years, the HIV-infected person may be completely symptom-free, despite the fact that the virus is still present, hidden in the tissues. However, the length of the latent period depends on many risk factors. Table 10.1 presents an estimate of the mean latent period for a series of risk factors. It is an estimate, as it is not possible to record the exact time from exposure to onset of symptoms for all people.

Although it is not known exactly how HIV causes suppression of the immune system, the virus preferentially enters T cells, depleting their concentration in blood. A substantial fall in T cell concentrations is diagnostic of AIDS. It is known that the risk of being infected with HIV is associated with some behavioral factors. Crack cocaine use and alcoholism greatly increase the risk of HIV infection due to loss of judgment, which leads to risky behavior. Intravenous use of drugs is also a risk factor, as transfer of blood from

Table 10.1 **Estimated Average Latent Period from Infection to Symptoms of AIDS.**

Risk Factor	Time from Infection to AIDS
Young hemophiliacs	1.5 years
Old hemophiliacs	10 years
Homosexual men	10 years
Heterosexual contact	10 years
HIV acquired through blood transfusion	6 years
Infection during organ transplantation or cancer chemotherapy	2–3 years

MILESTONE 10.1

Cancer

Cancer is essentially a genetic disease; chance mutations in any normal cell may turn it into a cancer cell. Rather than retaining its original characteristics, it develops the ability to reproduce abnormally and to **metastasize,** or spread to new sites. These mutations may occur by chance or be provoked by exposure to any of a large number of carcinogenic (cancer-producing) agents. Often it is not possible to know exactly why cancer strikes a particular part of the body. At other times, the reason seems clear, as in lung cancer of heavy smokers.

Cancer cells exhibit very rapid but undifferentiated growth in that they no longer act as they did in the tissue of origin. If the rapidly growing cells form a well-defined mass of tightly joined cells, the mass is usually considered benign and can be removed surgically with little problem. However, often the cells do not adhere to each other as they do in normal tissue. They tend to break away and are carried by blood and lymph to other parts of the body, where they continue to grow. These are the malignant, or dangerous, tumors known as cancer. The migrating cells, in turn, form new malignant tumors almost anywhere in the body.

If cancer is caught early, the malignant tumor can be removed surgically before metastasis has occurred. If it is diagnosed after metastasis, chemical methods or radiation may be used to destroy the cancerous cells. Powerful chemicals that act most strongly on rapidly growing cells, killing them, may be injected into the circulation. Unfortunately, these chemical agents may also kill some normal cells, although they act more potently on the cancer cells. For this reason, chemotherapy has many side effects, such as hair loss, nausea, and generalized weakness. However, untreated cancer is usually fatal because the cancerous cells crowd out normal cells and prevent normal tissue from carrying out its function.

Another method of treating cancer patients is with radiation therapy. A beam of ionizing radiation can be directed at specific sites in the body that contain the tumor cells. The ionizing radiation kills those cells, and often results in partial or complete regression of the tumor.

Although the immune system produces **cytotoxic T cells** that are able to destroy cancer cells, the complete destruction of cancer cells is seldom achieved without additional medical treatment. For reasons not well-understood, some cancer cells escape detection by the surveillance cells of the immune system. Early detection and treatment allow the best chance for recovery from cancer.

---

one individual to another by use of a common needle often transfers the virus.

Following a variable latent period, symptoms begin to appear as the virus concentration in blood rises. At this time, the person again becomes very infectious and is now considered to have AIDS. Because AIDS is a disease of the immune system, the symptoms are quite varied. The person has a very low resistance to most pathogens and may exhibit a wide constellation of symptoms. Often, bacteria that are seldom harmful become the source of a severe infection in people with AIDS. In addition, the immunodeficiency permits the development of cancers that are relatively rare in immunocompetent individuals. One such cancer is Karposi's sarcoma, a skin cancer rarely seen in people without AIDS.

Because the only way to contract HIV is through transfer of body fluids from an infected person, it is important to protect oneself during sexual intercourse. The only protection is either to avoid sexual contact or to use a condom during contact. AIDS affects both homosexual and heterosexual people. The virus may be transferred from male to female or from female to male, but women are more than twice as likely to become infected with HIV during sex with a man who has the virus as are men who have intercourse with women who have the virus. This probably is because the virus in the semen of the male remains in contact with the female for a longer period than does the virus of the female with the male.

Inflammation

Any injury to tissue, whether physical or chemical, may lead to **inflammation.** Inflammation is characterized by an increased blood flow to the affected area, increased capillary permeability with its attendant edema and swelling, and accumulation of granulocytes and monocytes. Pain is also symptomatic of inflammation. This set of symptoms is the result of liberation of a cascade of chemical substances from T cells, including histamine, prostaglandin, serotonin, and lymphokines.

Immediately after injury, macrophages begin phagocytic activity. This is followed by an increase in the neutrophil concentration in blood and invasion of the inflamed area with neutrophils. Some time later, many hours to days, bone marrow production of granulocytes and monocytes is increased. This stimulation of the bone marrow is caused by a large number of circulating factors released from activated macrophages. Among the more important initiating factors are interleukin 1 (Il 1) and tumor necrosis factor (TNF). These two factors appear to initiate the inflammatory process.

Hypersensitivity

Occasionally, an adaptive immune response occurs in reaction to an inappropriate situation, causing tissue damage. This is referred to as **hypersensitivity,** or **allergy.** When a person is exposed to an antigen such as pollen, dust, or some environmental chemical for the first time, there is little response. However, on subsequent exposures, chemical signals are released by sensitized cells of the immune system. In this case, they are the mast cells that are derivatives of basophils (See Figure 10.3), but are located throughout the body and do not circulate. They are associated with blood vessels of most tissues and exist in high concentrations in the mucosa of the lung and gut. Upon contact with an antigen, they release pharmacological mediators, one of which is histamine. Mast cell activation causes an inflammatory response that varies depending on the site of release. In general, blood vessels become more permeable, as noted in Chapter 9, and blood flow to the affected part is also increased. These changes contribute to the immune responses discussed above.

Some people are hypersensitive to certain antigens, and develop an allergy to what are usually benign substances. People with allergies to pollen know well the result of exposure: sneezing, reddening and perhaps itching eyes, as well as many other signs of a cold, but without the fever and body aches. These symptoms are due to hyperreactive mast cells that release inflammatory agents in improper amounts in response to an innocuous exposure. For some people this allergy may only be mildly irritating, but for others it might be life-threatening. Some people react so strongly to bee stings, or some other antigen, that a severe generalized response threatens their life. Fortunately, some simple remedies are quite effective in combating the generalized hypersensitivity response. Anti-inflammatory drugs as well as some that cause dilation of the trachea to ease breathing are available, often in aerosol form for rapid effects.

Another common reaction to nonspecific stimuli is **asthma,** which is an increased resistance to air flow in their respiratory tract on exposure to a variety of nonspecific stimuli. The asthmatic person experiences wheezing, coughing, and difficulty in breathing, all in response to some irritant in the air. The allergic response results in contraction of the smooth muscle in the trachea and bronchioles, restricting air flow. Although this is a relatively common disease and for most only mildly troublesome, the severity varies widely.

Summary

Numerous defense mechanisms have evolved to protect us from most pathogens that invade our bodies. Some are prevented from gaining access and others are destroyed in the body. Skin, cilia-lined mucosa, and stomach acids prevent a direct invasion or kill many of the pathogens. However, if this line of defense is breached, other mechanisms are available for warding off harm. Leukocytes with a nonspecific set of responses attack cells that are recognized as nonself. The ability to recognize such cells resides in specific receptor molecules on the surface of each leukocyte. Chemical signals given off by the leukocyte attract other leukocytes to the area of infection.

Shortly after a leukocyte encounters a foreign pathogen, the mobilization of leukocytes results in a cascade of responses. Both B cells and T cells proliferate, antibodies are released, and phagocytosis is stimulated. Additionally, memory cells proliferate. These cells contain the antibody specific against the invading pathogen. They are quiescent, however, until a subsequent exposure to the pathogen. The memory cells are then rapidly stimulated and the antibody response may be more than a hundred times greater than during the initial attack, conferring immunity against that pathogen.

At times, the immune system fails to function appropriately, as normal cells are not recognized. Antibodies are formed against the body's own organs, which are then attacked as if they were invading organisms. This autoimmune disease may be generalized or confined to one specific organ. Another form of an ill-adapted response is manifest as an allergy or hypersensitivity. On repeated exposure to pollen, dust, or certain foods, antibodies are released that cause mast cells to release a series of inflammatory mediators. Sensitized people may suffer symptoms ranging from sneezing to widespread rashes, or even experience life-threatening symptoms if there is severe respiratory involvement.

Questions

Conceptual and Factual Review

1. What are the first lines of defense against invading pathogens? (p. 263, p. 264)

2. When bacteria gain entrance to tissue, as through a cut or abrasion, what responses occur to destroy the bacteria? (p. 265)

3. How are specific receptors preformed for almost any invading organism? (pp. 268–269)

4. Name the types of lymphocytes and describe their functions. (pp. 269–272)

5. How is it possible for a vaccine to confer immunity to a disease? (pp. 270–271)

6. What is a memory cell? (pp. 270–271)

Applying What You Know

1. Why might you get many colds throughout your life, despite the immunity that should have been built up?

2. People who are going to travel in areas of the world where sanitary conditions are poor are often advised to get a gamma globulin shot before leaving. Why?

3. Why is it so difficult to produce a vaccine against HIV?

4. Some vaccines, such as those against typhoid fever and diphtheria, are made from dead organisms. How is this possible?

5. During a physical examination a physician may feel under your arm. Why do you think he or she does this?

Chapter 11
Respiration

Objectives

By the time you complete this chapter you should be able to

◆ Explain the functions of each part of the respiratory system.

◆ Describe the relationship between the alveoli and the pulmonary capillaries and explain the importance of this relationship.

◆ Describe the thoracic cavity and explain how the muscles of respiration change its shape.

◆ Define airway resistance and compliance and explain how each affects the work of breathing.

◆ Discuss the difference between gas partial pressure, concentration, and solubility.

◆ Explain how hemoglobin increases the effective O_2 solubility of blood and why this is advantageous.

◆ Describe the partial pressure gradients responsible for the movements of O_2 and CO_2.

◆ Explain how pulmonary blood flow and ventilation affect alveolar PO_2 and alveolar PCO_2.

◆ Discuss the intrinsic mechanisms used to maintain ventilation–perfusion matching.

◆ Describe the actions of the respiratory neurons and their modifying inputs.

◆ Discuss the effects that smoking has on the lungs and the health benefits of smoking cessation.

Introduction

Oxygen (O_2) is essential for our survival. We need it to generate most of the energy required for our cells to function. Our dependence on O_2 is so great that without it, we would lose consciousness within seconds and die after a few minutes. We call the process of supplying O_2 to our cells **respiration.** Respiration is so critical, all other physiological processes rely on it. The complex physiological system in charge of transporting O_2 from the atmosphere to the cells is called the **respiratory system.**

The air around us supplies all of our O_2. Oxygen makes up slightly more than 20% of the atmosphere, making air a very rich O_2 source. Although required for our life, atmospheric O_2 is actually a relatively recent addition to our planet. When life first appeared on Earth, over 3.5 billion years ago, the amount of O_2 in the atmosphere was zero! The first living organisms never came in contact with this gas, which is so vital to us today. They derived energy solely from biochemical pathways using other substances. For the next billion years, about ¼ of our evolutionary period, O_2 was completely absent on Earth. It was not until the first photosynthesizing cells appeared that O_2 was introduced into the atmosphere. These primitive plants converted the gas carbon dioxide (CO_2) and water into sugar and O_2. The O_2 produced by these plants accumulated in the atmosphere very slowly. It probably took another 2 billion years before O_2 levels in the atmosphere reached about 0.2%, just high enough for use in biochemical reactions generating energy. Shortly thereafter, the first organisms to use O_2 evolved. Thus, respiratory processes using O_2 were completely absent during more than 80% of our evolutionary period! As O_2 accumulated further, respiratory systems using O_2 evolved at an explosive rate. High atmospheric levels of O_2, similar to those of today, took another several hundred million years to develop, and probably coincided with the age of the dinosaurs. Since the time of the dinosaurs, O_2 levels have remained high. In this stable, O_2-rich environment, millions of animal species have evolved that are completely dependent on O_2.

Although transporting O_2 is the primary duty of our respiratory system, it has other vital functions. One is transporting CO_2 from the cells where it is produced to the environment. CO_2 is a byproduct of metabolism that is toxic at high levels, so eliminating it efficiently is essential. The respiratory system also regulates the hydrogen ion (H^+) concentration in blood. Because CO_2 combines with water to produce a bicarbonate ion (HCO_3^-) and H^+, the H^+ concentration of blood is directly related to CO_2 levels. The respiratory system responds to this chemistry, continuously modifying the transport of CO_2 to maintain H^+ concentrations at proper levels. This is a vital function! Practically all of the biochemical reactions occurring in our body are strongly affected by the concentration of H^+.

Gas transport between the environment and tissues is a very complex process involving a number of different organs and tissues. However, it can be divided into three stages (Figure 11.1): transport by the **respiratory tract,** transport by blood, and transport within tissue. The respiratory tract brings O_2-rich air from the environment into close contact with the blood in the lungs. The blood picks up the O_2 from the air in the lungs and carries it into the tissue capillaries. In the tissues, O_2 leaves the blood and moves to the metabolizing cells. CO_2 moves in the opposite direction, from its origin in the tissues to the blood in the capillaries. The blood transports CO_2 to the respiratory tract, where it is released into the environment. Although all of these stages are equally critical for gas transport, this chapter concentrates on events in the respiratory tract.

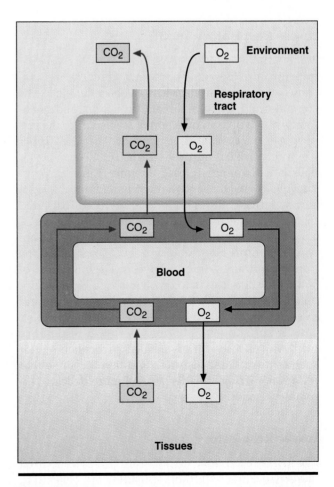

Figure 11.1 Gas transport between the environment and the tissues.

Structure of the Respiratory Tract

The primary functions of the respiratory tract are to transport gases between the environment and the body and to provide an exchange area between the air and blood. The distance between the environment just outside your nose and mouth and the blood flowing in your lungs is fairly long, about ⅓ of a meter (about 1 ft.). If you recall from Chapters 8 and 9, long-distance transport and the exchange of substances are also the primary functions of the circulatory system. It is not surprising that both the circulatory system and the respiratory tract share some very basic structural features. In the respiratory tract, gases are transported through tubes that branch into progressively smaller and smaller tubes, like blood flowing through the circulatory system. The smallest structures in the respiratory tract, like those in the circulatory system, are the most numerous, providing an extensive area for the exchange of material. Because of these structural similarities, the physical forces that move material through these systems act in similar ways.

Upper Respiratory Tract

Some structures of the respiratory tract are very familiar to us. A few are easily seen, and several may become all too apparent when we become sick. The common cold, flu, sore throats, bronchitis, and pneumonia all involve an infection of some part of the respiratory tract. The first structure of the upper respiratory tract is the nose (Figure 11.2). When we inhale with our mouths closed, air enters the respiratory tract by flowing through the nostrils and into the nasal cavity. Air then flows into the **pharynx** (throat), and from there into the **larynx,** which houses the vocal cords. The structures from the nose to the larynx constitute the upper respiratory tract. The lining of the upper respiratory tract receives a large amount of blood flow and it is coated with mucus. The blood not only serves its usual functions, but it also warms and moistens the mucus, which in turn warms and moistens the air before it flows into the lower respiratory tract. Mucus also entraps inhaled particulate matter such as dust. These features condition the air so that it does not irritate the deeper and more delicate respiratory tissues.

Lower Respiratory Tract

The lower respiratory tract (Figure 11.2) is structured like a tree trunk with roots spreading out beneath it. The trunk is the **trachea,** a short tube about 2.5 cm (about 1 inch) in diameter that extends from the larynx and splits into two slightly smaller tubes called main **bronchi.** One main bronchus goes to the right lung, the other to the left lung. The main bronchi split into smaller bronchi, like tree roots, which then divide into still smaller bronchi. The bronchi contain cartilage, making them fairly rigid, and smooth muscle, which can alter diameter. After the bronchi divide about 15 times, the air tubes are about 1 mm in diameter (about the width of a pencil point) and are called **bronchioles.** Unlike the bronchi, bronchioles have no cartilage, although they do contain a considerable amount of smooth muscle. The bronchioles, like the bronchi, split into even smaller tubes. After about a dozen divisions, very small sacs (about 0.5 mm in diameter), called **alveoli,** appear on the sides and at the ends of the smallest bronchioles. The alveoli are extremely important because they are the exchange sites where gases move between the air in the lungs and the blood. Consequently, this deep region of lung is called the **gas exchange region.** In the gas exchange region, there are no air currents and gases move only by diffusion. The other parts of the respiratory tract (the bronchioles, bronchi, trachea, and upper respiratory tract) are called the **conducting airways** because air moves through them by bulk flow, not diffusion.

Pulmonary Circulation

The lungs receive more blood than any other organ: In fact, they receive the entire cardiac output. The lung, or **pulmonary,** circulation begins with the pulmonary artery, which arises from the right ventricle. The pulmonary artery splits into smaller arteries that run along the sides of the conducting airways. The smallest of the pulmonary arteries give rise to arterioles, which supply the alveolar capillary network. These capillaries form a dense mesh that covers almost the entire outside of each alveolus (Figure 11.3(A)). This mesh is so dense that blood flows almost as a sheet over the alveolus. Once blood reaches the capillaries, only a small amount of tissue separates it from the air in the alveolus (Figure 11.3(B)). Because this tissue is so thin (about 0.5 μm, too small to see without a powerful microscope), gases diffuse very rapidly between the blood and the alveolar air. Blood leaving the alveolar capillary bed flows into the pulmonary venules, then into the pulmonary veins, and finally into the left atrium.

Thoracic Cavity

The lungs occupy the vast majority of the chest, or **thorax.** The ribs and the muscles between them form the walls of the thorax. The **diaphragm,** a flexible, plate-shaped muscle, forms the floor of the thoracic

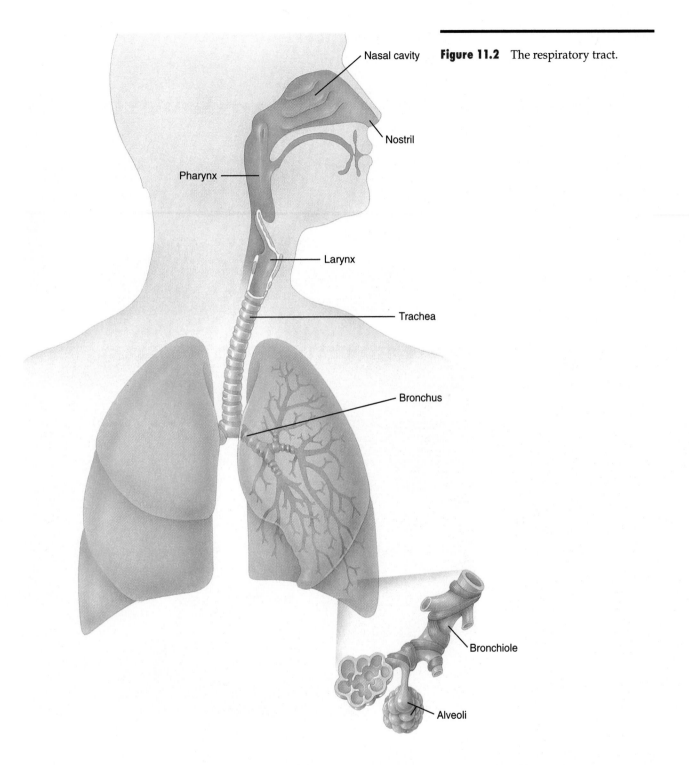

Figure 11.2 The respiratory tract.

Nasal cavity

Nostril

Pharynx

Larynx

Trachea

Bronchus

Bronchiole

Alveoli

cavity. (The diaphragm is the muscle that contracts rapidly and uncontrollably when we have the hiccups.) The shoulders and neck are at the top of the thorax. Extremely thin tissues, called **pleurae,** lie between the lungs and the walls of the thorax. They line both the outside of the lung and the inside walls of the thorax. Between these two layers of pleurae is a very thin, fluid-filled space called the **intrapleural space.** The liquid in it, the **intrapleural fluid,** allows the lungs to slide within the thoracic cavity with very little friction.

Muscles of Respiration

Like blood in the cardiovascular system, air is moved through the respiratory tract by a muscle pump. However, the respiratory pump functions

Figure 11.3 (A) Electron micrograph of lung capillaries (×1300). Note that the capillary bed is extremely dense. (B) Electron micrograph of a pulmonary capillary (×6000). Note that the tissue separating the alveolar air and the blood is extremely thin.

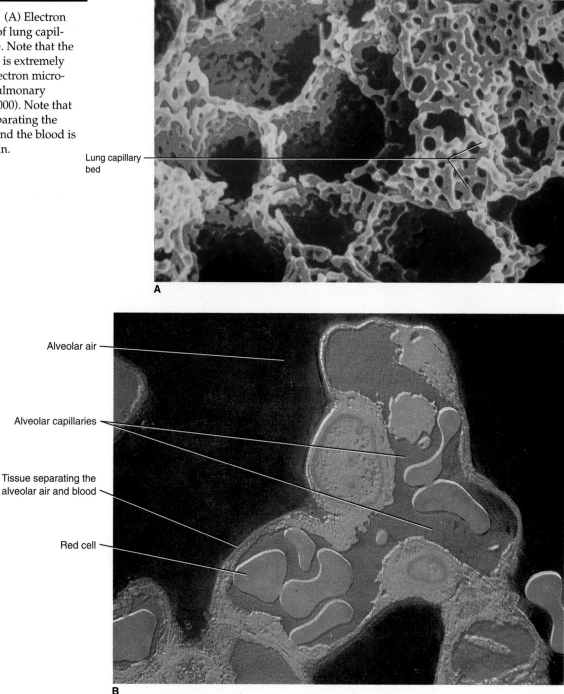

Lung capillary bed

A

Alveolar air

Alveolar capillaries

Tissue separating the alveolar air and blood

Red cell

B

quite differently from the heart. Instead of using a single muscle such as the heart, the respiratory pump uses a large number of skeletal muscles. At rest, when we inhale, the diaphragm at the base of the thorax contracts and pulls down, and the muscles between the ribs contract. This pulls the thoracic cage out and up, increasing the volume of the thoracic cavity. When we exhale at rest, these muscles relax, causing the diaphragm to move upward and the rib cage to move down and inward. Thus, at rest, inspiration requires muscular contraction (an active process), whereas expiration merely involves muscular relaxation (a passive process). During exercise, when we breathe faster and deeper, additional muscles in the abdomen and neck contract, which helps expand the thoracic cage for deeper inspiration. Active contraction of muscles in the abdomen during exercise assists expiration.

All of the muscles used during respiration change the shape of the thorax. However, none of these are connected to the lungs. We will now look at what makes the lungs expand and contract so that air can move through the respiratory tract.

Air Flow in the Respiratory Tract

In this section, you will learn how air flows in the respiratory tract. In addition, you will become familiar with the different parts of the air flow and the different functional air volumes in the lung. Finally, you will learn how much is required for breathing in both healthy and diseased lungs.

Respiratory Pump

If you blow up a balloon then leave it open to the atmosphere instead of tying it off, air rushes out of the balloon and it collapses. This is because the elastic forces acting on the balloon are directed inward. The lung is elastic like a balloon. As with a balloon, the elastic forces acting on the lung are directed inward, giving it a natural tendency to collapse. The thorax, on the other hand, has a natural tendency to move outward. It is similar to an archer's bow in this respect. Figure 11.4 illustrates the mechanism for getting air into and out of the lung. Thus, there are two structures, one within the other, with the elastic forces of the lung pulling in one direction and the elastic forces of the thorax pulling in the other direction. This situation creates a suction between the lung and the thorax in the intrapleural space. This suction causes the pressure exerted by the intrapleural fluid to be below atmospheric pressure.

During the time between inspiration and expiration, the intrapleural fluid pressure is approximately 5 mm Hg below atmospheric pressure. The air pressure in the lung is the same as the atmospheric pressure, about 760 mm Hg at sea level. There is a pressure difference across the lung wall: The pressure pushing against the inside of the lung (atmospheric pressure) is higher than that pushing on the outside of the lung (intrapleural pressure). The greater pressure inside the lung acts to push the lung walls outward, opposing the elastic inward forces of the lung. It is this pressure difference between the intrapleural fluid and the air in the lung that keeps the lungs expanded.

What happens when we inhale? The muscles of inspiration start pulling the thorax away from the lung wall even more. This creates a greater suction within the intrapleural space (that is, a further reduction in intrapleural fluid pressure), producing an even greater pressure difference across the lung wall. This increase in the pressure difference causes the lungs to expand.

As the lungs expand, the air pressure in the lungs falls slightly, usually by about 1 mm Hg, creating a pressure gradient between the air inside the lungs and the atmosphere. This pressure gradient sucks air in through the upper respiratory tract and then into the lungs.

During exhalation, the muscles of inspiration relax and the thoracic wall moves inward, toward the lungs. As the thoracic wall moves inward, the intrapleural pressure rises, reducing the pressure gradient across the lungs, allowing the lungs to collapse. The air pressure inside the lungs increases to about 1 mm Hg above atmospheric pressure as the lungs contract. Air then flows out of the lungs down this pressure gradient.

The flow of air through the respiratory tract is driven by pressure gradients, just like the flow of blood in the cardiovascular system. However, the pressure gradients moving air are only about 1% of those driving blood flow. Blood is much more viscous than air, so more energy and thus a higher pressure gradient is required to move it. That is why we cannot breathe water. Our respiratory muscles are not strong enough to generate the very high pressures needed to move enough water, which is almost 1000 times more viscous than air, into and out of the lungs. The reason why fish use gills instead of lungs is that much less energy is required to move water through gills.

Ventilation and Lung Volumes

Of utmost importance in respiration is the volume of air that flows into and out of the lungs each minute. This air flow is called **pulmonary ventilation.** As we shall see later, pulmonary ventilation helps determine the O_2 and CO_2 levels in the blood. Because the lung, like the heart, is a periodic pump, pulmonary ventilation is the product of two components: the number of breaths taken each minute, or the **ventilatory rate,** and the volume of air that is moved with each breath, or the **tidal volume.** At rest, the volume of air that we breathe (about 6 l/min) is close to the volume of blood that our hearts pump (about 5 l/min). However, our ventilatory rate is a lot slower than our heart rate. Whereas our hearts beat at least once each second, we take about one breath every 5 seconds or so. At rest, our tidal volume is approximately 500 ml. However, of that 500 ml, only about 70%, or 350 ml, reaches the alveoli and participates in gas exchange. That occurs because the conducting airways (with a volume of 150 ml) are filled with fresh air at the end of each inspiration. This fresh air never reaches the alveoli and is exhaled at the beginning of each expiration. Because the

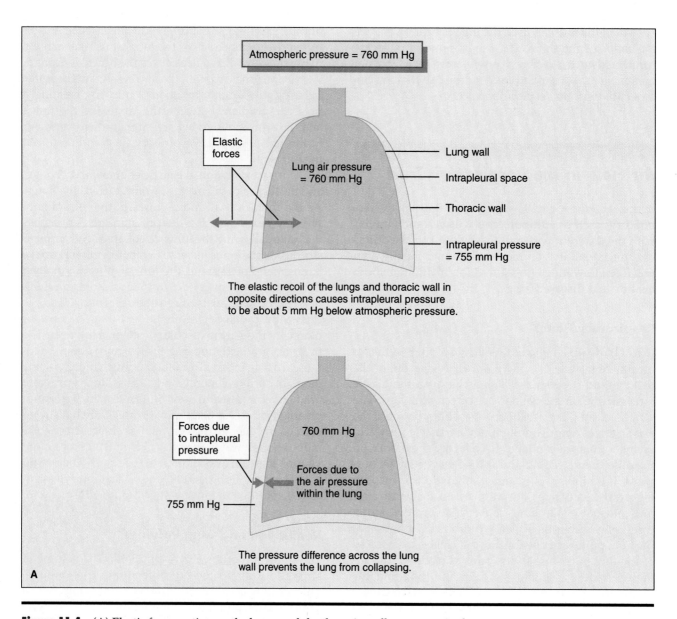

Atmospheric pressure = 760 mm Hg

Elastic forces

Lung air pressure = 760 mm Hg

Lung wall

Intrapleural space

Thoracic wall

Intrapleural pressure = 755 mm Hg

The elastic recoil of the lungs and thoracic wall in opposite directions causes intrapleural pressure to be about 5 mm Hg below atmospheric pressure.

Forces due to intrapleural pressure

760 mm Hg

Forces due to the air pressure within the lung

755 mm Hg

The pressure difference across the lung wall prevents the lung from collapsing.

A

Figure 11.4   (A) Elastic forces acting on the lungs and the thoracic wall, pressures in the intrapleural space and lung gas, and the pressure difference across the lung wall.

fresh air filling the airways at the end of inspiration does not participate in gas exchange, the volume of the conducting airways is called **dead space.**

When we breathe harder, such as during exercise, both ventilatory rate and tidal volume increase, although tidal volume tends to increase more than ventilatory rate. Because dead space does not change, a greater fraction of the tidal volume reaches the alveoli at higher tidal volumes.

The total volume of air in the lungs is considerably more than the tidal volume. After a normal exhalation, there is still about 2.5 l of air remaining in the lungs. If we exhale as much as we possibly can, about 1 liter of air still remains in the lungs. This air, which we can-

not get rid of, is called the **residual volume.** Six liters is about the most air with which we can fill our lungs. This volume is called **total lung volume.** The volume of the biggest breath we can take is called the **vital capacity.** This is equal to total lung volume minus residual volume (usually about 5 liters).

Work of Breathing

In healthy people, the work required to expand and contract the lungs is only 1–2% of the body's total energy expenditure. However, the work of breathing can increase considerably when a person is suffering from a lung disease such as asthma or pneumonia. In some

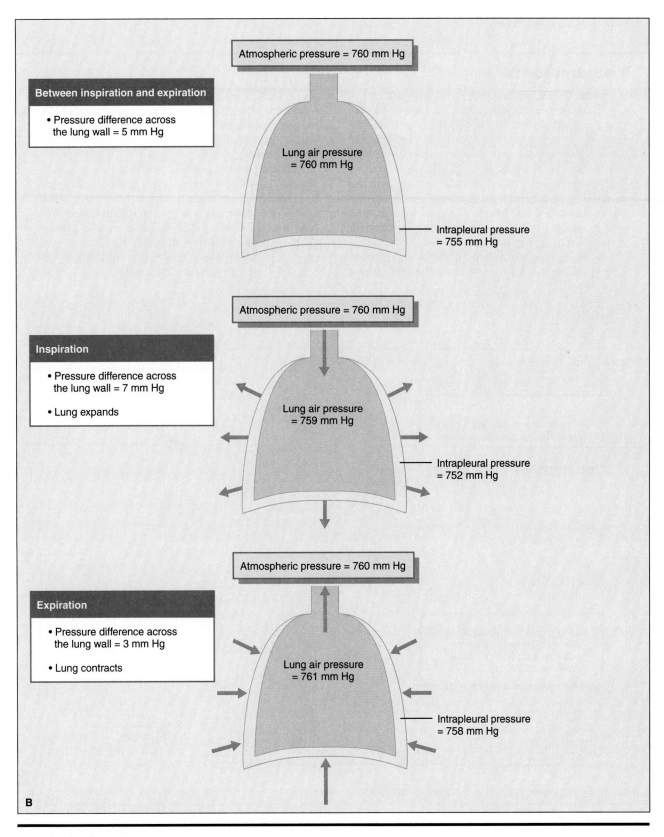

Figure 11.4 (continued)   (B) Pressures and air flow during breathing.

HIGHLIGHT 11.1

Pneumothorax

A **pneumothorax** (*pneumo = air, thorax = thoracic cavity*) is a dangerous condition where air enters the intrapleural space. This often results in the collapse of a lung. Air can enter the intrapleural space in a variety of ways. A puncture wound of the thoracic wall, such as a stab wound, is one example.

Normally, intrapleural pressure is lower than the pressure of the air contained in a lung. This difference in pressure across the lung wall opposes the inward elastic forces of the lung and keeps the lungs expanded. A puncture wound of the chest creates a hole that allows air to enter the intrapleural space through the thoracic wall. Entry of air into the intrapleural space causes the pressure of the intrapleural space to rise toward atmospheric pressure. As intrapleural pressure increases, the pressure difference across the lung diminishes. The elastic forces of the lung are then unopposed and the lung collapses.

To treat a pneumothorax created by a puncture wound, the wound must be sealed, preventing more air from entering the intrapleural space. Over the next day or two, the air trapped in the intrapleural space diffuses into the blood and is carried away as dissolved gas molecules. As the air is absorbed, intrapleural pressure drops below atmospheric pressure and reestablishes the normal pressure gradient across the lung.

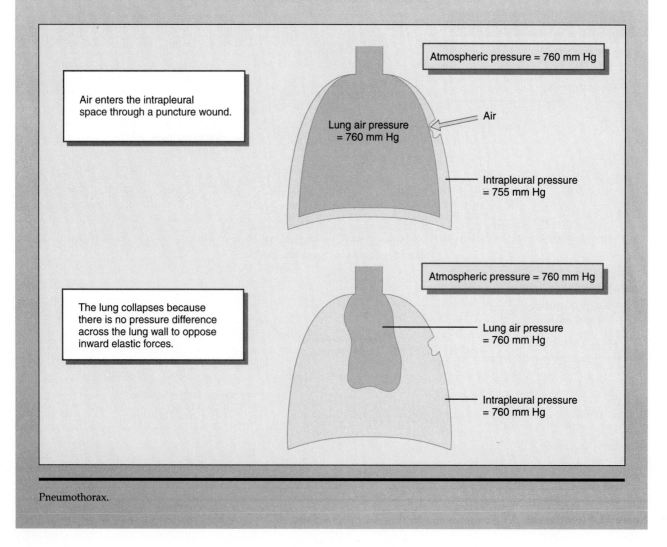

Pneumothorax.

extreme cases, breathing becomes so taxing that it exhausts the respiratory muscles and the person becomes incapable of breathing enough air to stay alive.

Two major factors determine the work of breathing: the compliance of the lungs and thorax and the resistance of the airways to air flow. The compliance of the lungs and thorax refers to how stretchable these tissues are. For example, if compliance is high, these structures stretch easily and, consequently, the work of breathing is small. If compliance is low, it takes a lot of energy to expand these structures.

Compliance There are two primary factors that determine the compliance of the lungs and thorax. One is the compliance of the actual tissue of these structures. Normally, these compliances are high, but a number of diseases can damage these structures and, as a result, decrease their compliance and increase the work of breathing.

The other major factor that determines compliance is alveolar surface tension. A fine layer of liquid covers the inner surface of each alveolus. When liquids contact air, they exert a force called **surface tension.** This can be a very strong force. In the alveolus, the force of surface tension pulls inward, opposing lung expansion. The stronger the alveolar surface tension, the more difficult it is to expand the lungs and the greater is the work of breathing. Normally, a substance called **surfactant,** present in the liquid lining of the alveoli, reduces surface tension. Special alveolar cells secrete surfactant.

Research involving premature babies first recognized the importance of surfactant. Premature babies often have serious health problems. A lung disorder called **respiratory distress syndrome,** characterized by a very low lung compliance, is the most common complication. Afflicted babies have great difficulty inflating their lungs and often they cannot obtain adequate O_2. Seventy percent of all deaths among premature babies result from this disorder. Babies who survive have a higher incidence of brain damage resulting from insufficient O_2 delivery to the brain. In the mid-1950s, experiments showed that a surfactant deficiency causes respiratory distress syndrome. In these experiments, lung extracts from babies with healthy lungs reduced considerably the surface tension of water, whereas similar extracts from babies with respiratory distress syndrome had very little effect on water surface tension. The reason premature birth can cause this deficiency is that surfactant production in the human fetus begins relatively late in development, usually between 24 and 34 weeks of gestation. Thus, babies born before this point are at

greater risk of having a surfactant deficiency. This discovery stimulated considerable research to develop a method for administering surfactant into the lungs of babies suffering from this disease. After about 30 years of research, first on animals then on babies, effective surfactant replacement therapy is available. Today, respiratory distress syndrome is often treated by administering aerosols of synthetic surfactant or surfactant isolated from humans or animals into the lungs of premature babies. Such treatment immediately improves lung function in babies with respiratory distress syndrome.

Resistance As mentioned above, the resistance of the airways to air flow is the other major factor determining the work of breathing. The higher the airway resistance, the more work is required to move air into and out of the lungs. The factors that determine airway resistance are the same factors that determine the resistance of the blood vessels to blood flow. Remember from Chapter 9 that the primary factor is the radius of the vessel; resistance varies inversely with the radius raised to the fourth power. Thus, a reduction in airway radius by $\frac{1}{2}$ will cause the resistance of the airway to increase by a factor of 16. Normally, the radii of the airways are sufficiently large that very little energy is required to move air through them. However, in many different types of lung disease, the airways constrict and intense muscular work is required to move sufficient air through them. To experience the extreme discomfort of airway constriction, try breathing through a drinking straw. The narrow radius of the straw produces a very high resistance to air flow, and a great deal of effort is needed to breathe.

Constriction of the airways occurs with asthma, the most common lung disease in nonsmokers. Approximately 5% of people in the United States have asthma, and it is the leading cause of chronic illness among children. Airway constriction due to asthma results from contraction of smooth muscle primarily in the bronchioles. Usually, an allergic response to very small particles (such as pollens, dust, and pollutants) causes constriction. Fortunately, there are now many drugs that can relax the bronchiolar smooth muscle and greatly reduce airway resistance. Many of these drugs are called beta-adrenergic agonists because they stimulate specific receptors, called beta-adrenergic receptors, on the bronchiolar smooth muscle. Stimulation of bronchiolar beta-adrenergic receptors causes the smooth muscle to relax very rapidly. Most of the inhalers that asthmatics use to relieve their symptoms contain beta-adrenergic agonists.

Gas Transport

This section describes the processes that move O_2 from the atmosphere to the tissues, where it is metabolized, and that move CO_2 from the tissues to the atmosphere. To appreciate these processes, it is important to understand some of the physical and chemical properties of gases in fluids and in air.

Properties of Gases

Before discussing the physiological mechanisms for moving gases between the environment and the tissues, it is important to understand the behavior of gases in both the gas phase and the liquid phase. To begin, suppose there is a box (Figure 11.5(A)), and the box is filled with air. The air in the box exerts pressure on all surfaces of the box. If the pressure in the box is the same as atmospheric pressure, 760 mm Hg at sea level, then 760 mm Hg is exerted on all sides of the box. If the gas in the box is half O_2 and half CO_2, then the O_2 exerts half the total pressure, or 380 mm Hg, and the CO_2 exerts the other half. The pressure exerted just by O_2 is called the **partial pressure** of O_2. It is abbreviated PO_2. The partial pressure of CO_2 is abbreviated PCO_2.

Gases can dissolve in water, and when dissolved, they also exert a pressure. As with air, this pressure is exerted in all directions. The pressure in a bottle of soda is generated by the very high partial pressure of CO_2 in the soda.

Suppose there is another box (Figure 11.5(B)) that contains both air and water. If O_2 is added to the air and none to the water, the O_2 in the air exerts a pressure on the surface of the water. The amount of pressure exerted by the O_2 depends on the PO_2. This pressure causes some O_2 to diffuse from the air into the water, where it dissolves.

If in another box (Figure 11.5(C)) containing both water and air, O_2 is added to the water but not to the air, the O_2 in the water exerts a pressure on the air surface. This causes O_2 to diffuse from the water into the air (Figure 11.5(D)). As O_2 diffuses from the water into the air, the PO_2 in the air rises and the PO_2 in the water falls. The PO_2 in the air continues to rise and the PO_2 in the water continues to fall until they are equal (Figure 11.5(E)). When that happens, the pressure of O_2 in the air exerted down on the water equals the O_2 pressure of the water exerted upward on the air. In this situation, the PO_2 in the air and the water are in equilibrium, and there is no net movement of O_2. Thus, the net diffusion of a gas occurs only when a difference in gas partial pressure exists. Notice that the driving force for gas diffusion, *partial pressure gradients,* is different from the driving force for the diffusion of dissolved salts and other nongaseous molecules, *concentration gradients.*

Gas partial pressure should not be confused with gas concentration, as they are quite different. The gas concentration, or the amount of gas in a certain volume, is the product of two factors: the partial pressure of the gas and the solubility of the gas. **Gas solubility** is a measure of how much gas is in a certain volume at a particular partial pressure. The greater the solubility of a gas, the higher the gas concentration at a particular partial pressure. Because of the relationship between partial pressure, solubility, and concentration, it is possible to increase gas concentration by increasing either the solubility or the partial pressure.

Aqueous solutions such as plasma have a low O_2 solubility. Consequently, they do not contain much O_2, even at reasonably high partial pressures. However, it is possible to increase the effective O_2 solubility of liquids, thereby allowing the liquid to hold more O_2. This can be done by adding hemoglobin to the solution. As discussed in Chapter 9, hemoglobin is the large, iron-containing protein in blood that forms chemical bonds with O_2. It is this binding of O_2 to hemoglobin that raises the effective solubility of O_2 in the blood.

This important property of hemoglobin can be appreciated if you consider the box in which O_2 is in equilibrium between the water and air (Figure 11.5(E)). Remember that at equilibrium, water PO_2 = air PO_2. If hemoglobin is added to the water, it begins to bind O_2 (Figure 11.6(F)). This effectively takes O_2 out of solution and causes the water PO_2 to fall. When water PO_2 falls, a PO_2 gradient is created between the air and water, causing a net diffusion of O_2 from the air to the water (Figure 11.5(G)). This movement of O_2 increases the total amount of O_2 contained in the liquid portion. Thus, hemoglobin acts as if it increases the O_2 solubility of a liquid, allowing it to contain more O_2.

Blood is a hemoglobin solution. If blood lacked hemoglobin, 100 ml of blood could hold only about 0.3 ml of O_2 at a PO_2 of 100 mm Hg. But because blood has hemoglobin, it holds approximately 19 ml of O_2 per 100 ml of blood at the same PO_2. That is about 65 times more O_2.

The precise relationship between PO_2 and O_2 concentration in blood is complex. This complex relationship is usually described by a curve, called the **hemoglobin–O_2 dissociation** curve, which results when O_2 concentration is plotted against PO_2, as in Figure 11.6. It is possible to determine the amount of O_2 in blood at any PO_2 with this curve. For example, line A shows that 100 ml of blood contains 14 ml of O_2 at a PO_2 of 40 mm Hg. Line B shows that 100 ml of blood contains 19 ml of O_2 at a PO_2 of 100 mm Hg. Note

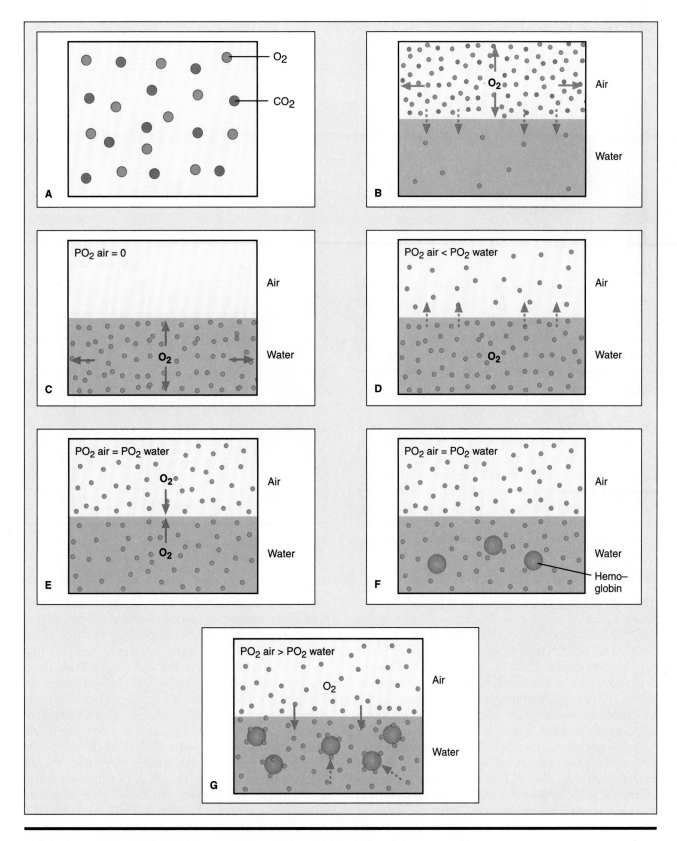

Figure 11.5 Properties of gases. (A) Total air pressure in box is 760 mm Hg (1 atm). Half of the gas is O_2 and half is CO_2. $PO_2 = \frac{1}{2} \times 760 = 380$ mm Hg; $PCO_2 = \frac{1}{2} \times 760 = 380$ mm Hg. (B) In a box with air and water, O_2 added to the air exerts pressure on the water's surface. (C) O_2 added to the water exerts pressure on the air's surface. (D) O_2 diffuses from water to air. (E) Equilibrium. (F) Hemoglobin is added to water. (G) O_2 binds to hemoglobin, lowering PO_2 of water.

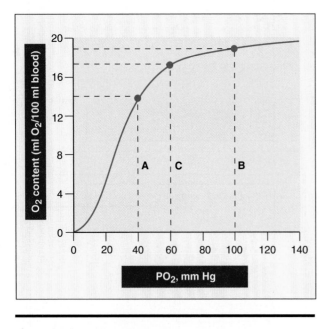

Figure 11.6 Hemoglobin–oxygen dissociation curve.

that above a PO_2 of approximately 80 mm Hg, increases in PO_2 produce only small increases in the O_2 concentration of blood. That is because above 80 mm Hg, the four O_2 binding sites on the hemoglobin molecule are nearly saturated with O_2 molecules and very little additional binding can occur. After all the O_2 binding sites are occupied, only a small additional amount of O_2 dissolves in the plasma as PO_2 is raised.

Oxygen Transport

The first step in transporting O_2 from the atmosphere to the tissues is inspiration. The amount of O_2 in the atmosphere is quite high—about 20% of all the gases is O_2. At sea level, where the total pressure of air is 760 mm Hg, atmospheric PO_2 is approximately 150 mm Hg (about 20% of 760 mm Hg). However, the PO_2 in the alveoli is considerably less, about 100 mm Hg.

There are several reasons why alveolar PO_2 is lower than atmospheric PO_2 (Figure 11.7). The primary reason is that when fresh air is inspired, it mixes with a considerable volume of old air still in the lungs and airways from the last expiration. (At rest, expiration only expels about 20% of the total air in the lungs and airways.) This old air is relatively deficient in O_2, and it dilutes O_2 in the inspired air. A second reason is that O_2 is continuously being removed from the alveolar gas by the blood. Third, CO_2 and water vapor are added to the lung gas. The addition of these gases further dilutes the O_2 in alveolar air to reduce alveolar PO_2.

After inspiration brings O_2 into the lungs, it moves from the alveoli into the blood (Figure 11.8). The blood entering the pulmonary capillaries is systemic venous

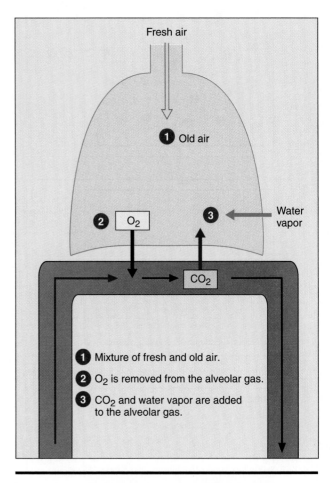

Figure 11.7 Three reasons why alveolar PO_2 is lower than atmospheric PO_2.

blood, which usually has a PO_2 of approximately 40 mm Hg. (Remember, blood flowing through the systemic veins has the same composition as blood flowing through the pulmonary arteries.) Consequently, as this blood flows into the alveolar capillaries, there is a PO_2 gradient of about 60 mm Hg between the alveolar air (100 mm Hg) and the incoming blood (40 mm Hg). Because gases diffuse down partial pressure gradients, there is a net diffusive flux of O_2 from the alveolar air to the blood across the extremely thin membrane separating the capillary blood from the air. Because this membrane is so thin, O_2 diffuses rapidly and the blood equilibrates with the alveolar air before it leaves the alveolar capillaries. Thus, the PO_2 of the blood leaving the alveolus, the arterial blood, has the same PO_2 as the alveolar gas, 100 mm Hg.

Because O_2 equilibrates between the alveolar air and blood, any changes in alveolar PO_2 cause the same changes in arterial PO_2, so the factors that alter alveolar PO_2 also affect the PO_2 of arterial blood.

One extremely important factor that can affect alveolar PO_2 is the total amount of pulmonary venti-

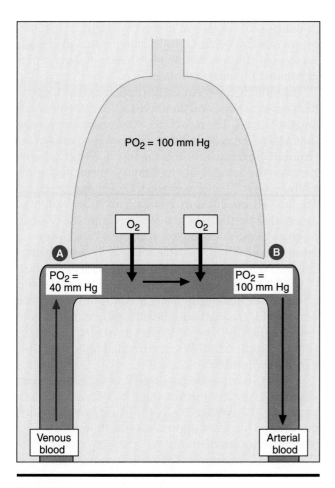

PO₂ = 100 mm Hg

O₂   O₂

A   B

PO₂ = 40 mm Hg   PO₂ = 100 mm Hg

Venous blood   Arterial blood

Figure 11.8 Transport of O_2 from the alveoli to the blood. Equilibrium occurs between points A and B.

lation, or the rate at which fresh air is moved into and out of the lungs. Increasing pulmonary ventilation, bringing more fresh air into the lungs, increases alveolar PO_2, and decreasing pulmonary ventilation does the opposite. Thus, arterial PO_2 is directly related to pulmonary ventilation.

Pulmonary blood flow, or **pulmonary perfusion,** also affects alveolar PO_2. As previously mentioned, one of the reasons why alveolar PO_2 is lower than the PO_2 of the atmosphere is that O_2 is always being extracted from the alveoli by the blood. This rate of O_2 extraction is directly proportional to pulmonary blood flow. The more blood flowing through the lungs, the more O_2 is removed from the alveolar air, causing alveolar PO_2 to decrease.

Considering these two factors, any increase in blood flow must be matched by a proportional increase in pulmonary ventilation in order to maintain a constant alveolar PO_2. This happens under most conditions. For example, the increased cardiac output, and thus pulmonary blood flow, needed to meet the increased blood flow requirement of exercising mus-

cle is matched by a similar increase in pulmonary ventilation, so the more you run, the more your heart pumps blood and the harder you breathe. The simultaneous increases in both pulmonary blood flow and ventilation maintain alveolar and arterial PO_2 at approximately 100 mm Hg. (The control mechanisms regulating cardiac output were presented in Chapter 8. Those regulating pulmonary ventilation are discussed in the next section of this chapter.)

A third important factor that affects alveolar PO_2 is altitude. For example, if a person travels from sea level to a mountain at 10,000 feet elevation (a common elevation for ski slopes in the Rocky Mountains), the PO_2 of inspired air declines from approximately 150 to 100 mm Hg. This decrease in inspired PO_2 causes the PO_2 in the alveoli, and thus in the arterial blood, to fall to about 60 mm Hg. Although traveling to the mountains decreases arterial PO_2 substantially, the O_2 concentration in the blood is relatively unaffected. Consider the hemoglobin–O_2 dissociation curve in Figure 11.6. At sea level with a PO_2 of 100 mm Hg, blood contains 19 ml of O_2 per 100 ml of blood. At 10,000 ft, with a PO_2 of 60 mm Hg, the O_2 concentration is 17 ml O_2 per 100 ml blood (line C in Figure 11.6), only moderately less than that at sea level. The reason for this relatively small drop in O_2 concentration, despite the rather large drop in PO_2, is that at 60 mm Hg, hemoglobin binds close to the maximum amount of O_2. This beneficial characteristic of hemoglobin is what allows us to travel fairly high into the mountains without experiencing a large decrease in the amount of O_2 carried by the blood. Of course, if we go high enough, reductions in alveolar PO_2, and thus arterial PO_2, eventually cause significant decreases in the O_2 concentration of blood. At the top of Mt. Everest (29,029 ft), the PO_2 of arterial blood falls to about 28 mm Hg. At this low PO_2, even the most experienced mountain climbers can survive for only a short period.

After O_2 has diffused from the alveolar air into the blood, the cardiovascular system transports the O_2 to the tissues. The O_2 concentration of the blood and organ blood flow determines the amount of O_2 transported to each organ. The factors governing tissue blood flow were discussed in Chapter 9.

The last step in the transport of O_2 is the diffusion of O_2 from the blood to the metabolizing cells (Figure 11.9). The PO_2 of the blood entering the capillaries almost equals the PO_2 of arterial blood, approximately 100 mm Hg. The PO_2 of the tissue surrounding the capillaries is always lower, but it varies. For example, in exercising skeletal muscle, PO_2 can be as low as 10 mm Hg. When we are hot, our skin is only a few mm Hg below arterial PO_2. An average value for tissue PO_2 is about 40 mm Hg. This difference in PO_2 between the blood and the tissue surrounding the

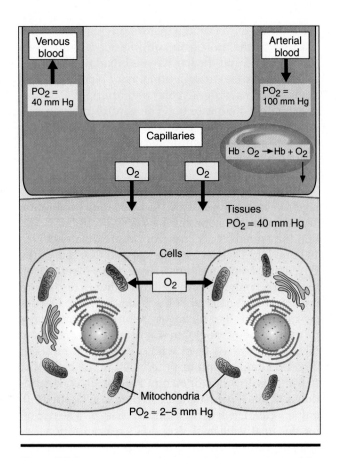

Figure 11.9 Oxygen transport from the blood to the cells.

capillaries drives O_2 out of the capillaries. However, before this can occur, O_2 must first be unloaded from hemoglobin; that is, the chemical bond between hemoglobin and O_2 must be broken.

Several things occur in the capillaries that weaken this chemical bond and therefore facilitate O_2 diffusion into the tissues. The most important are an increase in the CO_2 and H^+ concentrations in the blood and an increase in the temperature of the blood. These changes all act on the hemoglobin molecule to promote the unloading of O_2. The increases in CO_2, H^+, and temperature are all products of metabolism. Consequently, tissues that are the most active and require the most O_2 (such as muscles during exercise) produce the greatest increases. Thus, the unloading of O_2 from hemoglobin is facilitated the most in tissues with the highest O_2 need. Once O_2 leaves the hemoglobin molecule and diffuses out of the capillary, it continues to diffuse down a PO_2 gradient into the cells and finally into the mitochondria, where it is used in oxidative metabolism. Mitochondrial PO_2 is estimated to be very low, only a few mm Hg.

Before leaving the capillaries, blood equilibrates with the tissue just outside these blood vessels. Consequently, the average PO_2 of the blood leaving tissue, the venous blood, is approximately 40 mm Hg. This blood then returns to the lungs, where it picks up more O_2 from the alveolar air. This entire process of moving

HIGHLIGHT 11.2

What Is in Air?

Air that contains no water vapor (no humidity) is composed of 20.9% O_2, 78.1% nitrogen, 0.9% argon, 0.04% CO_2, and extremely small amounts of other gases such as neon and helium. The amount of these gases in air that contains water vapor is slightly less because water vapor dilutes these gases in the air. This effect of water vapor is small because the maximum amount of water vapor that can be contained in the air is only a few percent.

At sea level, the total pressure exerted by the atmosphere is usually about 760 mm Hg. Thus, if air has no humidity, the partial pressures of O_2 and CO_2 are 159 mm Hg (20.9% × 760 mm Hg) and 0.3 mm Hg (0.04% × 760 mm Hg), respectively. If air contains water vapor, these values are slightly less. At elevations above sea level, the total pressure of the atmosphere is less than 760 mm Hg, but the relative gas composition of the atmosphere is the same. For example, in Denver, Colorado, which is 5300 ft above sea level, total atmospheric pressure is approximately 630 mm Hg, but the percentage of the atmosphere that contains O_2 is still 20.9%. Because total pressure is less but the relative

composition is unchanged, the PO_2 of the air in Denver (assuming a humidity of zero) is 132 mm Hg (20.9% × 630 mm Hg). The table below shows the elevations, mean barometric pressures, and PO_2s of several places on Earth.

Table 11.1   Respiration

Location	Altitude (feet)	Barometric Pressure (mm Hg)	PO$_2$ (mm Hg)
Death Valley	−300	770	161
New York City	30	760	159
Chicago	600	740	155
Denver	5300	630	132
Mexico City	7350	575	120
Mt. Ranier	14,410	440	92
Mt. McKinley (Denali)	20,320	350	73
Mt. Everest	29,029	250	52

an O_2 molecule from the atmosphere to a mitochondrion takes less than 10 seconds.

Carbon Dioxide Transport

The respiratory system also transports another gas, CO_2. It moves in exactly the opposite direction to O_2.

Cells produce carbon dioxide as a byproduct of metabolism. The amount of CO_2 generated is usually directly related to the amount of O_2 consumed: The more O_2 used, the more CO_2 produced. As CO_2 forms in the cells, its partial pressure rises above that in the interstitial space, so it diffuses toward the blood in the capillaries (Figure 11.10). However, the PCO_2 gradient driving this diffusion is not as large as the gradient that drives O_2 diffusion. PCO_2 in the cells is usually slightly more than 46 mm Hg, whereas that of the blood entering the capillaries is 40 mm Hg. As blood moves through the capillaries, PCO_2 equilibrates between the blood and tissue just outside the capillaries. Because tissue PCO_2 is usually about 46 mm Hg, the blood leaving the tissues also has a PCO_2 of approximately 46 mm Hg. At higher metabolic rates, cell, tissue, and venous blood PCO_2 all rise a little due to the increased CO_2 production.

As with O_2, special adaptations increase the solubility of CO_2 in blood so that large amounts can be transported by the cardiovascular system. CO_2 is actually transported by the blood in three different forms (Figure 11.11). The simplest form is CO_2 molecules physically dissolved in the plasma and the blood cells. Only about 5% of the CO_2 in blood is transported this way.

As you can see Figure 11.11(B), approximately 7% of the CO_2 in blood is bound to the hemoglobin molecule. Hemoglobin has not only O_2 binding sites, but also CO_2 binding sites. These sites are different from the O_2 binding sites, and they do not involve the iron atom.

Most CO_2, about 88%, travels in the blood as bicarbonate ions, HCO_3^-, as shown in Figure 11.11(C). The conversion of CO_2 to HCO_3^- involves several steps. First, CO_2 diffuses from tissue into the plasma and then into the red cell. In the red cell, a chemical reaction occurs: CO_2 combines with water to form carbonic acid, H_2CO_3. Normally, this conversion is rather slow, taking many seconds. However, red cells contain

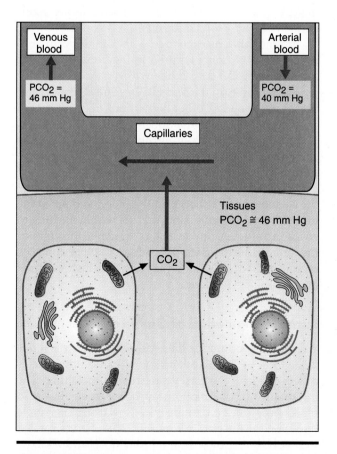

Figure 11.10 Carbon dioxide transport from the cells to the blood.

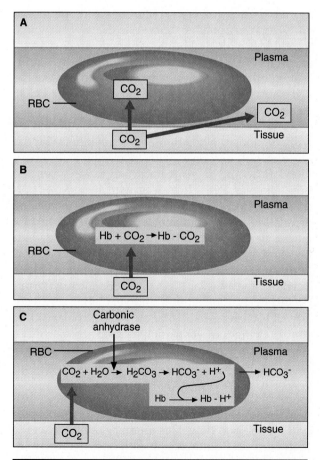

Figure 11.11 Carbon dioxide transport in blood. (A) Physically dissolved in blood. (B) Chemically bound to hemoglobin. (C) As bicarbonate.

an enzyme, carbonic anhydrase, that catalyzes this reaction so that it occurs 13,000 times faster than in its absence. H_2CO_3 is extremely unstable, and most of it immediately dissociates into HCO_3^- and H^+. At this point, the HCO_3^- ions leave the red cell and enter the plasma, whereas most of the H^+ ions generated in this reaction bind to specific H^+-binding sites on hemoglobin molecules. Thus, hemoglobin can bind three different chemicals involved in respiration: O_2, CO_2, and H^+. As mentioned in the previous section on O_2 transport, an increase in CO_2 or H^+ in the blood weakens the chemical bond between hemoglobin and O_2. Specifically, the binding of CO_2 and H^+ to the hemoglobin molecule weakens its bond with O_2, facilitating O_2 to movement from the blood to tissues.

In the lungs, there is another PCO_2 gradient. As stated above, the systemic venous blood entering the alveolar capillaries has a PCO_2 of approximately 46 mm Hg, whereas the alveolar gas has a PCO_2 of 40 mm Hg. This gradient causes CO_2 to diffuse from the capillary blood into the alveolar air. During this outward diffusion, CO_2 comes off of hemoglobin, and HCO_3^- and H^+ recombine into CO_2 and water. This conversion involves the same steps as in the generation of HCO_3^- in the tissues, except in the opposite direction. HCO_3^- moves from the plasma into the red cell, where it combines with H^+. H_2CO_3 is formed, which dissociates into CO_2 and H_2O. Blood PCO_2 rapidly equilibrates with alveolar gas so that the blood leaving the alveolar capillaries has a PCO_2 equal to alveolar PCO_2, 40 mm Hg.

Like alveolar PO_2, alveolar PCO_2 is affected by both pulmonary blood flow and ventilation. Whereas alveolar PO_2 decreases with increasing pulmonary blood flow and increases with increasing ventilation, alveolar PCO_2 increases with increasing pulmonary blood flow and decreases with increasing ventilation. With ventilation, the more fresh air that enters the lungs, the more lung CO_2 is diluted. This lowers alveolar PCO_2. With blood flow, the more blood perfusing the alveoli, the more CO_2 is delivered to the lung, causing alveolar PCO_2 to rise. Because of the opposite effects of blood flow and ventilation on alveolar PCO_2, changes in blood flow must be balanced by similar changes in ventilation to keep alveolar PCO_2 constant. Under normal conditions, control mechanisms regulating cardiac output and ventilation properly match blood and air flow to the lung to maintain arterial PCO_2 at 40 mm Hg. However, some conditions cause ventilation to change so that it is out of proportion to blood flow. This occurs when we travel to high altitudes. Pulmonary ventilation is greatly increased, but pulmonary blood flow is not. Consequently, alveolar PCO_2 falls, causing a reduction in arterial PCO_2. An extreme example of this comes from a fascinating physiological study done on climbers ascending Mt. Everest in 1981. One of the climbers, Christopher Pizzo, upon reaching the summit, stopped using his supplemental O_2 and collected samples of his expired air. Subsequent gas analysis revealed that his arterial PCO_2 was only 7.5 mm Hg, less than 20% of normal! This enormous decrease was due to his extremely high pulmonary ventilation (over 5 times normal) caused by the severe lack of O_2. In the next section, we shall see what mechanisms drive ventilation at high altitudes.

Regulation of Pulmonary Gas Exchange

Because supplying O_2 to and removing CO_2 from tissue is so vitally important, it must be precisely adjusted to meet our constantly changing metabolic needs. For this to occur, accurate regulation of pulmonary gas exchange is essential. Under most conditions, the body adjusts pulmonary gas exchange so that arterial PO_2 and PCO_2 remain constant, ensuring that adequate gas partial pressure gradients are maintained between blood and cells.

Pulmonary gas exchange is also adjusted to help regulate the H^+ concentration in the body fluids. Remember that CO_2 combines with water to produce H_2CO_3, and in turn, H^+ and HCO_3^-. The H^+ concentration of blood and all other body fluids depends on the PCO_2. Because of this relationship between PCO_2 and H^+ concentration, H^+ concentration can be adjusted by changing PCO_2. Our bodies regulate the H^+ concentration by controlling pulmonary gas exchange.

There are two types of mechanisms that regulate the exchange of gas between the blood and air in the lung: **intrinsic mechanisms** that act within the lung to ensure an efficient and even gas exchange throughout the lung, and **extrinsic mechanisms** that use complex neural pathways outside the lung to adjust the total amount of air ventilated.

Intrinsic Regulation of Gas Exchange

The most important regulatory mechanism in the lungs is the local matching of air flow to blood flow. This is called regional **ventilation–perfusion matching.** The importance of the relationship between air flow and blood flow can be demonstrated by using the single hypothetical alveolus in Figure 11.12(A). Air is supplied to this alveolus by a single bronchiole, and a single artery supplies its blood. If blood flow is properly

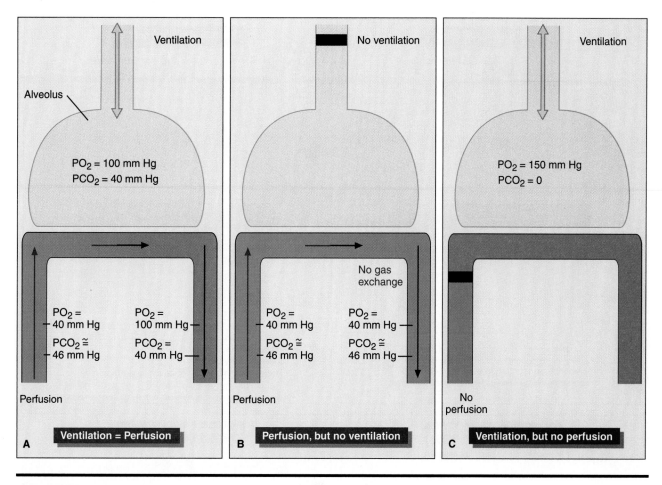

Figure 11.12 The effects of ventilation–perfusion matching on alveolar and arterial PO_2 and PCO_2.

matched to air flow, then both alveolar and arterial PO_2 and PCO_2 are 100 and 40 mm Hg, respectively. If something happens to block air flow into the alveolus (Figure 11.12(B)), such as a foreign particle lodging in the bronchiole, then the PO_2 in the alveolus falls and the PCO_2 rises. Very soon, alveolar PO_2 and PCO_2 become equal to the PO_2 and PCO_2 of the incoming venous blood, the partial pressure gradients driving diffusion disappear, and gas exchange ceases. Consequently, the blood flowing through the alveolus is wasted blood flow for the purposes of respiration. In addition, when this blood mixes with blood having already undergone proper gas exchange in the other parts of the lung, it lowers the PO_2 and raises the PCO_2 of the arterial mixture. This is detrimental to gas transport in tissues because such changes in arterial gas partial pressures decrease the partial pressure gradients driving diffusion in the tissue capillary beds. If the obstruction in the bronchiole is not complete, it still reduces air flow below normal, and blood gas partial pressures are affected in the same way, just not as much.

A decrease in blood flow to an alveolus is also detrimental. For example, in Figure 11.12(C), if the ar-

teriole supplying the alveolus becomes completely obstructed (blood clots can cause arteriolar obstruction) but air flow continues, the air flowing into the alveolus does not exchange any gas with the blood. Thus, this air and the energy required to move it are wasted.

Thus, for gas exchange to occur, both ventilation and perfusion of an alveolus are needed. For efficient gas exchange, perfusion must be matched to ventilation. If blood flow falls relative to ventilation, then air flow is wasted. If ventilation falls relative to perfusion, blood flow is wasted, arterial PO_2 decreases, and PCO_2 increases.

The lungs possess a number of intrinsic, or built-in, mechanisms that reduce mismatches of ventilation to perfusion. For example, if a partial obstruction occurs in a bronchiole, causing alveolar PO_2 to fall and alveolar PCO_2 to rise, the ventilation–perfusion mismatch elicits local responses that counteract it (Figure 11.13(A)). The primary response is a reduction in blood flow to the alveolus. The decrease in alveolar PO_2 stimulates the smooth muscle of the artery supplying the alveolus to contract, increasing its resistance to blood flow. Simultaneously, the flow of air into and out of the

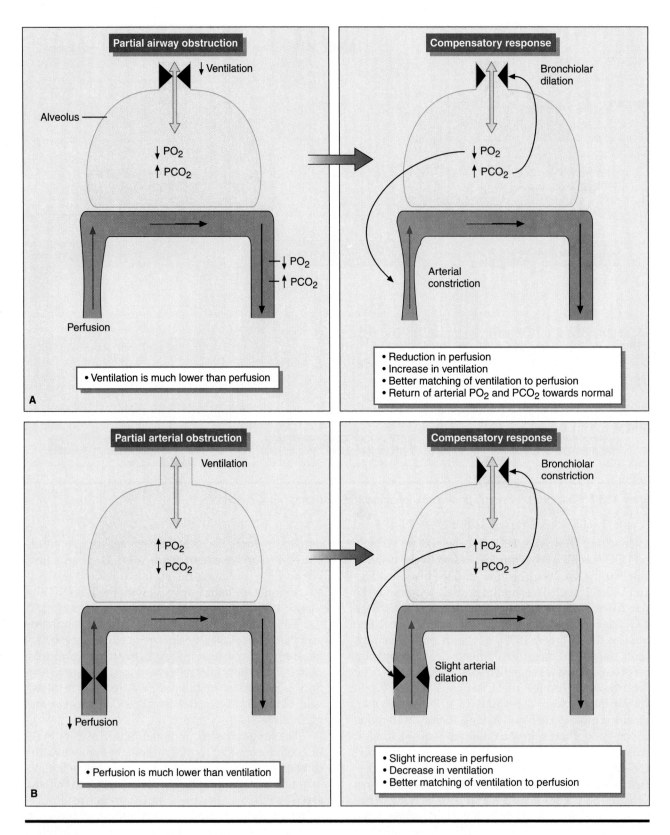

Figure 11.13 Intrinsic regulation of regional ventilation–perfusion matching.

alveolus increases. The increase in alveolar PCO_2 dilates the bronchiolar smooth muscle, reducing its resistance to air flow. Both of these responses change alveolar gas partial pressures, and thus arterial gas partial pressures, toward normal. In addition, less blood participates in inefficient gas exchange, reducing the amount of wasted blood flow.

Intrinsic compensatory responses also occur during the other type of ventilation-perfusion mismatch. A partial obstruction of an artery causes alveolar PO_2 to rise and PCO_2 to fall. The decrease in PCO_2 constricts the bronchiolar smooth muscle, thus decreasing air flow. The increase in alveolar PO_2 dilates the artery slightly, allowing a little more blood to perfuse the alveolar capillaries. Again, both of these changes in air and blood flow result in a better balance between ventilation and perfusion in the alveolus (Figure 11.13(B)).

The constriction of pulmonary arteries when alveolar PO_2 falls is extremely important. This response is called **hypoxic pulmonary vasoconstriction.** *Hypoxic* refers to decreased amount of O_2. However, hypoxic pulmonary vasoconstriction is not always a beneficial response. It is clear that local arterial constriction improves overall gas exchange when small regions of the lung become hypoxic, but what about conditions in which the alveolar PO_2 of the entire lung falls? This elicits arterial constriction throughout the whole lung. When arteries over the entire lung constrict, total pulmonary vascular resistance increases and arterial pressure in the lung rises. If arterial constriction is great enough, pulmonary arterial pressure will rise so high that many scientists believe that fluid will move out of the blood across the vessel walls and into the alveolar air spaces. This condition is called **pulmonary edema.** As fluid fills the air spaces, the distance between alveolar air and the capillary blood increases, slowing down diffusion of O_2 and CO_2. The blood flowing through the alveolar capillaries does not equilibrate with the alveolar gas, causing a decrease in arterial PO_2 and an increase in arterial PCO_2. If pulmonary edema is not treated quickly, arterial PO_2 can fall to lethal levels within one or two days. Occasionally, people ascending to very high elevations suffer from this serious condition. At high elevations, alveolar PO_2 falls throughout the lung because atmospheric PO_2 is reduced. In 1989, 23 people attempting to climb Mt. McKinley (elevation 20,320 ft) suffered severe pulmonary edema. Fortunately, most recovered when they were brought to lower elevations where atmospheric PO_2 is higher. The primary symptoms of pulmonary edema are shortness of breath, cough, mental confusion, and trouble beathing while lying flat.

Vasoconstriction in response to hypoxia also occurs in the lungs of most vertebrate animals, even frogs and lizards. In addition, hypoxic vasoconstriction occurs in other types of respiratory organs. For example, the blood vessels of fish gills show the same response to a decrease in water PO_2. Despite its importance and widespread occurrence, we still do not know how local hypoxia causes arterial constriction. No one understands how a decrease in alveolar PO_2 causes the arteries *upstream* from the alveoli to constrict because the arteries have no direct contact with the alveolar air. Animal experiments ruled out involvement by nerves because lungs with all of their nerves removed still exhibit hypoxic vasoconstriction. Other experiments showed that hormones are not involved either. When lungs are removed from animals and perfused with fluid that is free of all hormones, the pulmonary arteries still constrict when alveolar PO_2 falls. Perhaps signals traveling along blood vessels transmit information about alveolar PO_2 to the upstream arteries. However, this is merely an untested hypothesis.

Extrinsic Regulation of Gas Exchange

Breathing is a rhythmic process originating in the brain (Figure 11.14). The precise source of the breathing rhythm is not known, but it is controlled mostly by neurons in the brain stem. There are two basic types of respiratory neurons in the brain stem: inspiratory neurons and expiratory neurons. The inspiratory neurons make synaptic connections with inspiratory motoneurons. These motoneurons innervate the muscles of inspiration (primarily the diaphragm and the muscles between the ribs). The expiratory neurons in the brain stem communicate with expiratory motoneurons, which innervate the muscles of expiration (primarily the abdominal muscles). During inspiration, the inspiratory neurons are active and the expiratory neurons are quiet. As a result, inspiratory muscles contract and expiratory muscles relax and expand. During expiration, the reverse happens. The expiratory neurons become activated and the inspiratory neurons become inactive. When at rest, the expiratory muscles contract so slightly that the dimensions of the thorax are unaffected. During exercise, however, the muscles of expiration contract vigorously to help move air out of the lungs. Whether at rest or during exercise, there is an oscillation of neural activity between the inspiratory and expiratory neurons in the brain stem, and that oscillation of neural activity in the brain stem produces an oscillation of muscular activity between the inspiratory and expiratory muscles. The control of overall pulmonary ventilation involves changing the respiratory rhythm. This rhythm is modified through many different neural inputs to the brain stem that alter the activity of the respiratory neurons.

Figure 11.14 Extrinsic regulation of pulmonary ventilation.

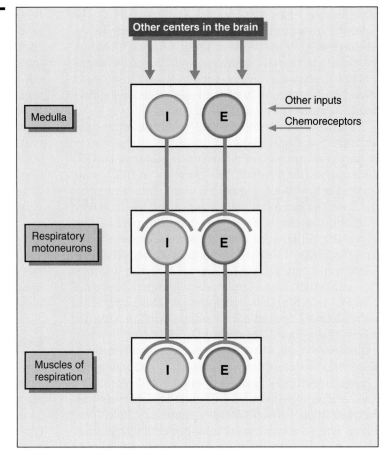

One important input to the brain stem is from special chemical sensing cells, called **O₂ chemoreceptors,** that are sensitive to the PO₂ of arterial blood. These O₂ chemoreceptors are located in the carotid and aortic bodies, which are small structures located on the sides of the carotid artery and the beginning part of the aorta, respectively. Because these chemoreceptors are located ouside the central nervous system, they are often called **peripheral chemoreceptors.** The carotid bodies are only about ½ cm (⅕ of an inch) in diameter, and the aortic bodies are even smaller. Nerves run from the carotid and aortic bodies to the respiratory neurons in the brain stem. If arterial PO₂ decreases, the O₂ chemoreceptors increase activity in these nerves, causing more activity in the respiratory neurons in the brain stem. This leads to greater contractions of the respiratory muscles and increased lung ventilation.

This creates a homeostatic control for arterial PO₂. The stimulus is a low arterial PO₂ and the response is increased lung ventilation. Because increased lung ventilation raises arterial PO₂, this response reduces the stimulus, and thus is a homeostatic control.

When might this response be elicited? If you have ever been in the mountains above 8000 ft, perhaps you

noticed that you breathe more than usual, particularly during exercise. Stimulation of the carotid and aortic chemoreceptors by the reduction in arterial PO₂ at high altitude causes a reflex increase in both the tidal volume and the rate of ventilation. Although increased lung ventilation raises arterial PO₂ in this situation, it does not restore it to normal, but it helps.

A second modifying input to the respiratory neurons comes from another type of chemoreceptor that is also located in the carotid bodies. These chemoreceptors are sensitive to the H⁺ concentration of arterial blood. Remember, the major source of H⁺ ions in blood is CO₂. An increase in blood PCO₂ generates more H⁺ ions. A decrease in PCO₂ reduces H⁺ concentration. Other processes besides the combination of CO₂ and water also produce H⁺ ions. For example, fat metabolism, ingestion of certain drugs such as aspirin, and kidney failure all add H⁺ to the blood. Thus, the arterial H⁺ concentration rises during starvation, when large amounts of fat are metabolized, or with an overdose of aspirin or kidney failure. In these situations, the increase in arterial H⁺ concentration stimulates the H⁺ chemoreceptors, which, like the stimulation of the O₂ chemoreceptors by low arterial PO₂, increases the activity of the brain stem respiratory neurons via neural connections. An increase in pulmonary ventilation results. This response also creates a homeostatic control. An increase in the H^+ concentration of blood stimulates the H^+ chemoreceptors of the carotid bodies, which leads to an increase in pulmonary ventilation. As pulmonary ventilation increases, alveolar and arterial PCO_2 declines. This reduces arterial H^+ concentration, which is the stimulus for this response. The same negative feedback system also decreases ventilation whenever arterial H^+ falls below normal.

A third type of chemoreceptor controlling lung ventilation senses the H^+ concentration of the cerebral spinal fluid (CSF, the fluid surrounding the brain and spinal cord). These chemoreceptors are called **central chemoreceptors** because they are located in the central nervous system. Although arterial H^+ concentration has almost no effect on cerebral spinal fluid H^+ concentration, arterial PCO_2 has a large effect on it. The reason for this paradox is that the blood–brain barrier (the tissue that separates the blood from the brain tissues) is nearly impermeable to H^+ but very permeable to CO_2 (Figure 11.15). An increase in arterial PCO_2 causes CO_2 to diffuse from the blood into the cerebral spinal fluid, raising its PCO_2. As in blood, an increase in cerebral spinal fluid PCO_2 increases the combination of CO_2 and H_2O, generating more H_2CO_3 and in turn more H^+. In contrast, an increase in arterial H^+ does not cause net movement of H^+ into the cerebral spinal fluid because H^+ cannot diffuse across the blood–brain barrier. Thus, whenever arterial PCO_2 increases, the cerebral spinal fluid H^+ concentration rises, stimulating the central H^+ chemoreceptors. Stimulation of these chemoreceptors increases neural activity to the brain stem respiratory neurons, causing an increase in pulmonary ventilation. Again, this response forms a negative feedback control: An increase in arterial PCO_2 elicits an increase in ventilation, which acts to decrease arterial PCO_2.

The peripheral and central H^+ chemoreceptors often function together to regulate pulmonary ventilation. Because both arterial and cerebral spinal fluid H^+ concentration are affected by arterial PCO_2, changes in arterial PCO_2 affect both receptor types. An easy way to affect both chemoreptors at once is to hyperventilate (breathe very rapidly and deeply) for approximately 1 minute. What does hyperventilation do to alveolar PCO_2, arterial PCO_2, cerebral spinal fluid PCO_2, arterial H^+ concentration, and cerebral spinal fluid H^+ concentration? After hyperventilating, your normal respiratory rate is noticeably reduced because of decreased activity of both types of chemoreceptors. Test this by counting the number of breaths you take during one minute while quietly sitting. Then hyperventilate for another minute, and again

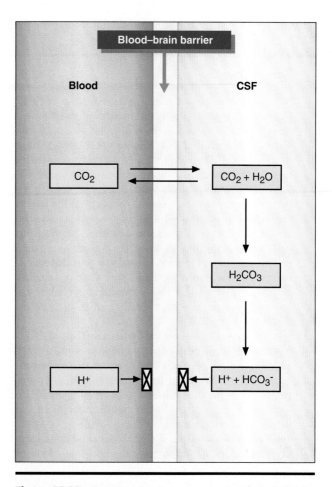

Figure 11.15 Relationship between cerebral spinal fluid, H^+ concentration, and blood PCO_2.

measure your respiratory rate after hyperventilating. In situations where arterial H^+ concentration increases due to the direct addition of H^+ to the blood (as in high fat metabolism, aspirin overdose, and kidney failure), not from an increase in PCO_2, only the H^+ chemoreceptors in the carotid bodies are stimulated.

The ventilatory response elicited by the H^+ chemoreceptors is extremely important because the functioning of most of our cells, especially brain cells, is affected by H^+ concentration. For example, a rapid and large increase in blood H^+ concentration interferes with proper nerve function, resulting in a marked decrease in mental function. This sometimes occurs in people with diabetes. When blood insulin concentration gets too low, the rate of fat metabolism increases, which leads to increased production of H^+. If blood H^+ concentration rises too much, confusion and sometimes coma or death result. The ventilatory response to increases in arterial H^+ is very effective in reducing the magnitude of such increases. In some cases, this response is lifesaving. Probably because of the extreme importance of maintaining H^+ within normal ranges, the H^+

MILESTONE 11.1

Smoking

In 1964, the Surgeon General released a report titled *Smoking and Health*. This landmark document reviewed all of the available data on the health effects of smoking. The report concluded that smoking contributes to lung cancer, chronic bronchitis and emphysema, cardiovascular diseases, and cancers of the lip, larynx, esophagus, and urinary bladder. Since that report, there have been 20 more Surgeon General's reports on the health hazards of tobacco use. These reports substantiate the findings of the 1964 report and describe a relationship between smoking and cancers of the pancreas, stomach, uterine cervix, and kidney. They also report that smoking may delay conception and that smoking during pregnancy increases the risk of miscarriage, fetal death, low birthweight, newborn death, and sudden infant death syndrome (SIDS).

Cigarette smoke is made up of more than 4000 different substances besides the commonly known "tar" and nicotine. Among these are chemicals that cause cancer, poison cells, change cell structure, elicit inappropriate immune responses, suppress appropriate immune responses, and alter neural activity in the brain. In addition, the carbon monoxide in cigarette smoke interferes with oxygen transport by displacing O_2 from the hemoglobin molecule. Thus, the blood of smokers carries significantly less O_2 than the blood of nonsmokers. These diverse biological actions contribute to the adverse effects of smoking throughout the body.

Lung cancer is the most common cause of cancer death in the United States and cigarette smoking is the leading cause of lung cancer (see Figure 1 in this Milestone). Men who smoke one pack of cigarettes per day have a 10-fold greater risk of lung cancer than do nonsmokers, and those who smoke two packs per day have a 25-fold increase in risk of lung cancer. Because cigarette use among women has increased markedly in the past 50 years, lung cancer has surpassed breast cancer as the leading cause of cancer death in women.

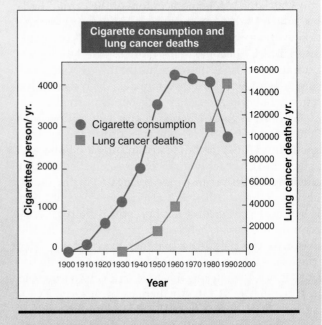

Figure 1   Cigarette consumption and lung cancer deaths in the United States during this century.

Cigarette smoking is also the leading cause of a debilitating and often fatal lung disease called **chronic obstructive pulmonary disease** (COPD), or chronic obstructive lung disease (COLD). People with COPD usually have a combination of chronic bronchitis and emphysema. Chronic bronchitis is characterized by excessive mucus production in the trachea, bronchi, and bronchioles. In addition, there is inflammation, edema, thickening of the bronchial wall, and an increase in smooth muscle in the airway walls. These changes disrupt normal air flow through the lungs and often block air flow to significant portions of the lungs. Emphysema destroys the tissue between the alveoli, which en-

chemoreceptors have evolved as the most important chemoreceptors in the control of ventilation under normal conditions. The O_2 chemoreceptors are also very important, but they exert large effects on ventilation only when blood PO_2 falls to relatively low levels.

Many other factors affect lung ventilation. Perhaps the most important is our voluntary control over breathing, because it enables us to speak, sing, play musical instruments, and hold our breath. Voluntary control involves numerous neural connections between the brain cortex and the respiratory neurons in the brain stem. Ventilation is also affected by reflexes,

such as sneezing and coughing, and by certain emotions. Fear and anger are potent ventilatory stimulants. Some hormones and drugs also affect ventilation. For example, epinephrine (adrenaline) increases ventilation. Have you ever noticed that your ventilation suddenly increases several seconds after you are startled? This results from the release of epinephrine by the adrenal glands. Ventilation is decreased by a number of commonly abused drugs. One of the primary symptoms of alcohol, heroin, or barbiturate overdose is a reduced ventilatory rate. These drugs act directly on the respiratory neurons of the brain stem, reducing their

MILESTONE 11.1 (CONTINUED)

larges the alveolar air spaces. These numerous changes caused by COPD result in shortness of breath, coughing, respiratory infections, an increase in arterial PCO_2, and a decrease in arterial PO_2. In severe cases, so little O_2 is able to move across the damaged lung into the blood that even the smallest amount of exercise, such as walking across a room, is impossible (see Figure 2 in this Milestone).

The addictive nature of smoking is due largely to nicotine, which is one of the components of tobacco. Nicotine acts on the central nervous system to increase arousal and attentiveness, decrease appetite and irritability, and relax the muscles. Because the nicotine inhaled in cigarette smoke reaches the brain within 8 seconds, its effects occur rapidly. Consequently, nicotine's effect positively reinforces each puff on a cigarette. A one-pack-per-day smoker puffs on a cigarette more than 70,000 times per year. That is a lot of reinforcement, which makes it very difficult to give up smoking. Perhaps understanding the dangers will prevent many from ever taking that first life-threatening puff.

For those who have already started smoking, there is some encouraging news tied to the benefits of smoking cessation. In 1990, the Surgeon General published a report titled *The Health Benefits of Smoking Cessation*. This report concluded that there are major health benefits of smoking cessation and they begin immediately; former smokers live longer than continuing smokers; smoking cessation reduces the risk of lung cancer, other cancers, heart attack, stroke, and chronic lung disease; and women who stop smoking before or during the first three months of pregnancy eliminate their increased risk of having a low-birthweight baby.

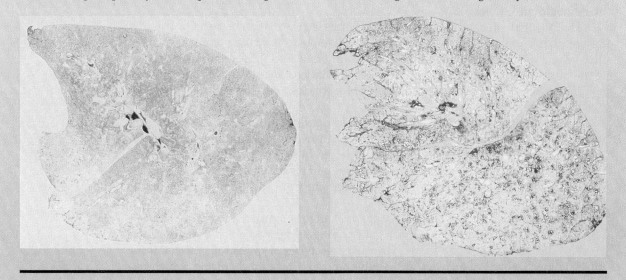

Figure 2   The lungs of a nonsmoker and a smoker. Notice the black color of the smoker's lungs and the light-colored cancerous tumors.

activity. Severe overdoses of these drugs can depress ventilation so much that death results.

There are also unknown factors affecting lung ventilation. Exercise is a very common stimulus for increasing lung ventilation, but the precise factors stimulating lung ventilation during most forms of exercise are not known. The chemoreceptors do not appear to be involved because arterial PO_2, PCO_2, and H^+ concentration do not change during exercise. Voluntary increases in ventilation, involving the neural connections between the cortex and the brain stem, also do not seem to be involved during exercise. The best proof of this comes from experiments on anesthetized animals, in which voluntary control of ventilation is absent. Direct electrical stimulation of their skeletal muscles causes muscular contraction, and pulmonary ventilation increases in response. In addition, neural connections between the exercising muscles and the brain stem do not appear to be important. Direct electrical stimulation of the leg muscles in paraplegics (people without neural connections between the legs and the brain) also causes increases in ventilation. Understanding the causes of increased ventilation during exercise must await further physiological studies.

Summary

Respiration is the process of supplying O_2 from the atmosphere to our cells. It is the most critical physiological process: Its disruption causes death within minutes. Although the primary duty of the respiratory system is to transport O_2, it has two other vital functions: to transport CO_2 from the cells to the environment and to regulate the H^+ concentration of blood.

The respiratory tract transports gases between the environment and the blood. The first step in O_2 transport is inspiration. Air with a PO_2 of 150 mm Hg (at sea level) is brought into the lung. Alveolar PO_2, 100 mm Hg at sea level, is lower than atmospheric PO_2 because the fresh air mixes with old residual air, blood continually removes O_2 from the alveolar gas, and CO_2 and water vapor are added to lung gas, diluting the O_2.

The blood entering the alveolar capillaries has a PO_2 of about 40 mm Hg. Thus, O_2 diffuses from the alveolar air into the blood. Blood equilibrates with the alveolar air before it leaves the capillaries, giving arterial blood the same PO_2 as the alveolar gas.

Three factors affect arterial PO_2: pulmonary ventilation, pulmonary blood flow, and altitude. Blood carries O_2 from the lungs to the metabolizing tissues. Practically all of the O_2 in blood is bound to the hemoglobin in the red blood cells. The binding of O_2 to hemoglobin enables blood to carry about 65 times more O_2 than if blood had no hemoglobin.

The PO_2 of blood entering the tissue capillaries is greater than tissue PO_2. This PO_2 difference drives O_2 out of the capillaries. However, before this can happen, O_2 must first be unloaded from hemoglobin. Increases in blood PCO_2, H^+ concentration, and temperature facilitate unloading of O_2 in the capillaries. Once O_2 leaves hemoglobin, it diffuses down a PO_2 gradient into the mitochondria, where it is used in oxidative metabolism.

The respiratory system transports CO_2 in the opposite direction. CO_2 is produced in cells as a byproduct of metabolism and diffuses down a PCO_2 gradient into the capillaries. Three different forms of CO_2 are carried in blood: physically dissolved CO_2 molecules, CO_2 chemically bound to hemoglobin, and bicarbonate ions (HCO_3^-). In the lung, CO_2 diffuses down a partial pressure gradient from the blood into the alveolar air.

Under most conditions, pulmonary gas exchange is regulated to maintain arterial PO_2 and PCO_2 constant. Pulmonary gas exchange is also adjusted to regulate the H^+ concentration in the body fluids. There are two types of regulatory mechanisms: intrinsic mechanisms, which act in the lung, and extrinsic mechanisms, which use neural pathways outside of the lung.

Questions

Conceptual and Factual Review

1. What are the three main functions of the respiratory system? (p. 279)
2. What happens to the air we breathe as it flows through the upper respiratory tract? Why is this important? (p. 280)
3. Where does gas exchange between the air and the blood occur? (p. 280)
4. Why is the pressure exerted by the intrapleural fluid lower than atmospheric pressure? (p. 283)
5. Why is the pressure difference across the lung wall important? (p. 283)
6. Why does air flow into the lungs during inhalation? (p. 283)
7. What happens to the intrapleural fluid pressure when the muscles of inspiration relax? What happens to the pressure difference across the lung wall? (p. 283)

8. What are the major factors that determine the work of breathing? (p. 287)
9. What does surfactant do? Why is surfactant important? What happens if surfactant is not present? (p. 287)
10. How does hemoglobin allow blood to carry more O_2? (pp. 288–289)
11. Why do increases in atmospheric PO_2 above 80 mm Hg produce only small changes in blood O_2 concentration? (pp. 288–290)
12. Why is alveolar PO_2 lower than atmospheric PO_2? (pp. 290–291)
13. What determines arterial PO_2? (pp. 290–291)
14. What factors affect alveolar PO_2? How do these factors affect arterial PO_2? (pp. 290–291)
15. What happens to facilitate O_2 diffusion from capillary blood to the tissues? (p. 292)
16. Describe the three forms of CO_2 that are transported by the blood. (p. 293)
17. How do changes in pulmonary blood flow and ventilation affect alveolar PCO_2? (p. 294)
18. What happens to alveolar PCO_2 at high altitudes? Why? (p. 294)

19. What happens to alveolar PO_2 and PCO_2 and to arterial PO_2 and PCO_2 when there is a decrease in ventilation to an alveolus (for example, when a foreign particle lodges in the bronchiole)? (pp. 294–297)
20. What compensatory responses counteract a ventilation–perfusion mismatch caused by partial obstruction of a bronchiole? (pp. 295–297)
21. When is hypoxic vasoconstriction not a beneficial response? Why? (p. 297)
22. What neural inputs alter the activity of the inspiratory and expiratory neurons to modify the respiratory rhythm? What effect do these neural inputs have on the respiratory neurons? (pp. 297–301)
23. Why is the ventilatory response elicited by the H^+ chemoreceptors so important? (pp. 298–300)
24. Besides neural inputs, what other factors can affect lung ventilation? (pp. 300–301)
25. What diseases can cigarette smoking cause? (pp. 300–301)
26. What happens when a cigarette smoker quits smoking? (pp. 300–301)

Applying What You Know

1. When many animals become hypoxic (i.e., unable to obtain enough O_2), their body temperature falls. How might a fall in body temperature be beneficial when O_2 is in short supply?
2. The brontosaurus was a dinosaur with a neck estimated at 25 feet in length. What respiratory problem would this present (besides the seriousness of a sore throat)?
3. Sometimes people experience a pneumothorax on just one side of the thorax, causing one lung to collapse. What do you think happens to blood flow to the collapsed lung? What happens to blood flow to the other lung? Why?
4. Many species of salamanders have no lungs or gills. Their gas exchange organ is the skin. Oxygen and CO_2 move directly between the environment and the blood across the outermost layer of skin. What problems and limitations do these animals experience by being skin-breathers? What benefit does skin-breathing offer over breathing with lungs?

Chapter 12
Renal Excretory System

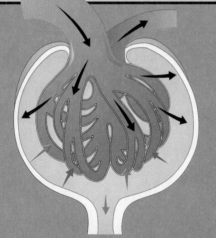

Objectives

By the time you complete this chapter you should be able to

◆ Describe the microanatomy of the kidney and understand the function of the different parts of the nephron.

◆ Understand that the kidney is the major organ responsible for maintaining the constancy of the internal environment regardless of wide variations in intake of salts and water.

◆ Explain how the kidney responds to changes in intake by altering the rate of excretion of those plasma constituents so that output always matches input.

◆ Describe how the kidney excretes many of the waste products of metabolism.

◆ Explain how the kidney reabsorbs some constituents from the tubular fluid and secretes other substances into that urine.

◆ Discuss factors that cause the kidney to alter the rate at which its transport systems work.

◆ Understand the role of hormones in regulating those processes as well as the mechanism by which the kidney is able to produce a concentrated urine.

◆ Understand the concept of clearance and how to calculate the rate at which the kidney filters fluid and reabsorbs and secretes plasma constituents.

◆ Describe the role of the renin–angiotensin system in the homeostatic control of salt and water balance.

◆ Explain the general concept of the kidney as a regulatory organ.

Introduction

A famous kidney physiologist once said that urine is the stuff of philosophy itself. Taken alone, that sentence sounds foolish. But what he meant was that the evolution of the mammalian kidney was responsible for allowing fish to evolve into humans. It wasn't that some prehistoric fish suddenly became human; rather, evolution of land-dwelling animals was possible only with the evolution of the fish kidney to the mammalian type. It is the organ that allowed animals to free themselves from the watery environment of the primordial sea, for the kidneys maintain the constancy of the internal environment of animals, or the interstitial fluid. However, that is not what most people think of when they think of kidney function. They usually think of the kidneys removing waste products of metabolism from the body. Although that is true, kidney function is far more complex than a simple waste disposal system.

Salt and water intake varies considerably among individuals. Some people feast on salted peanuts, pretzels, and potato chips, and salt their food even before tasting it. Others may believe, "Lips that touch salt shall not touch mine." Yet if one were to sample the blood of both groups of people, the NaCl (salt) concentration in plasma of the two groups would be nearly indistinguishable. You shall see how the kidneys act to maintain constancy of the body fluids, despite wide variations in salt and water intake. They are the organs of homeostatic control of the internal environment. Only when kidney (renal) function fails does the internal environment change significantly, threatening the organism's life.

Anatomy of the Renal System

The kidneys are located at the back of the abdomen, under the ribs, on either side of the spine (Figure 12.1). Although each kidney is about the size of a closed fist, and the two put together account for less than 1% of the total body weight, they receive about 20% of the blood pumped by the heart, an indication of their importance in homeostasis. Thus, the total renal blood flow is about 1 liter per minute in an adult. Arterial blood enters the kidneys, and venous blood and urine leave; during this process, the kidneys somehow make 1 ml per minute of urine out of arterial blood. The act of urination, however, is much less frequent than that. Peristalsis of the smooth muscle of the

ureters continuously forces urine from the kidneys into the bladder. Urine continues to flow to the bladder and is stored there, stretching the wall of the bladder. Stretch receptors in the bladder wall set up a spinal reflex that causes parasympathetic stimulation of the bladder muscles and simultaneously, via sensory nerves, a feeling of fullness and the urge to urinate. Voluntary urination, learned during childhood, is initiated by conscious control over relaxation of pelvic muscles that act as a urethral sphincter to control voiding of urine. Although the act of voiding (micturition) is a reflex, that reflex may be initiated or stopped voluntarily because of cortical control over the sphincter muscles.

In order to understand the processes involved in urine formation, it is necessary to consider both the gross and microscopic anatomy of the kidney (Figure 12.2). Figure 12.2(A) shows that the renal artery divides into several branches after entering the kidney. Each branch radiates toward the **renal cortex,** or the outer region of the kidney. In the cortex, the branches subdivide repeatedly until they reach the arteriolar level. At this level, there are approximately one million arterioles in each kidney, and each of these **afferent arterioles** supplies arterial blood to a **nephron.** Nephrons are the tubular structural and functional units of the kidney.

An enlarged diagram of a nephron, with its vascular and tubular components, is shown in Figure 12.2(B). Blood flows from the *afferent arteriole* into a tuft of capillaries, called the **glomerular capillaries,** and out of that tuft into another arteriole called the **efferent arteriole.** The blind end of the tubular nephron forms a cup-like structure, **Bowman's capsule.** Bowman's capsule surrounds the glomerular capillaries much like a balloon would surround a fist that is pushed into it. The tuft of glomerular capillaries is called the **glomerulus.** It is here that the process of urine formation begins with filtration of fluid out of the plasma into the Bowman's space.

The anatomy of the glomerular capillaries is unique. Surrounding the basement membrane of the capillaries are large octopus-like cells called **podocytes,** which project foot-like processes that enclose the capillary basement membrane. Figure 12.3 shows an electron micrograph and a scanning electron micrograph of the capillary structure. The first pore boundary is created by the capillaries, which restrict the cellular elements of blood from being filtered. This boundary is followed by the basement membrane, which blocks larger molecular solutes such as proteins. The final restriction barrier is formed by the interdigitating foot processes of the podocytes. The spaces between the foot processes are called the slit pores, which can be seen in Figure 12.3. The scanning electron micrograph

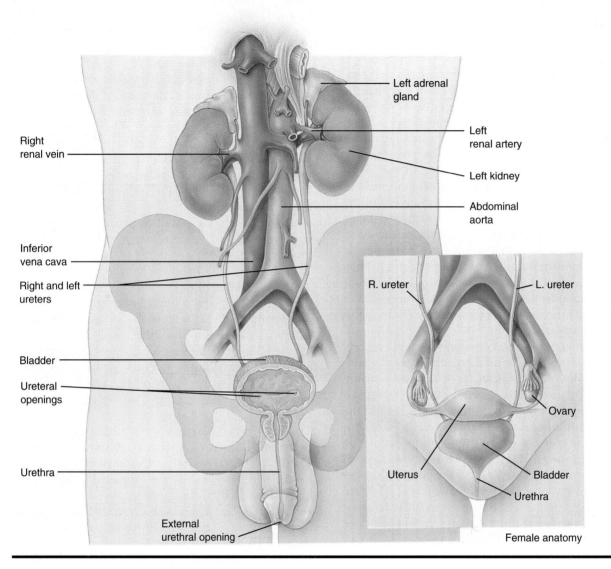

Figure 12.1 Anatomical location of the kidneys.

in Figure 12.3(B) shows the podocytes as they appear looking from inside Bowman's space.

Looking again at Figure 12.2(B), you can see that blood flows from the efferent arteriole into capillaries that surround the tubular portions of the nephron. For that reason, they are called **peritubular capillaries.** Blood from several peritubular capillaries flows into venules, which merge into larger and larger venous branches in the kidney until, finally, a single renal vein exits the kidney.

The tubular nephron is composed of a single layer of epithelial cells, which are specialized for the transport of water and solute molecules between the uri-

nary filtrate, or **tubular fluid** (fluid contained in the tubules), and the interstitial fluid surrounding the tubules. In sequence, the major regions or segments of the tubule are Bowman's space (the interior of the balloon indented by the fist), the **proximal convoluted tubule** (proximal to the glomerulus), the **loop of Henle** (which is divided into descending and ascending limbs), the **distal convoluted tubule** (distal to the glomerulus), and the **collecting duct.** Many nephrons ultimately merge into the same collecting duct. Whereas there are approximately one million nephrons per adult human kidney, there are only several thousand collecting ducts. Note the locations of these

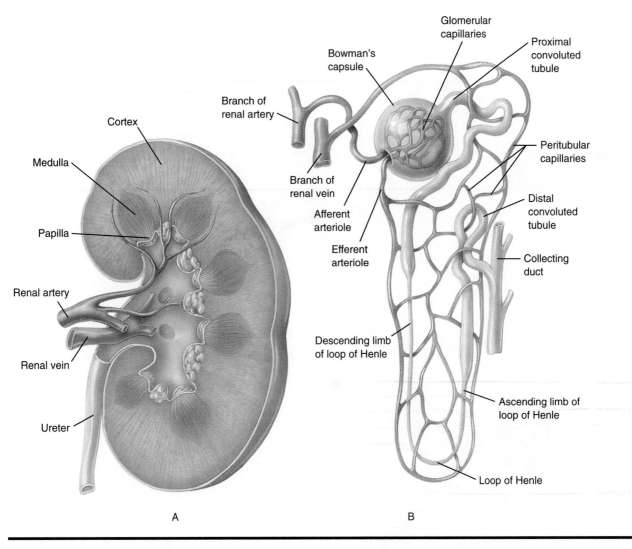

Figure 12.2 (A) Gross anatomy of the kidney. (B) Microanatomy of the kidney.

segments in the renal cortex and **renal medulla** (Figure 12.2(A)).

Function of the Nephron

Although physiologists recognized the structural relationships as far back as the mid-nineteenth century, they did not know what processes were involved in the formation of urine. Many physiologists still thought that vital processes were involved, meaning that living cells had the ability to perform certain processes without adhering to the known physical and chemical laws. Living tissue was thought to be able to use some mysterious power to accomplish an outcome. This thinking, of course, tends to end any search for mechanisms, as vital processes can never be understood. However, many physiologists believed that all life processes conform to physical and chemical laws, making them amenable to study. They actively sought to determine how urine is formed.

Glomerular Filtration

Physiologist Carl Ludwig, in 1847, was the first to identify a possible explanation. He noted that the glomerular capillaries seemed to be ideally suited to press fluid out of the capillaries into Bowman's space.

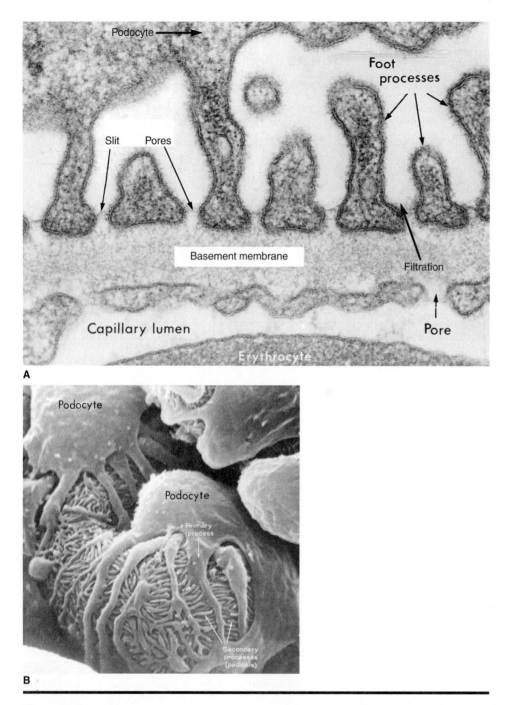

A

B

Figure 12.3 Glomerular capillary filtration barrier. (A) Electron micrograph. (B) Scanning electron micrograph.

The glomerular capillaries are interposed between two arterioles so that the resistance to flow out of the capillaries must be greater than in all other capillaries. He realized that the blood pressure in the capillaries should exceed the hydrostatic pressure in Bowman's space, and as a result force the filtration of a large amount of fluid out of the capillaries. Although this could account for the first step in urine formation, it could not account for the change that must take place in composition of the filtrate as it turns into final urine.

If Ludwig was correct in his assumption concerning glomerular function, a prediction could be made. Fluid filtered out of the glomerular capillaries into Bowman's space would be an ultrafiltrate of plasma; that is, it would resemble plasma in all respects, except that protein molecules would be lacking as they would be held back by the small pores in the capillaries. The next step was figuring out how to test this assumption.

In the early twentieth century, physiologists noted that the kidney of an amphibian, the mud puppy *Necturus*, contained very large glomeruli. They were so large that they were visible to the naked eye. Once this discovery was made, it was possible to develop microcapillary glass tubes and, with the aid of a microscope, insert them into the Bowman's space of a single glomerulus. They withdrew a small amount of fluid and analyzed its constituents. As Ludwig predicted, the composition of that fluid exactly matched an ultrafiltrate of plasma; it contained all the constituents of plasma in the same concentration as found in plasma, except for plasma proteins. The first riddle of urine formation was settled: Urine is formed from an ultrafiltrate of plasma moving out of the glomerular capillaries into Bowman's space in a process called **glomerular filtration.** The rate of formation of that fluid (the volume filtered each minute) is called the **glomerular filtration rate (GFR).**

Filtration Forces

Glomerular filtration is much like the process of filtration across the walls of typical systemic capillaries, described in Chapter 9, except that the glomerular filtrate ends up in tubules rather than in the interstitium. Figure 12.4 illustrates the three processes involved in urine formation. The size of each arrow represents the relative volumes of fluid flowing in the direction of the arrow. A very large volume of plasma flows past the glomerulus, and about 20% of that volume is filtered. As the fluid flows through the nephrons, most of it is reabsorbed back into the capillary network. The volume of blood passing the glomerular capillaries is reduced by the volume filtered, but as fluid is reabsorbed from along the nephron, all but a very small fraction of the volume is secreted back into the blood.

Despite the thickness of the filtration barrier, it is very permeable. It behaves as if it were perforated with holes or pores large enough to accommodate water molecules and all the small solute molecules in plasma. In Figure 12.5, the different forces acting on plasma are shown; these account for the filtration process. The major force is the blood hydrostatic pressure in the glomerular capillaries. On average, that pressure is about 60 mm Hg. Note that this pressure is considerably higher than that existing in all other capillaries in the body. In this respect, the kidney is unique.

Opposing the capillary hydrostatic pressure are two other forces. One is the colloid osmotic pressure of the plasma, illustrated in Chapter 9. As in all other capillaries, this force is about 30 mm Hg. In addition to this is the hydrostatic pressure in the capsule, ap-

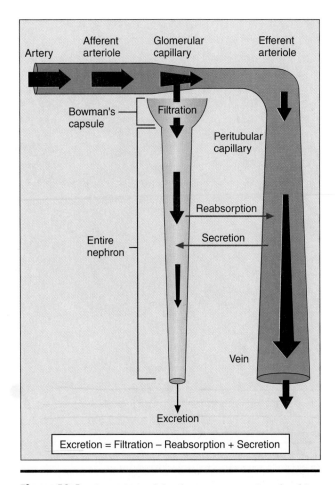

Figure 12.4 Overview of the three processes involved in the formation of urine.

proximately 20 mm Hg. The net filtration pressure is equal to the hydrostatic pressure along the glomerular capillaries minus the forces opposing filtration (the hydrostatic pressure in the capsule plus the colloidal osmotic pressure in the capillaries):

Net filtration pressure = Capillary blood pressure – (Plasma colloid osmotic pressure + Capsule hydrostatic pressure)

$$= 60 \text{ mm Hg} - (30 \text{ mm Hg} + 20 \text{ mm Hg})$$
$$= 10 \text{ mm Hg}$$

Note that this net filtration pressure of 10 mm Hg along the glomerular capillaries is unique in the body. It is the only place in which fluid is filtered out of the capillaries along its entire length.

Tubular Transport

Glomerular filtration is only the first process in urine formation, as shown in Figure 12.6. The glomerular

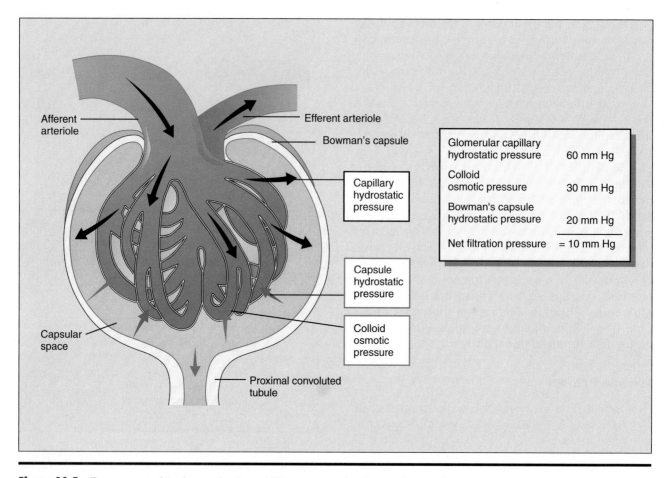

Afferent arteriole

Efferent arteriole

Bowman's capsule

Capillary hydrostatic pressure

Capsule hydrostatic pressure

Colloid osmotic pressure

Capsular space

Proximal convoluted tubule

Glomerular capillary hydrostatic pressure	60 mm Hg
Colloid osmotic pressure	30 mm Hg
Bowman's capsule hydrostatic pressure	20 mm Hg
Net filtration pressure	= 10 mm Hg

Figure 12.5   Forces exerted in the production of filtrate across the glomerular capillary membrane.

filtrate is virtually identical to protein-free plasma, but as this fluid flows along the tubule toward the end of the collecting duct, two additional processes transform the plasma-like filtrate into urine. Tubular cells transport some molecules that have diffused out of the peritubular capillaries into the interstitial space, and from that space carry them into the tubular fluid (B). This process is called **tubular secretion.** These secreted molecules may move across either the outer or the inner membranes of the tubular cells (indicated by the straight, thicker arrows), depending on the molecule in question. But regardless of the location of the active transport component, the secreted molecules ultimately come from the plasma in peritubular capillaries.

As secretion of a particular type of molecule occurs, its concentration in the interstitial fluid decreases, creating a concentration gradient for continued diffusion from the plasma to the interstitial fluid, as noted by the blue arrows. Tubular cells may transport some filtered molecules in the opposite direction, from the tubular fluid into the renal interstitium (C). This process is called **tubular reabsorption.** The term

reabsorption is used because all the solutes absorbed by the tubules were first filtered out of the blood and then reabsorbed back into the blood. Reabsorbed molecules in the interstitial fluid diffuse passively into the plasma in the peritubular capillaries, and ultimately leave the kidney in renal venous blood. Thus, although the glomerular filtrate in Bowman's space is very similar to plasma, secretory and reabsorptive processes dramatically alter both the volume and the composition of the tubular fluid as it flows along the nephron. These secretory and reabsorptive processes are complete by the end of the collecting duct, at which point the tubular fluid has become urine.

Overall, the rate of the reabsorptive processes exceeds that of the secretory ones. Removal of solute from the tubular fluid tends to reduce its osmotic pressure. However, the tubular epithelium, except for the ascending limb of the loop of Henle, is permeable to water, and water follows the solute passively in response to the small osmotic gradient set up by solute removal. Thus the volume of the filtrate is reduced as it moves along the renal tubules.

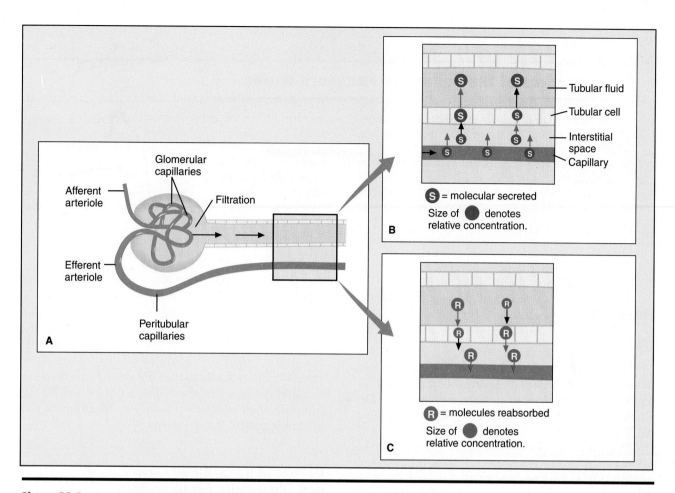

Figure 12.6 A closer look at the three processes involved in the formation of urine. (A) Glomerular filtration. (B) Tubular secretion. (C) Tubular reabsorption. Black arrows denote active transport; blue arrows represent either passive transport or facilitated diffusion.

Once urine leaves the kidneys via the ureters, no further changes in volume or composition occur. The rest of the urinary tract consists of the "plumbing" structures of ureters, bladder, and urethra (see Figure 12.1). From this description, you see that only three processes are involved in urine formation: the rate at which fluid is filtered out of the glomeruli, or the glomerular filtration rate; the movement of solute from capillaries into the tubular fluid, or secretion; and the movement of water and solute from tubular fluid back into the capillaries, or reabsorption. Urinary excretion is the sum of those processes. This may be expressed as a simple equation:

Urinary excretion = Glomerular filtration + Tubular secretion – Tubular reabsorption

Renal function is unique in that it can be quantified with relative ease. All excreted molecules, whether water or solute, are handled by some combi-nation of the three processes. In order to understand this central concept better, consider two examples. For an adult female, the GFR averages about 110 ml/min. The GFR of an adult male is slightly higher, about 125 ml/min. The difference in GFR is a reflection in the average difference in body weight between the sexes. For a female, 110 ml of water and associated solutes are filtered into the tubules every minute. Recall also that the rate of urine formation averages 1 ml/min. Because 110 ml of water are filtered into the tubules every minute but only 1 ml of urine is formed and excreted, it follows that, on average, the renal tubules must reabsorb 109 ml of water every minute; well over 99% of the filtered fluid is reabsorbed.

As another example of reabsorptive processes, consider the renal handling of glucose. The concentration of glucose in plasma averages 1 mg/ml, and the glomerular capillaries are freely permeable to glucose. If 110 ml of plasma is filtered into the tubules every minute, and each ml contains 1 mg of glucose, then

HIGHLIGHT 12.1

Consequences of the Failure to Reabsorb Water

If the renal tubules failed to reabsorb any of the filtrate, and the GFR remained constant at 110 ml/min, the consequences would be dire. (Of course this is impossible, but the calculations are informative.) A GFR of 110 ml/min equals 6600 ml/hour. If none of that fluid were reabsorbed, the body fluids would decrease by that amount each hour. Because about 60% of body weight is water, a woman weighing 50 kg would contain about 30 liters of water. Thus, within four and a half hours, no water would be left in her body. Actually, the person would die well before that. A loss of 20% of body water is fatal. In this hypothetical example, death would occur within one hour. This example illustrates the fact that exquisitely fine control over renal function is required for maintenance of life.

GFR × Concentration of glucose in the filtrate =
        Amount of glucose filtered/min
110 ml/min × 1 mg glucose/ml =
        110 mg glucose filtered/min

If the urine is glucose-free, as it is in most people, the renal tubules must have reabsorbed all of the filtered glucose, or 110 mg of glucose every minute. (The glucose reabsorptive mechanism is described in more detail below.)

Many of the solutes normally present in the plasma are filtered and reabsorbed, some (glucose and amino acids) more completely than others (urea, Na^+, and Cl^-). Other solutes, such as **creatinine** (an end product of muscle metabolism) and H^+, are filtered and secreted. Other plasma solutes, such as K^+, are filtered, reabsorbed, and secreted.

How can it be determined whether a particular solute is reabsorbed or secreted? If the filtered amount of a solute is greater than that excreted, some of it must have been reabsorbed from the filtrate as it flowed through the nephron. If the excreted amount of a solute is greater than that filtered, some must have been added to the tubular urine as it flowed down the nephron. If a solute is both reabsorbed and secreted (K^+, for example), applying the equation allows one only to calculate the net result of the two processes.

Renal Hemodynamics

The term *renal hemodynamics* refers to renal blood flow and glomerular filtration rate. These two flows determine renal function. If **renal blood flow (RBF)** falls to low values, for any reason, so will the GFR. Because it is the fluid filtered through the glomerular capillaries into Bowman's space (the glomerular filtrate) that is acted on to form urine, a reduction in GFR compromises the ability of the kidney to regulate the composition of the final urine. This, in turn, reduces its ability to keep the composition of body fluids constant. In fact, when physicians speak of kidney failure, they generally mean that both the GFR and RBF have fallen to values so low that the kidneys can no longer accomplish their function properly. In these cases, the life of the patient is in jeopardy.

Control of Renal Blood Flow and Glomerular Filtration Rate

Capillary pressure is physiologically controlled and regulates the GFR by afferent and efferent arteriolar resistances. As described in Chapter 9, the blood flow through any organ is equal to the driving pressure (difference between arterial and venous blood pressures) divided by the resistance to blood flow. The afferent and efferent arterioles are the major sites of resistance to blood flow through the kidney. Constriction of either arteriole increases resistance and therefore decreases renal blood flow. Conversely, dilation of either arteriole decreases resistance and increases renal blood flow.

Afferent and efferent arteriolar resistances affect not only the renal blood flow but also the glomerular capillary hydrostatic pressure (P_c) and therefore the GFR. Recall from Chapter 9 that if the resistance of an outflow pipe is increased, the pressure after the resistance is reduced and the pressure before the resistance is increased. (Pressure tends to increase upstream from the constriction and decrease downstream from the constriction.) It follows that constriction of the afferent arteriole decreases P_c, and with it the GFR. Constriction of the efferent arteriole increases P_c and the GFR. Note that constriction of the efferent arteriole reduces renal blood flow and increases GFR, whereas constriction of the afferent arteriole decreases both renal blood flow and GFR.

This is analogous to a garden hose with a small hole in the midsection. If the tap is turned on, water exits the hose at the end, but at the same time some leaks out of the hole. The leak is analogous to the GFR and the outflow is analogous to the renal blood flow (see Figure 12.7). If the gardener steps on the hose between the hole and the end, pressure increases in the hose between the constriction and the hole, forcing the leak to increase (B). If instead the gardener steps on the hose between the hole and the tap (C), pressure decreases at the point of the leak, and less water exits the hole. If the gardener places both feet on the hose so that it is constricted before and after the hole, the leak may be as great as in Figure 12.7(A), but the outflow at the end of the hose is lower. If one thinks of the hole as the glomerular capillaries, and the sections of hose before and after the hole as the afferent and efferent arterioles respectively, the analogy fits the kidney. Thus, both renal blood flow and glomerular filtration rate are controlled by altering the resistances of afferent and efferent arterioles.

Renal Autoregulation

Under most conditions, renal blood flow and glomerular filtration rate are held relatively constant, despite daily fluctuations of blood pressure. This phenomenon is called **renal autoregulation.** Independently of the renal nerves or of any substance in the blood, the afferent arteriole automatically adjusts to changes in the local blood pressure. If systemic arterial blood pressure decreases, the afferent arterioles dilate so that RBF and GFR remain relatively constant rather than decreasing. Conversely, if systemic arterial blood pressure increases, the afferent arterioles constrict, thereby preventing increases in renal blood flow and GFR. Thus, renal autoregulation tends to keep renal blood flow and GFR constant, independent of changes in arterial blood pressure. Because GFR is constant, alterations in the composition of final urine must reflect changes in the reabsorptive and secretory rates of different solutes. However, a few physiological situations are capable of altering GFR and RBF briefly.

The arterioles are composed of vascular smooth muscle cells that can contract or dilate in response to a variety of extrinsic stimuli, or signals that originate outside the kidney. The primary extrinsic control of arteriolar resistances is exerted by the renal sympathetic (adrenergic) nerves, which terminate on afferent and efferent arteriolar smooth muscle cells. Increased frequency of action potentials leads to increased neurotransmitter release (norepinephrine), which in turn

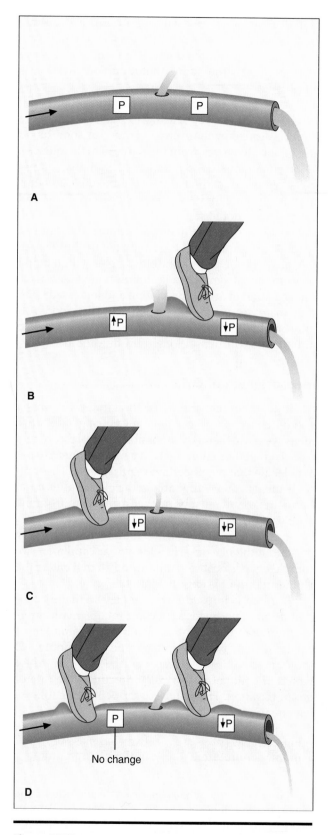

Figure 12.7 This analogy of a hose with a leak illustrates the effects of arteriolar constriction on glomerular filtration and renal blood flow.

leads to increased vascular smooth muscle contraction (increased resistance). This mechanism might explain the decreased renal blood flow during physical exercise, when the sympathetic nervous system is stimulated.

What purpose does a reduction in renal blood flow and GFR serve? During exercise, when more blood is needed by the exercising muscles, some blood can be shunted away from the kidneys to the muscles as a result of renal vasoconstriction. Similarly, following a hemorrhage, the reduction in renal blood helps to conserve flow for vital organs such as the heart and brain. This is not to say that the kidneys are not vital; rather, their function may be temporarily reduced without compromising the survival of the kidneys or the individual. During conditions of very high blood pressure, renal sympathetic stimulation is increased, causing afferent arteriolar dilation and increased GFR. This increases urine output and reduces blood volume and blood pressure. The kidney plays a key role in both short-term and long-term fluid homeostasis.

Measurement of GFR

A complete understanding of the kidney requires a good method for accurately measuring GFR. For many years it seemed impossible, until researchers realized that they could make the measurement using a molecule that meets certain requirements: The molecule must be freely filterable at the glomerulus, must be nontoxic, and must be neither reabsorbed from the tubules nor secreted into them. These criteria are met by inulin, a polysaccharide that is not normally present in the plasma. The glomerular capillaries are completely permeable to inulin, and all that enters the glomerular fluid is excreted into the urine.

If inulin is infused intravenously so that its concentration in plasma is kept constant, the mass entering the glomerular filtrate each minute (μg/min) must equal the mass excreted in the urine each minute. All the filtered inulin is excreted in the urine. In other words, the rate of inulin excretion equals the rate of inulin filtration. This can be described by the following equation:

$$\text{Filtered mass of inulin per unit time} = \text{Excreted mass of inulin per unit time}$$

Now consider what constitutes each side of the equation. The left side of the equation, the filtered mass, must equal the volume of filtrate formed per unit time (the GFR) times the concentration of inulin in that filtrate. Because the glomerular capillaries are freely permeable to inulin, the concentration of inulin in the filtrate is equal to its concentration in plasma (P_{inulin}).

In like manner, the right side of the equation, the excreted mass of inulin per unit time, equals the volume of urine formed in that time times the concentration of inulin in that urine. Thus:

$$\text{GFR} \times P_{inulin} = \text{Volume of urine} \times U_{inulin}$$

Rearranging the equation:

$$\text{GFR} = \frac{\text{Volume of urine} \times U_{inulin}}{P_{inulin}}$$

Each term on the right side of this rearranged equation is measurable. Suppose that an adult male is given an intravenous infusion of a sterile inulin solution. An hour later he empties his bladder and a blood sample is drawn 20 minutes later. Twenty minutes after blood sampling, the subject again empties his bladder into a container. Urine volume measures 60 ml; therefore the rate of urine flow is calculated as V = 60 ml/40 min = 1.5 ml/min. The blood sample is centrifuged to obtain plasma, and samples of plasma and urine are analyzed for inulin concentrations. The results are $P_{inulin} = 1$ μg/ml and $U_{inulin} = 80$ μg/ml. These numerical values of V, U_{inulin}, and P_{inulin} are brought into the equation, which can then be solved for the GFR:

$$\text{GFR} = \frac{(U_{inulin} \times V)}{P_{inulin}}$$
$$= \frac{80 \text{ μg/ml} \times 1.5 \text{ ml/min}}{1 \text{ μg/ml}}$$
$$= 120 \text{ ml/min}$$

The reason why U_{inulin} (80 μg/ml) is so much higher than P_{inulin} (1 μg/ml) is because most of the filtered water was reabsorbed from the tubules, but none of the filtered inulin was reabsorbed. Of the 120 ml of water filtered into the tubules every minute, 118.5 ml was reabsorbed and the remaining 1.5 ml of urine contained all of the inulin that was originally present in the 120 ml of filtered plasma.

The equation presented above is called the clearance equation. In the example given, each minute 120 ml of plasma was completely cleared of all its inulin. The clearance of inulin is a special case in that its clearance is equal to the GFR. The clearance of other substances can also be measured, but their clearance will rarely equal the GFR. Any substance that is reabsorbed has a clearance less than the GFR, as less of the sub-

HIGHLIGHT 12.2

Clinical Estimation of GFR

Measurement of the GFR by the inulin method is time-consuming and expensive because the physician must infuse inulin intravenously into the patient and perform timed, accurate urine collections. However, GFR can be estimated clinically by measuring the plasma concentration of creatinine or urea in plasma or blood. Both urea and creatinine are produced by the body at relatively constant rates and are removed from the body only by renal excretion. Thus, diagnosis of renal failure is based on the fact that the plasma concentrations of creatinine and urea are inversely proportional to the GFR.

To understand why this is so, assume for a moment that GFR in a hypothetical patient has fallen suddenly to 25% of normal, as it sometimes does in the condition known as acute renal failure. If the GFR falls to 25% of normal, the amount of urea and creatinine filtered and hence excreted will also be 25% of normal.

(Recall that the amount of a substance filtered each minute is equal to the product of the GFR times the concentration of that substance in plasma.) Because the rate of urea and creatinine production is unaffected by renal failure, the production of both compounds exceeds the rate of excretion, and the plasma concentrations of those substances rise until they are four times normal. When that happens, the rate of filtration and the rate of urea and creatinine excretion again exactly match the rate of production. Even though renal failure may develop slowly over many years, the inverse relation between plasma concentrations and GFR still holds.

Because the concentration of urea and creatinine in plasma increases as the GFR falls, a simple blood test allows for a good estimation of the GFR. For example, if both compounds are found to be three times the normal value, it is taken as evidence that the GFR is 1/3 normal.

---

stance is excreted than is filtered. Secreted substances have clearances greater than the GFR.

Transport Processes

As was outlined in Figure 12.4, the final composition of urine is the result of three processes: filtration at the glomerulus, reabsorption from the tubules, and secretion into the tubules. We shall consider the two transport systems separately.

Proximal Tubular Reabsorption

The glomerular capillaries are completely permeable to several metabolically useful molecules, including carbohydrates (such as glucose, fructose, xylose, and galactose), all of the amino acids used in protein synthesis, and intermediates in metabolism (such as lactic acid, citric acid, and acetoacetic acid). These low-molecular-weight molecules are freely filtered out of the glomerular capillaries into Bowman's space and flow into the proximal tubule. There they are completely reabsorbed by active processes by cells of the proximal tubule. They are not secreted by any nephron segment. Because the reabsorptive mechanisms for all these organic molecules have much in common, we

will discuss only one of them in detail, the mechanism for reabsorbing glucose from the proximal tubule.

Figure 12.8 illustrates how proximal tubular cells reabsorb filtered glucose molecules along with Na^+. The concentration of Na^+ in the tubular fluid is higher than its concentration in the cells lining the proximal tubule. This concentration difference, as well as the electrical difference across the cell membrane, favors the facilitated diffusion of Na^+ into the proximal tubular cells. That movement is aided by a carrier molecule, which also binds glucose, actively transporting it into the cell interior.

The carrier molecule is represented by the brown circle in the upper part of the diagram. This Na^+–glucose cotransport serves to raise the glucose concentration of the cell above that of the interstitial fluid on the opposing side, and so favors the facilitated diffusion of glucose out of the proximal tubular cells into the interstitial fluid. The peritubular capillaries then reabsorb the glucose back into the blood. At the same time, the peritubular cell membrane actively transports Na^+ out of the cell and K^+ into the cell by the Na^+/K^+ pump described in Chapter 3, which is shown by the darker brown circle. Active countertransport of Na^+ and K^+ across the peritubular side of the cell maintains the low Na^+ and high K^+ concentrations in the cell.

Thus, filtered glucose is reabsorbed from the proximal tubule by a secondary active transport mechanism, as described in Chapter 2. This mechanism is

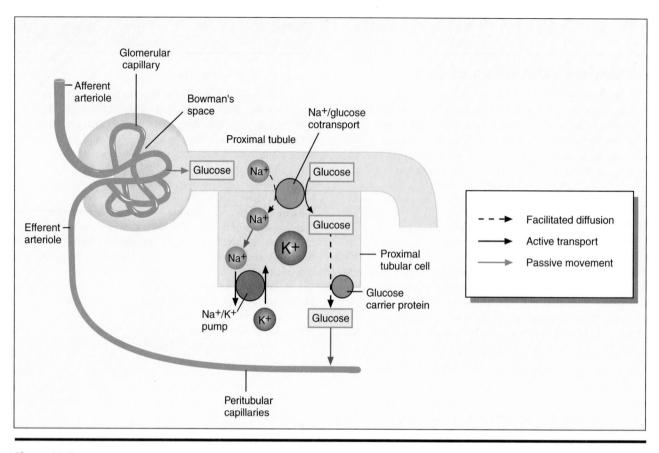

Figure 12.8 Cellular transport mechanisms for reabsorption of glucose from the proximal tubule.

very efficient because filtered glucose is completely reabsorbed (that is, the glucose concentration of the tubular fluid decreases to zero) even if the glucose concentration of plasma increases to values two or three times higher than the normal concentration of about 100 mg % (100 mg % = 100 mg/100 ml, or 1 mg/ml). However, because there is a fixed number of glucose–Na^+ carriers on the proximal tubule cells, and because each carrier can transport only a limited number of molecules per minute, it is possible to overwhelm or saturate the reabsorptive mechanism by increasing the amount of glucose filtered per minute, which is the product of the GFR and $P_{glucose}$. Once the mechanism is saturated, all of the carriers transport glucose at their maximal rate. If glucose is filtered into the tubules faster than the maximal rate of reabsorption by the proximal tubular cells, as is the case with untreated diabetics, as discussed in Chapter 5, then some of the filtered glucose escapes reabsorption and is excreted in the urine because there are no glucose–Na^+ carriers beyond the proximal tubule.

Figure 12.9 shows how the kidney handles glucose. Glucose is freely filtered into the tubules, and the amount filtered per minute (GFR $\times P_{glucose}$) increases linearly with increasing plasma glucose concentrations ($P_{glucose}$), as shown by the red line. Below a plasma concentration of about 3 mg/ml, glucose is not excreted in the urine ($U_{glucose} \times V = 0$). Because no glucose appears in the urine as the glucose concentration of plasma is increased from about 1 mg/ml to 3 mg/ml, it follows that the reabsorptive rate of glucose also increases. In Figure 12.9, the blue line represents the difference between the quantity filtered and that excreted; the amount reabsorbed. However, at a plasma concentration of approximately 3 mg/ml, called the **renal glucose threshold,** all of the Na^+–glucose carriers are saturated and are transporting glucose as fast as they can. No additional glucose is reabsorbed even though additional glucose may be filtered. The reabsorbed amount (filtered – excreted) remains constant. This constant or maximum value of glucose reabsorption is called the **glucose T_m,** which stands for transport maximum for glucose. As the concentration of glucose in plasma is increased above the threshold, all of that increment appears in the urine, as represented by the brown line.

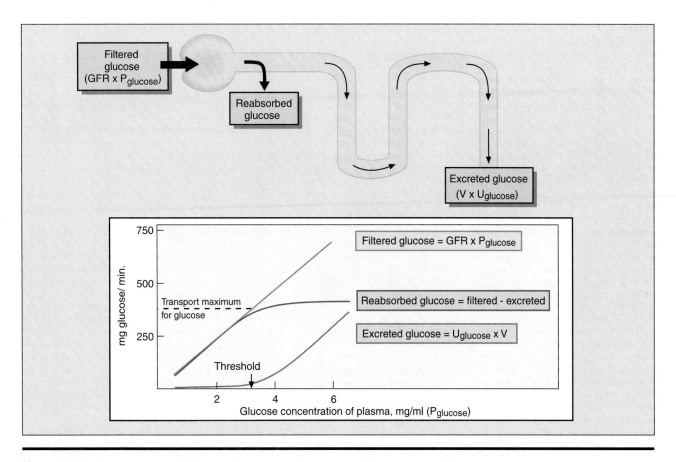

Figure 12.9 Representation of glucose filtration, reabsorption, and excretion as a function of its concentration in plasma.

Ordinarily the urine is glucose-free because plasma glucose concentrations rarely exceed the renal glucose threshold. In patients with untreated diabetes mellitus (sugar diabetes), however, plasma glucose concentrations may increase to values well above the threshold, and considerable amounts are excreted in the urine. In addition, urine flow increases because some of the filtered solutes are not reabsorbed, and some additional filtered water must also be excreted to carry off the extra solute molecules. Any increase in solute excretion necessitates excretion of additional water to carry out those solutes. This generalization has clinical implications. Any increase in solute excretion leads to an increase in urine flow known as **diuresis.** The diuresis resulting from this phenomenon is known as **osmotic diuresis** as it results directly from an increase in the number of osmotic particles appearing in the urine.

Proximal Tubular Secretion

In addition to their role in reabsorbing solutes, proximal tubular cells have the ability to secrete a variety of molecules. Some are normal plasma constituents

such as creatinine, uric acid, and histamine. Others are not, and must be ingested or injected, such as some antibiotics including penicillin and some herbicides and pesticides. The organic molecules secreted by proximal tubular cells are often classified on the basis of whether they dissociate a hydrogen ion (organic acids) or combine with a hydrogen ion (organic bases). Except for the fact that these substances are secreted rather than reabsorbed, the cellular transport mechanisms and characteristics are similar to those for proximal reabsorptive mechanisms, and are shown in Figure 12.10.

Some of the substance(s) is filtered at the glomerulus and some is secreted by the tubular cells into the tubular fluid. A carrier on the peritubular membrane of the cell binds both Na^+ and the secreted solute, much the same as was shown for the reabsorptive process for glucose in Figure 12.8. Na^+ moves into the cell by facilitated diffusion, down its electrochemical gradient, providing the energy for the active, uphill movement of S against its chemical gradient into the cell. The concentration of S builds up inside the cell, allowing it to move down its concentration gradient into the tubular

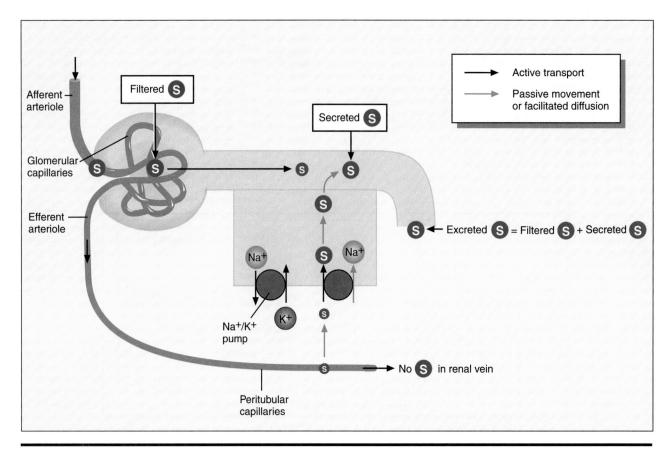

Figure 12.10 Schematic representation of cellular events in the secretory process of a proximal tubular cell.

fluid. Meanwhile, the Na⁺ transported by this carrier is actively transported back into the interstitium. The mechanism may be efficient enough to reduce the concentration of some secreted substances to near zero in the peritubular capillary plasma.

Acidification of the Urine

Normal metabolism generates a large quantity of acidic end products. Most of that acid results from metabolic CO_2, which combines with H_2O to form carbonic acid (H_2CO_3) and is removed from the body by the lungs, as described in Chapter 11. However, each day about 100 mM of nonvolatile acids are also produced and must be excreted by the kidney. Despite the fact that large quantities of acid are produced daily by the body, the acid–base balance of body fluids remains constant. That is because the kidneys are able to secrete that acid into the tubular fluid along the entire length of the nephron.

As was discussed in Chapter 2, the normal pH of blood is approximately 7.4. Urine may have a pH as low as 4.4. (This difference of 3 pH units means that the urine may have a H⁺ ion concentration 1000 times

as great as plasma.) As long as the kidneys function properly, they are able to excrete all the excess acid produced as a result of metabolism of food. However, if for any reason GFR falls to low values, the ability of the kidneys to secrete that acid is reduced, and acid is retained by the body. If this happens, the pH of body fluids falls and the individual is said to be **acidotic,** a potentially dangerous condition, as most of the cellular enzyme systems are very sensitive to pH changes.

Regulation of Na⁺, K⁺, and Cl⁻ Transport

Recall from Chapter 2 the reasons why intracellular and extracellular fluids are quite different in composition. All cells in the body actively transport Na⁺ from intracellular fluid to extracellular fluid, and K⁺ from extracellular fluid to intracellular fluid. This reciprocal movement of Na⁺ and K⁺ keeps the intracellular concentration of Na⁺ low and that of K⁺ high. Therefore, most of the Na⁺ contained within the body is dissolved in extracellular fluid, including the plasma. Most of the K⁺ is dissolved in intracellular fluid. Throughout life, the composition of both body compartments is

MILESTONE 12.1

Penicillin

During World War II, penicillin first came into general use, but it was expensive and difficult to make. Even more vexing was the fact that the kidney actively secretes it into the urine, so that much of the injected dose is lost and plasma levels decline rapidly following injection. Penicillin was in such short supply that hospitals were forced to save the patients' urine in order to reclaim the excreted penicillin. This was not a satisfactory arrangement.

Fortunately, experiments originally performed on laboratory animals showed that there are carriers for penicillin located in the cell membrane and that other substances might combine with those carriers. The phenomenon of two or more molecules being able to combine with the same carrier molecule is known as **competition.** That is, both molecules compete with each other for attachment to the carrier molecule. Knowledge of this fact proved to be very helpful and allowed use of much smaller doses of penicillin to accomplish the same effect. Penicillin was injected along with a substance that competed for attachment to the renal tubular carrier that transported penicillin. Although this didn't completely prevent the excretion of penicillin (filtered penicillin was still excreted in the urine), it greatly reduced the amount excreted because the competing molecule, rather than penicillin, bound to the carrier, preventing secretion of penicillin. By reducing penicillin excretion in this way, it was possible to inject smaller and less frequent doses of penicillin but still achieve effective penicillin concentrations in the blood.

kept relatively constant, despite the fact that we stress the system constantly. Our diets change, often from day to day. At times we eat a great deal of sodium (in salty foods) or potassium (found in certain fruits and meats). At other times we eat little of either, yet the Na^+ and K^+ concentrations of both cellular and extracellular fluids do not change.

The kidney regulates the amounts of K^+ and Na^+ in the body fluids by excreting any excess and retaining Na^+ and K^+ when their intake is low. Figure 12.11 is an overview of how the kidney handles Na^+, K^+, and Cl^-, for Cl^- is the major negative ion associated with Na^+ and K^+.

Proximal Tubule The largest fraction of the glomerular filtrate is reabsorbed from the proximal tubule. As already noted, approximately 65–75% of the filtrate is reabsorbed there, with only 25–35% remaining at the end of the proximal tubule. But how do we know that? Using the modern technique of micropuncture, it is possible to insert a fine micropipette into a single proximal tubule of an anesthetized animal. A small amount of the proximal fluid can be withdrawn and analyzed for its contents. If this is done while the animal receives an intravenous infusion of inulin, one can calculate the fraction of filtrate reabsorbed along the tubule.

In animals it is possible to measure accurately the fractional movement of water and associated solutes. For example, if the concentration of inulin in plasma is 1 µg/ml, and at the end of the proximal tubule the concentration of inulin is 3 µg/ml, that could only have occurred due to reabsorption of water along the length of the tubule. In order for the concentration of inulin to increase threefold, $\frac{2}{3}$ of the fluid must have been reabsorbed. If 3 ml of filtrate starts down the tubule, reabsorption of 2 ml leaves 1 ml behind, with all the inulin that was originally dissolved in 3 ml. Thus, the concentration of inulin in that 1 ml is 3 µg/ml. Micropuncture experiments have shown that at the end of the proximal tubule, about two-thirds of the filtrate is reabsorbed, but the concentration of Na^+ in that fluid is unchanged from that of plasma. (Only a rough estimate of fractional water and salt reabsorption is possible as these experiments cannot be done on humans.) But now to the mechanisms involved in reabsorption.

The first step in this process is passive, as indicated by the blue arrow in Figure 12.11. Filtered Na^+ enters the tubular cell and moves down the concentration gradient, which is set up and maintained primarily by active Na^+ and K^+ transport across the opposite side of the cell. The important point is, no matter how filtered Na^+ enters the cell, it is actively transported across the opposite, or peritubular, side of the renal cell into the renal interstitial fluid.

The rapid reabsorption of a large fraction of filtered Na^+ and the associated Cl^- (and HCO_3^-, carbohydrates, amino acids, and metabolic intermediates) accounts for the rapid removal of a large fraction of the total solute filtered into the tubules. Because the proximal tubular cells are very permeable to water, as filtered solutes are reabsorbed, filtered water is also reabsorbed by osmosis.

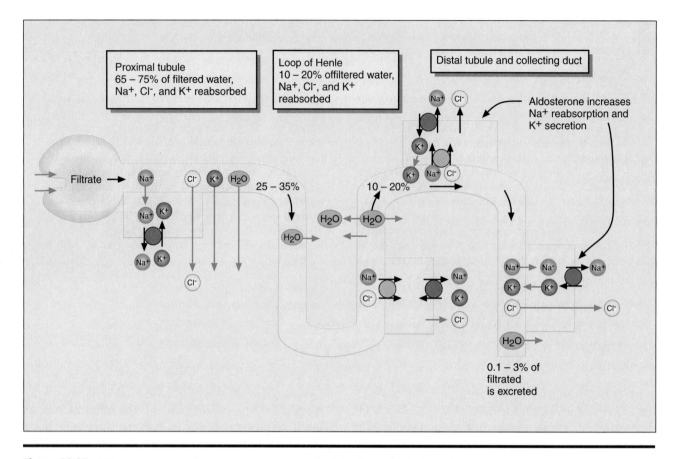

Figure 12.11 Diagram representing transport processes for Na+, K+, and Cl- along the nephron. Black arrows represent active transport; blue arrows represent either passive diffusion or facilitated diffusion.

Loop of Henle Additional Na+ and Cl- (salt) are reabsorbed from the loop of Henle. However, because the loop of Henle is unique in its function, those processes are discussed later in the chapter. At this point it is necessary only to note that additional solutes and water are reabsorbed from this segment.

Distal Tubule and Collecting Duct As shown in Figure 12.11, Na+ and Cl- continue to be reabsorbed from the distal tubule and collecting duct, whereas K+ is secreted into the tubule. The net movement of K+ in the distal tubule is opposite in direction to its movement in the proximal tubule. In many respects, the collecting duct cell shown in Figure 12.11 is similar to the proximal tubular cell. However, there is one large difference between the functions of the proximal and distal nephron areas. As already stated, proximal tubular reabsorption is essentially constant; about two-thirds of all the filtered water, Na+, and K+ are reabsorbed from this segment. If that is the case, then how can excretion of these substances be regulated? The answer, of course, is that the distal tubule and collecting ducts must act as the major regulators of salt

and water excretion. It is in these segments that the rates of transport of water, Na+, and K+ are regulated.

Aldosterone

Aldosterone is a hormone produced and released by the cortical layer of the adrenal gland that stimulates the activity of the Na+–K+ transport system in distal tubular and collecting duct cells. Increased concentrations of aldosterone in the plasma lead to increased Na+ reabsorption and therefore to decreased Na+ excretion. At the same time, aldosterone causes increased K+ secretion by the distal tubular cells and the collecting duct cells and therefore leads to increased K+ excretion. With maximal plasma concentrations of aldosterone, virtually all of the Na+ that reaches the distal tubule and collecting duct is reabsorbed, and little or no Na+ is excreted in the urine. Simultaneously, K+ secretion and excretion are very high. It should be noted that Na+ reabsorption and K+ secretion still occur in the complete absence of aldosterone, but at slower rates because aldosterone is the major hormone affecting the reabsorption of Na+ and secretion of K+.

HIGHLIGHT 12.3

Aldosterone

The ability of aldosterone to regulate the Na^+ balance of the body is indicative of the exquisitely fine control of transport by the kidney. An error in the transport of Na^+ of only a fraction of 1% would result in death. This may be illustrated by the following example. Only about 10–20% of the filtered Na^+ normally reaches the distal tubule. Even in a person whose adrenals are nonfunctional, most of the Na^+ is reabsorbed by distal tubular and collecting duct cells, and only 3–4% of the filtered Na^+ is excreted in the urine.

If this seems like a small percentage, let's estimate how quickly the kidneys could excrete all of the Na^+ in the body, assuming aldosterone levels in plasma are zero. Because the intracellular concentration of Na^+ is very low, the total amount of Na^+ in the body can be estimated as the total extracellular volume (10.5 liters) times the concentration of Na^+ in the extracellular fluid (145 mmoles/liter), or 1500 mmoles. The kidneys filter 17.4 mmoles of Na^+ every minute (GFR $\times P_{Na^+}$ = 0.120 l/min $\times$ 145 mmoles/l.) If 4% of this, or 0.70 mmoles, is excreted every minute, then it would take 2100 minutes, or only 36 hours, to excrete all of the Na^+ in the body! This exercise illustrates that the kidneys are capable of rapidly excreting excess Na^+ from the body, even if they reabsorb most of the filtered Na^+. Aldosterone is responsible for adjusting reabsorption of that 3–4%, and that is responsible for homeostatic control of salt balance.

Renin–Angiotensin System

At the beginning of the twentieth century, physiologists in Sweden did a simple but important experiment. They made a water extract of a rabbit kidney and injected some of that extract into another rabbit. The extract caused a rapid increase in blood pressure of the recipient rabbit, which slowly returned to normal. This experiment indicated that the kidney contained a substance that had a profound effect on blood pressure. Almost a century later, researchers have unraveled the sequence of events producing that effect, and have learned the physiological meaning of those early experiments.

The kidney contains and secretes into blood a protein called **renin,** which initiates a series of reactions outlined in Figure 12.12. Renin itself has no effect on blood pressure, but in plasma it causes the formation of a substance called **angiotensin I (Ang I).** Ang I is a peptide made up of ten amino acids and is split off from a large protein secreted by the liver called **angiotensinogen.** Ang I, in turn, is split into a smaller molecule, **angiotensin II (Ang II),** by the action of an enzyme present in most cells and blood called **angiotensin converting enzyme (ACE).** The highest concentration of ACE is in the lungs, so in a single pass through the lungs, most of the Ang I in blood is converted into Ang II. Ang II has a long list of important physiological effects.

Ang II causes blood pressure to increase by acting directly on small blood vessels, causing them to constrict. As mentioned in Chapter 5, it is also a potent stimulator of the adrenal gland, causing it to secrete aldosterone into the blood. Thus, when Ang II levels are high, aldosterone levels are also high in plasma and Na^+ reabsorption and K^+ secretion by the kidney increase.

In addition to the two effects noted above, Ang II stimulates a hypothalamic area of the brain that increases secretion of ADH, a hormone that regulates water reabsorption. (ADH is discussed in detail below.) Ang II has one other important action. It is a very powerful stimulator of thirst. Researchers have shown that if water is presented to animals that are deprived of water for 12–14 hours, they immediately drink large volumes of water. However, if they are injected with an Ang II inhibitor so that circulating Ang II is ineffective, the animals ignore the water. These experiments showed that circulating levels of Ang II regulate thirst.

Researchers noted that if a single renal artery of an animal is constricted slightly so that blood pressure to the kidney is reduced, renin levels in blood increase dramatically, indicating increased secretion by the kidney. The increase in plasma renin causes blood pressure to be elevated. This finding led to the realization that many humans suffer severe hypertension as a result of a small constriction, or **stenosis,** of one renal artery. Understanding the cause allowed for a cure. It was found that the constricted area of the artery could be surgically removed and the two ends reunited. This procedure reestablishes flow and increases the blood pressure to the kidney. Renin secretion falls and systemic blood pressure returns to normal.

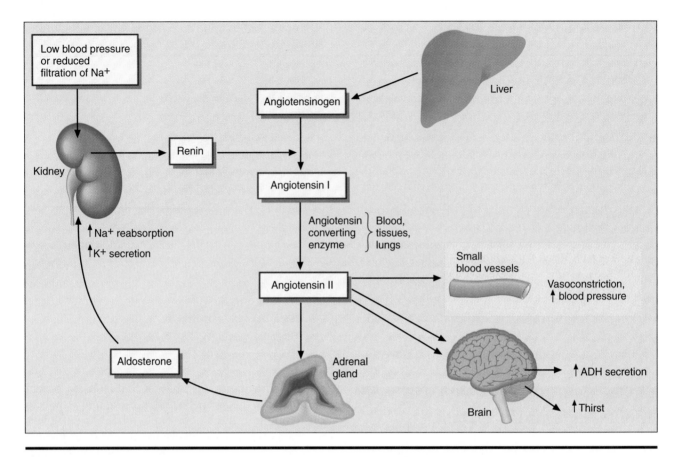

Figure 12.12 Diagram of the renin–angiotensin system and its multiple effects.

For some time it was believed that the only function of renin was to help regulate blood pressure. This made sense. If a person suffered a hemorrhage such that blood pressure fell, the kidney would respond by secreting more renin. This would cause generalized vasoconstriction and return blood pressure back toward normal. This seems to be a good, simple example of a negative feedback system. Although the kidney does respond to reductions in blood pressure as outlined above, researchers discovered other receptors that control renin secretion.

Researchers showed that the amount of Na+ entering the distal tubule is also a potent signal for renin secretion. If Na+ intake is low, its concentration in plasma is slightly lower than normal, thus reducing the amount of Na+ entering the distal tubule. Renin secretion is stimulated, causing aldosterone levels to rise due to the increased production of Ang II. This in turn stimulates Na+ reabsorption in the distal tubule, and the person loses less Na+ in the urine. (Refer to Figure 12.12.) This mechanism acts to keep the concentration of Na+ in plasma and body fluids relatively constant.

The kidney continuously adjusts the reabsorptive rate of Na+ so as to accommodate the changes in intake and maintain Na+ homeostasis.

The filtered amount of Na+ is much greater than the daily intake. You have seen that the filtered amount of Na+ is approximately equal to 17,000 mmoles/day; that is equivalent to about 1000 g, or 2.2 pounds, of NaCl. No one can eat that much salt in one day. The average amount of salt eaten per day is about 10 grams, but the range of intake is great. Yamomamo Indians of Brazil may eat less than 0.1 grams per day, whereas some groups in northern Japan may eat as much as 60–70 grams per day, yet all healthy people stay in balance. By simply adjusting the reabsorptive rate of Na+, the kidney is able to match output with intake. The same holds for all the other ions other than K+, which is both reabsorbed and secreted by the renal tubules.

Although we have outlined the renal response to changes only in Na+ intake, the same or similar mechanisms come into play regulating the excretion of other small charged ions such as calcium, potassium, phosphorous, and all other ions necessary for life.

Urine Concentration

Thus far we have considered only the question of how the kidney treats solutes; we have not considered how it handles water. In one respect, it is reasonable to suggest that the ability of the kidney to match water output to water intake is its most important function. When land animals evolved, the most difficult problem to overcome was intermittent access to water. They had to be able to conserve water during periods of deprivation. Furthermore, it was also necessary to rid the body of extra water when intake was high. For example, if you drink a large volume of water, as you would find in an extra large soda, shortly afterward, you would pass a large volume of dilute, light-colored urine. On the other hand, if you play a game of tennis on a hot day and do not drink, urine flow is low, and it is a dark-colored, concentrated urine. Why does this happen?

First let us consider the signal to the kidney, which dictates whether the urine is more or less concentrated. Next we will discuss the mechanism by which this is brought about. Consider an experiment in which researchers gave a dog a large amount of water and measured the rate of urine flow. As expected, urine flow rose above normal, falling back to starting values only when the excess water was excreted.

It had been known for some time that a few people have a serious illness known as **diabetes insipidus,** not to be confused with diabetes mellitus. In both types of diabetes, urine flow is very high. However, people with diabetes insipidus do not show any irregularity in glucose metabolism. The glucose concentration of their plasma is normal, no glucose appears in their urine, and there is no evidence of impairment of insulin secretion, yet these people are in serious danger solely because of the inordinately large volume of urine they produce. They may almost urinate themselves to death. The rate of urine formation exceeds their ability to drink enough water to keep pace with the water loss through their kidneys. The constant loss of water so dehydrates them that their blood pressure may fall low enough to produce shock. In severe cases, urine volume may exceed 20 liters per day! It was noted that on autopsy many showed lesions, or abnormal anatomy, of their posterior pituitary gland. Physiologist E. B. Verney was the first to realize the significance.

Verney performed a series of experiments on dogs that had consumed a large volume of water but were injected with an extract taken from a posterior pituitary gland. When he did this, urine flow did not increase when the animals drank water; instead, they continued to excrete a small volume of very concentrated urine. These experiments indicated that the posterior pituitary gland contains a substance that is able to prevent the normal increase in urine flow, or diuresis. That substance, called the **antidiuretic hormone (ADH)** because it inhibits diuresis, was later isolated, purified, and found to be a polypeptide hormone. (Another name for ADH is **vasopressin** because it also causes arteriolar constriction, increased peripheral resistance, and increased blood pressure.) The structure of ADH is known and it can be produced commercially. This is a boon to those suffering from diabetes insipidus, as injections of ADH or similar manufactured drugs may be used to bring their diuresis under control.

But what controls the secretion of ADH? This question could not be answered fully for many years, until the structure of ADH had been determined and it became possible to measure its concentration in plasma. We now know that although ADH is stored and secreted by the posterior pituitary gland, it is not made there. ADH is synthesized by the hypothalamic nuclei. It then is transported down the stalk of the pituitary to the posterior pituitary gland, where it is stored until released into the circulation.

Control of ADH Secretion

ADH is the major hormone that controls the volume of urine excreted. The posterior pituitary releases ADH into the circulation during periods of relative dehydration and lowers the rate of release in response to rehydration. For this reason, urine flow is very low following a period of exercise in hot weather, and very high after drinking a large soft drink. What is the stimulus for ADH secretion?

Osmoreceptors Verney continued his experiments by injecting a small volume of hypertonic salt solution into the blood vessels going to the brain of dogs undergoing a severe water diuresis. The diuresis immediately ceased. This increased the osmotic pressure of blood going to the brain, but had no significant effect on peripheral blood. The results of these experiments led to the conclusion that somewhere in the brain there must be a receptor that senses the osmolality of plasma; that receptor is called an **osmoreceptor.** We now know that ADH secretion is a direct function of plasma osmolality. If plasma osmolality is higher than normal, the secretory rate of ADH is increased, causing an increase in urine osmolality and a decrease in the rate of urine flow. Excretion of a concentrated urine dilutes body fluids a bit and so tends to restore plasma osmolality to the normal range. Conversely, when a person drinks a large soft drink, plasma osmolality falls. The decrease in plasma osmolality decreases the secretion of ADH and urine flow increases until once again plasma osmolality rises.

Stretch Receptors In addition to osmoreceptors, ADH secretion is also controlled by **stretch receptors** located in the heart. If blood volume falls for any reason, the volume (stretch) of the atria is reduced (see Chapter 8). Under this condition, the stretch receptors respond by stimulating neurons in the hypothalamus to signal ADH release by the pituitary gland, causing a reduction in urine flow. The reduction in urine output helps to conserve body fluids and to minimize the decrease in blood volume.

Stretch receptors in the atria have another important function. If the atria are stretched, atrial tissue releases a hormone that causes the kidney to excrete more Na+ and water. The hormone is called **atrial natriuretic peptide (ANP).** It has the ability to inhibit Na+ reabsorption by the renal tubules, which leads to an increased loss of Na+ into the urine, obligating more water to carry off the extra Na+. This endocrine function of the heart aids in the homeostatic control of salt and water balance. A snack of salted peanuts and your favorite drink increases the volume of your extracellular fluid, including your blood volume. This stretches your atria, signaling release of ANP and renal excretion of the excess salt and water.

Renal Concentrating Mechanism

It is one thing to determine the controlling factors for ADH release, and another to determine the mechanism by which they work. The quest to solve this problem began with the recognition that the interstitial fluid of the papilla of the kidney is very concentrated (refer to Figure 12.2). Furthermore, researchers noted that animals such as the kangaroo rat, which lives in the desert, produce a very concentrated urine and have a very long papilla. The kangaroo rat is able to concentrate its urine as much as 10 to 20 times that of plasma. (Human urine can be concentrated about 4.5 times that of plasma.) Researchers noticed that animals with easy access to fresh water, such as the beaver, cannot form very concentrated urine, and they have a very short papilla. These observations led to hypotheses regarding the concentrating mechanism, and these hypotheses led to a long series of studies in which the renal concentrating mechanism was worked out.

Countercurrent System

In order to understand the action of ADH on the renal concentrating system, it is helpful to discuss why a whale does not freeze to death while swimming in Arctic waters. This example is quite relevant

to the topic at hand. True, the whale has a very thick coat of blubber that insulates it from the ice cold water, but that in itself should not be enough to ensure survival, for its large fins do not have that thick coat of insulation. The situation of a whale is analogous to a person sitting on an ice floe, wrapped in a heavy parka, so that heat loss is very low. But the person has her legs dangling over the edge into the cold water. Heat loss would be so great from her legs that they would shortly suffer severe frostbite, and the person would lose so much heat she would die of hypothermia. How, then, can the whale survive if the situations are similar? A human would lose great amounts of heat, as shown in Figure 12.13(A), because surface blood vessels run in no particular pattern along the skin. Warm blood comes to the skin surface and loses its heat rapidly to the cold water, quickly chilling the body.

The whale's vascular system is quite different. Its surface vessels are formed as loops, with the long axis of the loop perpendicular to the skin surface. Let's see what difference this makes. Assume for a moment that it is possible to lift the whale to the same ice floe as our hypothetical person and have the whale bask in the sun completely out of the water. Heat loss would be minimal. That situation is shown in Figure 12.13(B). You are all familiar with the fact that heat loss is greater in water than in air. Now consider what would happen if we gently slide the whale into the water. Its warm skin would be opposed by very cold water and heat would be lost from the blood near the surface of the skin. Figure 12.13(C) shows what would happen after a very short time in the cold water. Blood at the turn of the loop would lose heat to the water, causing the temperature of the blood to fall.

Notice that the cold blood in the ascending limb of the loop is close to the descending vessel. Downward-flowing warm blood is close to cold upward-flowing blood. Under these conditions, heat is transferred from descending blood to ascending blood. As this happens, blood approaching the skin's surface cools even before being exposed to the cold water. As the system progresses (steps C–F), descending blood arrives at the turn in the loop only slightly above the temperature of the cold water. Very little heat is lost to the environment, as most of the heat transfer occurs across the loop instead of between blood and water. The whale loses only a fraction of the heat that a human would and survives in an environment lethal to us. Note, however, that the skin of the whale shows a steep gradient of temperature, coldest at the surface and warmest at the beginning of the capillary loops. This system is called a **countercurrent system** because the two limbs of the capillary run counter to each other.

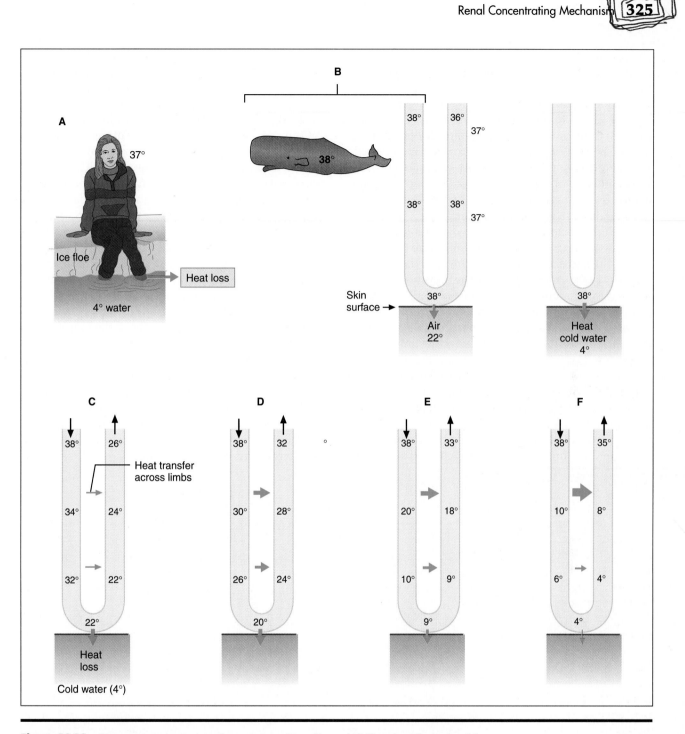

Figure 12.13 Countercurrent system for conservation of heat. (A) Heat loss from the skin of a human. (B–F) Heat conservation across the skin of a whale. The arrows indicate the direction of the heat transfer. The width of the arrows is proportional to the amount of heat transferred at that point.

Countercurrent Multiplier System

Perhaps you recognize a similarity between the anatomy described above and that of the inner part of the kidney (see Figure 12.2). The loop of Henle also forms a countercurrent system; however, an active process has been incorporated into the countercurrent system of the kidney. The descending limb of the loop serves as a passive conduit for urine, containing no measurable transport system for any urinary solute. However, it is permeable to water. The ascending limb of Henle's loop is functionally quite different from the descending limb. It is impermeable to water, but contains active transport systems that move Na^+ and Cl^- out of the tubular urine into the renal interstitium. Let us consider the consequences of this arrangement.

It is easiest to understand the system by taking it in steps. In Figure 12.14(A), the osmolality of the loop fluid is shown as it would be if loop flow were stopped and there were no NaCl transport system operating. The osmolality of fluid throughout the loop of Henle would be the same as plasma, approximately 300 milliosmoles (mOsm) per kg water.

In Figure 12.14(B), we picture what would happen if the Na^+ pump in the ascending limb were turned on but the fluid was not allowed to flow. The dashed line at the base of the loop represents the transition point at which the Na^+ pump would begin and continue along the ascending limb. In addition, we arbitrarily set the pump so that it can build up no more than a 200 mOsm difference across the limbs. The ascending limb would pump NaCl out into the interstitium, raising its osmotic pressure, and water would leave the descending limb in response to the osmolality difference set up by the pump. (For simplicity's sake, we shall ignore the changes in interstitial osmolality, although it is fair to say that interstitial osmolality would be similar to that of descending limb osmolality at each level along the

loop of Henle.) Figure 12.14(B) indicates what the situation would be after the pump reaches the maximal concentration gradient to be established. The ascending limb fluid would have fallen to 200 mOsm and that in the descending limb would have increased to 400 mOsm. Now we allow the fluid to flow again, but only for 1/2 turn along the loop. The osmotic concentrations would be as pictured in Figure 12.14(C).

Once again the pump is turned on without allowing flow. Figure 12.14(D) shows the concentrations that would develop. Again we allow fluid to flow, but only for a 1/4 turn (Figure 12.14(E)). The Na^+ pump is turned on again; Figure 12.14(F) shows the osmotic gradient as it would appear along the loop at this stage. One can go on with this simulation, but it should be enough at this stage to recognize that a concentration gradient is being built up along the long axis of the loop of Henle. The tip of the loop has the most concentrated fluid and the base of the loop the most dilute fluid. Furthermore, ascending limb fluid is actually hypotonic, less than 300 mOsm, as it enters the distal tubule. Because this system requires active transport, it differs from the whale analogy, which does not require active transport by the whale blood vessels. The system in the renal papilla is called a **countercurrent multiplier system.**

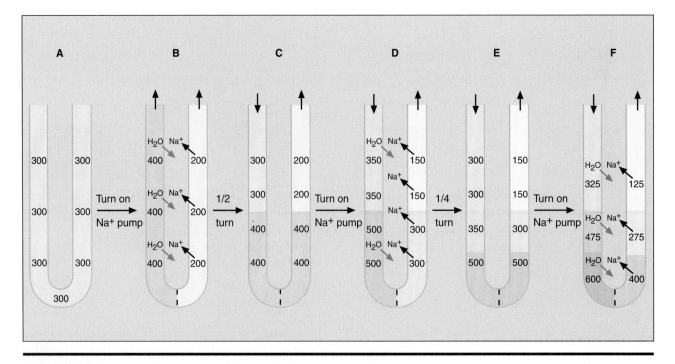

Figure 12.14 Countercurrent multiplier system of the kidney. The blue, downward-pointed arrows represent passive movement. Black, upward-pointed arrows represent active transport against a chemical gradient.

Having described the countercurrent multiplier system, one might still ask, "Of what use is it to the kidney in producing concentrated urine?" The answer lies in the collecting ducts. Figure 12.15 is a diagram of the complete picture of concentration gradients set up by the countercurrent multiplier system. Note that the collecting ducts carry tubular fluid through the region of the kidney in which the interstitial fluid is very concentrated. If the collecting duct epithelium is permeable to water but not solute, water leaves the collecting ducts and enters the renal interstitium to be reabsorbed. The volume of collecting duct fluid is reduced, and the collecting duct urine becomes hypertonic, or more concentrated than 300 mOsm/kg water. How concentrated? If collecting duct permeability to water is high, the urine osmolality is equal to the osmolality of the interstitial fluids. Kidneys functioning normally are able to establish an interstitial concentration of approximately 1400 mOsm/kg water. That is the upper limit of urine osmolality from a human kidney.

Permeability of the Collecting Ducts

Experiments with ADH have demonstrated that its major action on many epithelial tissues is to increase their permeability to water. Micropuncture experiments on animals have shown that in all animals studied, the collecting duct permeability to water is increased in the presence of ADH in a dose-dependent manner. The higher the level of ADH in plasma, the greater the permeability of tissues to water. Thus the circulating concentration of ADH determines the permeability of both the distal tubules and the collecting ducts. In the absence of ADH, both nephron areas do not allow water to leave the tubular urine, and the kidneys excrete a large volume of dilute urine, the situation that exists in patients with diabetes insipidus.

As the concentration of ADH increases, more and more water may leave the tubular urine, reducing the volume of final urine produced and increasing its concentration. In this manner, urine flow may vary between 0.2 ml/min and 20 ml/min. Urine osmolality may vary between 50 and 1400 mOsm/kg water. All

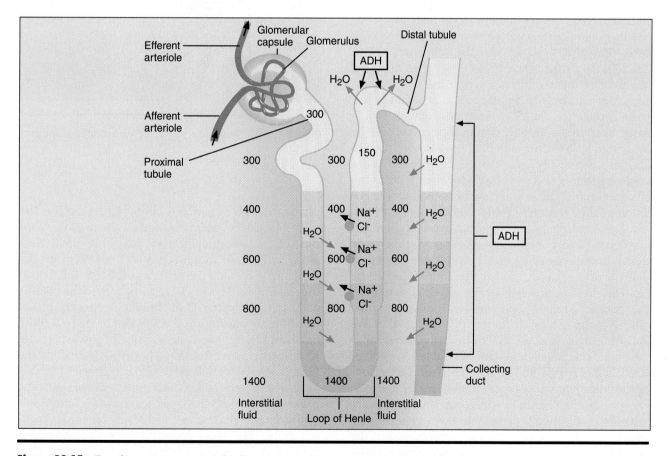

Figure 12.15   Renal countercurrent multiplier system with representative figures for the osmotic pressure at each level of the kidney. The numbers in the diagram refer to mOsm/kg water.

mammals studied have this system for forming a concentrated urine.

Renal Failure and the Artificial Kidney

The hallmark of many forms of renal disease is severely reduced GFR and renal blood flow. In some cases the progression of renal failure is severe enough to be life-threatening. Because the kidneys are responsible for regulating the salt and water balance of the body, if their function is severely impaired, it is not possible for salt and water excretion to match intake. In addition, end products of metabolism cannot be excreted in sufficient quantities to equal those produced. The most important metabolic end products in this respect are acid and K^+. These failings in function cause multiple changes in the composition of body fluids, the most harmful being an increase in the K^+ concentration of plasma.

Recall from Chapter 2 that the ratio of K^+ concentrations across the cell membrane is responsible for setting the membrane potential of cells. If the K^+ concentration of plasma increases, the membrane potential falls. As the K^+ concentration rises with time, the membrane potential of muscle cells may decrease to such low values that repolarization is severely impaired. The effect on the heart muscle may be lethal, as the heart may suddenly stop beating. Although the change in K^+ concentration is the most dangerous to life, other potentially lethal changes also occur. The important point, however, is that with alterations in the K^+ concentration there is no longer homeostatic regulation of the electrolyte composition of body fluids.

The hazards of renal failure were recognized many years ago, but only in the last few decades have researchers developed an **artificial kidney** that effectively managed to circumvent alterations imposed by renal failure. The problem for the physician is to keep the concentration of electrolytes in the patient's plasma constant, despite the loss of renal function. We know the ideal concentration of electrolytes and other metabolic substrates in plasma (see Table 9.1 in Chapter 9). In theory it should be possible to set plasma concentrations of any substance at any level by **dialysis.** Dialysis is a procedure in which two fluids are separated by a selectively permeable membrane so that water and small solutes may pass freely between the two solutions. Figure 12.16 is a photograph of a dialysis system used with humans.

The dialysis of plasma is accomplished by causing the patient's blood to flow through tubing made out of a selectively permeable membrane such as cellophane. The tubing is coiled, simply to save space, and placed in a large volume of an ideal plasma substitute. Blood from the patient's artery flows into the tubing, aided by a pump, and is returned to a vein in the patient's arm. Because the volume of the bath into which the coil is placed is very large compared to the volume of the patient's extracellular fluid, the concentration of all solutes in plasma approaches that of the bath fluid. For example, if the K^+ concentration of plasma is high, K^+ diffuses across the selectively permeable membrane from plasma to bath fluid. If the urea concentration is high in the patient's plasma, it too moves into the bathing fluid. Both these solutes show declining concentrations in plasma as the dialysis proceeds. Without going into detail concerning the makeup of the bath fluid, one can dialyze off any solute that may have accumulated in the plasma. In theory, this is a simple method of replacing kidney function in a patient with no renal function. The apparatus is naturally called an artificial kidney.

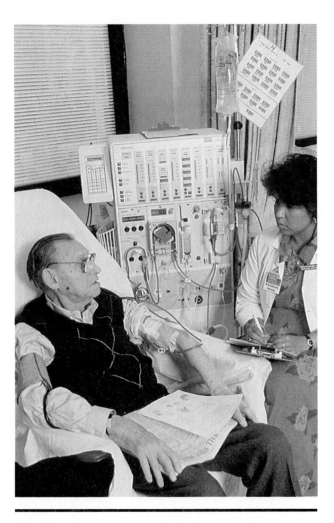

Figure 12.16 Photo of a patient on an artificial kidney machine.

MILESTONE 12.2

Kidney Transplantation

Although kidney dialysis can be used to save lives, it has the disadvantages pointed out. A better treatment would be transplantation of a donor kidney. Fortunately, renal transplantation is not technically difficult, and kidneys were the first organ to be transplanted successfully from one person to another. However, the supply of donor kidneys is limited, and immunologic rejection of foreign tissue is a major clinical problem to be overcome. At first the only source of kidneys was from living donors who were close relatives (most often a parent or sibling) because the closer the genetic match, the better the chance of acceptance of the transplanted kidney by the patient. More recently, however, the development of immunosuppressive drugs, techniques for short-term preservation of living kidneys, and advances in tissue typing have all led to the successful use of kidneys from recently deceased donors. Today most kidneys available for transplantation come from such donors who were unrelated to the recipient.

Although the site of the kidneys is just under the rib cage along the back wall of the abdominal cavity, the transplanted kidney is placed in the groin area for technical reasons. The figure in this Milestone illustrates the surgical technique used. The renal vein and artery of the transplanted kidney are attached to the patient's femoral vein and femoral artery. The ureter of the transplanted kidney is attached to the patient's bladder. Generally the remaining kidneys are left in place.

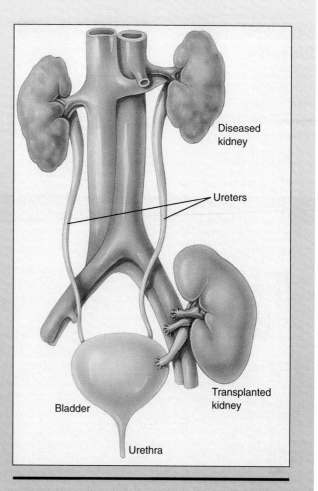

Drawing showing positioning of a transplanted kidney.

A few difficulties may be noted. Because one cannot walk around towing a 100-liter tank of fluid, the dialysis procedure must be intermittent; usually three times per week is sufficient to maintain health. Each dialysis procedure takes a few hours, the exact time depending on any residual renal function, the characteristics of the dialysis tubing, the size of the bath fluid, and the patient's ability to adhere to a diet designed to minimize changes in the body fluids.

Another difficulty is that of preparing an artery and vein in the patient so that they can be connected to the dialysis machine. The solution is to surgically place a blood vessel just under the skin so that it can be punctured easily, without trauma, and connected to the artificial kidney. These machines are responsible for saving many thousands of lives and allowing people who would otherwise die to live productive lives, with minimal discomfort.

Alcohol

Although many poisons exist that may cause irreversible renal damage, only one common one is consumed in our society: alcohol. Alcohol has two notable effects on the kidney. One is hardly serious, but may be annoying. Alcohol has the ability to reduce the rate at which ADH is secreted. Thus, following alcohol ingestion ADH levels in plasma fall and urine flow rises. This is in part the reason why urine flow is notoriously high after drinking beer. Chronic alcohol consumption, however, may lead to far more serious problems. Long-term alcoholism adversely affects liver function, often resulting in a condition known as cirrhosis of the liver. This alone is a serious health problem, but in addition, for reasons not well-understood, liver failure may lead to renal failure. The combination of liver failure and renal failure is critical.

Diuretics

As noted in Chapter 9, some clinical problems such as hypertension or congestive heart failure may be treated with **diuretics.** These are drugs that cause the kidney to excrete a larger than normal volume of urine and more salt than normal. The result of the diuresis and increase in salt excretion, called **natriuresis,** is a reduction in blood volume. In Chapter 9 you saw that reducing the circulating blood volume is often an effective treatment for lowering blood pressure. Many diuretics are available, but all rely on the same general principle. They inhibit salt transport somewhere along the tubules. Regardless of where they operate to inhibit salt transport, the result is the same. More salt and associated water are excreted. Any substance that inhibits the reabsorption of any urinary constituent causes diuresis.

Summary

The first step in the production of urine is filtration at the glomerulus. About 180 liters per day are formed. The proximal tubules reabsorb all the filtered glucose and other important plasma constituents. About 70–80% of the filtered water and associated salts are reabsorbed there. Some solutes, particularly organic molecules, may be secreted by the proximal tubular cells.

The loop of Henle is responsible for establishing a gradient of solute concentrations along the loop, so the tip of the loop is very concentrated and the base is dilute. ADH acts on the permeability of the distal tubules and collecting ducts and determines how much water leaves those nephron areas in response to the osmotic gradient set up by the countercurrent multiplier system. ADH concentrations are regulated by the osmolality of plasma, stretch receptors in the heart, and secretion of renin, which results in the generation of Ang II in plasma.

The distal tubule is the site of regulation for the amount of Na^+, K^+, and H^+ excretion. Aldosterone, a hormone secreted by the adrenal gland, increases sodium reabsorption by the distal tubule and potassium secretion by the same cells. Aldosterone levels in plasma are set largely by the end products of renin secretion by the kidney. Renin acts on a substrate produced by the liver, liberating angiotensin I, which in turn is cleaved into angiotensin II, the active ingredient of the renin–angiotensin system. The distal nephron segment can be considered the final regulator of salt and water balance.

Renal function can be measured by using the clearance equation:

$$\text{Clearance} = \frac{UV}{P}$$

The clearance of inulin is equal to the GFR. A more simple method is to measure the plasma concentration of creatinine or urea. Because creatinine and urea production are essentially constant in any individual and both solutes leave the body only via the kidneys, their concentrations in plasma are a reflection of the GFR. The lower the GFR, the higher the creatinine and urea concentrations of plasma.

Formation of a concentrated urine is dependent on the anatomy of the nephrons in the papilla of the kidney and the function of ADH. The renal countercurrent system requires hairpin-shaped loops, the loop of Henle, transport of Na^+ out of the ascending limb of the loop of Henle, and control of water permeability of the collecting ducts by ADH.

Questions

Conceptual and Factual Review

1. What are the different parts of the nephron? (p. 307)
2. What forces operate on the blood plasma to cause filtration at the glomerulus? How much plasma is filtered each day? (p. 309, p. 311)
3. How is the GFR measured? How is it estimated clinically? (p. 314)
4. Which plasma solutes are completely reabsorbed from the proximal tubular urine? Which are partially reabsorbed? (p. 315, p. 320)
5. Why do people with untreated diabetes mellitus and diabetes insipidus excrete large volumes of urine? (p. 317, p. 323)
6. In which part of the nephron are Na^+ and K^+ transport regulated? (p. 320)
7. Where along the nephron is most salt and water reabsorbed? (p. 320)
8. What hormone plays the major role in regulating Na^+ and K^+ concentrations of plasma? (p. 320, p. 321)
9. What substance is the end product of renin secretion? What are its functions? (p. 321)
10. What hormone regulates the volume and concentration of urine? What causes altered secretion of that hormone? (p. 321, p. 323)
11. What factors regulate the secretion of renin by the kidney? (p. 322)
12. What stimulates the pituitary to increase the secretion of ADH? (p. 323)
13. Where is ADH synthesized? (p. 323)
14. What would be the consequence of a loss of renal response to ADH? (p. 323)
15. How does ADH function to control urine osmolality? Where along the tubule does it act? (p. 327)
16. What changes would occur in the composition of plasma if the kidneys were to fail? (p. 328)
17. What is the result of inhibiting salt transport anywhere along the nephron? (p. 330)
18. Why might a physician prescribe diuretics for a patient? (p. 330)

Applying What You Know

1. Sailors stranded on a raft at sea do not drink sea water, as they know it will hasten death from dehydration. However, the osmotic concentration of sea water is less than that of urine, about 1100 mOsm/kg water versus 1400 mOsm/kg water. Why, then, can they not take advantage of this difference in osmolality and survive by drinking sea water?
2. What would be the consequences of a 1% error in the renal transport of Na^+? Of K^+? How long could this go on before a life-threatening situation arises?
3. A patient with untreated Addison's disease, in which the adrenal gland does not secrete sufficient aldosterone, may complain that he or she constantly craves salt. Why is that? Can you think of any other symptoms that might be typical of a patient with Addison's disease?
4. Some people have the reverse problem of a patient with Addison's disease; they secrete inappropriately high amounts of aldosterone. What do you think might be the consequences of that error in secretion?
5. If the clearance of substance X is found to be 155 ml/min, and the simultaneous clearance of inulin is 120 ml/min, what can you say with certainty about how the kidney handles substance X?
6. People with diabetes mellitus (sugar diabetes) and diabetes insipidus exhibit high urine flow rates. They often complain of thirst. Why? How might you distinguish between the two patients?
7. A person with poor renal function may be advised to eat little meat. Can you think of reasons for this advice?

Chapter 13
Digestive System

Objectives

By the time you complete this chapter you should be able to

◆ Describe the basic structure of the digestive system and list the organs and accessory organs associated with digestion.

◆ Define *digestion, absorption, secretion,* and *motility.*

◆ Describe the composition of saliva.

◆ Explain how the motility of the stomach contributes to the delivery of chyme to the small intestine.

◆ Understand the contribution of the stomach to food storage, digestion, and absorption.

◆ List the primary exocrine secretions of the pancreas, their functions, and their origins within the pancreas.

◆ Explain how bile is secreted, concentrated, stored, and delivered to the small intestine.

◆ Distinguish between peristalsis and segmentation, and compare their roles in movement of chyme.

◆ Review how each of the following is digested and absorbed in the small intestine: proteins, carbohydrates, fats, water and ions, vitamins, and bile.

◆ Know how digestion in the small intestine is regulated according to need and understand the mechanisms of that regulation.

◆ Compare the small intestine and large intestine in terms of their contribution to digestion and absorption.

◆ Explain the importance of bacteria in the large intestine to digestion.

Introduction

Food and water are vital ingredients for all living creatures. Food provides the necessary sources of energy and the materials needed for an organism to grow. In fact, the basic structural units of the food are the same as those of the human body.

As you will learn in more detail in Chapter 14, the food we eat is composed of large complex organic molecules (carbohydrates, proteins, and fats) as well as water and inorganic ions. It is the function of the digestive system, also called the **gastrointestinal (GI) system,** to break food down to sizes that can be absorbed, and then to absorb them into the body. (The prefix *gastro* comes from the Greek word *gaster,* meaning *stomach*).

The breakdown of food into its constituent parts requires fairly harsh conditions, including strong acid and enzymes. Thus, the body is faced with the general problem of how to break down the food into particles small enough to be absorbed into the body, without breaking down living cells of the body. In addition, the food we eat may contain bacteria that would be harmful if they were allowed to enter the body. Both of these problems are solved by having the breakdown of food occur essentially "outside" the body, in a continuous tube called the **GI tract,** that passes *through* the body from mouth to anus. The harsh conditions required for breaking down food and for killing ingested bacteria are created only at certain points in the tube, away from other living cells of the body. A schematic representation of how the GI tract passes through the body is shown in Figure 13.1.

Structure of the Digestive System

The digestive system is composed of several different organs, all of which are involved in digesting and absorbing food. It consists primarily of the GI tract; this tube includes the mouth (with its accessory structures, the teeth, tongue, and salivary glands), the pharynx, esophagus, stomach, small intestine, large intestine, rectum, and anal canal (Figure 13.2). The digestive system also includes the pancreas, the liver, and the gallbladder. These are referred to as accessory organs because they are not part of the tube structure previously described.

Examining more closely any region of the gastrointestinal tract from the esophagus to the large intestine (Figure 13.3), we see that it is composed of four distinct layers. The actual structure of the layers may vary somewhat from organ to organ; important differences are described later as we discuss the specific structure and function of each organ. The **mucosa** is

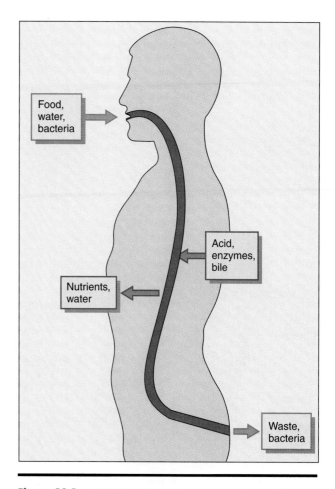

Figure 13.1   A highly schematic representation of the gastrointestinal tract, showing that the contents of the GI tract essentially lie "outside" the body.

the innermost layer; it is in close contact with the partially digested food in the **lumen,** which is the space within the tube. The mucosa consists of a thin sheet of epithelial cells supported by connective tissue and smooth muscle. The connective tissue contains blood vessels and lymph tissue, which is important for fighting disease (remember that the food may contain bacteria). The thin underlying smooth muscle supports the connective tissue and the epithelial sheet. In most regions of the GI tract, the mucosal layer exhibits extensive folding, which creates a much larger surface area than would be found in a smooth-surfaced layer of the same diameter.

Beneath the mucosal layer is the **submucosa,** a loose layer of connective tissue that joins the mucosa to the muscle layer beneath the submucosa. This layer contains numerous blood and lymph vessels that supply the mucosa. It also contains nerve cells that share in the control of gastrointestinal secretions. The third layer, the **muscularis,** consists primarily of smooth muscle arranged in two or three distinct layers. The muscularis also contains groups of nerve cells that are

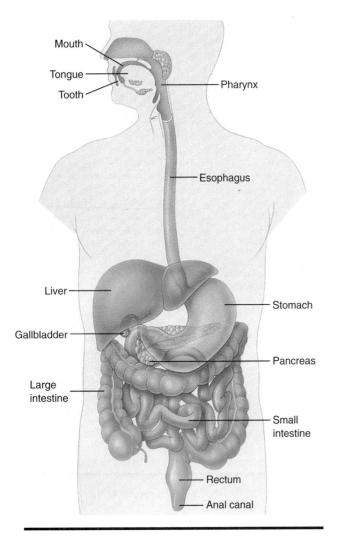

Figure 13.2 The human gastrointestinal system.

important in the local control of contraction of the muscle layers. The fourth layer, the **serosa,** is a connective tissue layer that surrounds the other three layers and attaches the entire structure to the body wall.

We are most interested in the mucosal layer because the epithelial cells of the mucosa are responsible for endocrine and exocrine secretions as well as the absorption of all substances from the GI tract. (Bear in mind, however, that endocrine and exocrine secretions are also produced by some of the accessory organs.) We also focus on the third layer, the muscularis, because it is responsible for causing movement through the GI tract.

Basic Processes

As mentioned earlier, the general function of the digestive system is to provide the body with the nutrients it needs. There are four basic processes we must consider in order to understand how the digestive system carries out its general functions. The first is **diges-**

tion. Obviously, whole chunks of food are too large to enter the bloodstream directly. Digestion is the process of breaking down the food into smaller units so it can undergo the second process, **absorption.** Absorption is the process whereby the final products of digestion cross the cells of the mucosa and enter the bloodstream. The units are thus absorbed into the body.

To carry out digestion and absorption, two additional processes must also be present. One is **motility** (or movement). Movement (chewing) occurs in the mouth to break up large pieces of food. The chewed food is then swallowed, and the food moves through the esophagus to the stomach. The stomach and intestines mix and churn the contents and slowly move the partially digested food down the digestive tract toward the colon. Motility is controlled by both nerves and hormones.

The other process critical to digestion and absorption is **secretion.** The GI tract produces both exocrine secretions (into the lumen of the GI tract) and endocrine secretions (into the blood). Exocrine secretions include saliva in the mouth, mucus in the esophagus, stomach, and intestines, acid in the stomach, bile by the liver, and pancreatic juice by the pancreas. Exocrine secretions contribute to lubrication of food during chewing and swallowing, protection of the lining of the GI tract from damage, killing of bacteria, and, most importantly, the breakdown (digestion) of the components of food into units small enough to be absorbed. The endocrine secretions (hormones) regulate the timing of some of the exocrine secretions, as we shall see later.

We will now examine the structure and function of each organ as food moves through the GI tract. For each organ, consider what contribution that organ makes to the overall process of providing the body with nutrients and take note of how the activity of that organ is controlled.

Mouth, Pharynx, and Esophagus

Mechanical breakdown, chemical digestion, and moistening of the food begin in the mouth. Once the food has been sufficiently broken down and moistened, it is transferred to the pharynx, swallowed, and passed down the esophagus to the stomach.

Mechanical Breakdown of Food

The primary role of the mouth in digestion is to break the chunks of food into ever-smaller sizes. The structures of the mouth and pharynx are shown in Figure 13.4. The teeth grind and pulverize the food as the tongue moves the food around and positions it for

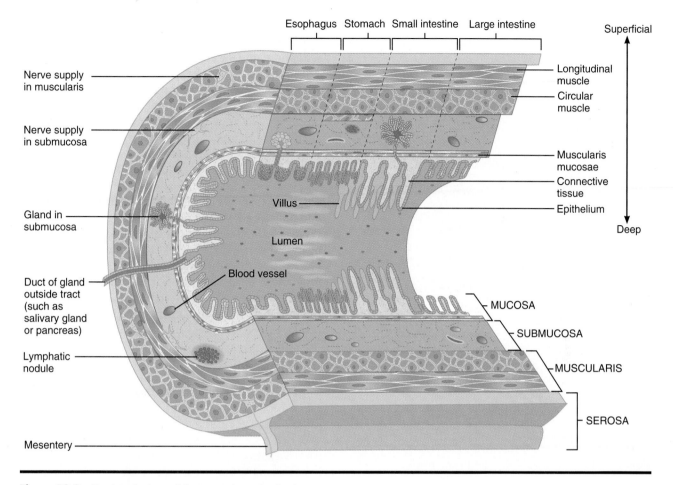

Figure 13.3 Sectional view of the gastrointestinal tube.

grinding. The mechanical breakdown of food is important because it greatly increases the surface area of food, enhancing the speed with which it can be digested and absorbed. This is one reason that it is a good idea to chew your food carefully; another, of course, is to avoid choking while swallowing. Chewing is under voluntary control because the muscles of the jaw and tongue are skeletal muscle, but it is so ingrained through practice that one generally doesn't think about it much. The tongue also contains the taste buds, discussed in Chapter 7.

Chemical Digestion and Lubrication by Saliva

Mechanical breakdown of food is not all that happens in the mouth. The mouth also adds **saliva** to the food. Saliva is a watery fluid containing bicarbonate (an acid-neutralizing molecule), enzymes, and a small amount of mucus. It is secreted by three main pairs of **salivary glands** (the parotid, the sublingual, and the submandibular glands) as well as by numerous smaller salivary glands. The salivary glands empty saliva into the mouth by way of ducts.

The primary function of saliva is to lubricate and moisten dry food. To demonstrate this, put a dry soda cracker into your mouth, chew it very quickly, and then try to swallow it immediately. If you have not allowed enough time for the chewed mass to become fully moistened by saliva, you may find the cracker extremely difficult to swallow.

A second function of saliva, though not an essential one, is to assist in the digestion of starch and triglycerides. Starch is a complex carbohydrate of plants that is composed of long, branching chains of monosaccharide units. Saliva contains **amylase,** an enzyme that breaks starch into disaccharides, trisaccharides, and other short chains of monosaccharide units. These relatively short-chain carbohydrates eventually are broken down completely to monosaccharides in the intestine, as only the monosaccharides can be absorbed. To demonstrate carbohydrate digestion in the mouth, again hold a soda cracker in your mouth, but this time keep it there for as long as possible. After several minutes you may notice that the moistened cracker begins to taste sweet as you digest the starch into short-chain carbohydrates. A second salivary enzyme, **lipase,** contributes to the digestion of triglycerides. Salivary lipase does not contribute

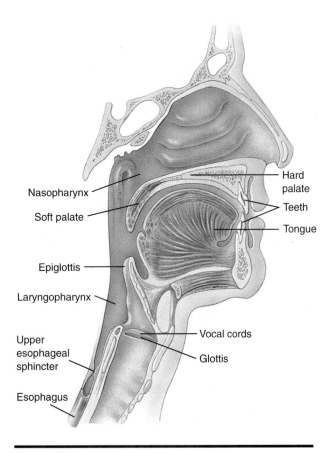

Figure 13.4 Structures of the mouth and pharynx.

in response to certain stimuli associated with eating. The composition of the saliva may change as well, with more mucus, enzymes, and bicarbonate secreted during times of high salivary secretion. The primary stimuli for salivary secretion are the taste and smell of the food and the physical movements associated with chewing. The most potent stimulus is a sour taste, such as lemon juice.

Swallowing

Once food has been chewed, reduced to a soft mass, and moistened with saliva, it is ready to be swallowed (Figure 13.5(A)). Swallowing can be initiated voluntarily; however, once begun, it is reflexive and cannot be stopped. Swallowing begins with the voluntary movement of food to the back of the mouth by the tongue, forcing the food against the upper part of the throat. The pressure against the upper part of the throat, called the **nasopharynx,** stimulates sensory

to triglyceride digestion in the mouth because it is not very active at the neutral pH of saliva. It is much more active in the acid environment of the stomach, and thus contributes to triglyceride digestion once the food and saliva reach the stomach.

A third function of saliva is to protect against tooth decay and gum disease. Very acid conditions in the mouth tend to enhance the removal of the minerals from teeth. Although acids are present in the food we eat, more damaging in the long-run are the acids produced by bacteria in the mouth between meals. These bacteria are killed by a third enzyme in saliva, called **lysozyme.** The combination of lysozyme and bicarbonate in saliva tends to protect against excessive prolonged acidity in the mouth.

Salivary secretion is controlled by the autonomic nervous system. Under normal conditions when you are awake, small amounts of saliva are continuously secreted to keep the mouth moistened and to lubricate the movement of the lips and tongue. The rate of salivary secretion is quite variable. It declines when you are asleep, when you are frightened, or when you are dehydrated, and it may increase more than twentyfold

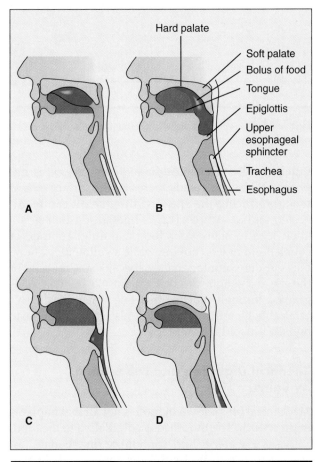

Figure 13.5 Swallowing. (A) Note the open passageway for air before swallowing. (B) and (C) The air passageways close during swallowing. (D) Food passes into the esophagus.

MILESTONE 13.1

Psychological Factors and Salivary Secretion

It is a common belief that one's mouth waters at the thought or sight of food. This belief persists, partially due to the classic experiments conducted by the Russian psychologist Ivan Pavlov in the late nineteenth century. Pavlov demonstrated that dogs can be trained to salivate in response to the sound of a bell (called a conditioned stimulus by psychologists), once the bell has been coupled with the delivery of food (an unconditioned stimulus) for a period of time. Thus, it seems only natural to assume that the same phenomenon occurs in humans. However, in 1961 a British dental surgeon and researcher, Dr. A.C. Kerr, reviewed the literature and concluded that this belief "is not substantiated by any objective experimental evidence in man." Subsequent experiments by Dr. Kerr showed that the rate of salivary secretion increased little or not at all in hungry human subjects who watched a lemon being sliced or who watched a meal of bacon and eggs being prepared. According to Kerr, the notion that our mouth waters at the thought of food may be ascribed to an increased *awareness* of the saliva, not to an increased rate of secretion. These experiments cast doubt on the role of psychological factors in controlling salivary secretion in humans, even though psychological factors clearly do contribute to salivary secretion in the dog. They also point out an important caveat: Although animal experiments may be applicable to humans, and indeed they usually are, there is no reason that they must be applicable to humans under all circumstances.

---

nerves, which in turn send sensory information to the brain, initiating the swallowing reflex. Breathing is halted temporarily because the respiratory centers of the brain are inhibited briefly. At the same time, the **soft palate** (the soft upper part of the back of the mouth) moves upward, closing the passageway for air from the nasal cavities to the nasopharynx (Figure 13.5(B)), and the **epiglottis** (the lid-like structure that covers the entrance to the trachea) tips downward, closing the entrance to the trachea. The **glottis,** or airway opening between the vocal cords, also closes. Finally, a ring-like band of muscle at the entrance to the esophagus called the **upper esophageal sphincter** relaxes (Figure 13.5(C)), allowing the mass or **bolus** of food to pass into the esophagus.

The term *sphincter* refers to a ring-like band of muscle that normally maintains considerable basal muscle tone, thereby partially or completely closing off an orifice or passage under normal circumstances. Sphincters must relax in order to allow passage of materials. They are usually named for their location, and the upper esophageal sphincter is just the first of seven that regulate flow into or through the GI tract. Table 13.1 summarizes the location, function, and control of GI sphincters. As they are mentioned in future sections, note that the mechanisms of their control make sense once you understand their functions.

Once the food leaves the **laryngopharynx,** it is in the esophagus. The esophagus moves the bolus of food quickly toward the stomach; it does not play a role in digestion or absorption. Mucus secreted by the mucosa of the esophagus lubricates the bolus of food as it passes, preventing mechanical damage.

Contraction of the muscle of the esophagus is initiated as part of the swallowing reflex. Muscular contractions of the esophagus occur in a particular pattern known as **peristalsis,** a wave-like contraction that moves in a sequential fashion along a tubular or hollow organ (Figure 13.6). Peristalsis is a common form of contraction in the GI tract, but it also occurs in the ureters, where it propels urine from the kidneys to the bladder, and in the uterus during childbirth. The function of peristalsis is to move the contents of hollow organs forward, even against the force of gravity if necessary. In the esophagus, peristalsis is produced by the coordinated action of nerves that stimulate successive rings of circular muscle to contract sequentially, from the top of the esophagus to the bottom. The entire peristaltic wave of contraction, and hence the movement of a bolus of food from the mouth to the stomach, takes about 6–10 seconds.

Sometimes a bolus of food may not be pushed all the way into the stomach by a single peristaltic wave. In such cases it may become lodged in the esophagus only partway down. This is particularly common with sticky or thick foods such as peanut butter or mashed potatoes. When this occurs, a second, more forceful peristaltic contraction is initiated directly behind the bolus. This secondary contraction is often sufficiently forceful that one becomes aware of a sense of mild pain or discomfort associated with the contraction of the esophagus against the bolus. Generally one can dislodge the bolus, and sometimes even feel it move into the stomach, by drinking a small amount of liquid.

At the lower end of the esophagus, the **lower esophageal sphincter** relaxes as the bolus of food

Table 13.1   **Sphincters in the GI tract.**

Name	Location	Function	Regulatory Control
Upper esophageal sphincter	At the entrance to the esophagus	Prevents air from entering the esophagus, yet allows food to enter	Relaxation of the upper esophageal sphincter is one component of the swallowing reflex. It is controlled by parasympathetic nerves activated by the brain.
Lower esophageal sphincter	Between the esophagus and the stomach	Controls access of food to the stomach, yet prevents reflux of stomach contents back into the esophagus	The lower esophageal sphincter relaxes as a bolus of food approaches. Lower esophageal sphincter relaxation is mediated by a local neural reflex initiated by the stretch of the esophagus by the approaching bolus.
Pyloric sphincter	Between the stomach and small intestine	Partly responsible for regulating the rate of emptying of the stomach into the small intestine	This sphincter is normally only partially constricted. Duodenal distention or the presence of acid or fats in the duodenum increases pyloric sphincter tone. Tone is increased by neural reflexes and by the secretion of cholecystokinin.
Sphincter of Oddi	At the end of the common bile duct, where it joins the small intestine	Regulates the rate of entry of bile into the small intestine	The presence of fat in the chyme in the duodenum stimulates the secretion of cholecystokinin into the bloodstream. Cholecystokinin relaxes the sphincter of Oddi.
Ileocecal sphincter	Between the ileum of the small intestine and the cecum of the large intestine	Regulates the passage of material from the small intestine into the large intestine	Mediated by neural reflexes. Distention of the terminal portion of the small intestine (the ileum) causes the ileocecal sphincter to relax, whereas distention of the large intestine causes it to contract.
Anal sphincters: Internal anal sphincter	At the terminus of the rectum	Regulate defecation	Internal anal sphincter: smooth muscle under involuntary control. Distention of the rectum produces reflex relaxation.
External anal sphincter			External anal sphincter: skeletal muscle primarily under voluntary control. Voluntary contraction can prevent defecation despite the presence of a defecation reflex.

HIGHLIGHT 13.1

Heartburn

The lower esophageal sphincter generally remains closed except during swallowing. This is important because the contents of the stomach are very acidic and the esophagus is not as well-protected from the acid as is the stomach. If the lower esophageal sphincter fails to remain closed or if there is excessive pressure on the stomach, reflux of stomach contents back into the esophagus can cause irritation and a burning sensation in the lower portion of the esophagus. This condition is called heartburn because the location of the pain is perceived as being near the heart. For people prone to this condition, the likelihood of heartburn can be reduced by not lying down after a meal and by eating smaller meals, both of which reduce the chance of reflux of stomach contents into the esophagus. In addition, antacids may reduce the severity of heartburn by neutralizing the stomach acid.

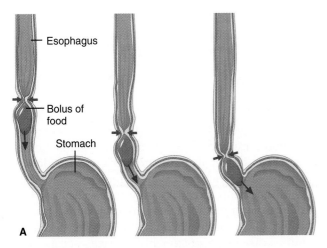

Figure 13.6 Peristalsis in the esophagus. (A) Sequential diagrams of how peristalsis moves a bolus down the esophagus toward the stomach. (B) An X ray of peristalsis in the esophagus.

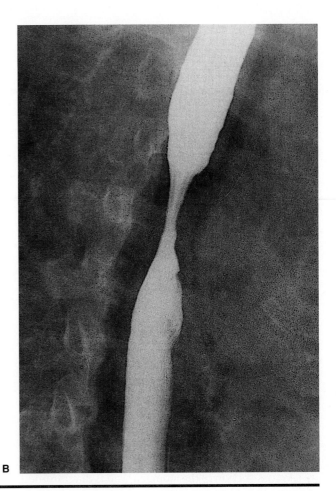

approaches. This allows the bolus to pass into the stomach, after which the sphincter closes again.

Stomach

Just below the esophagus is a large, J-shaped organ, the stomach. The first and most important function of the stomach is storage of food. A full meal can be eaten much more quickly than it can be digested and absorbed, and the stomach is the storage vessel for food until it can be processed. The muscles of the stomach undergo a generalized relaxation as food is ingested, in effect allowing the stomach to function as a large storage bag. A second important function is mixing. Superimposed on the overall generalized relaxation are rhythmic peristaltic contractions of the stomach that mix the partially digested contents and push them toward the small intestine, where most of the digestion and absorption takes place. The third function of the stomach is digestion of proteins. In order to ac-

complish this digestive function, the stomach secretes both acid and an enzyme into the lumen of the stomach. Note that digestion of carbohydrates or fats is *not* an important function of the stomach. Furthermore, with only a few exceptions to be noted later, the stomach plays essentially no role in absorption.

Structure

Anatomists generally define three regions of the stomach: the **fundus,** a rounded part that serves primarily a storage function; the body, or central portion; and the **pylorus,** which joins the small intestine. A diagram of the stomach is shown in Figure 13.7. At the point of union between the stomach and the small intestine is another sphincter, the **pyloric sphincter.** The wall of the stomach is composed of the same four general layers described previously, with some modifications. The muscularis is composed of three muscle layers rather than two, with the cells of each layer oriented on a different plane (oblique, circular, and longitudinal). These muscle layers are considerably thicker in the pylorus than in the fundus. The mucosal lining of

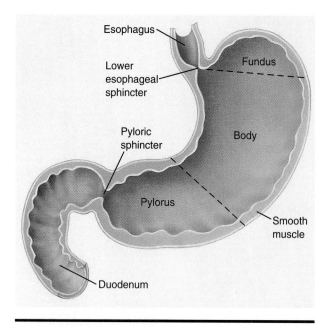

Figure 13.7 Anatomy of the stomach. Note that the thickness of the wall of the stomach increases from fundus to pylorus.

the empty stomach is folded into many ridges called **rugae,** allowing for considerable expansion when the stomach is full.

A cross-section of the stomach wall is diagrammed in Figure 13.8(A) and a closer view of the mucosa alone is shown in Figure 13.8(B). An electron micrograph of the mucosal surface is shown in Figure 13.8(C). The mucosal surface is dotted with gastric pits, which are the openings of deep thin folds of the gastric mucosa, the **gastric glands.** Four different cell types are present in the gastric glands. **Mucous cells** cover the surface of the mucosa and line the upper portion of the glands. These cells secrete mucus that protects the cells that line the stomach from digestion by the gastric contents. **Parietal cells** produce a watery fluid containing a strong acid. Parietal cells also produce **intrinsic factor,** a substance necessary for absorbing vitamin B_{12}. **Chief cells** produce **pepsinogen,** an inactive form of the protein-degrading enzyme **pepsin.**

Together, the secretions of the parietal cells and the chief cells make up **gastric juice,** a watery fluid containing hydrochloric acid (HCl), intrinsic factor, and pepsinogen. Gastric juice is secreted from the glands into the lumen of the stomach, where the acid and pepsinogen participate in the digestion of protein. Although intrinsic factor is necessary for the normal absorption of vitamin B_{12}, vitamin B_{12} is not actually absorbed until later, in the small intestine. The fourth type of cell of the gastric glands is the G-cell. **G-cells** do not secrete their product into the

gastric gland; instead they secrete a hormone, **gastrin,** into the bloodstream. Gastrin is involved in controlling both gastric motility and the secretion of gastric juice, as we shall see shortly.

Motility

Under resting conditions when the stomach is empty, the smooth muscle of the stomach maintains a continuous basal level of contraction, or tone, in order to keep the stomach small. When food is eaten, the smooth muscle relaxes, allowing the stomach to expand to accommodate the increased volume. This relaxation of the stomach is a neural reflex brought about by distention, or stretch, of the stomach as a result of the presence of food.

Unlike the basal tone of the smooth muscle, peristaltic contractions are virtually absent when the stomach is empty, but increase markedly when food is present in the stomach. Peristaltic waves of contraction start at the fundus and sweep down the body of the stomach at a regular rate of one wave every 15–25 seconds (Figure 13.9). Each wave takes about 5–10 seconds to travel the full length of the stomach, so that periods of contraction of any region of the stomach are preceded and followed by periods of relaxation.

The smooth muscle is relatively thin in the fundus, and the peristaltic waves of contraction that pass over this region have very little mixing effect on the fundus' contents. The fundus functions primarily as the storage region of the stomach, and food may remain unmixed in the fundus for an hour or more.

Both the force and the velocity of peristaltic contractions increase as the wave moves down the body of the stomach. Peristalsis is particularly strong in the pylorus. These stronger contractions mix the contents with gastric juice, mechanically reduce the size of the food particles, and propel the contents toward the small intestine. The watery fluid of partially digested food and gastric juice in the pylorus is called **chyme.**

The pyloric sphincter controls the flow of chyme from the stomach to the intestine. At rest, the pyloric sphincter is partially open. With each wave of peristalsis in the stomach, the chyme is pushed forward toward the pyloric sphincter, and then back into the body of the stomach again as the peristaltic wave of contraction passes by. As the peristaltic wave moves down the stomach, a small amount of chyme is forced through the sphincter, but then the sphincter contracts, completely closing the entry to the small intestine just before the peristaltic wave arrives at the end of the pylorus. Only a small amount of chyme enters the small intestine with each contraction. This cycle is repeated once every 15–25 seconds over a period of about 2–6 hours until all of the food in the stomach has been

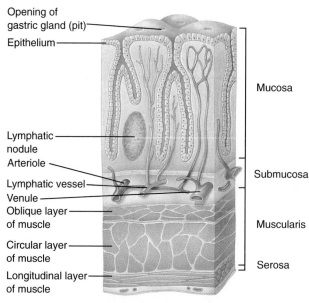

Opening of
gastric gland (pit)
Epithelium

Mucosa

Lymphatic
nodule
Arteriole
Lymphatic vessel
Venule

Submucosa

Oblique layer
of muscle
Circular layer
of muscle
Longitudinal layer
of muscle

Muscularis

Serosa

A Sectional views of layers of stomach

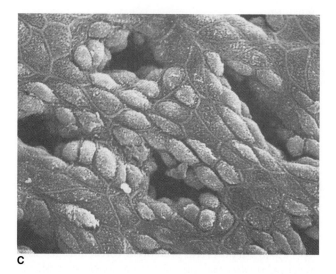

C

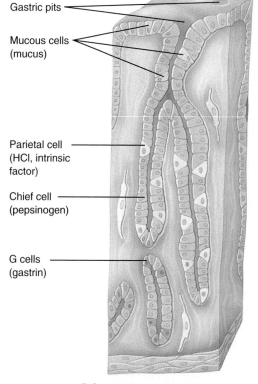

Gastric pits

Mucous cells
(mucus)

Parietal cell
(HCl, intrinsic
factor)

Chief cell
(pepsinogen)

G cells
(gastrin)

B Sectional view of the stomach mucosa
showing gastric glands

Figure 13.8 The stomach wall. (A) Cross-sectional view
of the four layers. (B) Expanded view of the mucosa
showing the gastric glands. The secretory products of each
cell type are noted in parentheses. (C) Scanning electron
micrograph (1400×) of the gastric pits.

mixed with gastric juice, becomes chyme, and is propelled into the small intestine. When the stomach is empty the peristaltic contractions cease, the basal tone of the smooth muscle returns, and the stomach shrinks to its empty size.

But what regulates these peristaltic waves? What makes them begin, and why do they end once the stomach is empty? Regulation of gastric motility can be divided into three phases. First, the sight, taste, smell, touch, and even the thought of food initiate neural reflexes that increase gastric motility. This anticipatory response, called the cephalic phase, is not as important as the gastric and intestinal phases that follow, and you may not even be aware of the increased motility of the stomach. However, many people say that their stomach growls when they experience a feel-

ing of hunger on smelling or seeing a food that they especially enjoy. Second, during the gastric phase, neural reflexes are initiated by stretch of the stomach wall (distention by food). These neural reflexes contribute to the initiation of peristalsis, which lasts for as long as food is present in the stomach. The hormone gastrin also stimulates motility. Secretion of gastrin is stimulated by distention of the stomach and by the presence of protein in the food. In general, the cephalic and gastric phases of regulation are designed to prepare the stomach for emptying even before it is filled, and then to facilitate gastric emptying once the stomach is filled.

The third phase of control of gastric motility is called the intestinal phase. This phase, which opposes the action of the cephalic and gastric phases, prevents the stomach from delivering chyme to the small

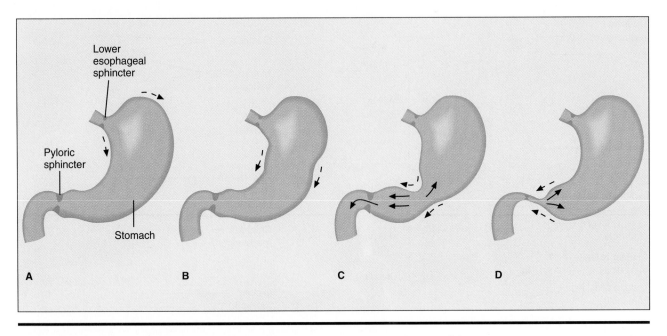

Figure 13.9 Peristalsis of the stomach. Dashed arrows indicate direction of travel of the peristaltic wave, solid arrows indicate movement of the stomach contents.

intestine more quickly than it can be digested and absorbed. There are at least two mechanisms for intestinal inhibition of gastric emptying. First, stretch of the small intestine inhibits gastric motility, thereby delaying gastric emptying. This, too, is a reflex mediated by nerves in the wall of the intestine. Second, the presence of acid and food breakdown products in the intestine leads to the secretion of three intestinal hormones: **secretin, cholecystokinin (CCK),** and **gastric inhibitory peptide.** All three of these hormones inhibit gastric motility, although they also have other actions that are described later. The intestinal control mechanisms, then, act to delay gastric emptying until the small intestine has digested and absorbed what it has already received. Ultimately, a balance is struck between factors that stimulate gastric emptying (volume and composition of the contents of the stomach), and those that inhibit it (volume and composition of the contents of the small intestine).

Gastric Secretions and Their Role in Digestion

The arrival of food in the stomach stimulates the stomach to secrete a watery gastric juice containing HCl and pepsinogen. Although strongly acid conditions are not beneficial in the mouth, the addition of acid in the stomach is essential to digestion. The strong acidity of gastric juice contributes to digestion in several ways. First, it assists in the breakdown of proteins and connective tissue in the ingested food. Second, the

acid causes the inactive enzyme pepsinogen, itself a protein, to be activated to pepsin. The strong acidity of the stomach has a third function: to kill most of the bacteria that are ingested, thereby preventing us from suffering from what we commonly call food poisoning. The term *food poisoning* is somewhat incorrect when used in this way, because what most of us actually experience is the result of ingestion of live bacteria; thus, it should more properly be called food infection. Technically speaking, food poisoning is the result of ingesting the toxic byproducts of bacteria or other microorganisms.

Pepsin is a powerful proteolytic (protein-digesting) enzyme, particularly in the very acidic conditions present in the stomach. Pepsin is largely responsible for breaking large protein molecules into smaller fragments. Indeed, because the inactive precursor of pepsin is itself a protein, pepsin can actually cause more pepsin to be produced from pepsinogen. These relationships are diagrammed in Figure 13.10.

Perhaps you are wondering how the stomach can contain a strong acid and an enzyme powerful enough to digest the proteins in the food without being digested itself. The first solution is to secrete the digestive enzyme in inactive form (as pepsinogen), which is converted to the active form (pepsin) in the lumen of the stomach following exposure to acid. Thus, the enzyme is not capable of digesting proteins until it is out in the lumen and ready to be used. The second solution lies in reducing the exposure of the mucosal cells to enzyme and strong acid. This is the primary

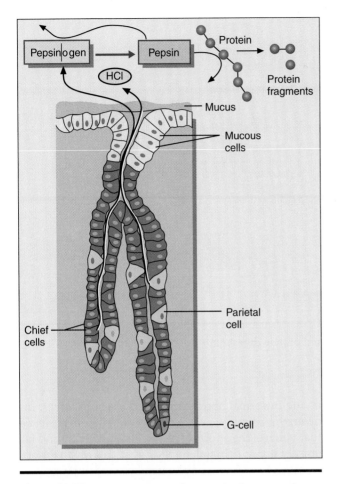

Figure 13.10 The formation of pepsin in the stomach and the digestion of proteins.

function of mucus. Mucus is a gel containing bicarbonate, which forms a flexible physical barrier over the mucosal surface. Although mucus is degraded continually by acid and pepsin, it is also continually replaced with fresh mucus by the underlying mucous cells.

Although mucus offers some protection from acid, bacteria, and proteolytic enzymes, some cells do become damaged and ultimately need to be replaced by new cells. The rate of cell replacement is extremely rapid in the gastrointestinal system; it has been estimated that the entire lining of the gastric mucosa is replaced every three days. These protective mechanisms notwithstanding, damage to the wall of the gastrointestinal system may occur in some people.

The secretory activity of the stomach is regulated according to need. The overall goal of regulation is to provide gastric juice to the stomach very quickly following a meal so that digestion can proceed, and then to reduce the rate of acid secretion when food is not present in order to prevent damage to the stomach itself.

The human stomach secretes approximately 1–2 liters of gastric juice per day, most of which is secreted when the stomach contains food. The regulatory mechanisms look very much like those for gastric motility. First, there is at least some anticipatory cephalic control by higher brain centers (remember, however, that evidence for such a neural reflex in the anticipatory control of salivary secretion is weak, at best). Gastric secretion increases in response to the taste and smell of food and the physical act of chewing. However, cephalic control does not always serve a true regulatory function in that it sometimes occurs in response to events totally unrelated to food. For example, neural influences are thought to cause an increase in gastric secretion and motility in response to stressful situations.

The primary physiological control of gastric secretion is gastric control. Distention of the stomach and the presence of protein in the stomach both stimulate the secretion of gastric juice via neural reflexes. In addition, distention and the presence of protein stimulate the secretion of gastrin from the G-cells of the gastric mucosa. Gastrin, in turn, circulates in the blood, ultimately causing the parietal and chief cells of the gastric mucosa to secrete more HCl and pepsinogen, respectively. Note that it makes sense that the most potent stimuli for gastric secretion are stretch of the stomach and the presence of protein because the actions of HCl and pepsin are to assist in the digestion of protein. This is yet another example of a negative feedback system; both stretch of the stomach by food and the presence of protein in the food set in motion events that result in the eventual digestion and removal of that food (Figure 13.11).

As was the case for gastric motility, events in the small intestine inhibit gastric secretions. Secretion of gastric juice is inhibited by stretch of the small intestine and by the hormones secretin, cholecystokinin, and gastric inhibitory peptide, all secreted by the small intestine into the bloodstream. Table 13.2 summarizes the control of gastric motility and secretion.

Absorption

In general, the stomach is relatively unimportant in terms of absorption of the partially digested food into the bloodstream. A small amount of water and electrolytes are absorbed, but the nutrient constituents of the food (proteins, carbohydrates, and fats) are not absorbed until they reach the small intestine, both because the stomach lacks the special cellular transport mechanisms necessary for absorption and because digestion is not yet complete. In general, the constituents of chyme leaving the stomach must undergo further digestion in the small intestine before

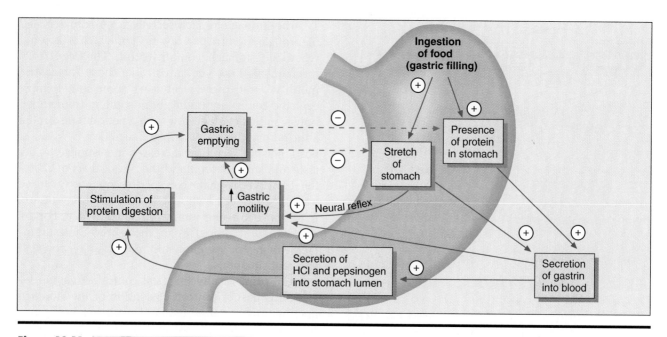

Figure 13.11 Gastric phase of the regulation of gastric motility and protein digestion.

absorption can take place. Most absorption occurs in the small intestine.

Among the few substances absorbed by the stomach are aspirin and alcohol. Both are fat-soluble and already small enough to be absorbed, and thus they can cross the cell membranes of the gastric mucosa rather easily by diffusion. Both aspirin and alcohol are absorbed even better by the small intestine, however. The presence of food in the stomach tends to delay their absorption in the stomach by virtue of the fact that their concentration in the stomach is diluted by the food and accompanying fluids. In addition, the presence of food in the small intestine tends to delay gastric emptying into the small intestine, thereby de-

laying intestinal absorption of these substances. Indeed, you may be aware from your own experience that you feel the effects of alcohol more quickly and intensely if you drink alcoholic beverages on an empty stomach than if you have eaten before drinking.

Pancreas

By the time the chyme reaches the small intestine it is an acidic, watery soup containing partially digested protein that is delivered from the stomach at a fairly

Table 13.2   **Control of gastric motility and secretion.**

Control Phases and Factors	Neural or Hormonal Response	Effects on Motility and Secretion
Cephalic phase		
Taste, sight, smell, touch, or thought of food	Increased neural activity to stomach from brain	Increased gastric motility and increased secretion of gastric juice
Gastric phase		
Distention of the stomach	Increased neural activity to stomach	Increased gastric motility and increased secretion of gastric juice
Protein and protein-digestive products in the stomach	Increased endocrine secretion of gastrin	
Intestinal phase		
Intestinal distention	Deceased neural activity to stomach	Decreased gastric motility and decreased secretion of gastric juice
Presence of fat in the duodenum	Increased endocrine secretion of secretin, cholecystokinin, and gastric inhibitory peptide from the duodenum	
Increased acidity in the duodenum		

HIGHLIGHT 13.2

Peptic Ulcers

An ulcer is a particular kind of damage to skin or internal mucous membranes in which the tissue dies and disintegrates. **Peptic ulcers** are damaged regions of the gastrointestinal tract caused by gastric juice. Symptoms of peptic ulcers may include pain and bleeding into the gastrointestinal lumen. If sufficiently severe, ulcers can result in complete perforation of the stomach or intestine, release of the luminal contents containing ingested bacteria into the abdominal cavity, and infection.

Peptic ulcers can form in the lower portion of the esophagus, but most often they form in the stomach or in the first part of the small intestine. The esophagus and the small intestine are not as well-protected from the effects of gastric juice as is the stomach. Thus, prolonged reflux of stomach acid into the esophagus can cause esophageal ulcers, and hypersecretion of stomach acid with the subsequent delivery of too much acid to the small intestine can cause ulcers in the duodenum.

Peptic ulcers of the stomach (gastric ulcers) may result either from hypersecretion of acid (possibly exacerbated by hypersecretion of pepsin) or from hyposecretion of the protective mucus. Hypersecretion of stomach acid can be caused in some people by emotional stress, cigarette smoking, and certain drugs or foods including aspirin, coffee, and alcohol. In addition to their direct effects, alcohol can damage the mucosal barrier directly and aspirin inhibits the formation of prostaglandins, which have a protective function.

In 1920 it was discovered that histamine, a molecule involved in allergic reactions and present in virtually all tissues, could stimulate gastric acid secretion. Subsequent research and testing, first on animals and then on humans, led to a series of drugs that block the action of histamine on gastric acid secretion. Several of these drugs, including cimetidine and ranitidine, are now licensed for the treatment of ulcers in humans.

However, that is not the end of the story. In just the past several years, a whole new approach to the treatment of peptic ulcers has become available with the recognition that patients with peptic ulcers generally are infected with a bacterium, *Helicobacter pylori*. Unlike most bacteria, *Helicobacter pylori* can survive in the strongly acid conditions present in the human stomach. Exactly *how* it contributes to the development of peptic ulcers is not yet known, but the evidence is clear that eradication of the bacterial infection results in very low rates of ulcer relapse compared with relapse rates in patients treated with conventional histamine-blocking therapy alone. Ongoing research in this field is likely to provide us with the answer in the very near future.

steady rate of perhaps several tablespoons every 15 seconds. The primary role of the small intestines is to digest and absorb this chyme. First, however, the chyme must be neutralized, both to protect the mucosa of the small intestine from damage and to allow digestion to proceed. The digestive enzymes present in the small intestine, unlike pepsin from the stomach, work best at a more neutral pH. Neutralization of the chyme and addition of digestive enzymes is the function of the pancreas.

Structure of the Pancreas

The pancreas is an elongated, somewhat lobular organ that lies just below the pylorus of the stomach (Figure 13.12(A)). It is connected to the **duodenum,** the first segment of the small intestine, by the pancreatic duct, which joins the common bile duct just before their common point of entry into the small intestine. In humans, there is also a second, accessory duct direct from the pancreas to the duodenum. The pancreas is both an exocrine and an endocrine gland. **Pancreatic juice,** a watery solution containing bicarbonate and digestive enzymes, is secreted into the pancreatic and accessory ducts (exocrine secretion), and several hormones are produced and secreted directly into the blood (endocrine secretion). The endocrine functions of the pancreas were discussed in Chapter 5.

The pancreatic and accessory ducts divide several times within the pancreas, forming progressively smaller and smaller ducts (Figure 13.12(B)). Cells lining the smaller ducts, called **duct cells,** secrete a watery fluid containing bicarbonate that forms the bulk of the pancreatic juice. Other cells lining the bulb-like terminal regions of the ducts, called **acinar cells,** produce various enzymes. The secretions of the duct cells and acinar cells mix in the larger ducts before secretion into the small intestine.

Functions of Pancreatic Secretions

The bicarbonate in pancreatic juice contributes to digestion in the small intestine by neutralizing the acidic chyme from the stomach. This provides the proper

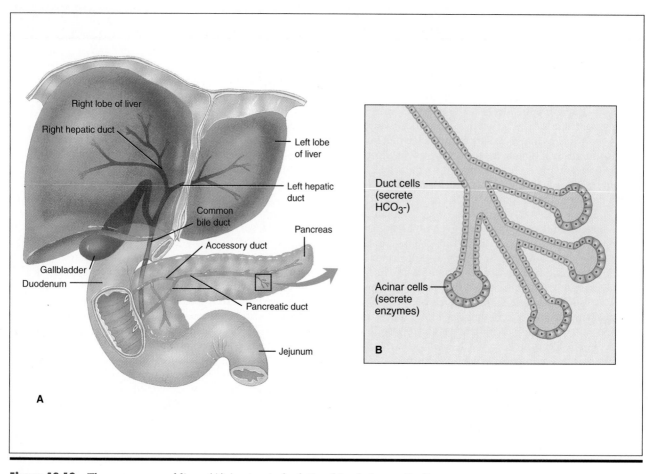

Figure 13.12   The pancreas and liver. (A) Anatomical relationships between the liver, gallbladder, pancreas, and small intestine. (B) A closer view of a section of the pancreas showing the duct cells lining the smaller ducts and the acinar cells lining the terminal sacs.

environment for the various enzymes to carry out their digestive functions. Each enzyme in pancreatic juice (and there are quite a few) is highly specific for the breakdown of certain constituents of the chyme. Many of these enzymes are inactive in the alkaline pancreatic juice, becoming active only when exposed to the more neutral conditions of the small intestine. Pancreatic enzymes include several different proteolytic enzymes that continue the process of protein digestion begun by pepsin in the stomach; pancreatic amylase, which continues the job of carbohydrate digestion begun by salivary amylase; and pancreatic lipase. Pancreatic lipase is the primary enzyme responsible for digesting fats, with salivary lipase playing only a minor role.

As was the case for the secretion of gastric juice, the secretion of pancreatic juice must be regulated so that the alkaline fluid and enzymes are delivered to the intestine only when needed. This topic will be considered further after we have discussed the small intestine.

Liver and Gallbladder

The liver and gallbladder play a pivotal role in the production, storage, and secretion of bile, important for the absorption of fats. In addition, the liver contributes

MILESTONE 13.2

Alexis St. Martin and Dr. William Beaumont

In the summer of 1822, a French Canadian named Alexis St. Martin was accidentally shot in the stomach at close range with a shotgun. Dr. William Beaumont, a young army surgeon stationed at Fort Mackinac in northern Michigan, was called to the scene. In his journal he later wrote, "In this dilemma I considered any attempt to save his life entirely useless. But as I had ever considered it a duty to use every means in my power to preserve life when called to administer relief, I proceeded to cleanse the wound and give it superficial dressing, not believing it possible for him to survive twenty minutes." Dr. Beaumont did manage to save Mr. St. Martin despite the limitations of medical and surgical techniques of the day (neither modern antibiotics nor anesthetics were available), nursing him back to health in his own home for two years because Mr. St. Martin had no money. However, Mr. St. Martin's wound healed with a permanent hole, or fistula, between the lumen of the stomach and the abdominal wall. Dr. Beaumont recognized the potential research value of being able to sample the contents of the stomach directly, and from 1825 to 1833 he hired Mr. St. Martin to be a research subject.

Dr. Beaumont placed food directly into the stomach and then withdrew it later to study the digestive process. He also sampled gastric juice and was the first to describe the content of the juice and its bacteriostatic (bacteria-inhibiting) and digestive functions. Dr. Beaumont is credited with discovering that the stomach acid is hydrochloric acid, that mucus is a separate secretion that coats and protects the stomach, and that a patient's emotional state can affect gastric motility and secretion. He also reported the effects of alcohol on the stomach: "The use of ardent spirits always produces disease of the stomach, if persevered in." Most of Dr. Beaumont's findings and conclusions remain unchallenged even today. Although one might not consider a young army surgeon stationed in the wilderness in 1822 to be a likely candidate for greatness in research, the story of his contributions to science illustrates the importance of curiosity, imagination, careful observation, and perhaps even serendipity in advancing scientific knowledge. Today Dr. Beaumont is generally recognized as the father of American physiology. Alexis St. Martin, the subject of his research, lived to age 83.

to energy exchange and metabolism, the subject of Chapter 14, and to detoxification and waste removal.

Anatomical Relationships

The liver, the largest of the abdominal organs, is actually made up of two separate lobes that lie in the upper part of the abdominal cavity just beneath the diaphragm (Figure 13.12). The liver synthesizes **bile,** a watery fluid containing bicarbonate, and **bile salts,** which are derivatives of cholesterol that facilitate fat digestion and absorption. After synthesis in the liver, bile is delivered via the hepatic ducts (*hepatic* is the adjective referring to the liver) to the gallbladder, where it is stored. During digestion the gallbladder contracts, expelling bile into the small intestine via the common bile duct.

The blood supply to the liver comes from two sources. The first is arterial blood via the hepatic artery, which arises ultimately from the aorta. The second is actually the venous blood draining most of the GI tract as well as the spleen. Venous blood from

these organs enters the liver via the hepatic portal vein, the word *portal* indicating that it is a gateway, or portal, into the liver from the other gastrointestinal organs. The arterial and hepatic portal venous blood mix in the small blood vessels of the liver. Thus, the cells of the liver are exposed to blood that has a substantial quantity of oxygen (arterial blood) and to blood that contains all of the absorbed nutrients and all of the secreted hormones from the entire GI tract (portal venous blood). After passing through the liver, the mixed blood returns to the vena cava by way of the hepatic vein. These relationships are shown in Figure 13.13.

Functions of the Liver

Secretion of Bile The liver produces and secretes approximately 1 liter of bile per day. The bile salts in bile are important for digestion and absorption of fats. The bicarbonate in bile contributes to the neutralization of chyme in the small intestine,

Figure 13.13 Diagram of the blood circulation to and from the liver in relation to other organs. Note that all of the blood leaving the portions of the GI tract involved in the absorption of nutrients must pass through the liver before gaining access to the general circulation.

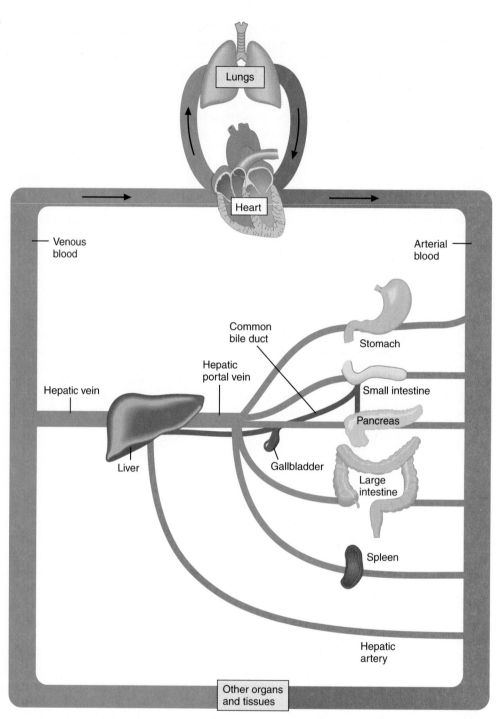

although it is less important than the bicarbonate in pancreatic juice. When food is not present in the small intestine, newly secreted bile is prevented from entering the intestine by a sphincter at the junction of the common bile duct and the small intestine, called the **sphincter of Oddi.** Thus, newly produced bile is diverted into the gallbladder, where it is concentrated and stored.

Metabolic Processing and Storage of Nutri-ents The liver plays a major role in processing and storing nutrients absorbed by the intestines. Indeed, the liver is ideally located to serve as a processing center for absorbed nutrients because the venous blood from the GI tract goes directly to the liver via the portal vein. Two key functions of the liver are that it helps control the blood glucose concentration, and it is a storage depot for glycogen, fats, and some vitamins and minerals. Complete descriptions of the metabolic pathways for newly absorbed nutrients are beyond the scope of this book, and many of these are covered in another subject area, biochemistry.

Detoxification and Waste Removal Another key role of the liver is the destruction and removal of certain body waste products and foreign substances from the body. There are two routes for their removal. Some are secreted into the bile and then excreted from the body with the feces; others are simply destroyed as they pass through the liver on their way to the general circulation.

Bilirubin and cholesterol are examples of substances that are secreted into bile and then excreted in the feces. Bilirubin is a breakdown product of hemoglobin. Most of the bilirubin secreted into the bile is in a form that cannot be absorbed by the intestines; thus, it cannot regain access to the body. In the intestines, bilirubin is further degraded to other compounds, which give the feces their characteristic yellowish-brown color. Cholesterol is also secreted into the bile, but only a small fraction is ultimately excreted; most is reabsorbed by the small intestine.

Other substances destroyed by the liver include alcohol, hormones, many drugs, and innumerable foreign compounds. Some of these may be toxic to the liver; alcohol, in particular, is known to be associated with liver disease. Although the precise mechanism for how alcohol damages the liver is not clear, current thinking is that one or more of the degradation products of alcohol is more toxic to the liver than the alcohol itself.

Synthesis of Plasma Proteins From Chapter 9 you know that the presence of proteins in the blood plasma is important in determining the movement of plasma fluid between capillaries and the interstitial fluid. The liver is responsible for the production of albumin, the primary plasma protein. Liver disease often leads to imbalances of fluid in the body because the diseased liver is unable to produce albumin at the normal rate. Another protein produced by the liver, fibrinogen, is important in blood clotting. In addition, many proteins produced by the liver serve as carriers for the transport of hormones in the blood.

The Gallbladder

The gallbladder stores and concentrates bile. As we shall see when we discuss digestion in the small intestine, bile is released into the intestine only in response to specific stimuli related to the presence of chyme. When chyme is not present, newly secreted bile from the liver is stored in the gallbladder. During storage, the gallbladder reabsorbs water and sodium chloride from the bile, effectively concentrating the bile salts. Even though the gallbladder can store only about 60 ml of bile, it can hold all of the bile salts secreted by the liver in about 12 hours (nearly all the bile salts in 500 ml of secreted bile) simply by removing water and other molecules. As a consequence, bile in storage in the gallbladder may have a concentration of bile salts eight to ten times higher than newly formed bile from the liver.

Under certain conditions, the ability to concentrate the bile can lead to problems. The excreted components of bile, most notably cholesterol and bilirubin, are relatively insoluble in watery solutions. If too much water is removed from the bile, or if cholesterol or bilirubin are secreted in excessive amounts, crystals of cholesterol or calcium bilirubin may precipitate out of the bile. Over time, the slow deposition of successive layers of cholesterol or bilirubin may result in the formation of large, hard objects in the gallbladder known as **gallstones** (Figure 13.14). Gallstones sometimes become so large that they partially or completely obstruct the outflow of bile from the gallbladder. When that happens, the person may suffer intense pain, especially after a meal.

Small Intestine

By the time the food enters the small intestine from the stomach, it has been broken down to smaller particles, mixed with stomach acid and saliva, and partially digested. The small intestine has two major roles. First, it continues the digestion of food. Second, it bears primary responsibility for absorption of nearly all of the digested products. In this section, we examine how the

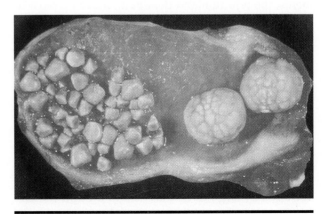

Figure 13.14 A human gallbladder and gallstones.

HIGHLIGHT 13.3

The Length of the Small Intestine

Some textbooks describe the small intestine as about 20 feet long, in contrast with the figure of about 11 feet used in this book. What accounts for such a substantial difference in what should be a simple measurement? The answer has to do with the functional activity of intestinal smooth muscle, and with the fact that intestinal length usually is measured on cadavers rather than on living persons.

Intestinal smooth muscle is composed of two distinct layers, one with the fibers aligned around the circumference of the intestine (the circular layer) and the other with the fibers aligned parallel to the long axis (the longitudinal layer). Contraction of the circular layer reduces the *diameter* of the small intestine, whereas contraction of the longitudinal layer reduces the *length*. In a living person, continuous partial contraction of the longitudinal layer keeps the small intestine at about 11 feet long. After death, the smooth muscle relaxes and the intestine lengthens to about 20 feet.

This example points out the need for a clear understanding of any underlying assumptions the investigators may have made when they designed their experiment. In this case the unspoken assumption implicit in using cadavers to measure intestinal length, and then applying the data to living human beings, is that the length of the human intestine is not altered by death. Note that it is the *assumption* that is wrong, not the measurement. In this book you may assume we are discussing the physiology of living human beings.

---

chyme is mixed and propelled forward (motility), how the small intestine digests the chyme into the smallest possible units, and how those units are absorbed across the mucosal cell layer and into the bloodstream. Finally, we address how the digestive and absorptive functions of the small intestine are regulated. In so doing, we discuss how the rates of delivery of pancreatic juice and bile are regulated.

Structure

It is customary to consider the small intestine as three segments that are defined by their location and length. The first section, the duodenum, is only about 10 inches long; it lies just beyond the pyloric sphincter. The common bile duct and accessory pancreatic duct empty into the duodenum. The remaining two segments, the **jejunum** and the **ileum,** make up most of the intestinal length. Together they may be more than 10 feet long, requiring a highly coiled arrangement to fit into the abdominal cavity. Ultimately, the ileum empties its contents into the large intestine. The three segments of the small intestine have slightly different functions as well, but we will consider them as one continuous unit.

The small intestine consists of the same four basic layers of tissue that were described for the stomach, with modifications that reflect the differences in function. In particular, the mucosa of the small intestine is arranged in large folds that extend around the circumference of the small intestine and are large enough to be seen with the naked eye (Figure 13.15(A)). The surfaces of the folds are covered with numerous microscopic projections called **villi** (Figure 13.15(B)). A close view of a villus reveals that it is composed of a layer of epithelial cells surrounding a central region containing blood and lymph vessels (Figure 13.15(C)). The luminal cell membranes of the epithelial cells (the portion of the cell membrane exposed to the lumen of the GI tract) are folded into many smaller projections called microvilli, increasing surface area even further. Together, the folds, villi, and microvilli give the small intestine an absorptive surface area that may be over 500 times as large as a tube of the same size with a smooth surface. Unlike the stomach, where the anatomical arrangement of the surface of the mucosa is designed to protect the mucosal cells from the harsh environment in the lumen, the mucosa of the small intestine is designed for maximum contact between the chyme and the epithelial cells. This greatly increases the efficiency of the absorptive process.

Note that there are only occasional mucous cells in the villi of the small intestine. Although mucus still serves to lubricate and protect the mucosal surface, less mucus is required than in the stomach because most of the stomach acid has been neutralized by pancreatic juice and bile. Indeed, too much mucus might impede the absorptive process.

Motility

For digestion to proceed in the small intestine, the chyme must be thoroughly mixed with pancreatic juice and bile and moved forward toward the large intestine. Two types of muscular contractions in the small intestine achieve these requirements: segmentation and peristalsis. **Segmentation** consists of seemingly random muscular contractions of short regions of the circular muscle. It is best visualized by imagining a long, thin, water-filled balloon that several people are squeezing randomly at the same time. Segmentation mixes the chyme, moving it in both directions from the point of contraction (Figure 13.16).

The rate of segmental contractions is about 12 contractions per minute in the duodenum, slightly less in the jejunum, and about 8 contractions per minute in the ileum. This difference in rates causes the chyme to be moved slowly forward over time because the chance that the chyme will be moved forward by the more frequent contractions in the duodenum is slightly greater than the chance that chyme will be moved backward by the less frequent contractions in the ileum. Segmentation is the primary cause of the overall forward movement of chyme in the small intestine.

Segmental contractions of the small intestine are weak or nearly absent when the stomach and intestines are empty, but they increase in force shortly after eating, even before the chyme reaches the small intestine. Increased contractile activity of the ileum shortly after eating is called the **gastroileal reflex.** This reflex is thought to be mediated by the hormone gastrin and by parasympathetic nerve activation induced by stretch of the stomach. Once chyme arrives in the small intestine, distention of the intestines induces a local reflex that also increases contractile activity.

In contrast to segmentation, peristaltic contractions in the small intestine are relatively weak during the active phase of digestion, contributing very little to the forward movement of chyme. However, peristaltic contractions become more prominent toward the end of the digestive period when there is very little chyme remaining. Thus, weak peristaltic contractions begin at just about the same time that segmentation ends. Each peristaltic wave travels only a short distance before dying out, to be replaced by another wave that begins slightly farther down the intestine. The predominant effect of peristalsis is to sweep the intestine clean of all remaining chyme, pushing the last of it

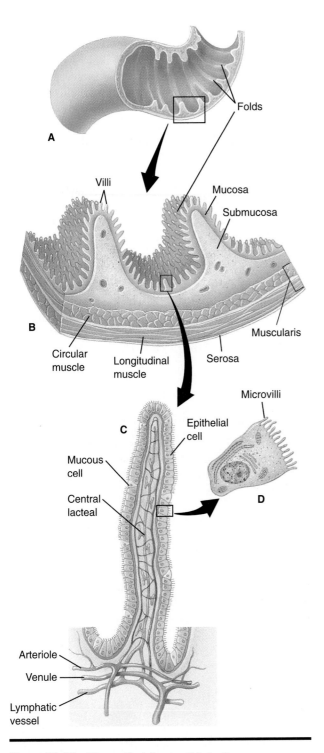

Figure 13.15 The wall of the small intestine. (A) Longitudinal section showing the folding of the mucosa. (B) Closer view of the folds of the submucosa showing the villi. (C) Closeup view of a villus. (D) A single epithelial cell on the surface of the mucosa.

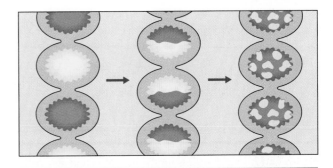

Figure 13.16 Segmentation in the small intestine.

toward the large intestine. Intestinal peristalsis is analogous to the sweeping motion a person uses to sweep a hallway.

Digestion and Absorption

The primary job of the small intestine is to absorb the bulk of the chyme. To accomplish this task, the particles in chyme must be broken down (digested) to components small enough to cross the cell membranes of the mucosal layer and move into the blood (absorption). Digestion, in turn, requires the exocrine secretions of the pancreas and liver, as well as enzymes associated with the mucosal cells themselves. Thus, we will consider secretion, digestion, and absorption together for each of several different constituents of chyme.

Proteins Recall that protein digestion begins in the stomach with the action of pepsin. In the small intestine, proteolytic enzymes from the pancreas continue the breakdown of proteins that have not been completely digested in the stomach (Figure 13.17). Pan-

creatic enzymes work best in a more neutral chemical environment than that of the very acidic chyme delivered from the stomach or of the pancreatic juice itself. Thus, it is important that the pancreatic juice contain bicarbonate in order to neutralize the stomach acid so that these enzymes can work. The pancreatic enzymes, of which there are many, break down the proteins to smaller and smaller peptides and some amino acids.

Not all of the enzymes responsible for protein breakdown come from the pancreas. In fact, the luminal cell membrane of intestinal mucosal cells is loaded with enzymes that are part of the structure of the cell membrane itself. This is an example of the general phenomenon described in Chapter 2; cell membranes contain proteins that have various functions. In this case the membrane proteins function as enzymes. These enzymes are not able to mix with the chyme as the free-floating enzymes from the pancreas do, but they are so numerous that they can very efficiently degrade the last of the peptide chains that are still too big to be absorbed.

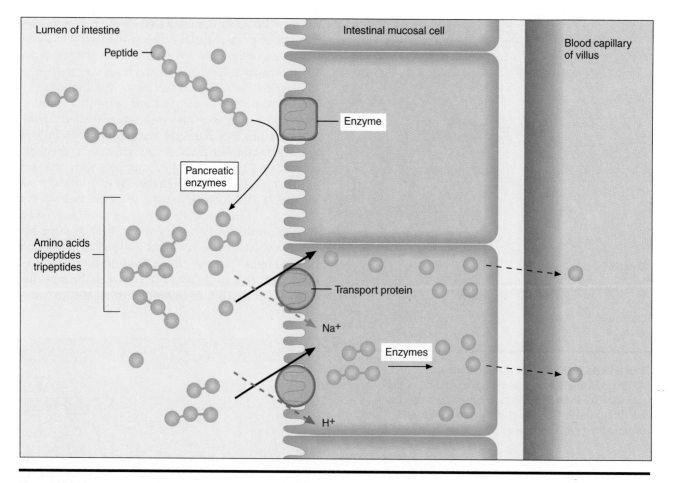

Figure 13.17 Mechanisms for the digestion and absorption of protein in the small intestine.

Once the proteins are reduced to dipeptides, tripeptides, or single amino acids, they are small enough to be transported into the mucosal cells. Absorption into the cell from the lumen requires a transport protein that is part of the cell membrane. Absorption is generally by a secondary active transport mechanism. The energy for the secondary active transport of amino acids, dipeptides, and tripeptides is provided by the facilitated diffusion of certain ions, most notably sodium and hydrogen ions. Once inside the cell, the dipeptides and tripeptides are further digested to amino acids by enzymes located inside the cell. Finally, the amino acids inside the cell diffuse out of the cell to the blood capillaries, where they are carried away to the liver by the hepatic portal circulation.

Carbohydrates The carbohydrates we eat are a mixture of complex carbohydrates, trisaccharides, and disaccharides. The digestion of the complex carbohydrates technically begins in the mouth with the secretion of salivary amylase. However, the food spends so little time in the mouth that digestion in the mouth itself is inconsequential. Carbohydrate digestion continues in the stomach at a slow pace, not because of anything that the stomach does, but as a result of the continued presence of salivary amylase that was swallowed along with the food. In the small intestine, digestion of carbohydrates is accelerated by the addition of amylases in the pancreatic juice. Pancreatic and salivary amylase are similar; both break down complex carbohydrates such as starch and glycogen to disaccharides and trisaccharides.

Disaccharides and trisaccharides are still too large to be absorbed, however. The job of breaking these small sugars into their constituent monosaccharides again falls to numerous specific enzymes that are part of the luminal cell membrane of small intestinal mucosal cells. Like the transport of amino acids, the transport of monosaccharides into the cell is generally accomplished by secondary active transport, followed by diffusion of the monosaccharides out of the cell into the blood (Figure 13.18).

Fats Digestion and absorption of fats is more difficult than digestion and absorption of proteins or carbohydrates because fats and the products of fat

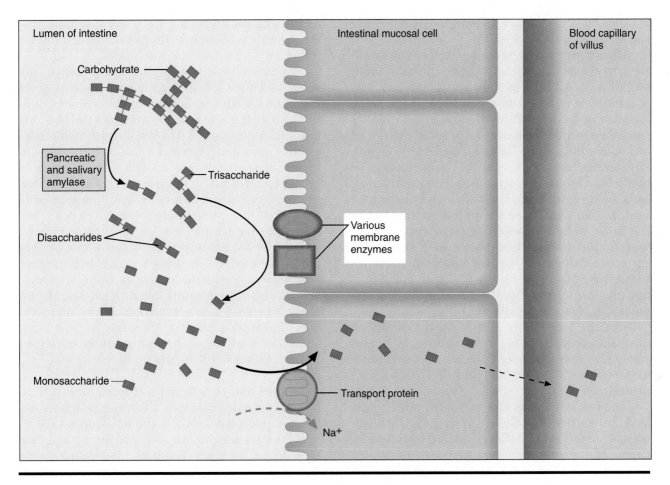

Figure 13.18 Mechanisms for the digestion and absorption of carbohydrates in the small intestine.

HIGHLIGHT 13.4

Lactose Intolerance

The carbohydrates in our diet generally consist of complex plant starches, animal glycogens, and a mixture of disaccharides and monosaccharides. All carbohydrates must be digested to monosaccharides before they can be absorbed. Lactose, the predominant disaccharide in milk, is composed of one glucose and one galactose. The bond between the monosaccharides can only be broken by the enzyme **lactase,** one of the many enzymes found in the luminal cell membrane of the small intestinal mucosal cells.

Although nearly all human infants have the lactase enzyme, approximately 50% of all humans lose the enzyme as they mature; these people are then deficient in the lactase enzyme as adults, and thus they are **lactose-intolerant.** The symptoms of lactose intolerance include intestinal distention, gas, and diarrhea upon ingestion of milk. The symptoms occur because undigested lactose reaches the large intestine, where it serves as an energy source for a flourishing colony of bacteria that produce gas and cause diarrhea. Most southern European populations, blacks of African descent, and American Indians are relatively lactose-intolerant. Populations that tolerate lactose relatively well include most northern and central Europeans, north Indians, and Arabians.

There are a number of ways milk products can be rendered useful for lactose-intolerant people. Cheese and yogurt, for example, no longer contain lactose because the lactose has already been digested by lactase-producing bacteria that were purposefully added as part of the production process. In many parts of the world, it is possible to buy milk to which lactase has been added directly.

digestion are not very soluble in water. Although peristalsis in the stomach and segmentation in the small intestine break the fats into droplets, these droplets are still far too large to be absorbed.

Digestion of fats begins in the small intestine with the arrival of bile salts from the liver and gallbladder (Figure 13.19(A)). Bile salts themselves are not very water soluble, so they tend to be delivered to the small intestine in the form of **micelles,** or small droplets composed of bile salts along with some cholesterol and other constituents. The bile salts in micelles **emulsify** large fat droplets, that is, they cause them to be broken into much smaller droplets. Emulsification allows the very small fat droplets to remain suspended in water, and it also increases the total surface area of the fat droplets. Emulsification is the same process whereby dishwashing liquid disperses grease on the surface of dishwater. Note that emulsification is different from enzymatic digestion; during emulsification the size of the droplets gets smaller, but the molecules of fat are not altered.

Once the large fat droplets have been emulsified to smaller ones, the fats can be digested by lipases present in pancreatic juice. Pancreatic lipases reduce the fats primarily to free fatty acids and monoglycerides (Figure 13.19(B)).

The free fatty acids and monoglycerides generated by the action of lipases are small enough to be absorbed, but they still are not very soluble in the watery chyme. However, they are quite soluble in the internal environment of the micelles, so that as soon as they are formed, they have a natural tendency to enter the micelle. The micelles carry the monoglycerides and free fatty acids to the surface of the mucosal cells of the small intestine, where absorption occurs (Figure 13.20). The micelles thus serve two functions: the emulsification of large fat droplets and the transport of free fatty acids and monoglycerides.

At the cell surface, monoglycerides and free fatty acids can enter the cell by diffusion because they are soluble in the lipid environment of the cell membrane. Once inside the cell, the monoglycerides and free fatty acids are resynthesized into triglycerides in the endoplasmic reticulum. Then they are coated with proteins in the Golgi apparatus to form a protein-coated packet of triglycerides called a **chylomicron.** Chylomicrons are packaged into secretory vesicles, transported to the cell surface, and released outside the cell. They then enter the lymph vessels in the villus, through which they are returned to the blood. Chylomicrons are transported via the lymph vessels because they are too large to enter the blood capillaries directly.

The luminal contents of the small intestine also include cholesterol derived from both dietary intake and bile. Cholesterol is relatively insoluble in water, so not all of it is available for absorption. Thus, some cholesterol is ultimately excreted in the feces.

Water and Ions Water and small ions (such as sodium, potassium, chloride, hydrogen, calcium, magnesium, phosphate, sulfate, and bicarbonate) are already small enough to be absorbed by the intestine; they do not undergo digestion. Most of these substances are absorbed rather quickly across the small intestine into the bloodstream. A good example is water. No matter how much water you drink, the intestines

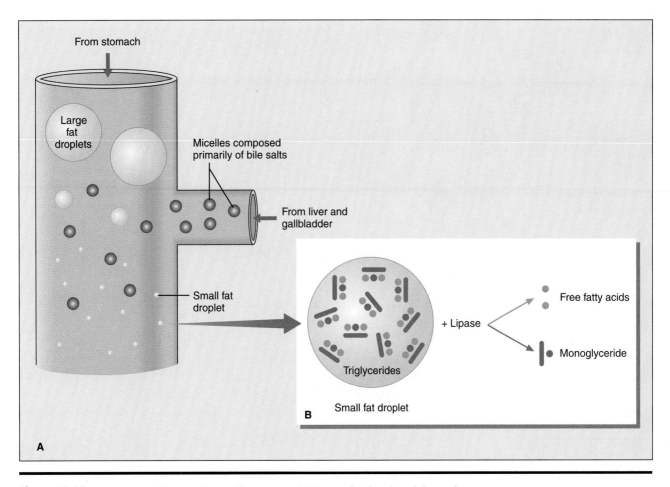

Figure 13.19   Digestion of fats in the small intestine. (A) Large fat droplets delivered to the duodenum from the stomach are emulsified to smaller droplets by micelles composed of bile salts from the liver and gallbladder. (B) The triglycerides that make up the small fat droplets are digested by lipases to free fatty acids and monoglycerides.

are normally capable of absorbing nearly all of it; otherwise, you would get diarrhea every time you drank. This observation illustrates a point made earlier, that the gastrointestinal system generally absorbs all that it can, leaving the regulation of the body's stores of absorbed substances to other systems (in this case the kidneys). Water simply diffuses through channels in the membrane, crossing the membrane at will. Here as in all cells, movement of water is governed strictly by osmotic forces, and thus it tends to follow the absorption of nutrients and ions. Small ions generally are absorbed either by facilitated diffusion via membrane protein carriers, or by active transport using membrane transport pumps. Recall that the energy generated by the inward facilitated diffusion of sodium into the mucosal cell is used to transport many amino acids and monosaccharides as well.

Bile Salts and Vitamins   Most of the bile salts that are secreted into the small intestine from the gall-

bladder are recycled and reused. Reabsorption of bile salts occurs primarily in the last portion of the small intestine (the ileum) and the first part of the large intestine. Once absorbed, bile salts are returned to the liver via the hepatic portal vein, where they are removed from the blood and again secreted into the bile. This process occurs so rapidly that bile salts are actually used twice in the course of digesting a single meal.

Vitamins are also absorbed in the small intestine. Water-soluble vitamins are absorbed by diffusion or active transport. Fat-soluble vitamins are carried in micelles to the surface of the mucosal cells, just like the products of fat digestion, where they are absorbed by diffusion. A special case is vitamin B_{12}, which is unique in that it cannot be absorbed unless it is attached to intrinsic factor. Recall that intrinsic factor is secreted into the lumen of the stomach by the parietal cells of the stomach's gastric glands. Once they are bound together, vitamin B_{12} and its associated intrinsic factor are absorbed in the terminal ileum. Without sufficient

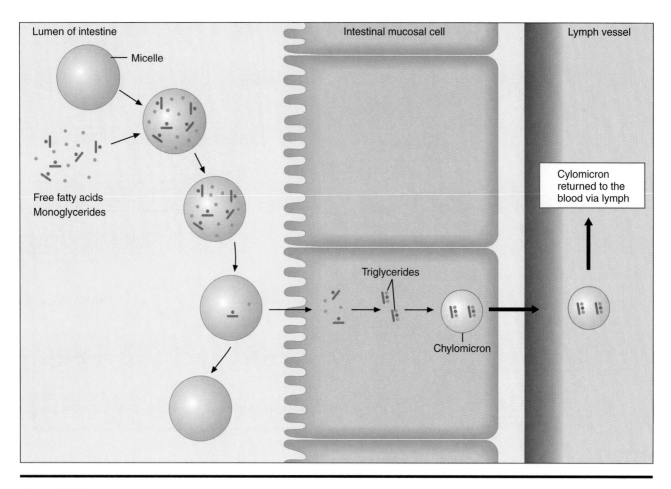

Figure 13.20 Absorption of fats in the small intestine.

intrinsic factor a person can become vitamin B_{12}-deficient. This may lead to anemia (a reduced number of red blood cells in the blood) because vitamin B_{12} is essential for the normal formation of red blood cells.

Regulation of Digestion and Absorption

Part of the overall regulation of the digestive process resides with the regulation of motility, for motility determines how quickly the stomach empties and how quickly chyme is moved down the small intestine. But what about the regulation of digestion and absorption? Of these two processes, only the rate of digestion is regulated, for under normal circumstances all of the digestive products that are small enough to be absorbed are absorbed quickly. The rate of digestion is controlled primarily by the timing and rate of delivery of pancreatic juice and bile to the small intestine. We now consider these processes.

Regulation of the Secretion of Pancreatic Juice The secretion of the water and bicarbonate components of pancreatic juice is controlled largely by the hormone secretin, which is secreted into the blood by the small intestine (Figure 13.21). The stimulus for secretion of secretin, in turn, is acid in the duodenum. Thus, the presence of acidic chyme in the small intestine stimulates the secretion of secretin into the blood, which in turn causes the pancreas to secrete more water and bicarbonate ions that effectively neutralize that acid. A similar control system exists for the enzyme component of pancreatic juice as well. The primary stimulus for the secretion of pancreatic enzymes is another hormone from the duodenum, cholecystokinin (CCK), which is secreted into the bloodstream in response to the presence of fat and protein digestive products in the duodenum.

The control of pancreatic secretion that we have just described is perhaps an oversimplification. First, the rate of pancreatic secretion (like that of gastric secretion) can also be controlled by neural reflexes involving the brain. Thus, secretion of pancreatic juice can begin in anticipation of the actual arrival of acidic chyme. Second, although secretin is the primary stimulus for pancreatic bicarbonate secretion, the simultaneous presence of CCK enhances the magnitude of

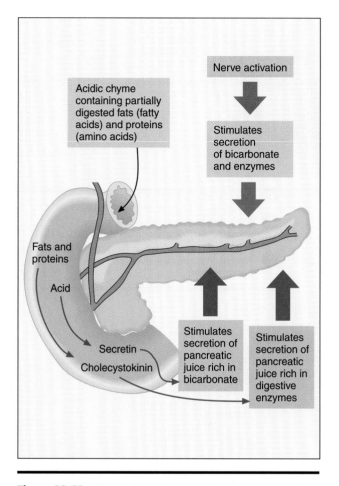

Figure 13.21 Regulation of pancreatic exocrine secretion.

the bicarbonate response to secretin. In other words, although CCK is not a direct stimulus of pancreatic bicarbonate secretion, secretin and CCK work synergistically to produce a greater rate of bicarbonate secretion than secretin alone does.

Regulation of the Secretion and Delivery of Bile Recall that the liver secretes bile at a continuous low rate even during fasting, but that the newly secreted bile is not delivered to the small intestine because the sphincter of Oddi remains closed between meals. Instead, newly secreted bile is diverted to the gallbladder, where it is stored and concentrated.

When chyme enters the small intestine, however, distention of the small intestine activates neural reflexes that stimulate contraction of the gallbladder. At the same time, the presence of fat in the chyme causes CCK secretion by the small intestine. CCK also causes gallbladder contraction, but in addition it relaxes the sphincter of Oddi. Together, contraction of the gallbladder and relaxation of the sphincter of Oddi result in an increased delivery of bile into the small intestine (Figure 13.22). Note that this is yet another example of

a negative feedback control system; the presence of fatty chyme in the small intestine causes bile to be released, which in turn assists in the digestion and absorption of fats.

Both secretin and neural reflexes stimulate the overall rate of hepatic bile secretion as well, but this has little effect on the rate of delivery of bile to the small intestine. Secretin and neural reflexes increase primarily the water and bicarbonate components of bile, not the bile salts.

Large Intestine

The large intestine completes the job of digestion and absorption and stores the remaining undigestible material (the feces) until the feces are passed from the body. The U-shaped large intestine (Figure 13.23) is shorter and less tortuous than the small intestine. Unabsorbed chyme from the small intestine enters the large intestine through the **ileocecal valve,** which remains closed except when chyme is present in the final portion of the ileum. The ileocecal valve prevents flow of chyme back into the small intestine. The **appendix** is a small outpouching of the **cecum,** which is the blind pouch at the proximal end of the large intestine. The appendix has no digestive function, but it does contain lymphatic nodules that tend to become inflamed, a condition known as **appendicitis.** If the inflammation is severe, the appendix may have to be removed surgically, as rupture of the appendix spills fecal matter containing bacteria into the abdominal cavity.

Most of the length of the large intestine is called the colon. The most striking anatomical feature of the colon is its arrangement into sac-like pouches called **haustra** (singular *haustrum*) separated by areas of constriction. Nevertheless, the mucosal cell layer of the colon is relatively smooth; it is not arranged into folds and lacks villi. Thus, the colon has much less surface area than the small intestine, consistent with its reduced role in absorption. The final segment of the large intestine is called the rectum.

Digestion and Absorption

Compared to the small intestine, the large intestine plays only a minor role in digestion and absorption. Most of the ions and water and essentially all of the easily digestible constituents of food were absorbed by the small intestine. Nevertheless, the large intestine does absorb most of the remaining water and ions, reducing the more liquid chyme to the semisolid feces.

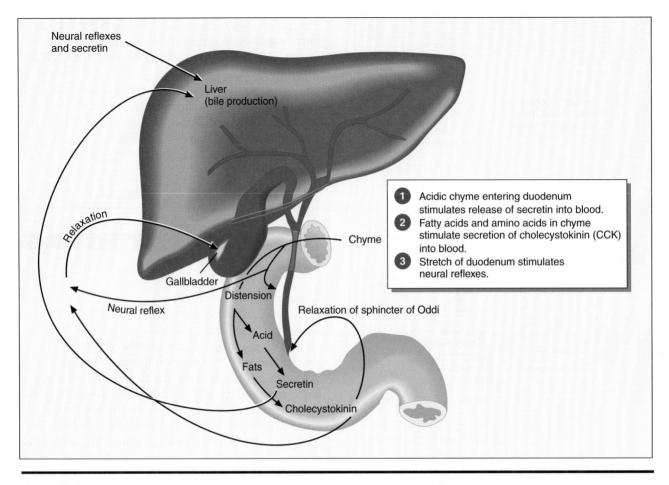

Figure 13.22 Regulation of the production and storage of bile and secretion of bile into the small intestine.

An important feature of the large intestine is its content of over 400 species of bacteria, far more than in the stomach and small intestine combined. Indeed, nearly a third of the dry weight of feces is bacteria.

Most of these bacteria are beneficial because they digest certain substances that are not digestible by the enzymes of the pancreas, thereby allowing them to be absorbed by diffusion. Examples of substances di-

Highlight 13.5

Dietary Fiber

Some carbohydrates cannot be digested and absorbed by the human intestine and thus they remain in the feces as indigestible dietary fiber. A good example is cellulose, a complex carbohydrate of plants that provides strength to plant cell walls. Humans cannot digest cellulose because they lack cellulase, an enzyme that breaks down cellulose to simpler carbohydrates. Fiber is an important element of our diets because it is the factor that most affects the transit time of fecal material through the large intestine; high-fiber diets increase the

fecal volume and decrease the transit time, whereas low-fiber diets do the opposite. Studies in animals and in humans suggest that a diet high in fiber is associated with a reduced risk of cancers of the colon and rectum. The mechanism of this beneficial effect is not well understood, but may be related to reduced time for cancer-causing agents to be absorbed in the colon and rectum. Good sources of fiber are whole-grain cereals (especially the bran of grains), peanut butter, and some vegetables.

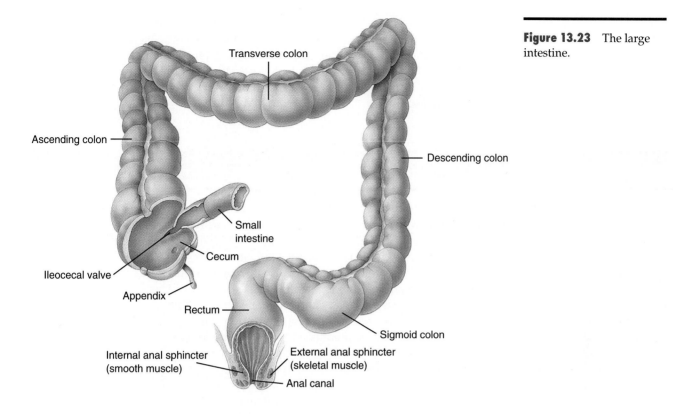

Figure 13.23 The large intestine.

gested by bacteria include some dietary fiber and certain fats. Bacterial digestion is also responsible for the production of gas in the large intestine.

In addition, certain bacteria can synthesize the K vitamins, which are then absorbed by the large intestine. Under normal circumstances these bacteria can synthesize sufficient K vitamins to meet our minimal daily requirements. This is not the only source of K vitamins, however; they are also ingested as part of our diet and absorbed in the small intestine along with the other fat-soluble vitamins. The K vitamins are important in the production of certain blood-clotting factors.

Motility and Defecation

The motility of the colon is considerably less than that of the stomach or the small intestine, consistent with its function as a temporary storage site for the undigestible material in food. Nevertheless, the colon does undergo segmental contractions. These contractions are so slow and so prolonged (each contraction is maintained for minutes rather than seconds) that at first glance, they seem to be an anatomical feature of the large intestine, rather than a functional one. In fact, the apparent arrangement of the large intestine into haustra is due to slow segmental contractions that occur at random positions along the colon. Segmental contractions in the colon accomplish slow mixing and movement known as haustral shuttling, similar to seg-

mentation in the small intestine. In addition to haustral shuttling, the colon occasionally exhibits more sustained contractions called mass movements, in which an entire section of the colon contracts at once, driving the colonic contents toward the rectum.

Motility of the colon increases soon after a meal enters the stomach, a phenomenon known as the **gastrocolic reflex.** The gastrocolic reflex is initiated by stretch of the stomach, and although the exact mechanism is not clear, it is thought to be mediated by the parasympathetic nerves and by the secretion of gastrin. Whatever the actual mechanism, the effect is to increase the movement of the colonic contents toward the rectum. The gastrocolic reflex accounts for the observation that the most common time to feel the urge to defecate is about one-half to one hour after the first meal of the day.

If sufficient fecal mass is present, either mass movements of the colon or the gastrocolic reflex may lead to distention of the rectum. Rectal distention, in turn, initiates the **defecation reflex.** Defecation is controlled by two sphincters: the **internal anal sphincter,** composed of smooth muscle under involuntary control, and the **external anal sphincter,** composed of skeletal muscle under voluntary control. During the defecation reflex the rectum contracts, the internal anal sphincter relaxes, and one becomes aware of the urge to defecate. If the time is appropriate, the individual allows the skeletal muscle of the external anal sphincter to relax, and defecation occurs. Defecation can be assisted by voluntary contraction of the abdominal and respiratory muscles with the glottis closed; this greatly increases intra-abdominal pressure and helps propel the feces forward. If the time is not appropriate for defecation, the external anal sphincter can be voluntarily contracted and defecation can be delayed. Eventually the gastrocolic and defecation reflexes will decline in intensity, and the urge to defecate may subside entirely, either until the next mass movement of the colon or until the next gastrocolic reflex.

Summary

The gastrointestinal system is designed to digest and absorb food. It is composed of a continuous tube from mouth to anus, and several accessory organs and structures including the salivary glands, teeth, and tongue in the mouth and the liver, gallbladder, and pancreas in the abdominal cavity. In general, the tube is composed of four distinct layers of tissue: the mucosa (where secretion and absorption occur), the submucosa, the muscularis (responsible for motility), and the serosa.

In the mouth the food is broken up into smaller parts and moistened with saliva, which contains enzymes that begin the process of carbohydrate digestion. The food then moves to the pharynx to be swallowed. The esophagus serves as a muscular conduit, transferring the food to the stomach. The stomach stores the food, mixes a portion of it with gastric juice containing acid and proteolytic enzymes, and passes the watery, partially digested food (now called chyme) to the small intestine at a steady rate over several hours. Very little absorption takes place in the stomach. Most of the digestion and nearly all of the absorption occurs in the small intestine, where secretions that facilitate digestion are added from the liver and pancreas. The liver secretes bile, which is stored between meals in the gallbladder and concentrated by the removal of water. Bile contains bile salts for emulsifying fats and for forming micelles that keep the fat digestive products in solution. The pancreas contributes enzymes for the digestion of fats, proteins, and carbohydrates, and both the pancreas and liver contribute bicarbonate ions that neutralize the acid from the stomach. Once the food has been digested to its smallest parts, the components are absorbed across the mucosal cells and into the blood. Digestion and absorption are completed in the large intestine; here, digestion is facilitated by bacteria. The primary role of the large intestine, however, is storage of the indigestible fraction of the chyme (now called feces) until it can be eliminated.

Digestive processes are regulated by negative feedback control systems involving nerves and exocrine and endocrine secretions. Muscular contraction of the stomach and the secretion of gastric juice, for example, are initiated by nerves and by the hormone gastrin, both of which are stimulated by the volume and composition of the chyme and by signals from the brain in response to the sight, taste, smell, or thought of food. In the small intestine, the presence of chyme stimulates neural reflexes and the endocrine secretion of hormones (secretin, CCK) that cause the release of bile and pancreatic juice and increase intestinal motility. These same nerves and hormones, along with another hormone, gastric inhibitory peptide, also inhibit gastric motility and secretion, slowing delivery of food to the small intestine until that which is already present in the small intestine can be digested and absorbed. Motility of the large intestine and even the urge to defecate are regulated by the presence of the products that remain. Thus, the processes that control digestion and absorption in each region of the digestive system are adjusted according to need.

Questions

Conceptual and Factual Review

1. What is saliva and what are its functions? (p. 335, p. 336)
2. What are the three primary functions of the stomach? (p. 339)
3. What substances (other than a watery fluid) are secreted into the stomach? (p. 340)
4. What is gastrin? What are its actions on the stomach? (p. 340, p. 341, p. 343)
5. What effect does the thought of food have on gastric motility? (p. 341)
6. What are the primary constituents of pancreatic juice? (p. 345)
7. What three hormones are secreted by the small intestine? What are their functions? (pp. 342, 343, 356, 357)
8. From what sources does the liver receive blood? (p. 347)
9. Where is bile produced? (p. 347)
10. What are the functions of the gallbladder? (p. 349)
11. What are the two main functions of the small intestine? (p. 349)
12. Of what functional importance are the folds, villi, and microvilli of the small intestine? (p. 350)
13. What is segmentation? What is peristalsis? (p. 350, p. 351)
14. How small must a chain of amino acids be before it can be absorbed across the luminal membrane of the cells lining the small intestine? (p. 353)
15. What is the fate of the bile salts that are secreted into the duodenum? (p. 355)
16. How and where is vitamin B_{12} absorbed? (p. 355)
17. Why are some bacteria beneficial for digestion? (p. 359)
18. What is haustral shuttling? (p. 359)

Applying What You Know

1. Both pepsinogen and HCl are secreted into the gastric pits in regions that are not well-protected by mucus (Figure 13.10). What explanation might you propose for why the cells of the gastric pits do not seem to be damaged as readily as the mucosal cells at the luminal surface of the stomach, even though they lack the protection of mucus?
2. What changes in digestion and absorption would you expect in a person who has had his or her gallbladder removed? Explain.
3. What happens to the enzymes secreted into the GI tract, such as salivary amylases, pepsin, and pancreatic lipases? Would you expect to find them in the feces? Why or why not?
4. What might be the advantages and disadvantages of attaching digestive enzymes to the luminal surface of an intestinal cell, as opposed to secreting them into the lumen?
5. Why is it not a good idea to take aspirin and alcohol together on an empty stomach?
6. In a person with gallstones, why are painful "gallbladder attacks" most likely to occur after eating a meal?
7. Why would you suppose that heartburn is more common in pregnant women than in nonpregnant women?

Chapter 14

Energy Exchange and Temperature Regulation

Objectives

By the time you complete this chapter you should be able to

◆ Describe how the human body performs obvious work such as muscular movement and less obvious work such as the synthesis of cellular molecules, tissue growth and repair, transport of molecules across membranes, and heat production.

◆ Explain how the energy needed for all of these functions comes from the food you eat.

◆ Understand that to supply the necessary energy, food must first be broken down into specific fragments and then absorbed into the blood from the gastrointestinal tract before it can be used directly as an energy source or stored for future use.

◆ Know the body's many mechanisms for the storage and use of energy sources.

◆ Understand that the body uses several different energy sources in different ways.

◆ Understand how the body's control mechanisms allow energy use to be changed quickly to meet rapidly changing demands.

◆ Know that many non–energy-supplying molecules (minerals and vitamins) are needed to keep the metabolic machinery of the body functioning properly.

◆ Learn how and why body temperature is controlled.

Introduction

The most basic necessity of all life is energy. Without it, all vital functions cease and organisms become merely objects. Because energy is at life's core, numerous biological mechanisms have evolved to obtain and store energy. Even more have evolved to regulate energy use with exquisite precision. In this chapter, you will learn how we obtain, store, and use this precious commodity.

But what exactly is energy? In our daily use of the word, it is synonymous with vitality, strength, and stamina. As we get older, we sometimes wish for more energy, or parents may wish for their child to have less energy at bedtime. However, to appreciate the physiology of energy, a more precise definition is needed. Energy is the ability to perform work or to generate heat.

Work is our primary activity. In a physiological sense, even the laziest person in the world is constantly working. Every single cell in your body continuously works to pump ions across its membrane and synthesize new compounds. The heart continually contracts, moving large volumes of blood throughout the body. Respiratory muscles never stop working, nor do the muscles of the gastrointestinal system. Even while sitting quietly, skeletal muscles work to keep us upright in our seats. If we lie motionless in bed for an entire day, we use more energy in that day than a trained runner does during a 15-mile race. In an average lifetime, each of us uses enough energy to send a rocket into space!

Our energy comes from the food we eat. More precisely, it comes from the energy contained in the chemical bonds of that food. Considerable energy is required to hold the atoms of molecules together. When molecular bonds are broken, this energy is released. Our bodies take advantage of that. We have very intricate processes that break the chemical bonds of food molecules, capture and store the energy, and then carefully direct that energy into processes requiring work. The released energy not only allows us to perform work, but also generates heat, which keeps our bodies warm and enables us to maintain a constant temperature.

Energy is usually measured in units called **kilocalories** (kcal), or simply calories, the term used on food packaging. One calorie or kilocalorie is defined as the amount of heat needed to raise the temperature of 1 liter of water 1°C. You can get some idea of the energy required to do that if you visualize a pot containing a quart of water at room temperature. The energy needed to boil it is approximately 80 calories.

The number of calories in different foods varies enormously. For example, a typical hamburger contains about 540 calories whereas a salad with vinegar dressing contains as few as 80 to 100 calories. However, these figures do not describe the total amount of energy contained in all of the chemical bonds; rather, they refer only to the energy the body can use.

Cellular Metabolism

Metabolism refers to both anabolism and catabolism. This section discusses the metabolic processes in cells that are most important for energy exchange.

Cellular Fuel Molecules

Cells are "picky eaters." Most use only a few types of molecules for fuel. The most important fuels are glucose, fatty acids, and ketones. As discussed in Chapter 2, glucose is a small, simple sugar molecule. Fatty acids are small, semisoluble fat molecules. The ketones comprise three small, soluble molecules. Glucose and fatty acids that supply our cells can come directly from the foods we eat. They can also be manufactured readily from smaller molecules within the body. Ketones, on the other hand, are only synthesized in the body by the liver. As you will see, the types of fuel molecules that cells take up vary from tissue to tissue. In addition, both the quantity and type of fuels used depend on specific physiological conditions.

Energy Transfer in Cells

Once fuel molecules enter cells, they are broken down and the energy released from the breakage of molecular bonds is captured and used for energy-requiring processes such as active ion transport and muscular contraction. However, energy from the breakdown of fuel cannot be transferred directly to energy-requiring processes. Instead, energy is transferred by way of a special molecule, called adenosine triphosphate (ATP), which stores the energy (Figure 14.1). As mentioned in Chapter 2, this molecule is composed of an adenosine core with three phosphate groups attached. Two of the phosphate bonds of this molecule are very special ones because they contain a large amount of energy. Consequently, when a phosphate bond is broken, considerable energy is released. However, not all the energy released can be converted into useful work. Your body, like any machine, is not 100% efficient. Approximately 40% of the energy available may be used for work and the other 60% dissipates as heat.

Figure 14.1 Chemical nature of the high-energy molecule ATP. Note that the molecule is a combination of adenine, ribose, and three phosphate groups. The bonds between the last two phosphate groups are the high-energy bonds, designated by ~. Breaking of these bonds releases energy for use by the cell.

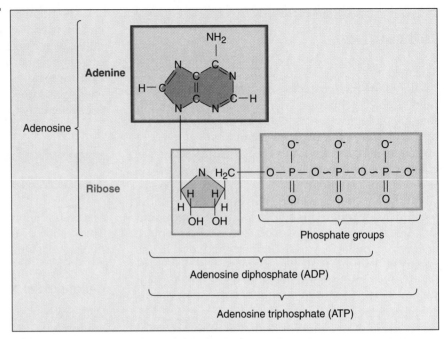

Most energy-requiring processes split a phosphate bond on ATP and use the energy released. The remains of the ATP molecule are adenosine diphosphate (ADP) and a phosphate ion. Other processes put the ATP molecule back together. This reverse process requires energy, which comes from breaking molecular bonds of fuel molecules. Thus, ATP is continuously being broken down into ADP and a phosphate ion by energy-requiring processes; it is then put back together by recovery processes that break down fuel molecules. ATP is thus an energy carrier, transferring energy from cellular fuels to the energy-requiring processes. This cycle of energy transfer is shown in Figure 14.2. The high-energy ATP molecule is at the top of the figure and its catabolic end products, ADP and a phosphate ion, are below. The ATP-generating processes are on the right of the figure and the cellular processes requiring the energy contained in ATP are on the left.

ATP Production

Although ATP is used as the energy source for literally thousands of energy-requiring processes, there are surprisingly few ways that ATP is generated. Nearly all of the body's cells use the same ATP-generating processes. In fact, these processes are used by practically all animal cells, and are even found in primitive organisms such as bacteria. This indicates that the ATP-producing processes evolved very early, probably two to three billion years ago. Their conservation through billions of years of evolution demonstrates

how critical they are for life. Most ATP is generated by two major biochemical processes: anaerobic respiration, which does not require oxygen, and aerobic respiration, which does require oxygen. The use of the word *respiration* here may be somewhat confusing. In Chapter 11, *respiration* refers to gas transport. However, *respiration* also refers to the biochemical processes in cells that produce ATP.

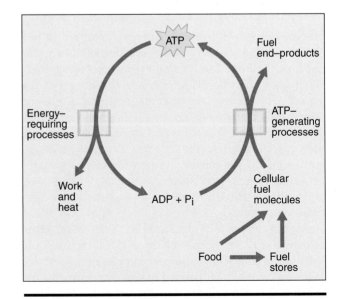

Figure 14.2 The generation and use of ATP. P_i is inorganic phosphate, a phosphate ion that is not attached to another compound.

Anaerobic Respiration ATP can be generated in the absence of oxygen by a series of biochemical reactions called **glycolysis.** Glycolysis is the oldest ATP-generating pathway, evolving even before O_2 was present in the atmosphere. Glycolysis in yeast produces the alcohol in beer and wine and the CO_2 that causes bread dough to rise. Glycolysis is a series of chemical reactions which breaks down a glucose molecule into two smaller molecules of pyruvate in the cytosol (see Figure 14.3). The energy released in the degradation of one molecule of glucose generates two molecules of ATP. No oxygen is needed for glycolysis to take place, so this method of producing ATP becomes extremely important when the O_2 supply to cells is limited or absent.

Aerobic Respiration Aerobic respiration is a much more complex process than anaerobic respiration and, as its name implies, requires O_2. In contrast to glycolysis, in which only glucose is broken down, numerous fuel molecules supply the energy for aerobic respiration. The fuels most often used are derived from fatty acids, ketones, and the pyruvate generated in glycolysis. Although glycolysis can occur when oxygen is absent, it also occurs when oxygen is present. Two major biochemical processes are involved in aerobic respiration: the **citric acid cycle** and **oxidative phosphorylation.**

These pathways are very complex and, for our purposes, need not be described in detail. However, several important aspects of aerobic respiration should be noted. First, fragments of fuel molecules enter the citric acid cycle, where bonds between carbon atoms are broken. The energy released from these bonds transfers electrons to specialized electron carrier molecules. These carriers then transfer the electrons to

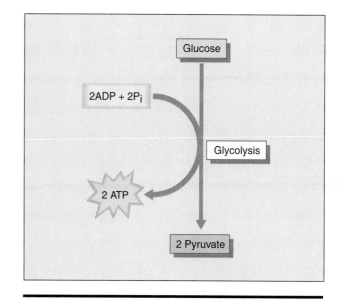

Figure 14.3 Anaerobic respiration (glycolysis).

the **electron transport chain,** a group of membrane-bound compounds in the mitochondria. Electrons flow through the electron transport chain until they combine with oxygen. As electrons flow down the chain, energy is released and used to manufacture ATP from ADP and phosphate. The collection of processes comprising the electron flow down the electron transport chain, the release of energy from that flow, and the subsequent generation of ATP is called oxidative phosphorylation. The major components of aerobic respiration are diagrammed in Figure 14.4.

For each fuel molecule degraded in aerobic respiration, multiple ATP molecules are generated. For example, the breakdown of one molecule of pyruvate results in the generation of approximately 18 ATP

MILESTONE 14.1

Hans Krebs

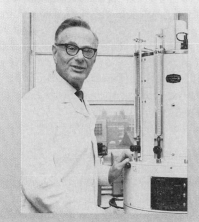

Hans Krebs.

The citric acid cycle derives its name from citric acid, a major substrate in this pathway. This process also goes by two other names: the tricarboxylic acid cycle and the Krebs cycle. Tricarboxylic acid is another name for citric acid, and the name *Krebs cycle* honors Dr. Hans Krebs, the scientist who first described this pathway. Dr. Krebs won the Nobel Prize in physiology and medicine in 1953.

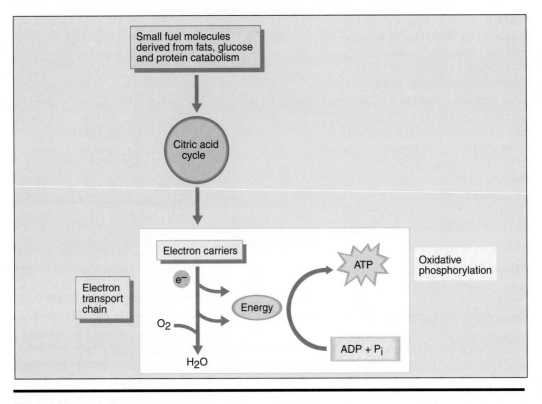

Figure 14.4 Aerobic respiration.

molecules. Thus, compared with glycolysis, aerobic respiration produces considerably more energy in the form of ATP and is the dominant ATP-generating process in most cells. The energy contained in the ATP generated is then used in the countless energy-requiring processes that occur in cells.

A critical aspect of aerobic respiration is that oxygen is required; without it, electrons will not move down the electron transport chain and release the necessary energy to produce ATP. Consequently, any disruption of the oxygen supply to cells immediately turns off this process. If this occurs, most cells compensate by greatly increasing their rate of glycolysis to meet their ATP requirements. However, high rates of glycolysis cause an increased production of pyruvate, which, when oxygen is scarce, is rapidly converted to lactic acid. High concentrations of lactic acid are toxic to cells, so most cells can rely on glycolysis only for short periods.

Energy Delivery to Cells and Energy Storage

Because of the large and constant need for energy, our bodies have a multitude of mechanisms for delivering

a continuous and rich supply of fuel to our cells. During a meal and several hours after, the molecules absorbed from the gastrointestinal tract provide most of the fuel used. After digestion is completed, fuel stores in the body are used. Two types of mechanisms supply and store cellular fuels: those that are used during the absorption of food from the gastrointestinal tract and those that are used after absorption is complete. During these two periods, our bodies are in the absorptive state and postabsorptive state, respectively.

The Molecules of Fuel Transport and Storage

Literally thousands of different types of molecules are absorbed into the body following a meal; even more are synthesized in our bodies. However, only a limited number are used specifically for fuel transport and storage. These molecules fall into three categories: carbohydrates, lipids (fats), and proteins. In Chapter 2, you were introduced to these molecular categories as structural building blocks of the cell; here we discuss their use as cellular fuels.

Carbohydrates As you learned in Chapter 2, carbohydrates are composed of carbon, hydrogen, and oxygen atoms. They are sugars and starches ranging from simple sugars, which have a molecular weight of about 130, to large starch molecules having molecular weights of several hundred thousand. In the gastrointestinal tract, most carbohydrates are broken down into simple sugars and then absorbed. These simple sugars are glucose, fructose, and galactose. Common table sugar is sucrose, a disaccharide, which means two sugars (glucose and fructose) combined into one molecule. Both fructose and galactose directly enter the hepatic portal blood and are rapidly converted to glucose as this blood flows through the liver. Thus, consumption of carbohydrates increases the glucose concentration of blood.

Carbohydrates are also synthesized in the body. The liver can manufacture glucose from smaller molecules for release into the blood. This process is called **gluconeogenesis,** or the formation of new glucose. Several tissues can manufacture a large carbohydrate called glycogen; the liver and skeletal muscle are the most important sources. Glycogen is a very large molecule composed of many thousands of glucose molecules joined together, and so does not circulate in the blood but remains trapped in the cells where it is produced. Within the cell, glycogen can release glucose on demand, serving as an important glucose store. When glycogen is broken down, a process called **glycogenolysis,** individual glucose molecules are released. Although glucose produced from the breakdown of liver glycogen can enter the blood, glucose within muscle cells cannot. Consequently, the breakdown of muscle glycogen produces glucose available only to the muscle cells in which it is stored.

Fats As you learned in Chapter 2, fats are composed of carbon, hydrogen, and oxygen atoms, and are classed into three types: simple fats, compound fats, and derived fats. Most simple fats are triglycerides and form the major energy storage pool in the body. One triglyceride molecule consists of three free fatty acids attached to one glycerol molecule.

Derived fats are manufactured from other fats. The most well-known derived fat is cholesterol, a necessary component of all cells. Cholesterol can come directly from the diet or it can be synthesized within cells from fatty acids. Organ meats such as liver, kidney, and brain are rich sources of cholesterol, as are dairy products such as cream, cheese, and egg yolk. Although cholesterol is a normal constituent of cellular structures and is always present in the blood, too much cholesterol can be harmful, as you learned in Chapter 9. For this reason, people with very high cholesterol levels in their blood are advised to eat less of those foods and rely more on vegetable products for their energy supply.

Fatty acids that are absorbed by the gastrointestinal tract and thus enter the blood can be taken up by cells and used as fuel in oxidative phosphory-

lation. However, most of the fatty acids in blood are taken up and stored by fat cells, which are also called **adipocytes.** In storing fatty acids, adipocytes synthesize triglycerides, a beneficial synthesis because triglycerides can be packed more closely than free fatty acids, allowing more compact storage. The very high energy content of triglycerides makes them well-suited for this role. A gram of fat contains over twice the energy of a gram of glucose or protein. In a process called **lipolysis,** adipocytes rapidly degrade triglycerides, releasing fatty acids into the blood. Although fatty acids are stored in adipose tissue, the liver can also manufacture fatty acids from a variety of smaller molecules such as glucose and amino acids in a process called **lipogenesis.**

Proteins Proteins are chains of amino acids, which are small molecules containing nitrogen. There are 20 different amino acids, usually grouped into two categories: essential amino acids and nonessential amino acids. These can be confusing terms because all 20 amino acids are essential constituents of most proteins. However, the 11 nonessential amino acids can be synthesized by the body and are not an essential part of the diet, whereas the 9 essential amino acids cannot be synthesized and must be supplied in the food we eat.

Although the most important function of amino acids is in protein synthesis (see Chapter 2), they are also an important source of energy. All dietary protein is broken down in the gastrointestinal tract into its constituent amino acids, which are then absorbed into the blood. These amino acids can be broken down in the liver and their carbon containing parts used to make both glucose and fatty acids. It is also possible for cellular proteins, such as actin and myosin in skeletal muscle cells, to be degraded into amino acids, which are released into the blood.

The Absorptive State

People living in industrialized countries tend to eat meals rich in carbohydrate, fat, and protein. Consequently, during and for several hours following a meal, considerable amounts of glucose, fatty acids, glycerides, and amino acids enter the blood from the gastrointestinal tract. During this absorptive period, cells preferentially take up glucose from the blood for use as intracellular fuel. The fats, amino acids, and any dietary glucose that are not immediately used for fuel are added to the body's fuel stores. Energy storage and the use of glucose as the primary energy source are the hallmarks of the absorptive state.

As noted above, glycogen is manufactured mostly in the liver and skeletal muscles from blood glucose. Within reasonable limits, the more carbohydrates consumed, the more glucose enters the blood and the more glycogen is produced. This is the reason that athletes often "carbohydrate load" by eating a large spaghetti dinner the day before an event. However, there is a limit to the amount of glycogen that can be made and stored in advance. No matter how much carbohydrate we eat, we can store only enough glycogen to run our bodies for a few hours.

Fat is by far the largest energy store in the body. Most people store enough fat to fuel their bodies for more than one month without resupply. Thus, the energy content of our triglyceride stores is at least 100 times greater than our glycogen stores. Most of the triglycerides are produced in the liver and subsequently transported in the blood to adipose cells, where they are stored. The materials for triglyceride synthesis can come not only from dietary fats, but also from glucose and amino acids. If the carbohydrate content of a meal is large, more glucose enters the body than can be used for immediate cell metabolism and glycogen production. The excess glucose is converted to fatty acids, which are then used in triglyceride synthesis. Likewise, a high-protein meal produces more amino acids than are needed for protein synthesis. The excess amino acids are then converted into fat for energy storage. Note that triglycerides can be produced from fats, glucose, and amino acids, whereas glycogen can be made only from glucose.

Most protein synthesis occurs in the absorptive state, when amino acids from dietary protein enter the body. However, the amount of protein synthesized is small and is not stimulated by eating more protein. Only enough protein is produced to replace the small amounts routinely broken down as a result of the metabolic wear and tear of cells. Thus, your diet does not affect your muscle mass, no matter how much steak you might eat. However, the reverse is possible; muscle protein declines when dietary energy supply is insufficient. During prolonged fasting or starvation, proteins become a vital source of energy. However, the use of protein for energy is deleterious because it reduces the amount of protein available for other vital functions such as muscle contraction and enzyme synthesis. Proteins are thus the energy store of last resort. The pattern of energy flow in the absorptive state is diagrammed in Figure 14.5.

The Postabsorptive State

In the postabsorptive state, the most important mechanisms for supplying glucose are glycogenolysis and gluconeogenesis. These processes, introduced earlier,

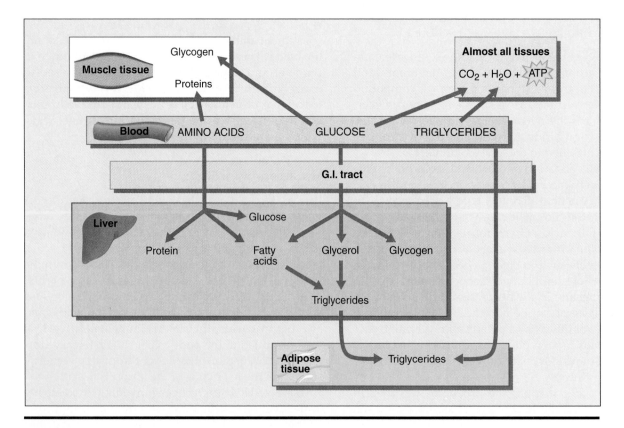

Figure 14.5 Major metabolic pathways of the absorptive (fed) state.

are essential because the nervous system requires a continuous supply of glucose. If the glucose supply to the brain is insufficient, coma and even death may result. Because glycogen stores are limited, glycogenolysis can provide glucose for only short periods. Consequently, without a fresh supply of glucose from food, gluconeogenesis becomes increasingly important. In the liver, glucose is formed mainly from the smaller molecules of glycerol and amino acids. Much of the glycerol for gluconeogenesis comes from lipolysis in adipose tissue. However, during long periods of fasting, as adipose tissue shrinks and limits the triglyceride supply, amino acids from protein degradation become a more important supply for gluconeogenesis.

Besides glucose and fatty acids, ketones are produced during a fast. These substances, which include acetone, are produced in the liver from fatty acids and can be metabolized by most tissues, including the nervous system. Normally, ketones provide only a very small fraction of the body's fuel molecules. However, during prolonged fasts, their production increases, reducing the body's requirement for glucose. Consequently, gluconeogenesis is reduced, slowing the breakdown of protein. The pattern of energy flow in the postabsorptive state is diagrammed in Figure 14.6.

Regulation of Energy Exchange in the Absorptive and Postabsorptive States

The patterns of energy exchange in the fed and fasting states are regulated with great precision by hormones. Many hormones are involved in this regulation, but the most important are two hormones secreted by the pancreas, insulin and glucagon, and the catecholamines, epinephrine and norepinephrine. These hormones have been discussed in Chapter 5 and are discussed only briefly here.

Insulin We saw in Chapter 5 that insulin concentrations in blood vary directly with the glucose concentration and that insulin directs the absorption of glucose into most cells of the body. But insulin also has other effects that regulate metabolism. Insulin stimulates key enzymes in the anabolic processes that synthesize glycogen, protein, and triglycerides.

Insulin also inhibits the catabolic pathways that mobilize fuel molecules from the energy stores. In addition, gluconeogenesis and synthesis of ketones are inhibited by insulin. Thus, the metabolic pattern of the fasting state is turned off by the insulin secreted in response to a meal. In the postabsorptive state, glucose concentrations fall, lowering insulin concentrations in the blood. This reduces

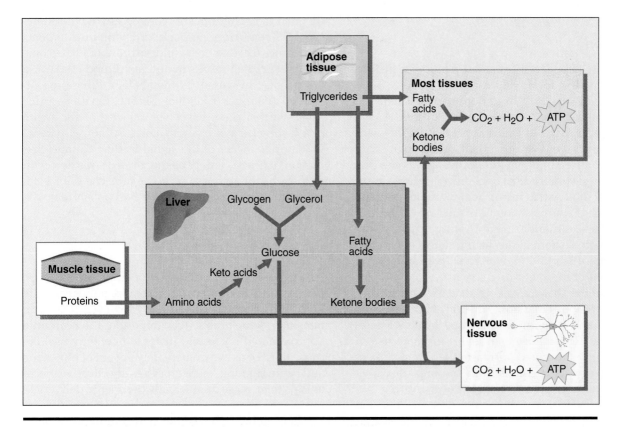

Figure 14.6 Major metabolic pathways of the postabsorptive (fasting) state.

the stimulation of anabolic processes and the inhibition of catabolic processes.

Glucagon In contrast to insulin, glucagon's action is restricted to the liver, where it inhibits glycolysis and stimulates gluconeogenesis and ketone synthesis. These actions help prevent the blood glucose concentration from falling too low during a fast. Ketone synthesis helps maintain blood glucose concentrations because ketones enter tissue and substitute for glucose as an energy source, preventing unnecessary glucose uptake from the blood.

Catecholamines Although catecholamines do not normally regulate metabolism, they become critically important when the blood level of glucose falls to very low values. This can occur in pancreatic diseases that impair glucagon secretion. This also occurs if a person with diabetes takes too much insulin, or if someone is starving. When plasma glucose levels fall below about 50% of normal levels, glucose receptors in the brain are activated, increasing the secretion of catecholamines. This increase in circulating catecholamine levels stimulates liver glycogenolysis and gluconeogenesis, as well as lipolysis in adipose tissue and glycogenolysis in skeletal muscle. Thus, the actions of catecholamines are generally opposite to those of insulin and similar to those of glucagon: They promote the metabolic pattern of the postabsorptive state.

Starvation

People may experience extended periods of insufficient food intake for a wide variety of reasons. They may be poor or living in underdeveloped nations. Starvation also occurs as a result of some diseases such as cancer and AIDS, in which appetite is suppressed. Some people severely restrict their eating voluntarily in order to lose large amounts of weight. Some diseases of the gastrointestinal tract prevent adequate absorption of food so that insufficient nutrients enter the body although normal amounts of food are consumed. All of these situations significantly prolong the amount of time spent in the fasting state. Under such conditions of starvation or near starvation, several adaptations occur to improve survival.

Remember that the fasting state is characterized by low levels of circulating insulin, high levels of glucagon, and sometimes increased levels of catecholamines. These hormonal levels promote use of stored energy. During a short fast of 4–12 hours, most of the energy used comes from the breakdown of glycogen and triglycerides. However, during extended periods of fasting, glycogen stores become depleted, decreasing the availability of glucose. Although gluconeogenesis remains as a glucose source, this process is costly because it requires amino acids from the breakdown of protein. Most tissues easily adapt to the shortage of glucose by using fatty acids for fuel, but the brain cannot do this because fatty acids are unable to cross the blood–brain barrier. However, the brain can use ketones for fuel because they easily cross the blood–brain barrier, and ketone production by the liver increases enormously during a prolonged fast. After several days of fasting, ketones provide approximately 50% of the energy needs of the brain. Thus, increased ketone production reduces glucose use, which ultimately reduces protein breakdown and facilitates survival.

Another adaptation to prolonged fasting is a decrease in the body's energy requirements. One of the primary symptoms of starvation is lethargy. People who are inadequately nourished are much less active than well-fed people. This is beneficial because less energy is needed, slowing down the depletion of energy stores. Another important adaptation to fasting is a reduction in the basal metabolic rate. That is, the body uses less energy even in the absence of physical activity. The reason for this change is not known, but it is probably the result of the large alterations in metabolic hormones.

The increased ketone production during a fast causes ketones to build up in the blood. This accounts, at least in part, for the ability of some parents to predict that their children are "coming down with something." Even before noting overt symptoms, parents sometimes feel that their children are about to become ill. Disease often results in decreased food intake and an increase in stored fat metabolism, liberating increased quantities of ketones. Ketones have a very distinct, sweetish odor that can be detected on the breath of a fasting person or a sick child. Although parents may not be aware of the signal leading to their prediction of illness, it may be the unconscious detection of the odor of ketones on the breath. The odor is familiar to most people; common nail polish remover, acetone, is a ketone.

Exercise

At rest, our skeletal muscles consume little energy. Our leg muscles, for example, require only about 0.12 kcal/min while we are sitting. However, if we run fast, these same muscles consume about 16 kcal/min, approximately a 130-fold increase in energy use. Because most of us can immediately get out of our chairs and begin running, there must be very effective mechanisms to increase dramatically the energy delivery to muscle proteins.

During brief exercise, such as running up a flight of stairs or sprinting to catch a bus, most of the energy used comes from muscle glycogen stores. Glycogen in the muscle can be broken down rapidly and the resulting glucose used directly by the contracting muscle cells. However, during prolonged exercise, such as long-distance running, bicycling, or playing soccer, additional energy sources are needed. These additional fuels are glucose and fatty acids that are delivered to muscle by the blood. Several processes enrich the blood with these fuels and promote their delivery to the contracting muscle cells.

One very important event during exercise is a net breakdown of fuel stores, causing glucose and fatty acids to enter the blood. This breakdown is mediated primarily by hormonal changes. Exercise activates the sympathetic nervous system, causing the release of catecholamines. Glucagon secretion also increases during exercise. Remember, all of these hormones cause the breakdown of glycogen and triglycerides. They also promote gluconeogenesis in the liver, which adds more glucose to the blood.

Another very important event is an increase in blood flow to the working muscles. As you learned in Chapters 8 and 9, exercise induces a number of cardiovascular changes that greatly increase blood flow to contracting muscle. This increase in flow is essential in delivering fuel to the muscle cells. Another significant change during exercise is an increase in the rate of glucose transport into muscle cells. Although exercise does not appreciably affect the insulin levels of the blood, muscle cells become more sensitive to this hormone during exercise, which increases their glucose uptake from blood.

Thus, during exercise, the energy needed for muscle contraction comes first from energy stores in the muscle itself. During continued exercise, circulating glucose and fatty acids become the major fuels. These fuels enter the blood from the breakdown of energy stores and from gluconeogenesis, and are delivered in greater amounts by the high blood flow to working muscle. In muscle, they are readily taken up by the exercising cells. The pattern of energy flow during exercise is diagrammed in Figure 14.7.

As everyone knows, we cannot exercise indefinitely. At some point, fatigue sets in and exercise halts. For many years, it was thought that fatigue resulted from depletion of muscle glycogen. This led athletes to eat a high-carbohydrate meal the night before a contest. However, it has recently been shown that a high-carbohydrate meal increases muscle glycogen only a little, and that fatigue is not solely the result of depleted muscle stores of glycogen. If glucose is added to the blood by ingesting sugar during exercise, fatigue can be delayed with no effect on the rate of muscle glycogen depletion. Currently, the most effective

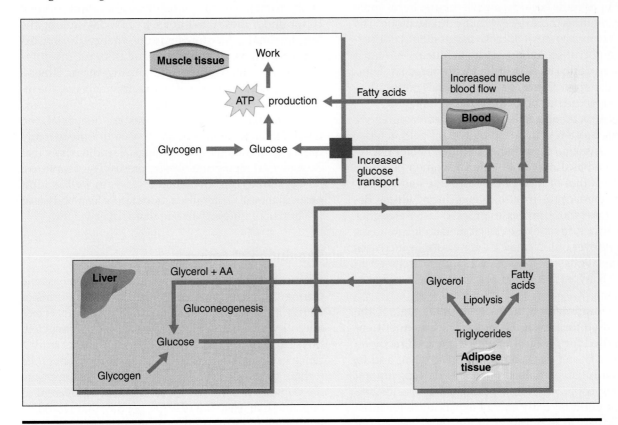

Figure 14.7   Major metabolic pathways during exercise.

method to increase endurance during sustained exercise is to drink or eat a moderate amount of sugar during the exercise period.

Whole Body Energy Balance

So far, we have discussed how cells obtain and use energy, how energy is stored in the body, and how energy moves from energy stores to cells. In this section, you will learn how we meet our total energy requirements.

The total amount of energy we use is determined by three factors: basal metabolic rate, physical activity, and **thermogenesis** (*thermo* means heat; *genesis* means creation). For most of us, the majority of our energy expenditure sustains our basal metabolic rate, the energy required for all vital functions such as the maintenance of ionic gradients across membranes, the production of new tissues, the replacement of old tissues, and the maintenance of normal body temperature. Although our basal metabolic rate is relatively constant, it is determined by a number of factors. Size is most important. A large person usually has a higher basal metabolic rate than a small person; the large person simply has more tissue to maintain. Body composition is also important because muscle tissue, even while resting, requires more energy than adipose tissue. The lower metabolic rate of adipose tissue should not surprise you because adipose tissue is little more than a collection of adipose cells stuffed with stored fat. Thus, the basal metabolic rate of a lean person is greater than that of an obese person of the same weight.

Basal metabolic rate gradually decreases as we age. This is often manifested by weight gain in older people despite no clear change in eating habits. On average, the basal metabolic rate of women is 5–10% lower then that of men of the same age and weight. Basal metabolic rate can also change during illness. Because all metabolic processes proceed faster at higher temperatures, fever causes basal metabolic rate to increase. A fever of 102°C (39°C), for example, increases basal metabolic rate by about 20%. As discussed in Chapter 5, hyperthyroidism also increases basal metabolic rate.

Superimposed on the basal metabolic rate is the energy used for physical activity. The amount of energy needed for physical activity depends directly on the intensity and duration of muscular work, and is, of course, highly variable from day to day and among different people. A sedentary person may use 15–20% of his or her total daily energy expenditure for muscular work; a marathon runner may use more than 75%.

The third and smallest component of energy output is thermogenesis. This is the heat produced in response to cold (usually by shivering) and for the digestion and absorption of a meal.

Over long periods, these energy requirements must be met by dietary energy input. If energy input is less than energy output, fuel stores are consumed and weight loss occurs. If energy input exceeds energy output, the excess energy is stored as fat and weight gain occurs. For most people, body weight stays remarkably constant as energy expenditure and input are balanced with exquisite precision. See Highlight 14.1 for a more detailed description of weight control.

What determines energy input? Certainly the amount of food eaten is one factor, but the type of food is also important. Both carbohydrates and proteins provide four calories for every gram ingested; fat, however, provides *nine* calories per gram. For this reason, a meal consisting largely of protein and carbohydrate generally supplies fewer calories than one rich in fats. A 1-pound, well-marbled steak supplies many more calories than 1 pound of bread.

Diets that are high in fat are also high in calories. This is potentially beneficial if large stores of energy are needed, as in fasting. This situation was common with ancient peoples, who often relied on hunting and primitive agriculture, both variable food sources. The ability to store fat was probably an important survival factor under these conditions. Even today, among the less fortunate, fat stores supply energy during extended periods without food. For all of us, fat stores can be extremely important during major illness, when we eat less and energy requirements may be increased for extended periods.

Although adequate fat stores are beneficial, too much fat intake or fat storage may contribute to many diseases. Numerous epidemiological studies show that excessive fat intake and obesity predispose people to a host of serious cardiovascular diseases as well as adult onset diabetes. Generally, a moderately lean body is at lower risk of illness than one that is overweight.

Regulation of Eating

For most people, body weight remains relatively stable during their adult lives. What this means in terms of energy balance is that the intake of food energy is balanced by the energy output, regardless of changes in physical activity throughout the years. This is truly a remarkable phenomenon. Consider the consequences if there were a systematic error in food consumption. If a person ate extra food with a caloric equivalent of only one-half teaspoon of sugar per day, that would be equivalent to an excess of only about 10 calories daily. Over one year, that would represent 3650 calories, or

HIGHLIGHT 14.1

Diets and Dieting

Although it is clear that excess weight presents a health risk, weight reduction often stems more from a desire to attain the current ideal look than from a desire to prolong life. Not long ago, the "full figure" was popular for both men and women. Now the lean look is popular. To this end, a host of diet plans are marketed, most promising weight reduction with little or no effort, and promising weight loss of up to 5 pounds a week. Are these promises reasonable?

We have seen that body weight is determined solely by the difference between calories consumed and calories expended. The average person expends approximately 2500 calories per day. This assumes average weight, average activity, and average basal metabolic rate.

The desire to lose weight translates into a wish to lose fat, not muscle or body water, which means that stored fat must be metabolized. The important question is, "How many calories are available in one pound of fat?" Each pound of fat supplies 3500 calories. Thus, to lose only 1 pound of fat in one day, an average person would have to eat nothing and exercise enough to burn an extra 1000 calories. To lose 5 pounds, it would be necessary to use 17,500 calories more than are ingested. That is clearly impossible for most people, but because the basal metabolic rate is proportional to body weight, it would be possible for a person weighing 1030 pounds to accomplish this feat.

Let us do some more simple calculations. We know that exercise burns calories, but how many for each type of exercise? Rapid bicycling or swimming requires about 500 additional calories per hour. Thus, to lose 5 pounds of fat in one day the individual would have to swim 35 hours a day and eat nothing. This is obviously impossible. Can any diet *appear* to deliver the promise of large rapid weight loss? Unfortunately, yes. It is possible to lose a few pounds of water rather quickly on various "diet" schedules. However, because the object is to lose fat, the scale gives false information to the naive dieter. If the weight loss is due to water loss, a person will retain water and gain weight as soon as the crash "diet" is abandoned. (Review Chapter 12 for the control of salt and water balance.) How, then, can one lose weight effectively and keep it off over the long term?

Another small series of calculations is helpful in answering the above question. Assume a person wishes to lose 5 pounds without swimming for 35 hours in one day. You have already seen that 5 pounds represents a fat loss equal to 17,500 calories. There are many routes to that end, all allowing the expenditure of 17,500 calories more than eaten during whatever time is reasonable. This could be accomplished by reducing food intake or increasing energy expenditure, or both. A modest exercise program, such as bicycling an hour every other day, coupled with a small reduction in food intake of only 250 calories/day, will achieve a weight loss of 1 pound per week, or 52 pounds per year! This type of diet plan requires a change in both food consumption and activity, but it is the most effective and long-lasting way to health.

about the number of calories stored in 1 pound of fat. In 50 years, the person would gain over 50 pounds. Yet this is a rare occurrence. Without thinking, most people maintain their weight within small bounds. The reasons are not well-known, but some hints are available.

Control of food intake seems to reside in the brain rather than in the stomach, as might seem probable. Although the stomach becomes distended as one eats, the *feeling* of hunger is largely independent of stomach distention. That drive to eat seems to originate in a hypothalamic area called the satiety center. If this center is destroyed in experimental animals, they eat almost continually, attaining collosal sizes. Thus, there is some central control over the amount of food we take in, which in most people balances the energy expended.

Other factors also play a role in eating, especially in humans. Often our eating habits are conditioned by social situations and cultural norms. We are more apt to gorge ourselves at a banquet than at home; we tend to eat between meals more often during "coffee hours" or some social occasion than while attending school or sitting in an office at work. We also consciously try to design our eating habits to allow us to achieve a body shape our culture sees as the ideal. Finally, some people overeat or undereat because of psychological problems that are not well-understood.

Other Necessary Nutrients

Minerals Thus far, we have talked only about food as an energy source, neglecting other aspects of food intake. To remain healthy, we need not only a sufficient intake of calories, but non–energy-producing elements, including certain common minerals such as sodium, potassium, and calcium, as well as some rarer ones such as iron and zinc. Most people have no problem consuming sufficient minerals to supply their needs. However, for many, obtaining enough calcium or iron may be a problem. Iron is required for red cell production; deficiency results in

HIGHLIGHT 14.2

Hypertension, Apples, and Pears

Researchers have clearly demonstrated that obesity places a person at a higher risk of hypertension than a lean person. Conversely, a person with hypertension who loses excess weight improves the chances of reducing his or her blood pressure. However, not all adipose tissue poses the same risk of hypertension. Long-term studies of large populations indicate that excess adipose tissue confined to the abdominal area places a person at greater risk of hypertension than the same amount of adipose tissue distributed elsewhere on the body. People with fat distributed largely around the hips and thighs are at a lower risk for hypertension.

The photos in this box illustrate the two body shapes. Accumulation of abdominal fat in a person creates an apple shape; accumulation in the hip and thigh area creates a pear shape. It is not quite clear why the fat distribution should have such an effect on blood pressure or diabetes, but some hints are available. The venous drainage of abdominal fat goes directly to the liver, whereas the venous drainage of other fat tissue enters the vena cava and is diluted by the entire blood volume. Consequently, fatty acids liberated by abdominal fat reach the liver in higher concentrations than fatty acids liberated by other fat tissue. Some researchers have shown that fatty acids injected directly into the hepatic portal vein increase blood pressure. This suggests that the increased risk of hypertension associated with the apple shape is, at least in part, the result of abdominal fat releasing fatty acids directly into the blood supply to the liver.

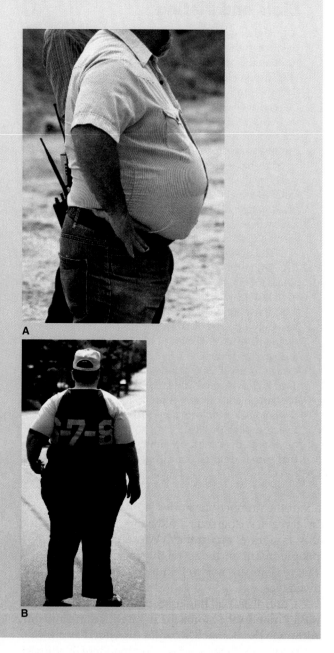

(A) Apple and (B) pear body shapes.

---

a condition called anemia. This is particularly common among menstruating women. Calcium aids bone metabolism, and women generally require a bit more calcium than men. In general, however, we are not faced with life-threatening mineral deficiencies in our society.

Vitamins A second type of nutrient required is vitamins. Although they do not supply energy di-

rectly, they are required in very minute amounts for the body to perform certain vital functions. Most vitamins cannot be synthesized by the body like proteins; rather, they must be provided in our food or by pills. Although it has been known for several hundred years that a diet of fresh food is required for good health, the reasons have remained a mystery until fairly recently. Hundreds of years ago, when sailing techniques developed to enable ships to sail the oceans for many months at a time, sailors invariably became ill with a disease called scurvy (see Milestone 14.2). This very debilitating disease, characterized by soft, bleeding gums, weakness, and bleeding under the skin, often killed more than half of the crew during a long sea voyage. Largely by trial and error, it became apparent that if fresh food, either vegetables or meat, were added to the traditional sailing menu of crackers and salt-cured meats, scurvy could be prevented. It took until early this century to learn that the missing ingredient in preserved foods was vitamin C, or ascorbic acid. We now know that there are many vitamins, all controlling some specialized and vital metabolic function, and that these vitamins must be continually supplied. But what do they do?

Vitamins are compounds that do not perform a function directly, but act with proteins called apoenzymes. An apoenzyme is almost a true enzyme, except that it lacks one essential component to function. The missing component is a coenzyme, or vitamin. When a vitamin combines with its apoenzyme, a functioning enzyme is formed that allows specific chemical reactions to proceed at an optimal rate. Without adequate vitamins, those reactions do not proceed normally, producing serious consequences. It should now be apparent that the necessary quantity of any vitamin depends on the amount of apoenzymes in the body. If the quantity of vitamin far exceeds the quantity of apoenzymes present, ingesting more of the vitamin is not beneficial. At best, the extra vitamin is stored, but often it is simply excreted.

Table 14.1 on pp. 377–378 is a list of vitamins, divided into water-soluble and fat-soluble groups. For example, the B vitamins and vitamin C are water-soluble, whereas vitamins A, D, E, and K are fat-soluble. In general, the water-soluble vitamins are stored in smaller quantities because they are easily excreted in urine. If one takes large doses of vitamin C, most of it is excreted in the urine within a few hours. This is not the case for vitamin A because most of it can be stored in the tissues, particularly the liver. The table also shows some of the major sources in food and some disorders that result from a deficiency of each vitamin.

It should be stressed at this point that if a little is good, more is not necessarily better. In fact, some vitamins are quite toxic when taken in large doses. The Center for Disease Control in Atlanta reports about 5000 cases of vitamin toxicity each year caused by overzealous ingestion of vitamins. Table 14.1 also lists some adverse effects that can result from self-treatment with megadoses of each vitamin. A megadose ranges from 100 to 1000 times the recommended daily allowance, which is the amount that meets most people's needs.

Temperature Regulation

Maintenance of a certain body temperature is one of our most important physiological tasks. In this section, we discuss why body temperature is so important, how it is controlled, and how it can be changed under certain conditions for our benefit.

Importance of Body Temperature

As mentioned above, the body temperature of a healthy person is quite stable, rarely changing by more than a few degrees, regardless of the environmental temperature. Why is this, and how is it accomplished?

You already know from Chapter 2 that most chemical reactions are temperature-dependent. In general, as temperature rises, chemical reactions proceed at a faster pace. However, in biological systems, enzymes control many of the chemical reactions in a cell. Most enzymes are proteins whose chemical configuration is altered by temperature. A most obvious example is a fried egg. The white of an egg is composed of the protein albumin dissolved in water. If the egg white is heated beyond some point, it changes from a viscous, translucent material to one that is harder and opaque. Although this is an extreme example, it should be clear that proteins are temperature-sensitive.

For proper body function, cellular reactions must proceed at specific rates. Most enzymes regulate our cellular reactions best at temperatures of about 98–99°F (37–37.5°C). At temperatures below 93°F (34°C), many cellular reactions are too slow, seriously impairing physiological functions. Above 107°F (42°C), irreversible tissue damage occurs. Thus, body temperature *must* be maintained within a limited range.

Regulation of Body Temperature

The critically important process of temperature regulation is controlled by a physiological system that in many ways is similar to the temperature control system in buildings. Like many buildings, the body regulates its temperature using a set of components that include a cooling system, a heating system, temperature sensors, and a thermostat. The body's thermostat is a collection of thermoregulatory cells in the hypothalamus of the brain. These cells set our body temperature at approximately 98°–99°F (37°C). If body temperature falls below this thermal set point, temperature sensors located in the skin, abdominal organs, and central nervous system detect

MILESTONE 14.2

Discovery of Vitamins

Although it had been known for a few centuries that a restricted diet such as sailors and explorers might eat during long expeditions led inexorably to scurvy, the cause was unknown. It seemed most probable to people of those days that scurvy was a result of spoiled food or some contaminant in the food. Eating fresh meat or vegetable products prevented or cured scurvy; this was taken as evidence that unspoiled food was the antidote. It was not until animal experiments were performed during the last few years of the nineteenth century that it became evident that chemicals contained in food were necessary to prevent certain diseases.

In 1893, Dr. Christian Eijkman traveled to Indonesia to study a group of chickens that exhibited a set of symptoms very similar to those of the human disease called beri beri. The animals showed serious neuromuscular problems, weakness, and early death. Eijkman believed that the disease was spread by some bacterium and searched diligently for it with no success. But suddenly, without warning, the disease disappeared from all his chickens, seemingly for no reason. As he searched for the answer, he found that the chickens exhibiting signs of beri beri were fed the leftover rice from the officers' mess on the military base in which he was working. Officers ate the more expensive polished rice, from which the outer covering of the grain was removed. However, the cook who fed the chickens was replaced with another cook who believed it irresponsible to give mere birds such expensive rice for food. He replaced their polished rice with brown rice containing the whole grain, not telling anyone of the switch. As soon as the chickens were fed whole grain rice, beri beri disappeared from the flock. Eijkman realized that beri beri must be a deficiency disease, not one caused by a bacterium. He then replaced the whole grain rice with polished rice and watched beri beri reap-

pear. Next he fed the affected chickens polished rice supplemented with the discarded polishings. The chickens were cured.

For the first time, it was clearly demonstrated that disease might result from inadequate supplies of a chemical found in certain foods. We now know that chemical to be vitamin B_1. Following his careful experiments, vitamins were isolated, the first being vitamin C, which prevented scurvy. These experiments also led the way for establishing minimum standards for vitamin consumption. Experimentally, this is determined by completely depriving an animal of a given vitamin until signs of a deficiency are observed. The vitamin is then added to the diet until all symptoms are gone. That amount could be taken as the minimum daily requirement; however, for safety's sake, the minimum daily requirement for humans is set at about twice the minimum required to prevent any symptoms of the deficiency.

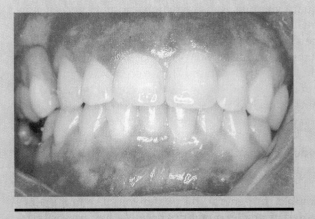

Bleeding gums, one of the first symptoms of scurvy.

this temperature change and activate our heating system. If the body temperature rises above the hypothalamic thermal set point, temperature sensors detect the change and activate our cooling system.

However, these components are considerably more complex in the body than in a building. For example, although a building is cooled by circulating cold air, our cooling system consists of several different mechanisms that increase the body's heat loss to

the environment. These cooling mechanisms include sweating, increased blood flow to the skin, and a variety of behaviors. By increasing blood flow to the skin, we transport heat from the body's core to its periphery, where it can rapidly move into the environment. The flush of facial skin on a hot day is the result of this cooling mechanism. The evaporation of sweat cools the skin's surface, facilitating heat loss from the blood flowing through the skin. Behavioral

Table 14.1   The Vitamins

Vitamins	Source	Function	Deficiency Disease	Dangers of Vitamin Megadose Self-Therapy
Water-Soluble Vitamins				
B₁ or thiamine	Whole grains, meat, liver, and eggs	Aids carbohydrate metabolism. Required for proper muscle and nerve function.	Beri beri, partial muscle paralysis, and neuritis	None reported.
B₂ or riboflavin	Whole grains, meat, liver, eggs, and milk	Aids carbohydrate and protein metabolism. Promotes tissue repair. Helps body use oxygen.	Poor use of oxygen, cataracts, dermatitis, and anemia	None reported.
B₃ or niacin	Whole grains, meat, liver, eggs, and milk	Essential for energy-releasing reactions and cell metabolism.	Pellagra, dermatitis, and mental disturbances	Transient flushing, itching, irregular heartbeat, and possible liver damage.
B₆ or pyridoxine	Whole grains, meat, liver, eggs, and milk	Required for normal metabolism of amino acids and fats.	Retarded growth, convulsions, dermatitis, and nausea	Possible nerve damage.
B₁₂ or cyanocobalamine	Eggs, milk products, and liver	Required for red blood cell formation and proper function of nervous system.	Pernicious anemia and mental disturbance	None reported.
Pantothenic acid	Most foods	Helps convert metabolic fuels to energy. Required for normal nerve and immune function.	Fatigue, poor adrenal function, and neuromuscular problems	None reported.
Biotin	G.I. bacteria, eggs, and liver	Aids fatty acid synthesis. Needed for maintaining healthy skin.	Dermatitis and mental depression	None reported.
Folic acid	G.I. bacteria, leafy vegetables, meat, poultry, and fish	Coenzyme for some metabolic processes. Required for red blood cell production.	Anemia	None reported.
C or ascorbic acid	Fresh fruits, vegetables, and meat	Essential for sound teeth and bones and tissue healing. It is an antioxidant.	Scurvy, swollen gums, loose teeth, and poor wound healing	Rebound scurvy. Newborn babies of mothers taking high doses of vitamin C may exhibit signs of scurvy. In adults, megadoses cause diarrhea, induce menstrual bleeding in pregnant women, and destroy vitamin B.

continued

Table 14.1 **The Vitamins (cont.)**

Vitamins	Source	Function	Deficiency Disease	Dangers of Vitamin Megadose Self-Therapy
Fat-Soluble Vitamins				
A or retinol	Yellow vegetables, milk products, and fish oil	Needed for good night vision. Helps maintain epithelial tissue.	Night blindness, dry skin, and urinary and gastro-intestinal infections	Anorexia, retardation of growth, irritability, and bone pain.
D or calciferol	Sunlight on skin, fish oils, eggs, and milk	Needed for calcium and phosphorus metabolism.	Rickets, poor bone forma-tion in children, and mal-formed skeleton	Nausea, weakness, weight loss, kidney stones, kidney failure, and hypertension.
E or tocopherol	Whole grains and vegetable oils	Antioxidant. Inhibits catabolism of fatty acids in cell membranes.	Unknown in humans	Nausea, diarrhea, and headache.
K	Leafy vegetables and intestinal bacteria	Needed for normal blood clotting.	Long clotting time and excessive bleeding	None reported.

cooling involves numerous actions such as wearing fewer clothes, drinking cold beverages, and finding a lake to swim in on a hot summer day.

Our heating system comprises mechanisms that generate more heat as well as those that conserve the heat we generate. Extra heat is generated by muscular movements, both voluntary and involuntary. Shiver-ing is an involuntary muscular movement, whereas jumping up and down outside on a cold day is vol-untary. We conserve heat through reduced blood flow to the skin and by many behaviors such as wearing more clothes, standing closer to a heater, and keeping our arms closer to our bodies.

All of these thermoregulatory components func-tion as a homeostatic control system. Body tempera-ture is detected at numerous sites by the temperature sensors, which send afferent information to the thermo-regulatory cells of the hypothalamus. This information is integrated by the hypothalamic cells, which in turn send efferent information to the numerous organs and behavioral centers in the brain that produce changes in heat production and heat dissipation to the envi-ronment. The components of our thermoregulatory system are shown in Figure 14.8.

For example, if you go to the beach on a warm summer day and lie in the sun, your body temperature begins to rise above your temperature set point. Tem-perature sensors detect this change in temperature and send afferent information to the hypothalamus, which reads this information and then sends efferent signals to the skin and the brain areas controlling be-havior. The efferent signals to the skin initiate sweat-ing and increase skin blood flow, thus promoting heat transfer from the body to the environment. The effer-ent signals eliciting behavioral responses cause you to get out of the sun and jump into the water.

If you stay in the water too long, your body tem-perature falls below your body's temperature set point. Again, your temperature sensors detect this change and send afferent information to the hypo-thalamus. Efferent signals are sent to the skin, skeletal muscles, and the brain's behavioral centers. Cuta-neous blood flow falls, reducing heat loss to the envi-ronment. If your body temperature falls low enough, shivering occurs, generating more heat. Your volun-tary behavioral responses may include swimming fast to generate more internal heat or getting out of the water and lying in the sun again.

Fever

The body's homeostatic control system for maintain-ing temperature is extremely effective, normally main-taining body temperature within a very narrow temperature range. However, a number of internal factors can also cause body temperature to move out of the normal range. One that we are all familiar with is a bacterial or viral infection. One of the most com-mon symptoms of infection is fever. When we begin to get sick, we feel cold even though our body tempera-ture is normal. A number of thermoregulatory re-sponses are then initiated to increase our body temperature. We put on more clothes, shiver, and reduce blood flow to the skin. The pallor of a sick person is a result of this latter response. These responses generate and conserve more heat, thus increasing body temperature.

But how does infection initiate these responses? When bacteria or viruses invade the body, immune system cells detect the infection. These cells, primarily macrophages, secrete a number of substances collectively called pyrogens into the blood. The pyrogens circulate to the hypothalamus, where they increase the production of prostaglandins (a class of fat molecules) in the hypothalamic thermoregulatory cells. These prostaglandins then reset the body's thermostat to a higher temperature. When that happens, the information transmitted by the temperature sensors tells the hypothalamus that the body temperature is now below the new thermostatic set point. The hypothalamic thermoregulatory cells respond by signaling the skeletal muscles, skin, and behavioral sensors in the brain to generate more heat and conserve it.

When a fever "breaks," the opposite set of reactions occurs. Fewer pyrogens are secreted by the white blood cells, reducing prostaglandin synthesis in the hypothalamus. The thermoregulatory set point consequently falls back to normal, below the feverish body temperature. This difference between body temperature and set point initiates the typical set of cooling responses: increased cutaneous blood flow, sweating, and behaviors to dissipate more body heat.

When many of us get sick, we "break" the fever with drugs. All of the common fever-reducing drugs, such as aspirin, ibuprofen, and acetaminophen inhibit prostaglandin synthesis. Thus, these agents prevent the pyrogens released by the immune system cells from stimulating the prostaglandin synthesis that raises the thermoregulatory set point.

Why do we get a fever when sick? Traditionally, fever was considered a deleterious side effect of illness and measures were taken to reduce or to eliminate fever. Before aspirin was discovered, the sick were given cold baths to reduce their body temperatures,

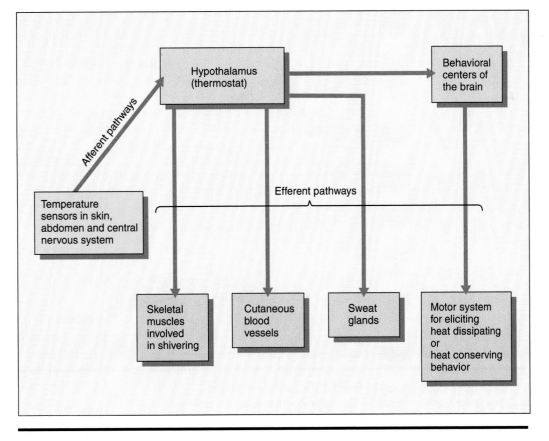

Figure 14.8 The thermoregulatory system.

and after the discovery of aspirin, this drug was administered liberally to lower body temperature. However, fairly recently it was discovered that fever is beneficial, improving the immune response to infection. This was first demonstrated in the early 1970s in a very clever experiment on lizards by Dr. Matthew Kluger. Unlike mammals, lizards are ectothermic ("cold-blooded"); that is, they are unable to maintain their bodies at a temperature different from that of their environment. However, even though lizards are always at their environmental temperature, they do regulate their temperatures with remarkable precision by seeking out environments that have a temperature they prefer. For example, if a lizard is too cool, it basks in the sun to raise its temperature. If its temperature rises too high, it gets out of the sun and seeks a shady area to cool off.

In his study, Dr. Kluger first put healthy lizards in an environmental chamber with many different temperatures. The lizards moved around in the chamber until they found a spot with the temperature they preferred, and then stayed there. This temperature was 38°C, slightly above our normal body temperature (Figure 14.9(A)). When he infected the lizards with bacteria, he observed that they all moved to warmer areas of the environmental chamber. The temperature they chose when sick was 40°C (Figure 14.9(B)). This first part of his experiment demonstrated that lizards give themselves a fever when sick. Dr. Kluger then tested the hypothesis that fever, at least in lizards, is beneficial. If this hypothesis is correct, then illness in the infected lizards should be less severe at higher temperatures. He tested this hypothesis by infecting several groups of lizards with the same bacteria. One group was kept at the temperature they normally prefer (38°C), another at the higher temperature they preferred when sick (40°C). One week later, 75% of the lizards maintained at their normal body temperature died, whereas only 25% of the lizards maintained at their preferred temperature when sick died. Thus,

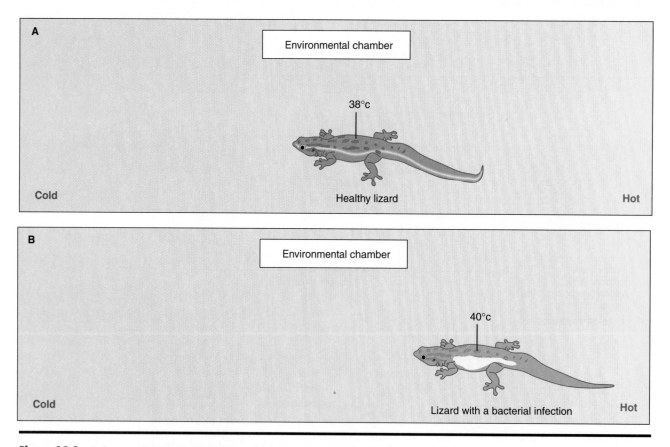

Figure 14.9 Behavioral thermoregulation. (A) Healthy lizard. (B) Lizard infected with bacteria. Lizards, like people, develop a fever when sick. This indicates that fever developed early in evolution.

fever, at least in lizards, is extremely beneficial. It allows these animals to survive infection much better.

Is fever beneficial for humans? Although this question is still being researched, much evidence indicates that the answer is "yes." Slightly higher body temperatures have been shown experimentally to improve a number of important immunological functions that allow us to fight bacterial and viral infections more effectively. What does this tell us of the wisdom of taking aspirin when sick? It suggests, but does not prove, that we may recover from an illness more quickly if we do not treat a mild fever. Very high fevers, however, are not beneficial and are a serious danger. Children often run extremely high fevers, which can lead to seizures if untreated. Generally, temperatures above 104°F (40°C) should be treated.

Summary

Life requires a continuous supply of energy to fuel a host of vital biochemical reactions. These reactions allow cells to synthesize molecules; to move, to maintain and to repair structures; and to generate heat. All of the energy we use comes from the food we eat. Numerous processes break the chemical bonds of the food molecules, capture the energy released, and then direct that energy into processes that require energy.

Energy is supplied directly to cells by three types of molecules: glucose, fatty acids, and ketones. The energy released from their breakdown is transferred to energy-requiring processes by ATP. Two major processes produce ATP: anaerobic respiration and aerobic respiration. Anaerobic respiration generates ATP by breaking down glucose into pyruvate. Aerobic respiration can use energy from all three types of cellular fuel molecules. Numerous complex processes generate ATP from ADP and phosphate ions.

To meet our constant need for energy, fuel molecules are continuously supplied to our cells. During the absorptive state, fuel molecules are derived directly from glucose, fats, and amino acids absorbed by the gastrointestinal tract. Cells preferentially take up glucose for energy use during this period. The fats, plus the glucose and amino acids not immediately used for energy and protein synthesis, are added to the body's energy stores. The excess glucose is converted into glycogen and fat.

In the postabsorptive state, fuel stores provide energy to our cells. Carbohydrates, fats, and protein are used for fuel storage and transport in the blood. Carbohydrates are transported in blood as glucose and are stored as glycogen. Fats are transported as fatty acids and stored in adipose tissues as triglycerides. Although proteins are not normally used to supply cellular energy fuels, they can be broken down into amino acids, which are then converted to glucose by the liver. This process usually occurs only during starvation, when carbohydrate and fat stores are depleted.

In the postabsorptive state, glycogen is broken down into glucose and triglycerides are degraded into fatty acids and glycerol. The glycerol is then converted to glucose in a process called gluconeogenesis. Some of the fatty acids are converted to ketones. Most of the glucose formed during the postabsorptive state is used by the nervous system; the other tissues use mostly fatty acids. Ketones are used by all tissues.

Energy exchange is regulated by hormones. The most important hormones involved are insulin, glucagon, and the catecholamines. Increased insulin secretion in the absorptive state directs glucose into most cells and stimulates glycogen, protein, and triglyceride synthesis. Insulin also inhibits the metabolic events of the postabsorptive state: glycogenolysis, lipolysis, gluconeogenesis, and synthesis of ketones. In the postabsorptive state, insulin secretion decreases while glucagon secretion increases. Glucagon prevents plasma glucose levels from falling too low by inhibiting glycolysis and stimulating gluconeogenesis and ketone synthesis. The catecholamines are important only when blood glucose levels fall very low. Their actions are generally opposite to those of insulin and similar to those of glucagon.

During prolonged starvation, the body undergoes several adaptations to reduce energy use and conserve glucose. The primary adaptation is an increased reliance on fatty acids and ketones for fuel. Energy use is reduced by decreasing basal metabolic rate and reducing physical activity.

Exercise elicits enormous increases in energy use by muscle cells. During brief exercise, most of the energy comes from the breakdown of muscle glycogen. During longer periods of exercise, triglycerides and liver glycogen are degraded and gluconeogenesis is stimulated. These metabolic events, in combination with increased muscle blood flow, deliver large amounts of fatty acids and glucose to the working muscles.

Our body weight is totally dependent on the difference between total energy input and output. Energy output is determined by basal metabolic rate, physical activity, and thermogenesis. Energy input is determined by the amount and types of food we eat. Eating in most people is regulated with remarkable precision by cells in the hypothalamus. However, for some people, energy input and output are not well-matched, resulting in weight gain or loss.

Normally, our body temperature is regulated within a narrow range, 98–99°F. Precise regulation of body temperature is essential because many enzymes cannot function properly above or below that range. Body temperature is regulated by a negative feedback control system comprising temperature sensors, a physiological thermostat in the hypothalamus, and a number of mechanisms to alter heat loss and gain.

When we get sick, infection is detected by immune system cells, which release substances called pyrogens. Pyrogens cause increased production of prostaglandins in the thermoregulatory cells of the hypothalamus, which in turn elevate our thermoregulatory set point. Our normal body temperature is then sensed as being too cold, and our heating mechanisms are activated, resulting in a fever.

Questions

Conceptual and Factual Review

1. What is the difference between the processes of catabolism and anabolism? (p. 363)
2. What is a calorie? (p. 363)
3. What molecule transfers energy from cellular fuel molecules to energy-requiring processes? (pp. 363–364)
4. Define aerobic respiration, anaerobic respiration, glycolysis, gluconeogenesis, lipolysis, lipogenesis, and glycogenolysis. (pp. 364–367)
5. Distinguish between a protein, a fatty acid, and a carbohydrate. (pp. 366–367)
6. What are the relative sizes of the energy stores of muscle glycogen, liver glycogen, and adipose tissue in an average person? (pp. 366–367)
7. What metabolic differences exist between the absorptive state and the postabsorptive state? (pp. 367–372)
8. In what form is most of the energy stored in the body, and where is this energy stored? (p. 368)
9. During a prolonged fast, what changes occur in the use of energy stores? (p. 370)
10. Of the total energy used in one day, about what fraction is used to support basal metabolic processes? (p. 372)
11. If an adult gains 20 pounds from age 30 to age 60, and if all that weight gain is fat tissue, what average percentage error in energy balance occurred during those 30 years? Assume the weight gain was equally distributed over the years. (p. 373)
12. What is an apoenzyme? A coenzyme? (p. 374)
13. What physiological mechanisms keep you warm on a cold day? What physiological mechanisms keep you cool on a hot day? (pp. 376–378)
14. What is fever? (pp. 378–381)
15. Does fever have a useful function? Is it ever dangerous? (p. 381)

Applying What You Know

1. Bookstores are filled with diet books, each advocating a different "secret" to weight loss, yet most people who embark on these diet plans have difficulty keeping the weight off. Why do you think this is so?
2. Endurance athletes often eat a large meal of carbohydrates the night before a contest. Why?
3. As a person develops a fever and body temperature rises, she feels cold. She shivers, curls up to conserve heat, and covers herself with blankets. Why does a sick person feel cold even though her body temperature is actually elevated?

4. When a fever breaks, the opposite occurs. The body responds as if it is overheated. The person feels warm, sweats, and kicks off the covers, even though body temperature is falling. Why?
5. Although there are some advocates of megavitamin regimens, even for people in good health, why are megadoses of most vitamins no more effective than the recommended daily allowance?

Chapter 15
Reproductive System

Objectives

By the time you complete this chapter you should be able to

◆ Understand the advantages of sexual reproduction.

◆ Describe cell division and the differences between mitosis and meiosis.

◆ Describe the anatomy of the reproductive tracts of both males and females.

◆ Understand the function of the primary sex organs.

◆ Describe the hormonal changes that take place at puberty.

◆ Explain the hormonal and physical changes that occur during a menstrual cycle.

◆ Describe the physiological events that lead to pregnancy, the sex act, emission of sperm, fertilization of the ovum, and implantation.

◆ Understand the nature of the sexual response in both male and female.

◆ Understand the changes in the developing fetus from the time of implantation to birth.

◆ Be familiar with the different methods of birth control and be able to discuss the advantages and disadvantages of each.

Introduction

Fossil bacteria are found in ancient strata, indicating that approximately four billion years ago bacteria evolved. These single cells developed the ability to divide and to produce other living cells that were essentially the same as the parent cell. Each cell, in turn, contained all the genetic information necessary to reproduce. This genetic information was gathered within the cytoplasm of each cell and was carried on a long molecule of DNA, as described in Chapter 2, that is called a chromosome.

During the process of division, or reproduction, shown in Figure 15.1, the genetic information of the cell first doubles in quantity. Next, the surface of the cell constricts. As it does, half of the genetic material flows to one half of the constricting cell, and half of the genetic material enters the other half of the cell. Finally the constriction pinches the single cell completely in half, making two identical cells with identical genetic material. If a cell transmits exactly the same information to the two

resulting halves, each must be an exact replica of the original cell. No matter how many times this happens, the future organisms should remain absolutely unchanged. However, we know change has occurred through the ages, and change is still with us. How?

About 1.5 billion years ago the first nucleated cells, or **eukaryotic cells,** evolved. This seemingly small change in structure allowed for a more rapid introduction of new species. Now genetic information was contained within the nucleus and individual chromosomes aligned the genes necessary for expression of the individual characteristics. Along with the nucleus came a unique method of reproduction. Cells began to exchange genetic information. The genetic identity of each individual of a species was unique; no longer was every organism an "identical twin." Furthermore, each reproductive episode led to another unique animal or plant. In short, a mechanism for rapid change was introduced into the reproductive cycle.

Sexual Reproduction

Humans have 23 pairs of chromosomes, one half of each pair coming from each parent. Thus, in each cell, there are usually two genes for every attribute. This is not to imply that the genes are identical. Rather, both genes code for a protein that has the same function, such as eye color. The genes contained on the chromosomes that determine sex are an exception, and will be discussed later in this chapter. The pairs of chromosomes are called **homologous chromosomes,** and because the chromosomes are paired we refer to the cells of the organism as being **diploid.**

We now recognize that **sexual reproduction** involves the coming together of genes from both parents to form a new individual with the same number of chromosomes and genes. This can happen only if each parent contributes half of its normal number of chromosomes to the progeny. Otherwise, each generation would have twice the number of chromosomes and genes as the parent generation.

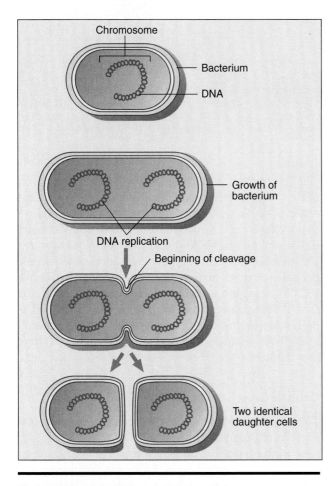

Chromosome

Bacterium

DNA

Growth of bacterium

DNA replication

Beginning of cleavage

Two identical daughter cells

Figure 15.1 Cell division of a bacterium.

Reproductive Physiology

Although researchers recognized for centuries that development of a new mammalian individual required the union of a male and a female, the nature of that union was not understood. How was it that the "secretions" of the male, when introduced into the female,

MILESTONE 15.1

Genetic Fingerprinting

Although one might like to think that single genes are arranged in perfect order, one after another on each chromosome, the truth is somewhat different. In 1985 an important discovery was made. Sections of the DNA molecule are repeated over and over, often one thousand or more times. This alone was intriguing, but even more interesting was the finding that the repeating sections of DNA are essentially unique for each individual, just as fingerprints are unique for each individual. Because of this variation in the makeup of the repeating fragments, they are called hypervariable regions.

Using modern techniques it is possible to remove these regions from chromosomes and sort them according to size. Approximately twenty such fragments are large enough to be sized from a single person's DNA. The fragments are placed on a thin sheet of gel and an electric current is applied to the top and bottom of the gel. The fragments move along the gel in response to the electric current, the larger fragments moving more slowly than the smaller ones. Thus, they arrange themselves in order of size along the gel. They then may be visualized when stained with a dye. The pattern of banding, as shown in the photo, is unique for each individual and is called a **genetic fingerprint.**

Columns 1 and 2 of the photo show the DNA fingerprints for one individual's blood and sperm, respectively. Columns 3 and 4 show DNA fingerprints from identical twins. Columns 5 and 6 are DNA fingerprints from two unrelated men. Note the correspon-

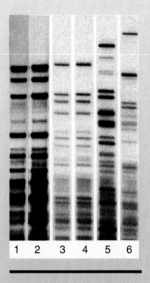

Genetic fingerprints.

dence of bands for identical twins and the noncorrespondence for the two unrelated men. It is this property of the hypervariable regions that allows one to determine whether a blood sample came from a particular individual, or if a semen specimen recovered from a rape victim came from the accused. Although there are some technical problems associated with the methodology, DNA fingerprinting has played, and probably will continue to play, an important role in legal cases.

At the present time DNA fingerprinting is admissible as evidence in many but not all states. There is always the possibility that two unrelated people will have the same DNA fingerprint, although the probability is very low. However, a clear difference in the DNA fingerprints indicates that two different people are represented. Thus, DNA fingerprinting can exclude suspects unequivocally, but not include a suspect with the same degree of certainty.

---

resulted in the formation of an embryo that had the capacity to emerge some time later as a new individual? Before the introduction of the high-power microscope it was believed that the male secretions coagulated when introduced into the female; this coagulum was thought to take form slowly and became the **fetus.** The invention of high-power microscopes has allowed us to view more clearly the processes involved.

Microscopic examination of **semen,** the whitish fluid secreted by the male reproductive tract, reveals that "swimming" in the fluid are small organisms called **sperm.** These are single cells containing a long, thin tail that beats so that it drives the sperm through the watery medium of the semen. It is the messenger bringing genetic material to its female counterpart, the **ovum** or egg. Figure 15.2(B) shows the details of a human sperm. It is divided into three sections: a head containing 23 chromosomes and a midpiece holding the mitochondria that supply energy for the last section, the tail.

The beating tail drives the sperm in the direction of the ovum. **Fertilization** occurs when a single sperm enters a single ovum and adds its genetic material to that of the ovum. Additional sperm that may contact the ovum are unable to penetrate the cell membrane. This fertilized ovum then divides, and subsequently becomes a new member of the species. In Figure 15.2(A), a single sperm has made contact with the surface of a single ovum and is about to penetrate it. This figure shows the coming union of the genetic material of mother and father. But how does that material come into being?

Cell Division

There are two kinds of human cell division. In one type, as noted in Chapter 2, the chromosome number remains the same during the process, which is known as mitosis. The other type of division is the type in which the chromosome number of the new cells is half

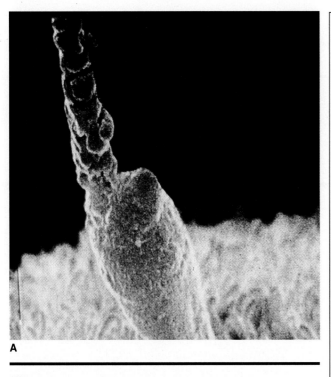

A

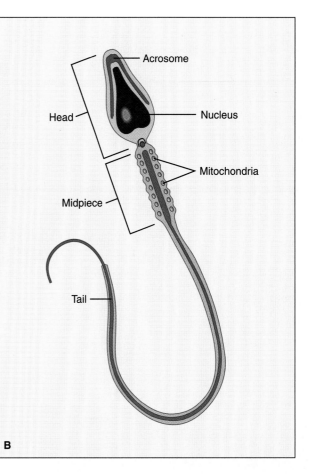

B

Figure 15.2 (A) Photomicrograph (×16,800) of a sperm contacting an ovum. (B) Enlarged diagram of a human sperm.

that of the parent cell; that is called **meiosis.** Meiosis occurs in the ovaries and testes as a means of generating reproductive cells.

    Mitosis The first step in mitosis is movement of centrioles to opposite ends of the cell. The centriole may be considered an anchoring point toward which migrating chromosomes move. Next, strands of DNA duplicate and condense, becoming visible as individual chromosomes, which then align themselves along an axis of the cell. A special attachment called the **spindle** forms, connecting each chromosome to a centriole. The point of attachment of the spindle to the chromosome is called the **centromere.** Next, the nuclear membrane disappears and the chromosomes begin to migrate along the spindle toward the two poles of the cell. One set of daughter chromosomes goes to one pole and the other set to the opposite pole. As the two sets of chromosomes approach the poles, the cell begins to divide, finally pinching into two new cells, each having the original number of chromosomes. This sequence of events is illustrated in Figure 15.3.

    The ability to sort out the complex stages of mitosis is due, in part, to the fortuitous characteristic of chromosomes. Each one has a unique form, so they can be followed individually. Figure 15.4 presents

photomicrographs of chromosomes in a dividing cell. Note that the paired chromosomes have different shapes that allow researchers to follow individual chromosomes through the process of cell division, and even to number each chromosome.

    Meiosis In both the male and female the meiotic divisions are similar, but occur at different times in their life cycles. The process of meiosis proceeds much as mitosis through the first cell division, with one important exception. Figure 15.5 illustrates this point. Following duplication, at the time they begin to separate and move to each pole of the dividing cell, for reasons not well-understood, the chromosomes may exchange homologous segments. The daughter chromosomes separate, as in mitosis, except that some of the chromosomes now have a different alignment of genes due to **crossing over,** an event in which segments of the maternal and paternal chromosomes break away, in some random manner, and exchange places with each other. This exchange of homologous segments serves to mix parental genes, allowing for even greater diversity of the genetic make-up of the reproductive cells, or **gametes,** and aiding continued evolutionary adaptation. At the end of the first mitotic cell division the diploid number of chromosomes

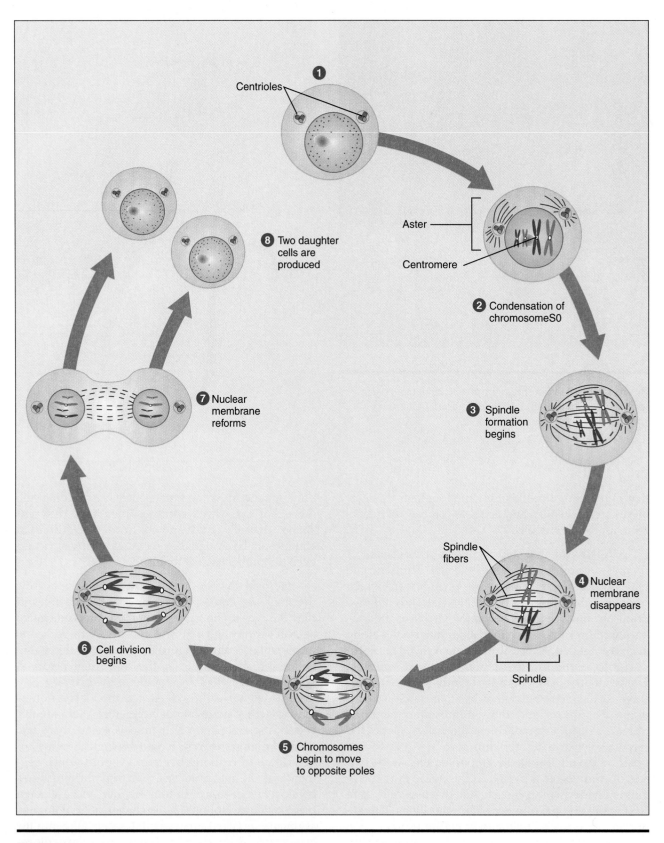

Figure 15.3 Single cell with two pairs of chromosomes before and during a complete mitotic division.

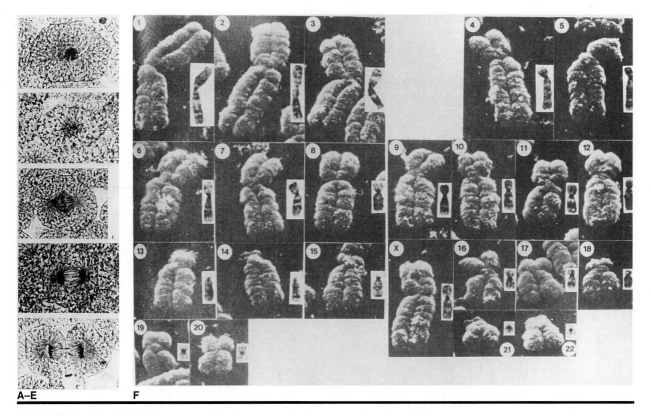

A–E   F

Figure 15.4 (A)–(E) Photomicrographs of an animal cell undergoing mitosis. The chromosomes are clearly visible, as are the spindle fibers along which the chromosomes will migrate. (F) Photomicrograph showing unique forms for each chromosome.

is maintained, but the genetic character of the individual chromosomes is different from that of the starting chromosomes. The cell now begins a second division, which is different from mitotic divisions.

The sequence of events from the primary germ cells, **spermatogonia** and **oogonia** (which give rise to sperm and ova), are similar, so they are pictured together in Figure 15.6. The spermatogonia and oogonia are produced by the testicles and ovaries, respectively. During the first meiotic division, homologous chromosomes remain joined rather than segregating in a random manner. As the cell divides, homologous pairs migrate to opposite poles as in mitosis, so that each daughter cell, called secondary **spermatocyte** or secondary **oocyte,** contains a half set of replicated chromosomes.

The second meiotic division now begins, giving rise to four daughter cells, each with one half the diploid number of chromosomes (haploid). Thus, a single meiotic sequence creates four haploid cells; each spermatogonium that completes meiosis gives rise to four sperm. However, if all primary spermatogonia underwent complete meiotic divisions, the spermatogonia would soon be used up and shortly after pu-

berty the male would become sterile. That does not happen. As shown in the left side of Figure 15.6, some of the spermatogonia do not complete meiotic divisions and therefore never become sperm, reverting back to primary spermatogonia. This allows the process to continue for many years.

The events described above begin at the onset of sexual maturity, or **puberty,** at about age 14 in the male. The timing and end results in the female are a bit different. In the female, the first meiotic division begins before birth, but it is not completed until ovulation begins many years later. At birth, the female contains about one million ova, and no new ova are produced during the rest of her life. Of those million ova, only about 300 to 400 reach maturity during a woman's reproductive life. The second meiotic division takes place only after fertilization of the secondary oocyte.

The final difference between male and female meiosis is that in the female, the sequence from oogonium to ovum does not result in an increase in the number of cells. At both meiotic divisions, one daughter cell is much smaller than the other and is nonfunctional; it cannot be fertilized and develop into an embryo. This cell is called a **polar body.**

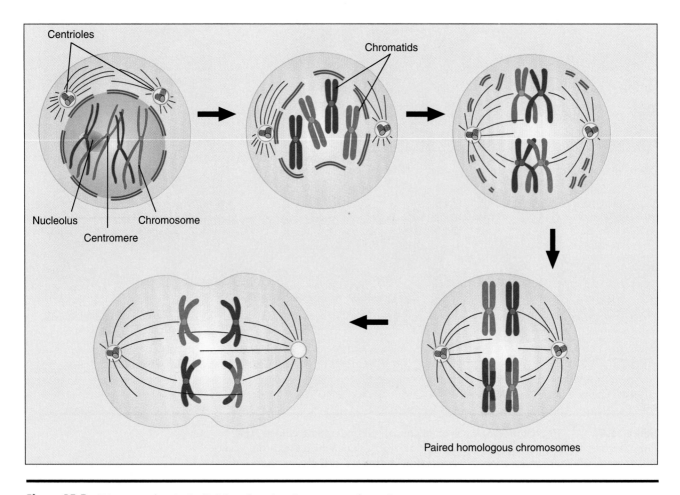

Figure 15.5 Diagram of meiotic division showing the process of crossing over.

Sexual Differentiation

Although humans have 23 pairs of chromosomes, only 22 pairs are homologous. The human chromosomes have been numbered arbitrarily by geneticists. The chromosome in position number 23 may be one of two different types. One chromosome of this pair is considerably larger than the other. The larger chromosome is named the **X chromosome** and the smaller one the **Y chromosome.** Examination of the chromosomes of males shows that their genetic makeup, or **genotype,** consists of one X and one Y chromosome in position 23. Females have the genotype XX. From these observations, it is evident that the sperm determines sex. Female ova contain only X chromosomes. Sperm, on the other hand, are divided equally between those bearing X and Y chromosomes. Thus, the likelihood of a newborn being male or female should be equal. However, for reasons not well-understood, slightly more than 50 percent of newborns are male.

Sex is determined by the genotype, but on occasion genotypes other than XX or XY occur. Some people are born with an extra chromosome, or even with one of the sex chromosomes missing. This type of error in chromosome distribution results in people having unusual sexual characteristics. A person born with an additional X chromosome, so that his genotype is XXY, develops as a male but has underdeveloped **testes,** the male organ that produces sperm. A female who inherits a single X chromosome with no accompanying X or Y chromosome develops as a female, but fails to develop the external **secondary sexual characteristics** that distinguish a male from a female, and is also sterile. It is as if we are all programmed to be female unless the appropriate male hormones are present.

A single gene on the Y chromosome determines the early development of the gonads. Presence of the Y chromosome results in the development of testes at about the seventh week of gestation. Absence of a Y chromosome causes the same embryonic tissue to develop into ovaries. The testes of the genetic XY embryo begin to secrete testosterone, which directs the development of the male reproductive organs. Absence of testes and of testosterone directs embryonic tissue to form the female reproductive organs.

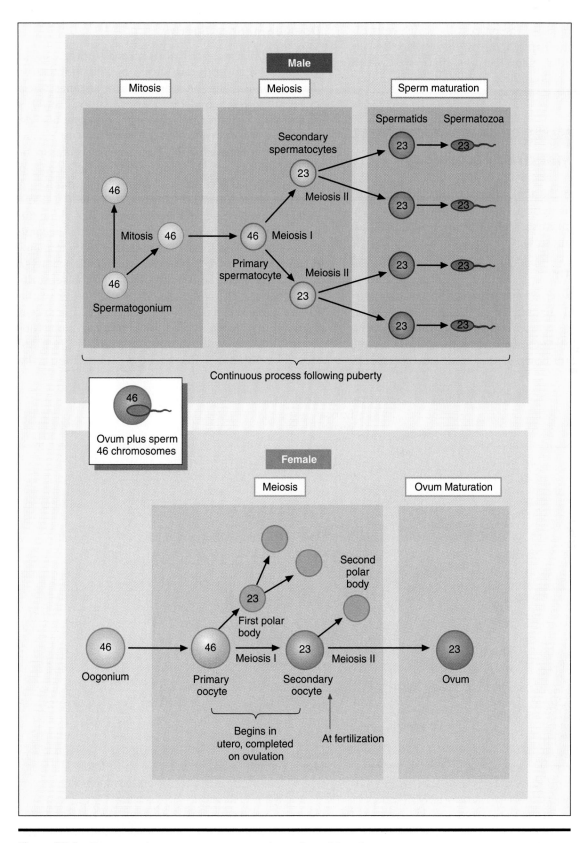

Figure 15.6 Diagram of meiosis as it occurs in the male and female gametes. The numbers in the figure represent the number of chromosomes in the cells at each stage of the cell divisions.

We have followed the process by which cells divide and become haploid, but we have not followed the process further, showing how sperm and ovum meet. In order to do so, it is first necessary to understand both male and female anatomy.

Male Reproductive Physiology

Although both the male and female partners contribute equal shares of genes to form a new human, the reproductive processes of the two sexes differ in anatomy and physiology.

Male Anatomy

Figure 15.7 shows the male reproductive structures, or **primary sexual characteristics.** The male genitals include the testes, encased by the scrotum, and the penis. The testes give rise to the sperm and also secrete the male sex hormone, testosterone, into the blood.

Development of Sperm

We have already described the process of meiosis, in which a cell is produced with a haploid number of chromosomes. This halving of chromosomes takes place in the testes and the ovaries. Figure 15.8(A) is a photomicrograph through a section of one **testicle** and illustrates sperm formation. That process is diagrammed in Figure 15.8(B). Note that the testicle contains tubular structures made up of large cells along an extracellular layer called the basement membrane. As the cells move toward the lumen, they change shape and get smaller. Spermatogonia are aligned along the basement membrane. As they divide, some become differentiated and follow the path that leads to meiosis and sperm formation. As the spermatocyte moves towards the lumen of the tubule, called the **seminiferous tubule,** it divides again, at the border of the lumen. Here it is released into the lumen. The sperm then move from the seminiferous tubules into an oblong body called the **epididymis** where they are stored and mature. This process is continuous; the

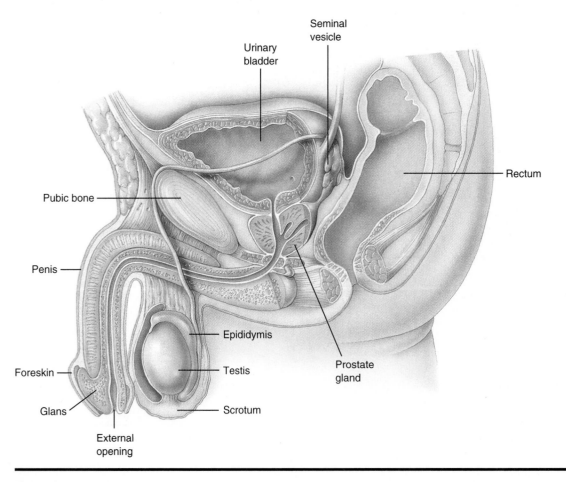

Figure 15.7 Cross-section of the male reproductive organs.

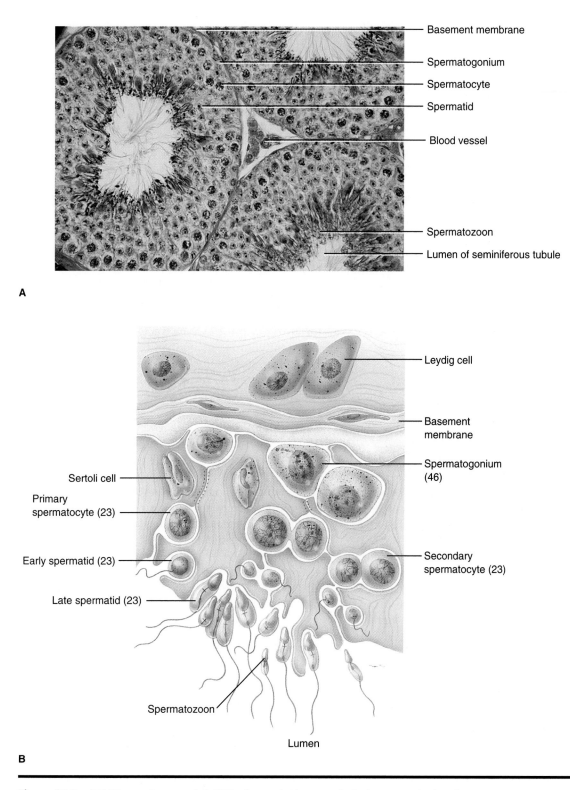

Figure 15.8 (A) Photomicrograph (×850) of a seminiferous tubule (cross-section) and surrounding tissue. (B) Formation of sperm in the testes.

sperm are released from the penis only periodically, during **ejaculation.**

Ejaculation

Ejaculation is a reflex discharge of semen from the penis. It is initiated by reflex stimulation of the parasympathetic nerves and inhibition of the sympathetic nerves. Contraction of the epididymis and seminal vesicles forces sperm into the urethra. At the same time other fluids are added to the stored sperm. The seminal vesicles produce a mucus-like fluid containing nutrients that empty into the ejaculatory duct and increase the volume of the ejaculate. The **prostate** gland also produces a fluid that is added to the sperm. During emission the prostate contracts along with the vas deferens, adding to the volume of the semen. The secretions of both the seminal vesicles and prostate supply nutrients and other substances that allow the sperm to remain active within the female for as long as two to three days.

Male Puberty

Thus far we have considered adult reproductive physiology. However, it is only at the age of puberty that the male secondary sexual characteristics develop. They are the external physical characteristics not associated with reproduction that define the male body: facial hair, deep voice, body shape, and muscle development. This generally begins in the male at about age 13 or 14, although the onset of puberty may be as early as age 11 or even as late as age 20. This great variability is normal.

At puberty, a number of changes take place in the male. At the onset of puberty, a hormone called **gonadotropin-releasing hormone (GnRH)** is secreted by the hypothalamus in the brain. GnRH in turn is transported by blood to the anterior pituitary gland in the base of the brain. Here it stimulates the anterior pituitary to secrete two hormones, **luteinizing hormone (LH)** and **follicle-stimulating hormone (FSH).** These hormones, in turn, act on special cells in the testes. FSH acts on **Sertoli cells,** causing the testes to secrete testosterone. LH stimulates **Leydig cells** of the testes to begin the maturation process that results in the production of viable sperm.

As puberty approaches, blood concentrations of the gonadotropins LH and FSH, released by the pituitary gland in the brain, all increase. Testosterone secretion increases dramatically and initiates a series of physiological and anatomical changes. Primary sexual characteristics develop as the testes, scrotum, and penis greatly enlarge. Additionally, the secondary sexual characteristics appear. The distribution of body hair changes; pubic hair and facial hair begin to grow.

The voice begins to deepen as the vocal chords and larynx enlarge. Muscular development begins to change so that the adult form becomes evident. During this same time, bone growth ceases and the male attains his adult height.

These changes are usually accompanied by an increased awareness of the sexual drive. We generally speak of a young man entering puberty as becoming aware of the other sex. It should be pointed out that until puberty, the male is sterile. Although he may be aroused, no sperm are produced, nor are sufficient fluids present to allow ejaculation, even though orgasm may be experienced.

Figure 15.9 shows the relationships between the hormonal and physiological changes during the onset of puberty. The hormones are secreted by the hypothalamus, the anterior pituitary, and the testes. FSH acts on the Sertoli cells of the testes, promoting spermatogenesis. At the same time, the Sertoli cells secrete the hormone **inhibin,** which acts on the anterior pituitary to inhibit the secretion of FSH.

LH stimulates the Leydig cells of the testes to increase testosterone secretion. Note the negative feedback system between testosterone and GnRH and testosterone and LH. Testosterone feeds back to the hypothalamus, inhibiting the secretion of GnRH. Thus, if testosterone levels rise very high, testosterone acts to decrease the secretion of GnRH, leading to a decrease in the secretion of LH. Falling levels of LH then cause a decrease in testosterone secretion. At the same time, testosterone also acts directly on the anterior pituitary to decrease LH secretion. In this way, testosterone levels are kept fairly constant. However, a question still remains. What triggers the release of GnRH at about the fourteenth year of life? We do not have the answer, and have yet to discover how the events from birth to puberty cause the changes outlined above.

Male Sexual Response

The **glans penis** (Figure 15.7) is richly supplied with sensory nerves that, when stimulated by touch, send impulses to the brain. This neural discharge results in the unique set of feelings, usually intensely pleasant, called sexual sensation. These feelings generally give rise to efferent nervous impulses that reach the penis through both the sympathetic and parasympathetic nerves. The autonomic stimulation causes dilation of the arterioles of the penis, compressing the venous sinuses in the penis and causing them to become engorged with blood. The increased arterial pressure and venous compression results in distention of the erectile tissue, which is known as an **erection.**

Although tactile stimuli are effective, other stimuli may also cause erection. Psychic stimulation, visual

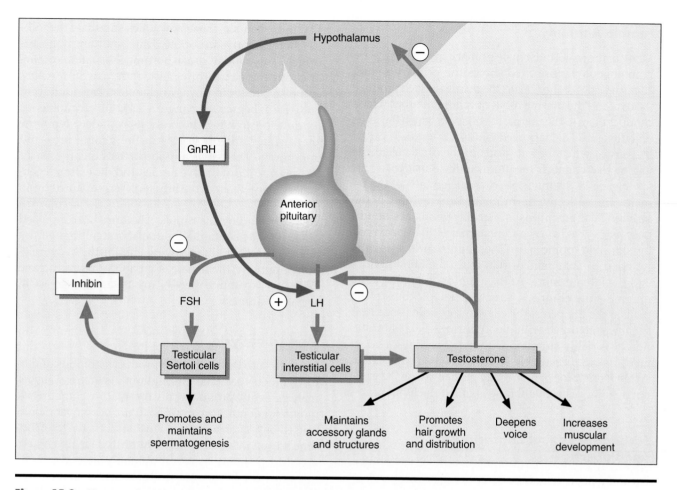

Figure 15.9 Hormonal relationships during puberty in the male.

stimulation, olfactory stimulation, and even auditory stimulation may lead to erection, and on occasion to ejaculation. Other stimuli may inhibit erection. The human sexual response is not only a very complex event, it is also unique for many individuals. No single stereotypical behavioral pattern can describe the human sexual drives. This normal variation of erogenous stimuli may be unique to humans, and adds a richness of experience and variety to the sexual encounter.

Intense stimulation initiates a series of reflexes that result in **emission,** or the contraction of the vas deferens, ampulla, seminal vesicles, and prostate. Semen is pushed into the base of the urethra, beginning a spinal reflex of rhythmic contraction of muscles within the penis that forces semen out of the urethra. Accompanying the act of ejaculation is an intense pleasurable sensation. The two events, emission and ejaculation, are called the male **orgasm.**

After orgasm, the male undergoes a period of resolution during which the erection is generally lost and a period of calm intervenes. A refractory period follows, when it is difficult for the male to be aroused.

This period may be minutes or hours. In general the refractory period usually lengthens with age.

Female Reproductive Physiology

Female reproductive function is somewhat more complex than that of the male. This follows from the fact that the male physiological contribution to the reproductive cycle is momentary, whereas the female contribution lasts nine months plus the time of nursing, if a woman chooses to breast feed. Additionally, the female reproductive process is cyclical. In general terms, however, the female shares equally with the male the important aspects of reproductive success of the species. She supplies half of the material for life in the ovum, which contains half the normal complement of chromosomes. Thus, the female and male each contribute half of the genetic material required for continuation of the species.

Female Anatomy

Many of the male and female primary sexual organs are homologous. Figure 15.10 shows the primary sexual characteristics of the female: the **ovaries,** which produce the ova, and the **uterine tubes** or **fallopian tubes** which conduct them to the **uterus,** the organ in which the fetus will develop. The narrowed end of the uterus is called the **cervix,** which in turn leads to the canal called the **vagina** that connects the uterus to the outside.

Figure 15.11 shows the external genitalia of the female. The collective term for these structures is the **vulva.** The **mons pubis** is a small prominence of adipose tissue that lies over the external genitalia, covered by skin and pubic hair. The vaginal orifice is surrounded by two pairs of skin folds, the **labia majora** and the **labia minora.** At the anterior end of the labia minora is the **clitoris,** a structure homologous to the penis. Like the penis, the clitoris is erectile tissue that becomes engorged with blood during sexual stimulation. It is richly supplied with nerve endings, and so gives rise to pleasurable sensations when stimulated. A thin membrane of tissue may partially close the vaginal opening. This membrane is called the **hymen.**

Secondary Sexual Characteristics and Puberty

As in the male, female puberty is triggered by the secretion of a cascade of hormones. GnRH secretion increases and, in turn, stimulates secretion of FSH and LH. In the female, these hormones stimulate the follicles in the ovaries to secrete their major hormone, **estrogen.** This sudden rise in hormonal secretions occurs at about the eleventh to thirteenth year in females, and also results in physical changes. The adolescent body form begins to change; a layer of subcutaneous fat develops, giving the body its rounded, feminine shape. Hair growth begins in the genital area and underarms. The pelvis and the breasts enlarge, as do the internal reproductive organs and the external genitalia. Enlargement of the breasts is accompanied by an increase in size and pigmentation of the **areola,** the area surrounding the nipple. The breast is composed of fatty tissue containing a ductal system that allows the secretion of milk following childbirth. At completion of puberty, the female is capable of reproduction. With this capability, there is an associated increase in sexual awareness.

Menstrual Cycle

In contrast to the male, who produces millions of mature gametes each day throughout his lifetime, the female generally produces only one mature gamete each month for about 30 years. (The figure of one per month is used only as an average.) Furthermore, the total lifetime supply of ova is present at birth. During embryonic life, as many as seven million potential ova, or oocytes, may be present in the ovaries. As development continues, that number decreases, so that at birth about 1–2 million ova are present. The number con-

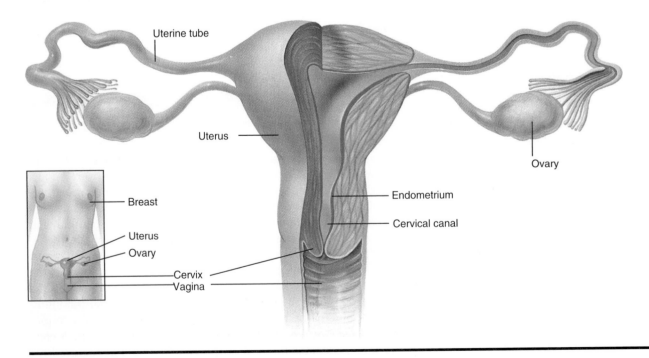

Figure 15.10 Anatomy of the female reproductive organs.

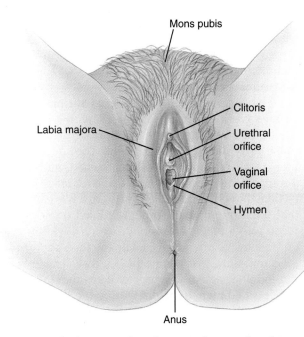

Mons pubis

Clitoris

Labia majora

Urethral orifice

Vaginal orifice

Hymen

Anus

Figure 15.11 External genital anatomy of the female.

tinues to decline until puberty, when only about 300,000–400,000 may approach maturity. Of these, only a small fraction, numbering only in the hundreds, ever develop to the stage at which they might receive a sperm. If one assumes a 30-year period of fertility, then less than 400 of the mature ova will be discharged from the ovaries and be available for fertilization. This periodic maturation of an oocyte in the ovary, coupled with physiological changes that occur in the uterus, is called the **menstrual cycle.** You will see that the menstrual cycle includes cycling of ovarian hormones and cycling of uterine anatomy.

Although puberty may begin as early as the tenth year of life, the first menstrual period usually occurs at about age 12 and is called **menarche.** The events during a single cycle are outlined below. We shall first consider the structural changes that occur during the cycle, and then consider the hormonal ones that accompany them.

Figure 15.12 illustrates the sequence of events of the maturation of an ovum and its discharge from the ovary, in clockwise manner, beginning with the primary follicle in the upper left. Many oocytes are embedded in a flattened layer of cells in the ovary. These structures are known as **primary follicles.** Each month as many as 20 of these follicles begin to mature. As the follicle enlarges, the primary oocyte undergoes a meiotic division, changing the primary oocyte to a secondary

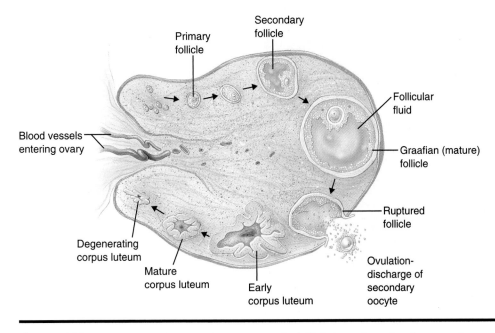

Primary follicle

Secondary follicle

Follicular fluid

Blood vessels entering ovary

Graafian (mature) follicle

Ruptured follicle

Degenerating corpus luteum

Mature corpus luteum

Early corpus luteum

Ovulation-discharge of secondary oocyte

Figure 15.12 Diagram of an ovary showing a single follicle in various developmental stages.

haploid oocyte and one polar body. As the follicle continues to enlarge, it fills with fluid; this is called a **Graafian follicle.** The photomicrograph in Figure 15.13 is a cross-section of a rabbit ovary showing the stages of maturation. Note that within the ovary, all stages of maturation from primary oocyte to Graafian follicle are present. Although many primary follicles begin the maturation process, all but one usually regress. In rare cases, more than one matures, accounting for the birth of non-identical twins or triplets.

The follicle ruptures when fully mature, releasing the secondary oocyte into and down the fallopian tube. The ovum is covered by a layer of granular cells, which is necessary for transport of the ovum along the fallopian tube, which conducts the ovum from the ovary to the uterus. Release of the ovum from the ovary is called **ovulation.**

Figure 15.14 shows a rabbit ovary removed from the animal and cultured in a salt solution. The ovum is seen being extruded from the ovary. When the ovum released into the fallopian tube is fertilized by a single sperm, the sperm adds its genetic material to the nucleus of the ovum, which rapidly undergoes the second meiotic division and casts off the second polar body. The fertilized ovum, known as a **zygote,** now has the diploid number of chromosomes, and begins development of a new member of the species.

Following rupture, the Graafian follicle fills with a yellowish material, hence the name **corpus luteum,** or yellow body. The corpus luteum secretes hormones necessary for maintenance of the uterus if the expelled

ovum is fertilized. However, most of the time fertilization does not take place. In this case, the corpus luteum atrophies, or gets smaller, and although remaining in the ovary, becomes a nonfunctional structure called a **corpus albicans.**

Changes also occur in the uterus during the menstrual cycle. In order for a fertilized ovum to attach to the uterine wall and develop normally, the uterus must be prepared for that event. The lining of the uterus, called the **endometrium,** thickens and exudes a nutrient-rich secretion that aids in nourishing the fertilized ovum if it reaches the uterus. It is estimated that only one in about three fertilized ova ever develop into a viable embryo. When fertilization does not take place, the uterus sheds the thickened endometrium along with some blood, discharging them through the vagina. This process takes about five days and is called **menstruation.** During the shedding period, about 100 ml of blood plus uterine tissue and tissue fluids are lost, although the volume of total blood and fluid lost is highly variable from woman to woman. During the menstrual period, cramps and discomfort may be experienced, but the severity is highly variable. Following menstruation, the menstrual cycle begins again, repeating for some 30 to 40 years.

Figure 15.15 plots both the ovarian changes and the endometrial changes that take place throughout the menstrual cycle. The hormone levels in blood that initiate those changes are shown in the mid-section of Figure 15.15. The anatomical changes in the ovary and

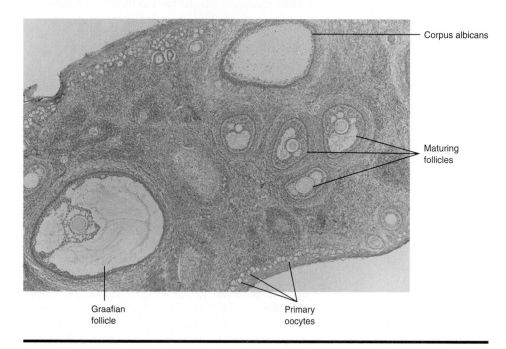

Figure 15.13 Photomicrograph of a rabbit ovary showing some of the same structures as in Figure 15.12.

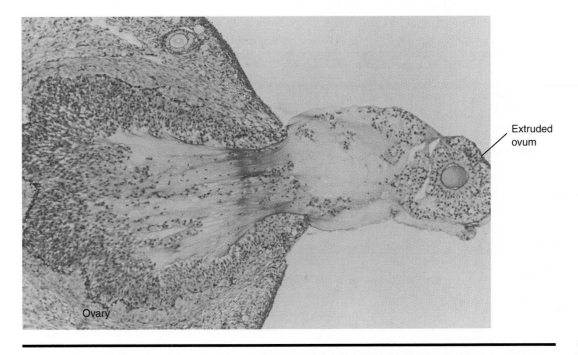

Extruded
ovum

Ovary

Figure 15.14 Photomicrograph capturing the moment of ovum expulsion from a rabbit ovary.

uterus are illustrated in the top portion of the figure. In Figure 15.15, the cycle begins on day one of menstruation. During this time, the major hormones involved in the menstrual cycle are at their low points, the endometrium is being expelled, and the corpus luteum is becoming nonfunctional and evolving into the corpus albicans.

Shortly after cessation of menstruation, about day five, increases in blood levels of FSH and LH begin, stimulating follicular growth. Although many follicles begin to enlarge, generally only one matures into a Graafian follicle. The stimulated follicles then begin to secrete increasing amounts of estrogen, acting on the uterus to cause rapid growth of the endometrium. At about the midpoint of the cycle (approximately day 12), the high blood levels of estrogen stimulate GnRH, which in turn causes the sudden release and rapid surge of LH and FSH levels in blood. This surge induces ovulation.

With ovulation, estrogen levels fall rapidly, as do FSH and LH. The Graafian follicle becomes the corpus luteum, reducing its secretion of estrogen, as it begins to secrete increasing levels of another hormone, **progesterone.** Progesterone aids in stimulating growth, integrity, and glandular function of the endometrium. About ten days later, the corpus luteum begins to degenerate, diminishing its secretions, and at about 28 days degeneration is completed. As the blood levels of estrogen and progesterone fall, their stimulatory effect on endometrial growth also falls, until finally men-

struation begins again. You should note that the scale of days on the horizontal axis of Figure 15.15 represents the average for women. Although the preovulatory phase is rather constant, for some women the entire cycle is longer, for some shorter, and for some the cycle is irregular.

In some individuals the appearance of the secondary sexual characteristics and menstruation is delayed significantly. Young women who train strenuously as athletes and women who diet excessively or who suffer from anorexia show delayed onset of puberty. After menarche, the menstrual cycles may become irregular and may even cease. The precise reasons for this are not yet known, but seem to be associated with physiological changes associated with the loss of body fat.

Premenstrual Syndrome (PMS) PMS is a cluster of symptoms often experienced by women preceding menstruation and sometimes continuing during menstruation. The intensity and number of symptoms are variable and may consist of one or more of the following: joint pain, abdominal pain, anxiety, irritability, headache, retention of fluids with attendant weight gain, swollen or tender breasts, and mood swings. Although the precise cause of PMS is not known, some treatments have been used to reduce the severity of symptoms, including aspirin, tranquilizers, antidepressants, progesterone, dietary changes, and diuretics. However, the particular treatment for any person is

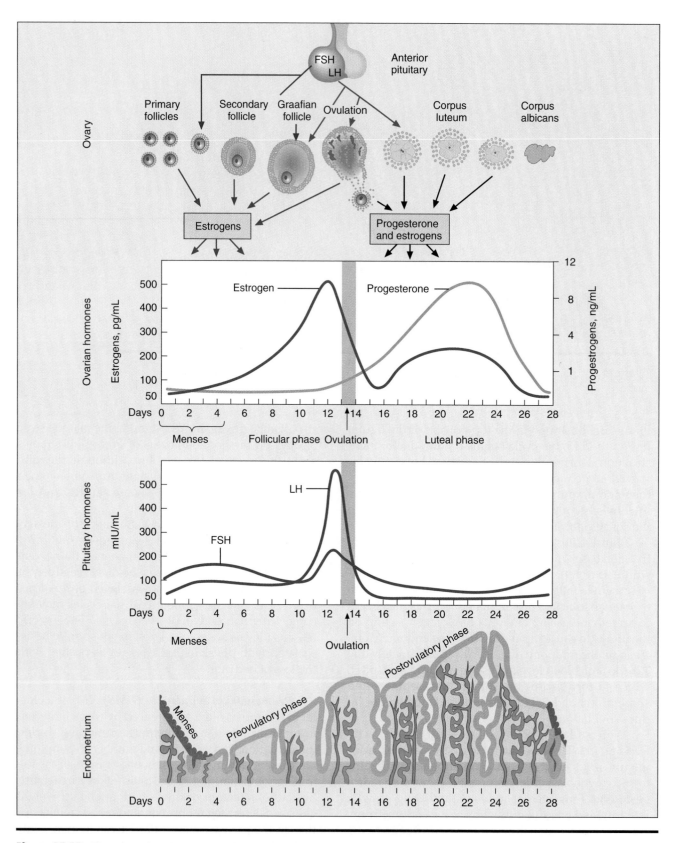

Figure 15.15   Diagram showing the changes in hormone levels and uterine and ovarian anatomy during a single menstrual cycle.

individualized. It is estimated that about one-third of women of childbearing age experience PMS.

Menopause Sometime between the ages of about 40 to 50, for most women, the menstrual cycle becomes irregular and finally ceases. Cessation of menstruation, or **menopause,** is accompanied by hormonal changes. Both estrogen and progesterone secretion are diminished. Some women experience one or more of the following: periodic hot flashes and sweating, headaches, mood swings, muscular aches, and depression. Fortunately for most women, these uncomfortable manifestations are usually transient, but for others they may last years; the length of the symptoms is very variable. Menopause is usually accompanied by a decrease in breast size and a reduction in the secretion of vaginal lubricants during sexual intercourse.

Female Sexual Response

In general, the male and female sexual responses are similar. During excitation, the clitoris and labia minora become engorged with blood and swell. The vagina becomes lubricated with vaginal secretions. Sexual arousal results from physical stimulation as well as psychic stimulation. At some stage of sexual contact orgasm (climax) may occur and is also described as intensely pleasurable. Although the male requires a variable period of rest before being able to reach another orgasm, the female may experience multiple orgasms with little or no interval between them. In this respect the two sexes differ in their sexual response.

Pregnancy

Pregnancy begins only after fertilization of the ovum by a sperm. It includes the events between fertilization and delivery of the baby. Although there is some variation in the length of time between the two events, the average time is 266 days, approximately nine months.

Fertilization

Discharge of the ovum into the fallopian tube is the first step required for pregnancy to occur. As many as 300–500 million sperm may enter the uterus during intercourse. The motility of the sperm and ciliary motion within the uterus propel the sperm along the reproductive tract. Of the many millions of sperm introduced into the female, perhaps only a few hundred survive and enter the fallopian tubes. The movement

from cervix to fallopian tubes takes a few hours. Although mature sperm are introduced into the female during intercourse, the sperm are incapable of fertilizing an ovum until they reside in the uterus for a few hours. This time is necessary for **capacitation** of the sperm. A layer of protein molecules is removed from the sperm. In the area of the ovum, additional changes in the sperm occur; they take on a lurching motion and become able to fuse with the ovum. This later process is called **activation.**

The mechanisms responsible for capacitation and activation are not well-understood. The best guess at the present time is that while sperm are stored in the male reproductive tract, a component of the semen prevents activation. Mature sperm residing in the male reproductive tract are exposed to various inhibitory factors that suppress their ability to fertilize an ovum. When sperm enter the uterus, uterine fluids wash away those factors, enabling one sperm to fertilize the mature ovum.

Embryonic Development

Figure 15.16 diagrams the sequence of events during the week following ovulation. The ovum slowly moves down the fallopian tube, where it comes into contact with sperm. One may fertilize the ovum, which then undergoes a series of mitotic divisions. From this stage until about eight weeks later, the developing organism is called an **embryo.** After about three days, the multiple cell divisions transform the single cell into a **morula,** a compact body of about sixteen cells. Further cell divisions cause the morula to enlarge and become a hollow, fluid-filled ball of cells, the **blastocyst,** which is ready for implantation upon reaching the uterus. At this point, about one week after ovulation, the lining of the uterine wall is maximally developed. Cells of the blastocyst grow into the endometrium, attaching the blastocyst firmly to the uterine wall. It is usually at this stage, called **implantation,** that pregnancy is considered to have begun.

Development of a Fetus

It is customary to divide the period of pregnancy into thirds, or **trimesters.** However, because the process from fertilization to birth is continuous, no clear-cut separations between trimesters exist. Figure 15.17 shows some stages of the development of the embryo during the first trimester. After two weeks the embryo begins to take form, in another two weeks the umbilical stalk becomes recognizable, and at five weeks the placenta is evident. The **placenta** is the structure that connects the developing embryo to

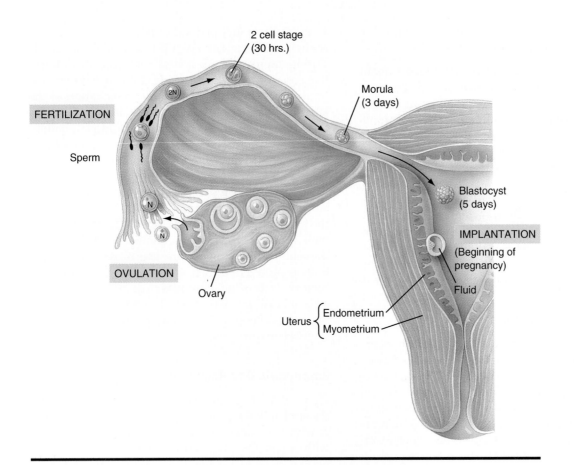

2 cell stage
(30 hrs.)

2N

FERTILIZATION

Sperm

Morula
(3 days)

N

N

OVULATION

Ovary

Blastocyst
(5 days)

IMPLANTATION

(Beginning of
pregnancy)

Fluid

Uterus { Endometrium
Myometrium

Figure 15.16 Schematic diagram showing the changes that occur in a fertilized ovum from ovulation through implantation.

MILESTONE 15.2

In Vitro Fertilization

In some cases, either the male or the female partner may be infertile. The male partner may not be able to supply enough sperm during normal intercourse for fertilization to take place. Recall that a few hundred million sperm are usually ejaculated during intercourse. For reasons not well-understood, if the number of sperm is too low, fertilization may not take place.

Some females may also be unable to conceive because of the presence of a barrier to the sperm somewhere in the reproductive tract. If for some reason the fallopian tubes are blocked, sperm may not be able to reach the mature ovum. In such cases it is possible to remove mature ova from the female and collect sperm from the male. The two samples are combined in a small dish or test tube filled with an appropriate fluid medium. Because sperm and ova are confined in a small space, far fewer sperm are necessary for fertilization to take place (only about 50,000 to 100,000 sperm per ovum). When the fertilized ova develop to about the 10-cell stage, they are transferred to the uterus of the female.

In the majority of cases the in vitro fertilized ovum goes on to develop normally. However, the rate of spontaneous abortion of the fetus is about 30–40 percent. Although this seems to be a high figure, it is approximately the rate of in vivo spontaneous abortions.

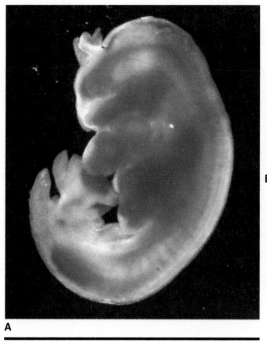

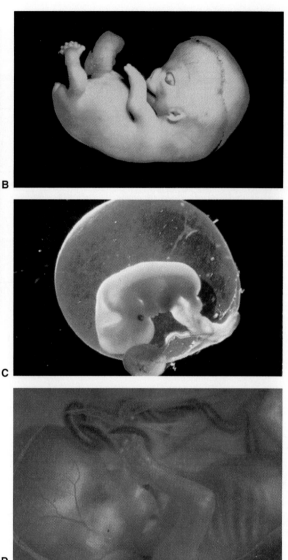

Figure 15.17 Photographs of a developing human embryo. (A) 4 weeks. (B) 6 weeks. (C) 7 weeks. (D) 12 weeks.

the mother and nourishes it from the mother's blood until birth. By the ninth week, the limbs are defined. As the first trimester ends, the developing embryo is called a **fetus.** At this point, the fetus is only about the size of a thumb and is many weeks away from being able to survive outside the uterus. Six more months are required for complete development.

Figure 15.18 shows how the placenta is attached to the uterus by the umbilical cord, which contains the fetal artery and vein. Blood from the mother supplies the placenta so the two streams permit the transfer of nutrients from mother to fetus. It is important that the mother eat a nutritious diet high in iron and other minerals. The mother must supply to the developing embryo all the substances required for building a new human.

If the mother ingests drugs or alcohol or if she smokes tobacco, toxic substances may cross the placenta and enter the blood of the developing embryo or fetus. Depending on the dose and the frequency of exposure, irreparable damage may be inflicted on the fetus.

Substances that produce birth defects are called **teratogens.** One of the most common teratogens is alcohol. As little as two alcoholic drinks per day may lead to irreversible damage to the fetus. Tragically, the child may be born with malformed limbs and face and severe mental deficiency. The constellation of symptoms resulting from alcohol consumption is known as **fetal alcohol syndrome (FAS).** A child born with FAS is shown in Figure 15.19.

Although it is clear that alcohol is a true risk factor for the developing embryo, it is not possible to quantify the volume of alcohol needed to effect a specific change in the development. Current evidence indicates that the fetus is most susceptible to damage during the first trimester. It is also clear that chronic use of alcohol constitutes the greatest risk. In light of

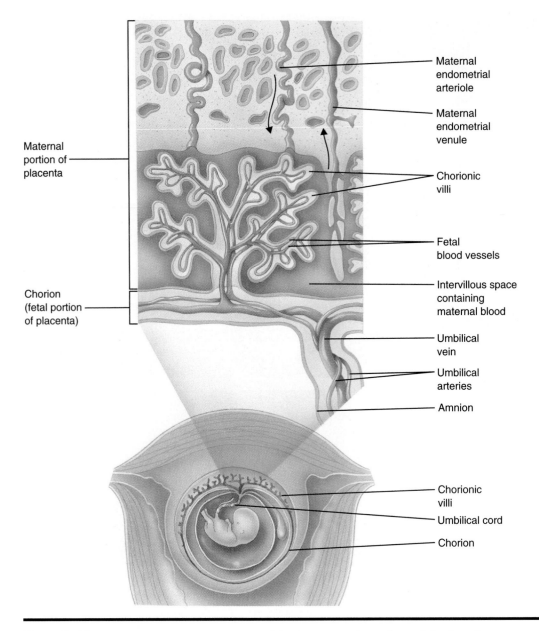

Figure 15.18 Diagram showing the anatomy of the developing fetus, the umbilical cord, and the placenta.

these findings, it is recommended that a woman abstain from drinking alcoholic beverages throughout her entire pregnancy. She should also consult a physician before taking any medication, or even high doses of vitamins, as these substances may have a deleterious effect on the developing embryo.

Endocrine Maintenance of Pregnancy

During the first 5–6 weeks following implantation, the ovaries are required for maintaining the pregnancy. During the early phase of pregnancy, estrogen

and progesterone levels in blood rise due to increased ovarian secretion of these hormones. If the ovaries are removed before the second month of pregnancy, the fetus will not survive and will be spontaneously aborted. However, as the placenta matures, it begins to secrete both estrogens and progesterone, as well as some other hormones of pregnancy. At this stage, hormone secretion by the placenta is sufficient to maintain pregnancy.

As the pregnancy continues, other physiological changes take place, preparing the mother for birth. The breasts enlarge as a result of increased secretion of

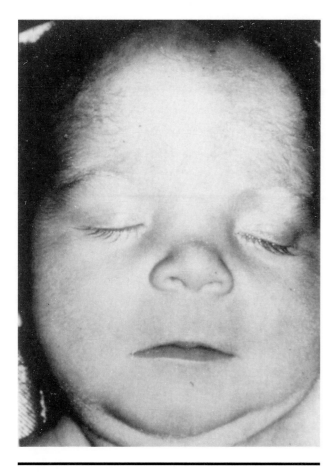

Figure 15.19   Child born with fetal alcohol syndrome.

progesterone. This enlargement increases the ductal system, preparing the breasts for milk secretion, which will allow her to nurse the baby.

Parturition

Development of the fetus continues until birth, or **parturition.** Although it is not entirely clear what initiates parturition, animal research has given some clues as to the initiating causes of labor. Sheep are large enough that it is possible to monitor hormone levels in their blood continuously throughout pregnancy. During the last few days of pregnancy, progesterone levels fall rapidly. At the same time, prostaglandin synthesis and release from the uterus increases. Among the complex hormonal changes occurring in the final stage of pregnancy is an increase in the pituitary hormone **oxytocin.** Oxytocin plays an active role along with prostaglandin in stimulating uterine contractions.

Uterine contractions are similar to the peristaltic waves in the intestines, as described in Chapter 13, that force a bolus along a tubular structure. Uterine contractions begin at the top of the uterus and move

downward, forcing the fetus toward the cervix (Figure 15.20). In the early phase of labor, estrogen levels in blood increase and initiate weak and irregular contractions, causing minor discomfort. As the fetus moves toward the cervix, it initiates a reflex increase in oxytocin secretion and prostaglandin release, both acting to increase the frequency and strength of uterine contractions. The increase in the strength of the contractions is felt as increasing discomfort, or labor pains, and the cervix dilates in preparation for delivery of the fetus. Further stimulation of the cervix causes positive feedback as still more prostaglandin is released, causing still stronger contractions.

At some point during labor, the contractions are sufficiently strong that the amniotic sac surrounding the fetus ruptures and the contained fluid leaks out of the vaginal opening. Figure 15.20 shows the progression of the fetus from the uterus, through the cervix, and finally into the external world.

Following birth of the newborn, the umbilical cord connecting the baby to the placenta is tied off and a short while later the placenta detaches from the uterine wall and is expelled by the continued contractions. This latter event is called the **afterbirth.** The cervix can now relax and prostaglandin and oxytocin concentrations in blood fall.

Lactation

During pregnancy, the high levels of estrogen and progesterone prepare the breasts for milk secretion or **lactation.** Estrogen stimulates duct development and progesterone stimulates alveolar development. The breasts enlarge and the developed alveoli become capable of producing milk, which may be stored in the newly enlarged alveolar and ductal system (Figure 15.21).

Although the breasts are well-prepared for producing milk by the middle of gestation, none is produced or secreted. This is because the high levels of progesterone and estrogen that prepare the breasts for milk production also inhibit the action of the hormone **prolactin,** which is secreted by the anterior pituitary. Prolactin is the hormone that stimulates the actual production of milk by the glandular tissue of the breasts. It is not until parturition, when the levels of estrogen and progesterone diminish and their inhibitory influence on prolactin is reduced, that prolactin is able to exert its stimulatory effect on the breasts.

When the baby begins to suckle, or in some cases, when a nursing mother hears the baby crying, a neural reflex initiates the secretion of oxytocin by the posterior pituitary. Oxytocin stimulates ejection of milk, a process called **milk let-down.** As long as the baby is breast-fed, both LH and FSH are inhibited. Because these hormones

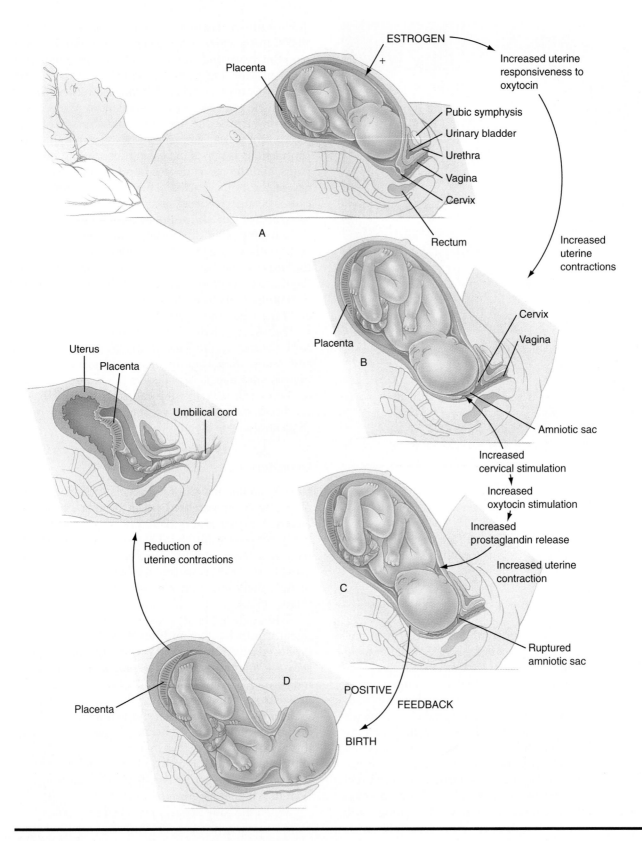

Figure 15.20 Sequence of events during parturition.

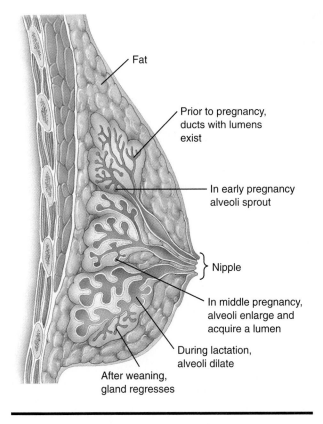

Fat

Prior to pregnancy, ducts with lumens exist

In early pregnancy alveoli sprout

} Nipple

In middle pregnancy, alveoli enlarge and acquire a lumen

During lactation, alveoli dilate

After weaning, gland regresses

Figure 15.21 Schematic diagram of changes in the breast before pregnancy, during lactation, and after weaning.

are required for ovulation, ovulation is also inhibited, although the inhibition is not 100% effective. The reduced levels of FSH and LH reduce the probability of conception during the breast-feeding period.

Fertility Control Options

To this point we have considered the mechanisms of reproduction; however, prevention of the reproductive process is often the goal of both married and unmarried couples. A variety of methods now exist to prevent ovulation, fertilization, or implantation following fertilization.

Barrier Methods

One general method of preventing fertilization is to place a barrier between the sperm and ovum and thus prevent their union. The barrier may be placed over the penis or within the vagina.

Condom As soon as it was recognized that the introduction of male semen into the female uterus was required for conception, a method of birth control became evident. If a barrier could be introduced between the penis and uterus, fertilization could not occur. The most effective method for this is using a **condom,** which is a very thin rubber sheath placed over the penis before intercourse. The ejaculate is trapped in the sheath, preventing fertilization. Although this is a very effective method, some find it objectionable. It requires planning because it must be applied before initiation of intercourse, and some men and women find that the tactile sensations are significantly altered.

Perhaps most importantly, the condom protects against sexually transmitted diseases. Neither bacteria nor viruses are able to cross the membrane of the condom, so both partners are protected from sexually transmitted diseases.

Diaphragm A diaphragm is a circular ring of rubber, individually fitted by a physician for the woman. It is placed so that it fits snugly over the cervix. The diaphragm must be in place before intercourse, and to be effective should be used with a spermicidal jelly. Following intercourse it should be left in place for a few hours so that no living sperm can contact an ovum. If used properly, both the condom and the diaphragm are effective in preventing pregnancy, although neither is 100% effective.

Cervical Cap The **cervical cap** is similar to a diaphragm, but smaller, and may be left in place for one to two days. Like the diaphragm, it must be fitted by a physician and used with a spermacide. It is placed over the cervix before intercourse and should be left in place for a few hours afterward.

Sterilization

There are a few methods of birth control that might be called **sterilization.** By that we mean any method by which a person is made incapable of reproducing. The various methods of sterilization may be reversible or nonreversible.

Tubal Ligation In the female, the fallopian tubes may be surgically cut and the cut ends tied off. This process, called **tubal ligation,** allows for completely normal menstrual cycles, but following ovulation, the ovum cannot traverse the fallopian tube. Conversely, sperm that have entered the uterus during intercourse are unable to travel up the fallopian tube to meet an ovum. Although excellent microsurgical techniques have been very successful, tubal ligation represents a small risk that some are unwilling to take. Any invasive operative procedure carries the possibility of

HIGHLIGHT 15.1

Amniocentesis

In recent years it has become possible to identify the genes responsible for certain genetic abnormalities. For example, the chromosomes or genes that are responsible for Down syndrome, cystic fibrosis, Tay–Sachs disease, sickle-cell anemia, and a few hundred other genetic defects are now known and have been characterized. Down syndrome is caused by a chromosomal error during meiosis. An extra chromosome is included in a gamete so that the child has three copies of chromosome number 21. A person with Down syndrome shows mental retardation and usually a shortened life span. Tay–Sachs disease results from a mutation of a single gene. A child born with this genetic defect suffers from brain degeneration and an early death at about age four or five.

Modern techniques of analyzing genetic material from the fetus allow a pregnant woman to get genetic counseling not previously available. The ability to counsel women results from the fact that the developing embryo or fetus sheds some of its cells into the amniotic fluid. During the second trimester, or occasionally in the later part of the first trimester, sufficient cells exist in the amniotic fluid for genetic analysis. The figure shows a schematic diagram of the method called **amniocentesis** used for obtaining amniotic fluid. A physician inserts a needle through the abdominal wall of the mother into the amniotic fluid, using only a local anesthetic. The cells in the fluid are then subjected to genetic analysis.

In this way, the pregnant woman can get excellent information regarding the genetic status of the fetus. In some cases this early knowledge allows for treatment of the newborn before overt symptoms become apparent. An example of this is the genetic disease **phenylketonuria (PKU).** Children with this defect are unable to metabolize properly the amino acid phenylalanine. If untreated, the child may be severely retarded. Fortunately, the condition can be treated successfully if it is detected before or shortly after birth.

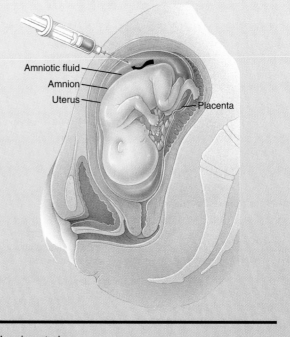

Amniocentesis.

In some cases, no known treatment is available and decisions must be made concerning the continuation of pregnancy. Genetic diseases such as Tay–Sachs and cystic fibrosis currently have no cure. Cystic fibrosis shortens life span by about half, but there is now more hope for people with this disease. Researchers have been able to insert the gene for cystic fibrosis into mice, causing these animals to show the symptoms of the disease. Use of an animal model will allow researchers to gather much information concerning cystic fibrosis that would not be possible otherwise. Perhaps in the near future a cure or preventative will be found.

---

infection or postoperative complications. There is also the possibility that the sterilization procedure will be permanent, even though techniques have been devised to reattach the ends of the tubes. The reattachment techniques are not always successful.

Vasectomy Male sterilization is an operation called a **vasectomy,** which is simpler than tubal ligation. A small incision is made in the scrotum and the vas deferens are cut and tied off. This prevents the sperm from reaching the seminal fluid, and thus they are not ejected with the semen. Because the sperm

make up only a small part of the ejaculate, the volume ejected during orgasm is essentially normal. As with tubal ligation, methods have been worked out to reattach the cut ends of the vas deferens should the individual wish to father children, but this method is not always reversible.

Oral Contraception A second method of contraception is chemical, rather than surgical. The oral contraceptive methods depend on altering the hormonal changes that occur naturally so that ovulation or implantation does not occur. The most commonly used method is a pill containing a high concentration of progesterone and a lesser one of estrogen. As noted earlier, high levels of progesterone and estrogen inhibit the FSH and LH secretion that are required for ovulation. Some people exhibit side effects such as weight gain, nausea, headaches, and other minor problems. Although this is an effective method of birth control for those who are able to use it, the pill is not recommended for women who have a history of hypertension, breast cancer, or liver disease.

Although the above hormonal method is widely used, other combinations of hormones are available. Long-acting solutions of hormones may be injected or even implanted subcutaneously (under the skin). Injected contraceptive hormones may prevent pregnancy for a few months; implanted hormones may prevent pregnancy for up to a few years.

Intrauterine Devices An **intrauterine device (IUD)** is placed in the uterus. The size and shape of IUDs are variable. Most are made of some plastic material, and have a tail attached to the body of the IUD. It is not quite clear why this is so, but researchers have found that introduction of an IUD causes a local inflammatory reaction in the endometrium of the uterus that may prevent implantation or hinder fertilization. Regardless of the type used, it must be inserted into the uterus by a physician. IUDs have been shown to cause pelvic inflammatory disease in some women, and even infertility. For this reason their use has declined.

Rhythm Method

Although the rhythm method is not strictly a method of contraception, it is included under that heading as it is designed to take advantage of the period during which a mature ovum is not ready for fertilization. Among the changes associated with the menstrual cycle is a change in basal body temperature. About a day or so before ovulation, body temperature falls slightly, and following ovulation it increases about 1°F. If daily temperature readings are kept for a period of months, some idea of the time of ovulation may be determined. The object is to abstain from intercourse during that period.

One difficulty here is that the temperature changes are small, and vary among women. Other factors can also change body temperature and may interfere with the determination of when ovulation will take place. Also, the sperm and ovum may live as long as two to four days, respectively, in the female reproductive tract, so it is necessary to have no intercourse for at least two days before and following ovulation. The rhythm method for most people is ineffective.

Efficacy and Safety of Contraceptive Methods

Although many different methods may be used to prevent pregnancy, they are not all equally effective. The choice of any one method depends on the individual's own belief system as well as knowledge of the probability of success with each contraceptive method. Table 15.1 presents the best estimates of the effectiveness of some of the currently used methods. The failure rate is estimated per 100 women over a period of a year. This means that there will be the number of pregnancies shown in 100 women of childbearing age who use the method in question for one year. The only method known to be 100% effective is abstinence.

Termination of Pregnancy

For a variety of reasons some people wish to terminate their pregnancies. A few such methods are available and usually entail small risk to the woman carrying the fetus.

Surgical Methods A few surgical methods are available for inducing abortion. Such methods as saline injection, suction of the uterus, or surgical scraping of the uterine lining all may be used. These methods entail some small surgical risk, but the health risk associated with medical abortions is less than that encountered in completing pregnancy and childbirth. These methods are also highly controversial as they entail ethical considerations.

Table 15.1	**Rates of failure for different methods of contraception.**
Method	**Estimated Failure Rate per 100 Women Years***
Oral contraceptives	0.3
Subdermal implant	
Year 1	0.2
Year 2	1.2
Year 3	2.7
IUD	2.5
Diaphragm	2.4
Condom	2.0
Tubal ligation	0.01
Vasectomy	0.2
Rhythm method	19
Abstinence	0.0

*Failure rate is estimated as the number of women who get pregnant for every 100 women using one of the methods above during their reproductive years. The estimate assumes each woman will use that method every time, and do so properly. However, the true failure rate is usually higher than the estimated rate, due to failure of compliance and differences in individual physiology.

Chemical Method A successful pregnancy requires that the uterus be prepared for implantation, and that throughout fetal development it nourish the fetus. A drug called **RU486,** given as a pill, blocks the action of progesterone in preparing the uterus for implantation. If RU486 is given shortly after fertilization, pregnancy is prevented because the zygote does not attach to the endometrium. If implantation has already taken place, RU486, followed by administration of prostaglandin, causes uterine contractions that expel the placenta and fetus. At this writing, RU486 is available only on a trial basis in the United States, but is used successfully in many other countries.

Summary

In order for any species to thrive, it is necessary for the genetic material of the individual to be passed on to his or her progeny. Animals such as humans require that this genetic material be recombined so that one set of genes from each parent is passed on to the child. The gametes of each parent undergo a cell division in which the chromosome number (and number of genes) is halved; this process is called meiosis. Thus, when sperm and ovum meet, the resulting offspring have the same number of chromosomes as each parent. Meiotic divisions ensure that genetic variability is maximized. Although both the sperm and ova undergo meiosis, there is one significant difference in their development. The male produces new sperm for most or all of his life. The female is born with a lifetime supply of ova.

Both sexes are sterile during their early years, becoming fertile at puberty. The physiological and anatomical changes that occur during this period are complex. Both male and female develop secondary sexual characteristics and become physiologically and mentally prepared for reproduction. The female begins menstruation, part of a monthly reproductive cycle that will continue for about 30–40 years. The menstrual cycle requires a cycling of different hormones that prepare the uterus for possible implantation of a fertilized ovum. If the ovum is not fertilized, the enlarged portion of the inner lining of the uterus is shed, causing some bleeding. This condition lasts a few days and is called menstruation.

If a viable ovum is fertilized, implantation may occur, followed by pregnancy and birth. However, pregnancy may be avoided by use of any one of many methods. Two general methods are reliable. Barrier methods prevent the union of sperm and ovum. Hormonal methods may prevent ovulation or implantation. It is also possible to cause rejection of the fertilized ovum with the use of appropriate hormones. Although many methods may be used to prevent pregnancy, only the use of a condom offers protection against sexually transmitted diseases.

Questions

Conceptual and Factual Review

1. What is a chromosome? How many does a human have? (p. 385)

2. What is the difference between a diploid cell and a haploid cell? (p. 389)

3. Which gamete, the ovum or sperm, determines the sex of the fetus? Why? (p. 390)

4. Identify the male and female primary and secondary sexual characteristics. (pp. 392, 394, 396)

5. What are the primary male and female anatomical sexual characteristics? The secondary characteristics? (p. 392, p. 396)

6. What hormonal changes initiate puberty? (p. 394, p. 396)

7. How do the male and female sexual responses differ? In what ways are they the same? (pp. 394–395, p. 401)

8. What is a blastocyst? What hormones are responsible for preparing the uterus to allow implantation of the blastocyst? (pp. 398–400, p. 401)

9. At what stage in the menstrual cycle can the ovum be fertilized? (p. 400, p. 409)

10. Following fertilization, what happens to the zygote? (pp. 401–404)
11. What is the function of the placenta? (p. 401, p. 403)
12. Name some ways in which fertilization may be prevented. (pp. 407–409)

13. What are some methods of birth control? Discuss the advantages and disadvantages of each. (pp. 407–409)
14. If fertilization takes place, what methods might be used to expel the fetus? (pp. 409–410)

Applying What You Know

1. What issues do you believe are necessary to discuss in order to come to some conclusion regarding when human life begins?
2. The chances of carrying a fetus to term with a defective genetic makeup is small. Most such fetuses are spontaneously aborted. However, women over age 35 experience a greater probability (although a small probability) of delivering a child with some genetic abnormality, regardless of the father's age. Why do you think this is so?
3. The birth rate in undeveloped societies, in which contraceptive methods are unknown, is lower than in more industrialized societies. Why do you think this is so?

Glossary

Absolute refractory period The time period when a recently activated section of cell membrane is completely unresponsive to further stimulation.

Absorption A general term for the taking up of fluids or other substances; in the GI system, absorption is the process whereby the products of digestion cross the cells of the mucosa and enter the bloodstream.

Accommodation The adjustment of the curvature of the lens of the eye that enables it to focus on either near or far objects.

Acetylcholine (as̩ĕ-til-ko'-lēn) A common neurotransmitter substance; located primarily at the neuromuscular junction and in the autonomic nervous system.

Acid A molecule that can donate a hydrogen ion.

Acidosis (as̩ĭ-do'sis) A condition of higher blood acidity than normal.

Acinar cells (as̩ĭ-nar) Cells lining the terminal bulb-like regions of the pancreatic ducts that produce primarily pancreatic enzymes.

Acquired immune deficiency syndrome (AIDS) An immune deficiency disease caused by human immunodeficiency virus.

Acromegaly (ak̩ro-meg'ah-le) A condition in which human growth hormone is secreted in large amounts after puberty when bone growth is no longer possible.

Actin (ak'tin) The protein of the thin filament that interacts with myosin to produce shortening of a sarcomere during muscle contraction.

Action potential A large, rapid, short-lived depolarizing membrane potential that travels down a nerve axon or across the surface of a muscle cell.

Activation Changes in the sperm as they take on a lurching motion and become able to fuse with the ovum.

Active transport The transport of a substance across the cell membrane against a concentration gradient.

Addison's disease A disease resulting from an adrenal insufficiency.

Adenosine diphosphate (ADP) (ah-den'o-sēn) A two-phosphate molecule formed from the splitting of adenosine triphosphate to yield energy.

Adenosine triphosphate (ATP) An adenosine molecule linked to three phosphate groups that yields energy when split to ADP and inorganic phosphate.

Adenylate cyclase (ah-den'i-lāt) A second messenger.

Adipocytes (ad'ĭ-po-sīt) Fat cells that take up and store most of the fatty acids in the blood and synthesize triglycerides.

Adrenal cortex (ah-dre'nal) The outer part of the adrenal gland.

Adrenal glands Endocrine glands, of which the outer portion secretes steroids and the inner part secretes catecholamines.

Adrenal insufficiency Inadequate secretion of adrenal hormones.

Adrenal medulla (me-dul'ah) The inner part of the adrenal gland.

Adrenergic (ad̩ren-er'jik) Describing nerves that release the neurotransmitter norepinephrine.

Adrenocorticotropic hormone (ACTH) (ah-dre'bo-kor̩ti-ko-tro'pik) A tropic hormone that stimulates the secretion of adrenal cortical hormones.

Adult-onset diabetes (Type II diabetes or non–insulin-dependent diabetes) (di̩ah-be'tez) The most common form of diabetes, in which the pancreas secretes normal amounts of insulin, but the individual does not respond normally to insulin levels in the blood.

Aerobic metabolism (a'er-ob-ik me-tab'o-lizm) The breakdown of chemical energy sources in the presence of oxygen.

Afferent arteriole (af'er-ent ar-ter'e-ol) The structure that supplies arterial blood to the glomerulus.

Afferent nerves Nerves that carry information to the central nervous system from the periphery.

Afterbirth Delivery of the placenta after birth of the baby.

Agglutination (ah-gloo̩tĭ-na'shun) The clumping of red blood cells when incompatible blood is mixed.

Albumin (al-bu'min) The small protein fraction of plasma.

Alcohol The intoxicating chemical in fermented drinks.

Aldosterone (al-dos'ter-on) A hormone produced and released by the cortical layer of the adrenal gland, which stimulates the activity of the Na$^+$/K$^+$ transport of the kidney.

All-or-none rule The rule stating that once initiated, all action potentials are alike in any given excitable cell; either they occur or they do not occur.

Allergy A response to an antigen resulting in a complex constellation of symptoms.

Alpha (α) cells Cells of the pancreas that secrete glucagon.

Alpha motor neuron A neuron that innervates skeletal muscle fibers.

Alveolus (al-ve'o-lus) A very small sac (~ ½ mm in diameter) on the side or at the end of the smallest bronchioles. Alveoli are the exchange sites where gases move between the air in the lungs and the blood.

Amines A class of hormones derived from the amino acid tyrosine.

Amino acid (ah-me'no) An organic molecule with an amino group on one end and a carboxyl group on the other; the building blocks of proteins.

Amniocentesis (am,ne-o-sen-te'sis) A method of obtaining amniotic fluid from a pregnant woman for genetic analysis.

Ampulla (am-pul'lah) A widened section in each semicircular canal; the ampulla contains the hair cells responsible for detecting movement of the head.

Amylase (am'ĭ-lăs) An enzyme present in saliva and pancreatic juice that breaks starch into short-chain carbohydrates.

Amyotrophic lateral sclerosis (ALS) (ah-mi,o-trof'ik lat'eral sklĕ-ro'sis) A disease characterized by hardening, and hence damage, of the lateral columns of the spinal cord, leading to progressive weakening and wasting of skeletal muscle.

Anabolic steroid A masculinizing hormone.

Anabolism (ah-nab'o-lizm) The chemical processes in cells that convert simple molecules to more complex ones.

Anaerobic metabolism Metabolism that continues in the absence of oxygen.

Anatomy (ah-nat'o-me) The study of the structure of the body.

Androgen (an'dro-jen) A hormone that acts as testosterone.

Anemia (ah-ne-me-ah) A deficiency of red blood cells or of hemoglobin, reducing the oxygen-carrying capacity of blood.

Angiotensin I (Ang I) (an'je-o-ten'sin) A peptide made up of 10 amino acids split from a large protein secreted by the liver.

Angiotensin II (Ang II) An octapeptide split from angiotensin I by the action of an enzyme present in most cells and blood.

Angiotensin-converting enzyme (ACE) The enzyme that splits angiotensin II from angiotensin I.

Angiotensinogen (an,je-o-ten'sin-o-jen) A protein secreted by the liver that is split by renin into angiotensin I.

Anterior pituitary (an-tēr'e-er) The frontmost lobe of the pituitary that secrets LH, FSH, TSH, GH, ad prolactin.

Antibody An immunoglobulin protein that produces an immune response to a specific (usually foreign) molecule.

Antidiuretic hormone (an,tī-dī,yer-eh'tik) A hormone secreted by the posterior pituitary that inhibits diuresis.

Antigen (an'tĭ-jen) A large, complex molecule that stimulates the immune system to produce a specific antibody.

Aorta The large, thick-walled vessel that arises from the heart.

Aphasic (ah-fa'zik) Describing people with deficits in their ability to speak.

Appendicitis (ap,ĕn-dĭ-si'tis) Inflammation of the lymph nodes of the appendix.

Appendicular skeleton (ap,en-dik'u-lar) The portion of the skeleton forming the shoulders, the hips, and the four limbs.

Appendix A small outpouching of the cecum.

Areola (ah-re'o-lah) The area surrounding the nipple of the breasts.

Arrhythmia (ah-rith'me-ah) Variation from the normal rhythm of the heartbeat.

Arterioles The smallest arteries, responsible for the major resistance to blood flow.

Arteriosclerosis (ar-tēr'e-o-skler-o'sis) Stiffening of the arteries, usually with increasing age.

Arthritis Inflammation of the joints.

Artificial kidney A machine that acts as a kidney.

Asthma (az'mah) The most common lung disease in non-smokers, caused by constriction of the airways.

Astrocyte (as'tro-sīt) A type of glial cell.

Atom The smallest unit of an element, consisting of a nucleus and electrons.

Atria (ay'tre-ah) The small, thin, muscular upper chambers of the heart.

Atrial fibrillation The action that occurs when the atrial wave of depolarization arises in multiple sites and sweeps over the atria in an almost continuous and rapid irregular path.

Atrial natriuretic peptide (na,tre-u-ret'ik) A peptide hormone secreted by the heart that has the ability to inhibit Na^+ reabsorption by the renal tubules.

Atrioventricular (AV) node (ay,tre-oh-ven-tri'kyoo-ler) A node similar to the sinoatrial node at the junction of the atria and ventricles.

Atrioventricular (AV) valves The valves between the atria and ventricles that are responsible for the unidirectional flow.

Autocrine factor (aw-tō-krin) A chemical messenger that acts directly on the cell that produces it.

Autoimmune Describing the formation of antibodies against one's own organs.

Autonomic nervous system (ANS) (aw,to-nom'ik) Nerves that carry information to and from the periphery and the brain; the ANS is divided into the sympathetic and parasympathetic divisions.

Axial skeleton (ak-sil) The portion of the skeleton forming a central axis.

Axon An elongated extension of a neuron that conducts action potentials; also known as a nerve fiber.

Axon collaterals Side branches that extend from the main axon.

Axon hillock The region of a nerve cell where the cell body narrows to form the axon; the site where action potentials are initiated.

Axon terminals The branched endings of a neuron, which release neurotransmitter.

Baroreceptor (bayr,i-re-sep-tor) A general term for a pressure-sensitive nerve terminal or region.

Basal cell A general term for an undifferentiated cell able to give rise to other, more specialized cells.

Basal metabolic rate (BMR) The minimal energy required for the maintenance of all bodily functions, usually calculated by oxygen consumption or by the number of calories expended per unit time.

Base A molecule that can accept a hydrogen ion.

Basilar membrane (bas'ĭ-lar) The lower membrane of the inner ear; the basilar membrane provides structural support for the hair cells and vibrates in response to sound waves.

Basophil (ba'zo-fil) A white blood cell involved in the inflammatory responses.

Benign tumor (be-nin') A tumor that is not malignant, and will not spread through the body.

Beta (β) blockers Drugs that have the ability to block the effect of norepinephrine on its receptors on the cell membrane, the beta receptors.

Beta (β) cells Pancreatic cells that secrete insulin.

Beta (β) receptors Cellular receptors that bind noradrenalin.

Betz cells The cells in the brain from which the pyramidal axons arise.

Bile A watery fluid produced by the liver and concentrated in the gallbladder, containing primarily bicarbonate and bile salts; bile also contains several waste products, including bilirubin and cholesterol.

Bile salts Substances derived from cholesterol that facilitate the digestion and absorption of fats.

Bilirubin (bil,i-roo'bin) One of the products of hemoglobin metabolism.

Bipolar cells Neurons located in the retina between the photoreceptor cells (rods and cones) and the ganglion cells.

Blastocyst (blas'to-sist) A hollow fluid-filled ball of cells formed by further cell divisions of the morula.

Blood–brain barrier The limitation of the brain capillaries to the movement of certain solutes.

Blood group A classification of blood according to its antigen characteristics.

Bowman's capsule A cup-like structure that surrounds the glomerular capillaries.

Brain stem The base of the brain.

Broca's area The primary speech area of the cortex.

Bronchiole a small conducting airway of the respiratory tract that has no cartilage, but a considerable amount of smooth muscle.

Bronchus (brong'kus) A conducting airway of the respiratory tract that contains cartilage, making it fairly rigid, and smooth muscle, which can alter diameter.

Bulk flow The term used to describe the movement of fluid and small ions across a membrane which has pores too small to allow the movement of protein.

Bundle of His The specialized group of conducting fibers leaving the AV node and running down either side of the inner wall of the heart.

Calcitonin (cal,sih-to'nin) A hormone produced by the thyroid glands that lowers the calcium level in plasma.

Capacitation The change in sperm that occurs in the female reproductive tract that is necessary for sperm to fertilize an ovum.

Capillaries (kap'il-layr,ēs) The smallest blood vessels.

Carbohydrate An organic molecule containing carbon, hydrogen, and oxygen in a ratio of 1:2:1; sugars and starches are carbohydrates.

Cardiac cycle The sequence of events that occur from one beat of the heart to the next.

Cardiac muscle A type of muscle found only in the heart.

Cardiac output The total volume of blood pumped into the aorta each minute.

Cardiopulmonary resuscitation (CPR) (kar,de-o-pul'mo-ner,e) A method of artificial respiration coupled with manual compression of the heart.

Cardiovascular center (kar,de-o-vas'ku-lar) A region in the medulla of the brain that controls blood pressure.

Carotid artery (kar-rot'id) One of the two major arteries supplying the head.

Carotid sinus A specific area in each carotid artery that contains baroreceptors.

Cartilage (kar'tĭ-lij) A specialized fibrous connective tissue that forms most of the temporary skeleton of the embryo.

Catabolism (kat-tab'oh-lizm) The process by which larger molecules are broken down into smaller ones to obtain energy.

Catalyst (kat'ah-list) A substance that accelerates the rate of a chemical reaction without being altered by the reaction.

Catheters (kath'ĕ-ter) Small hollow-bore tubes.

Cecum (se'kum) The blind pouch at the proximal end of the large intestine.

Cell The smallest living unit in multicellular organisms.

Cell body The portion of a neuron that contains the nucleus and organelles.

Cell membrane The selectively permeable membrane that surrounds all cells, composed primarily of phospholipids, proteins, and cholesterol.

Central chemoreceptors (ke'moh-re-sep,terz) Chemical sensing cells located in the central nervous system.

Central nervous system (CNS) The part of the nervous system made up of the brain and its associated spinal cord.

Centrioles (sen'tre-ol) A pair of rod-like tubular structures in the cell nucleus that are involved in cell division.

Centromere (sen'tro-mēr) The point of attachment of the spindle to the chromosome.

Cerebellum (ser,ĕ-bel'lum) The posterior part of the lower brain.

Cerebral cortex (se-re'bral) The outer, grayish portion of the brain.

Cerebral hemispheres The brain structures credited with the higher levels of function such as thought, intelligence, and ability to plan and solve problems.

Cerebrospinal fluid (CSF) The fluid surrounding the brain and spinal cord and filling the cerebral ventricles.

Cervical An area of the body pertaining to the level of the neck; also, the adjective referring to the cervix.

Cervical cap Contraceptive device similar to a diaphragm, but smaller and may be left in place for one to two days.

Cervix (ser'viks) The narrowed end of the uterus.

Chemoreceptor A receptor cell or region that responds to certain chemicals in its environment.

Chemotaxis (ke,mo-tak'sis) Chemical signals recruiting white cells to the area of invasion.

Chief cells Cells of the gastric glands of the stomach, which produce pepsinogen, the inactive precursor of the proteolytic enzyme pepsin.

Cholecystokinin (ko,le-sis,to-ki'nin) A hormone secreted by the small intestine into the bloodstream that inhibits gastric motility, promotes the secretion of bile, and increases the secretion of pancreatic enzymes.

Cholesterol (ko-les'ter-ol,) A type of lipid that stabilizes the cell membrane and is also a precursor of bile salts and steroid compounds; cholesterol may form plaques in blood vessels.

Chondroblast (kon'dro-blast) A cartilage-forming cell, present primarily during fetal development.

Chordae tendinae Tendons in the heart that help prevent inversion of the AV valves during ventricular contraction.

Choroid plexus (kor'roid plek'sus) Specialized tissue in the brain ventricles that produces CSF.

Chromosome (kro'mo-som) One of 23 paired structures (in humans) in the cell nucleus, composed of DNA and proteins; chromosomes are visible only during cell division.

Chronic obstructive pulmonary disease (COPD), or chronic obstructive lung disease (COLD) A debilitating and often fatal lung disease of which cigarette smoking is the primary cause. People with COPD usually have a combination of chronic bronchitis (excessive mucus production in the airways) and emphysema.

Chylomicron (ki,lo-mi'kron) A protein-coated packet of triglycerides formed by the Golgi apparatus in an intestinal mucosal cell; chylomicrons are exported from the cell into the lymph vessels.

Chyme (kīm) The watery fluid in the pylorus of the stomach, consisting of partially digested food and gastric juice.

Cilia (si'e-ah) Minute, vibratile, hair-like processes attached to the surface of some cells.

Citric acid cycle (Kreb's cycle) A major cytosolic biochemical process involved in aerobic respiration.

Clitoris (klit'o-ris) An erectile structure of the female homologous to the penis.

Clotting factors A series of blood factors required for normal clotting.

Cocaine An addicting chemical obtained from coca leaves.

Cochlea (kok'le-ah) The bony, coiled, fluid-filled structure of the inner ear that is the organ of hearing.

Colitis (ko-li'tis) Severe inflammation of the bowel and gastrointestinal tract.

Collagen (kol'ah-jen) A type of protein that is the primary constituent of ligaments and tendons.

Collecting duct The last section of the renal tubules leading to the ureters.

Colloid osmotic pressure (kol'oid-ahl ahz-ma'tik) The osmotic pressure contributed only by the protein molecules.

Compact bone A very hard, dense form of bone.

Competition The phenomenon of two or more molecules being able to combine with the same carrier molecule.

Complement (kom'plĕ-ment) Plasma proteins active in the immune responses.

Condom A very thin rubber sheath placed over the penis before intercourse to prevent fertilization.

Conducting airways The parts of the respiratory tract (the bronchioles, bronchi, trachea, upper respiratory tract) where air moves by bulk flow, not diffusion.

Cones Cone-shaped photoreceptor cells of the retina containing one of three different photopigments; the photoreceptor cells responsible for color vision.

Control group The group of subjects in an experiment that do not receive the treatment being tested.

Convergence Joining of more than one into one.

Cornea (kor'ne-ah) The transparent convex layer at the outer surface of the eye through which light must pass to reach the lens.

Corpus albicans (kor'pus al-bi-kanz) A small, atrophied, nonfunctional corpus luteum.

Corpus collosum (kor'pus kal-lō-sum) A large tract of nerve fibers connecting the two hemispheres.

Corpus luteum (kor'pus loo-tē-um) The ruptured Graafian follicle of the ovary filled with a yellowish fluid.

Corticosterone (kor,tĭ-kos'te-rōn) A glucocorticoid hormone.

Corticotropin-releasing hormone (CRH) (kor,tĭ-ko-tro'fin) A tropic hormone that stimulates secretion of ACTH.

Cortisol (kor'tih-sol) A glucocorticoid hormone.

Cortisone (kor'ti-son) A synthetic glucocorticoid.

Countercurrent multiplier system The system used by the kidney to form a concentrated urine.

Countercurrent system A system of blood flow in which the vessels are hairpin-shaped so that blood flowing in opposite directions is in proximity.

Covalent bond (ko-va'lent) A bond between atoms that is characterized by the sharing of electrons.

Cranial nerves Nerves that exit the CNS directly from the brain through small holes in the skull.

Creatinine (kre-at'ĭ-nin) An end product of muscle metabolism that may be used to estimate the glomerular filtration rate by measuring its concentration in plasma.

Cretinism (kre'tĭ-nizm) Severe retardation in a newborn due to hypothyroidism in the mother during pregnancy.

Cross-bridges The heads of the myosin molecule that interact with actin molecules of the thin filaments, leading to a shortening of a sarcomere.

Crossing over An event in which segments of the maternal and paternal chromosomes break away in a random manner and exchange places with each other.

Cupula (ku'pu-lah) A cap of gelatinous fluid over the hair cells of the ampulla, in the semicircular canals of the inner ear.

Cushing's syndrome A disease caused by the release of large amounts of adrenal cortical hormones.

Cyclic AMP (cAMP) An intracellular molecule that often serves as a second messenger.

Cytokines (si,to-ki-ne'sis) A chemical that stimulates macrophage activation and antibody responses.

Cytoplasm (si'to-plazm) The material inside a cell, other than the nucleus.

Cytoskeleton (si,to-skel'ĕ-ton) A protein network lending structural support to the cell membrane and organelles.

Cytosol (si'to-sol) A jelly-like material that fills the space within a cell that is not occupied by organelles or cytoskeleton.

Cytotoxic T cells (si-to-tok-sĭk) T cells that are able to destroy cancer cells.

Dead space The volume of air filling the airways at the end of inspiration, which does not participate in gas exchange. Dead space is approximately equal to the volume of the conducting airways.

Decussation (de,kus-sa'shun) The crossing of nerve fibers from one side of the spinal cord to the other.

Defecation reflex (def,ih-ka'shun) A neurally mediated reflex leading to defecation, initiated by distention of the rectum.

Dehydration synthesis The process of forming carbohydrates from simple sugars, in which the equivalent of a water molecule is removed.

Delirium tremens (dĕ-lēr'e-um tre'mens) A mental disturbance marked with trembling, excitement, and distress.

Delta (Δ) cells Pancreatic cells that secrete somatostatin.

Denaturation The permanent disruption of the tertiary structure of a protein.

Dendrites (den'drĭts) The extensions from the cell body of a nerve cell that receive input from other neurons.

Depolarization A reduction in membrane potential from resting potential; a change in membrane potential from resting potential toward zero mV.

Desmosome (dez'muh-sōm) A physical structure that joins two cardiac cells together.

Diabetes insipidus (di,ah-be'tez in-sih'pih-dus) A disorder in which inappropriately large amounts of urine are formed due to a failure of antidiuretic hormone production, or to the inability of the renal tubules to respond to antidiuretic hormone.

Diabetes mellitus (meh-li'tus) The disease resulting from too little secretion of insulin or the inability to respond to insulin.

Dialysis (di-al'ah-sis) A procedure in which two fluids are separated by a selectively permeable membrane so that water and small solutes may pass freely between the two solutions.

Diaphragm (di'ah-fram) *In anatomy,* a flexible, plate-shaped muscle that forms the floor of the thoracic cavity. It is the most important inspiratory muscle.

Diaphragm *As a method of birth control,* a circular ring of rubber, individually fitted by a physician for the female, placed so that it fits snugly over the cervix to prevent sperm from entering the uterus.

Diastole (di-as'to-le) The period of ventricular relaxation.

Diastolic pressure The lowest pressure reached during the cardiac cycle.

Diencephalon (di,ĕn-sŭ-lon) A division of the brain serving as a relay station that contains the thalamus and the hypothalamus.

Differentiation The taking on of different forms and functions.

Diffusion The net movement of particles from an area of higher concentration to an area of lower concentration, as a result of random thermal motion.

Diffusion potential The electrical charge that develops across a selectively permeable membrane when ions diffuse.

Digestion The process of breaking down foodstuffs in the GI tract into smaller units so they can undergo absorption.

Diploid A cell having two of each type of chromosome.

Disaccharide (di-sak'ah-rid) a sugar molecule consisting of two monosaccharides.

Distal convoluted tubule The section of the renal tubule between the loop of Henle and the collecting duct.

Diuresis (di,u-re'sis) An increase in the rate of urine formation.

Diuretics (di,u-ret'iks) Drugs that cause the kidney to excrete a larger than normal volume of urine.

Divergence Branching of one to more than one.

DNA (deoxyribonucleic acid) (de-ok,sĭ-ri,bo-nu-kle'ik) A double-helical molecule consisting of two strands of nucleotides, containing the genetic information necessary for the synthesis of proteins.

DNA replication The process of making an exact copy of a cell's DNA before cell division.

DNA transcription The process of making an RNA molecule from a gene.

Dopamine (do'pah-mēn) A common neurotransmitter of the central nervous system.

Dorsal root Nerves that exit the spinal cord from the dorsal side.

Dorsal root ganglion (gang'gle-on) The ganglion in the dorsal nerve root containing the cell bodies of the sensory fibers.

Duct cells Cells lining the ducts of the pancreas that secrete primarily a watery fluid containing bicarbonate.

Duodenum (du,o-de'num) The first segment of the small intestine; it is about 10 inches long.

Ectopic focus (ek-top'ik) An excitable piece of heart tissue, other than the SA node, at which an extrasystole originates.

Edema (ĕ-de'mah) An area of localized swelling.

Effector cell A specific B cell that creates many copies of itself in response to an antigen and secretes antibodies.

Efferent arteriole (ef'er-ent) The arteriole that leaves the glomerular capillaries.

Efferent nerves Nerves that carry information away from the CNS to the periphery.

Ejaculation (e-jak,u-la'shun) A reflex discharge of semen from the penis.

Electrocardiogram (ECG or EKG) (e-lek,tro-car'de-o-gram,) A recording of the electrical activity of the heart.

Electrocardiograph The machine used to record the electrical activity of the heart.

Electrochemical gradient The sum of the electrical and chemical gradients for a particular ion.

Electroencephalogram (EEG) (e-lek,tro-en-sef'ah-lo-gram,) Small cyclical alterations in the electrical voltages recorded at the surface of the skull.

Electron transport chain A group of membrane-bound mitochondrial electron carrier molecules that allow electrons to move down the chain until they combine with oxygen. As electrons move down the chain, energy is released and used to manufacture ATP from ADP and phosphate.

Embryo (em'bre-o) The developing organism from fertilization until about 8 weeks.

Emission A spinal reflex of rhythmic contraction of muscles within the penis that forces semen out of the urethra.

Emulsify (e-mul,sĭ-fi) To break large fat droplets into smaller ones, so that the smaller droplets can remain suspended in water.

End diastolic volume (di-as'to-lik) The volume of blood remaining in the ventricles at the end of systole.

Endocrine system (en'do-krin) The portion of the body comprising the glands that empty their secretions into the blood.

Endometrium (en,do-me,tre-um) The lining of the uterus.

Endoplasmic reticulum (en,do-plaz'mik rĕ-tik'u-lum) Folded, membranous structures within a cell that are responsible for producing most of the chemical compounds made by the cell.

Endorphins (en-dor'fins) CNS neurotransmitters.

Endoskeleton An internal skeleton, as in humans.

Endothelium-derived relaxing factor (EDRF) (en,do-the'le-um) A substance released by the endothelial cells of vessels that relaxes vascular smooth muscle.

Enkephlins (en-kef'ah-lin) CNS neurotransmitters.

Enzyme (en'zīm) A protein that functions as a catalyst.

Eosinophil (e,o-sin'o-fil) A white blood cell with some phagocytic action.

Epididymis (ep,ĭ-dĭ dĭ-mis) An organ associated with the testes in which maturation of sperm occurs.

Epiglottis (eh,puh-glah'tis) The lid-like structure that covers the entrance to the trachea.

Epinephrine (ep,ĭ-nef'rin) Also known as *adrenaline;* a neurotransmitter in the central nervous system, and also a hormone secreted by the adrenal gland.

Erection Distention of the erectile tissue of the penis.

Erythrocyte (e-rith'ro-sit) The red blood cell.

Erythropoietin (e-rith,ro-poi'e-tin) A hormone secreted by the kidney that stimulates bone marrow production of red cells.

Estrogen (es'tro-jen) A female sex hormone secreted by the ovaries.

Eukayrotic cells (u,kar-e-ot'ik) Cells having a true nucleus.

Eustachian tube (u-sta'ke-an) A narrow tube connecting the upper portion of the throat to the middle ear.

Exocrine glands (ek'so-krin) Exocrine glands empty their secretions into ducts that empty to the outside of the body.

Exoskeleton A jointed external skeleton, common among insects.

Experimental group The group of subjects in an experiment that receive the treatment being tested.

Extension A form of motion in which the angle between two bones is increased.

External anal sphincter A ring of skeletal muscle at the distal end of the rectum, under voluntary control; voluntary relaxation of the anal sphincter allows defecation.

Extrasystole (ek,strah-sis'to-le) An extra beat of the heart inserted between two normal ones.

Extrinsic mechanisms (ek-strin'sik) Regulatory mechanisms in an organ that use substances and pathways lying outside of the organ.

Facilitated diffusion A form of carrier-mediated transport across a cell membrane in which the substance moves down its concentration gradient without expending metabolic energy.

Fallopian tubes (fah-lō'pē-un) The uterine tubes.

Fatigue A decline in muscle performance during sustained heavy exercise.

Fatty acid A type of lipid consisting of a carbon–hydrogen chain that ends with a carboxyl group.

Fertilization The entry of a single sperm into an ovum.

Fetal alcohol syndrome (FAS) The constellation of symptoms resulting from alcohol consumption by the mother during pregnancy.

Fetus The developing embryo 8 weeks after fertilization.

Fibrin (fi'brin) Protein strands that form the basis of a clot.

Fibrinogen (fi-brin'o-jin) The precursor of fibrin.

First messenger A molecule that causes the release of another chemical responsible for a series of actions within the target cell.

Flexion (flek'shun) A form of motion in which the angle between two bones is decreased.

Follicle-stimulating hormone (FSH) (fah'lih-kul) A hormone that stimulates the Sertoli cells of the testes.

Fovea (fo've-ah) A small region at the center of the macula that contains cones but is devoid of rods.

Frontal lobe The frontmost cortical lobe.

Fundus (fun'dus) The first, rounded part of the stomach that serves primarily a storage function.

G-cells Cells of the gastric gland of the stomach that secrete the hormone gastrin into the bloodstream.

Gallstones Hard masses or stones in the gallbladder or bile duct, usually consisting of precipitated mineral salts, cholesterol, and bilirubin.

Gametes (gam'ēts) The reproductive cells of the male and female.

Gamma globulin (or immunoglobulin) (glob'u-lin) The protein components of blood that confer immunity.

Gamma motor neuron A neuron that innervates a contractile region of a muscle spindle.

Ganglia (gang'gle-ia) Small swellings containing the cell bodies of nerves.

Ganglion cells (gang'gle-on) Neurons of the innermost layer of the retina; their axons make up the optic nerve.

Gap junction A region of connection or union between two muscle cells, composed of membrane proteins common to both cells, permitting diffusion of ions and molecules between them.

Gas exchange region The deep region of lung containing the alveoli where gases move by diffusion between the air in the lungs and the blood.

Gas solubility The amount of gas in a certain volume per unit of partial pressure.

Gastric glands Glands of the stomach, located in deep folds of the gastric mucosa.

Gastric inhibitory peptide A hormone secreted by the small intestine into the bloodstream that inhibits gastric motility.

Gastric juice The watery fluid produced by gastric glands of the stomach, containing hydrochloric acid, intrinsic factor, and pepsinogen.

Gastrin A hormone produced by the G-cells of the gastric glands of the stomach that stimulates the secretion of gastric juice and increases gastric and intestinal motility.

Gastrocolic reflex (gas,tro-kol'ik) A peristaltic wave of contraction in the colon induced by the entrance of food into an empty stomach.

Gastroileal reflex (gas,tro-il'e-al) Increased segmental activity of the ileum when food is present in the stomach, even before any chyme reaches the ileum; the gastroileal reflex is thought to be mediated by the hormone gastrin.

Gastrointestinal (GI) system (gas,tro-in-tes'tĭ-nal) The digestive system.

Gastrointestinal (GI) tract The digestive tract.

Gated channels Cell membrane proteins that regulate the movement of particular molecules through the cell membrane by opening and closing, like a gate.

Gene A segment of DNA that contains the information needed to synthesize a particular protein.

Genetic fingerprint An array of DNA fragments, isolated and visualized on a gel, that is unique for each person.

Genotype (jen,o-tīp) Refers to the genetic makeup of an individual.

Glans penis (glanz pe'nis) The end portion of the penis.

Glia (gli'ah) A class of central nervous system neural elements whose function is not completely known.

Globin A large polypeptide chain of hemoglobin.

Globulin The large protein fraction of plasma.

Glomerular capillaries (glō-mer yoo-ler) The tuft of capillaries that supply blood to the glomerulus.

Glomerular filtration The formation of fluid that filters out of the glomerular capillaries into Bowman's space.

Glomerular filtration rate (GFR) The rate at which glomerular filtration occurs.

Glomerulus (glo-mer'u-lus) The tuft of glomerular capillaries.

Glottis (glah'tis) The airway opening between the vocal cords.

Glucagon (gloo'kuh-gon) A hormone secreted by the alpha cells of the pancreas.

Glucocorticoids (gloo,ko-kor'tĭ-koidz) Steroid hormones affecting glucose metabolism.

Gluconeogenesis (gloo,ko-ne,o-jen'ĕ-sis) The biochemical process that forms glucose from smaller molecules.

Glucose Tm (gloo'kos) The maximum rate of glucose reabsorption by the kidney.

Glycerol (glis'er-ol) A three-carbon organic molecule that is a subunit of triglycerides.

Glycogen (gli'kō-jin) A complex carbohydrate that is the primary form of stored carbohydrate in animals.

Glycogenolysis (gli,ko-je-nol'i-sis) The biochemical process that breaks down glycogen into glucose.

Glycolysis (gli-kol'ĭ-sis) The process that breaks down sugars into simpler compounds.

Glycoprotein (gli,ko-pro'te-in) Proteins that have carbohydrate groups attached to them.

Goiter (goi'ter) Large swelling of the thyroid gland due to a lack of iodine.

Golgi apparatus (gol'je) A membranous cell organelle responsible for final processing and packaging of substances synthesized by the endoplasmic reticulum.

Golgi tendon organ A specialized structure for sensing tension in a muscle.

Gonadotropin-releasing hormone (GnRH) (go-nad,o-trōp'in) A hypothalamic hormone that stimulates the anterior pituitary to secrete luteinizing hormone and follicle-stimulating hormone.

Gonadotropins (go-nad,o-trop'ins) Tropic hormones that regulate secretion of other hormones produced by the ovaries and testicles.

Graafian follicle (grāf'e-an) A fluid-filled follicle containing a mature ovum.

Graded potential A small change in membrane potential in a localized region of the cell membrane of an excitable cell; unlike an action potential, a graded potential can vary in magnitude.

Gray matter The outer, grayish portion of the brain composed of a series of folded convolutions.

Gyri (ji're) The elevated parts of the convolutions of the cerebral cortex.

Hair cells Receptor cells for hearing and for equilibrium, characterized by having hairlike projections that detect movement.

Haustra (haws-tra) Sac-like pouches of the large intestine that appear to be anatomical features, but are actually caused by very slow segmentation of the circular smooth muscle.

Heart failure Weakened ventricular muscle preventing the heart from beating with its normal force.

Heart rate The number of times the heart beats per minute.

Helper T cell A class of T cell that increases the activity of the B cell.

Hematocrit (hem,ah'to-krit) The fraction of whole blood that is composed of red blood cells.

Heme (hem) Smaller polypeptide chain of hemoglobin containing an atom of iron.

Hemoglobin (he'muh-glo-bin) The oxygen-carrying molecule of red blood cells.

Hemoglobin–O2 dissociation curve The complex curve that results when blood O_2 concentration is plotted against blood PO_2.

Hemophilia (he,mo-fil'e-ah) A disease in which the time needed for clot formation is prolonged excessively.

High-density lipoprotein (HDL) Sometimes called "good" cholesterol because HDL transports cholesterol from cells to the liver for elimination and offers some protection against plaque formation in the coronary arteries.

Histamine (his'tuh-mēn) A paracrine factor responsible for some of the local swelling of a damaged tissue and for some of the effects of allergic reactions.

Homeostasis (ho,me-o-sta'sis) The stability that the body acts to maintain in most physiological parameters.

Homologous chromosomes (ho-mol'ŏ-gus) Paired chromosomes.

Hormone A chemical messenger secreted by a gland into the bloodstream and conveyed by the blood to its target cells.

Human growth hormone (hGH) A hormone that regulates the rate of growth of body cells and bone.

Human immunodeficiency virus (HIV) The virus responsible for AIDS.

Hydrogen bond The weak attractive bond between two polar molecules, at least one of which contains hydrogen.

Hydrolysis (hi,drah-la'sis) The process of breaking down complex sugars and starches, in which the equivalent of a water molecule is added to the final products.

Hydrophilic (hi,dro-fil'ik) Tending to dissolve readily in water.

Hydrophobic (hi,dro-fo-bik) Tending not to dissolve readily in water.

Hymen (hi'men) A thin membrane that partially closes the vaginal opening.

Hyperglycemic (hi,per-gli-se'mik) High blood sugar concentration.

Hyperpolarization An increase in membrane potential from resting potential, making the membrane potential even more negative than at resting potential.

Hypersensitivity Allergy.

Hypertension High blood pressure.

Hyperthyroidism Oversecretion of thyroid hormones.

Hypoglycemia (hi,po-gli-se'me-ah) Low blood sugar concentration.

Hypothalamus (hi,po-thal'ah-mus) The brain area forming part of the floor of the third ventricle that serves to regulate many of the homeostatic mechanisms of the body.

Hypothesis (hi-poth'ĕ-sis) An idea, or thesis, as yet unproved.

Hypothyroidism Undersecretion of thyroid hormones.

Hypoxia (hi-pok'se-ah) A condition of inadequate oxygen supply to an organ, tissue, or whole body.

Hypoxic pulmonary vasoconstriction The constriction of pulmonary arteries that results from a decrease in alveolar PO_2.

Ileocecal valve (il,e-o-se'kal) Two folds of tissue at the junction of the ileum and the cecum that function as a valve to prevent backflow from the cecum to the ileum.

Ileum (il'e-um) The third segment of the small intestine; it is about 6 feet long.

Immune system the system that defends the body against pathogens and toxic substances.

Immunoglobulin (Ig) (im,u-no-glob'u-lin) A term for antibodies.

Implantation Attachment of the blastocyst to the uterine wall.

Incus (ing'kus) The second of the three bones in the middle ear that connect the eardrum to the oval window.

Inferior vena cava (ve'nah ka'vah) The very large vein that travels along the back wall of the body and returns blood from the lower body to the right atrium.

Inflammation An injury to tissue characterized by an increased blood flow to the affected area, increased capillary permeability with its attendant edema, and accumulation of granulocytes and monocytes.

Infundibulum (in,fun-dib'u-lum) The tissue connecting the pituitary with the hypothalamus.

Inhibin (in,hi'-bin) A hormone that feeds back on the anterior pituitary to inhibit the secretion of FSH.

Inoculation An injection of antibodies or altered antigens to prevent a specific illness.

Insertion The end of a muscle that is attached to the part of the skeleton that moves when the muscle contracts.

Insulin A protein hormone secreted by the pancreas that controls the concentration of sugar in blood.

Insulin-dependent diabetes Diabetes due to a deficiency of insulin.

Integral protein A protein that spans the width of the cell membrane.

Intercalated disc (in-ter'kah-la,ted) A region of contact between two cardiac muscle cells, bridged by desmosomes and gap junctions.

Internal anal sphincter A ring of smooth muscle at the distal end of the rectum; distention of the rectum produces involuntary relaxation.

Interstitial fluid (in,ter-stish'al) Extracellular fluid that is not contained in the vascular system; it surrounds and bathes all cells.

Interstitium (in,ter-stish'ĭ-um) The space between cells, occupied by interstitial fluid.

Intervertebral disc (in,ter-ver'teh-brul) A flattened, slightly compressible disc located between the vertebrae.

Intrapleural fluid (in,trah-ploo'ral) The liquid in the intrapleural space that allows the lungs to slide within the thoracic cavity with very little friction.

Intrapleural space A very thin fluid-filled space between the two layers of pleurae.

Intrauterine device (IUD) A device placed in the uterus to prevent pregnancy.

Intrinsic factor (in-trin'sik) A substance secreted by parietal cells of the gastric glands, necessary for the absorption of vitamin B12.

Intrinsic mechanisms Local regulatory mechanisms in an organ.

Iodine An element taken up by the thyroid gland that is required for the synthesis of thyroid hormones.

Ion (ī'on) A charged atom or molecule.

Ionic bond (ī-ah'nik) The bond that forms between oppositely charged ions.

Iris The muscular structure that regulates how much light enters the eye.

Islets of Langerhans (i'let lahng'er-hanz) Islands of cells in the pancreas that secrete insulin and glucogon.

Isometric (i,so-mĕ-trik) Meaning *same length;* a muscle contraction in which the development of tension occurs without a change in the position of bones to which the muscle is attached.

Isotonic (i,so-tah'nik) Meaning *same tension;* a muscle contraction in which a constant force (tension) is applied and the muscle changes in length.

Jaundice Yellow color imparted to the skin and whites of the eyes due to excess bilirubin in the blood.

Jejunum (jĕ-joo'num) The second segment of the small intestine; about 4 feet long.

Joint A point of union between two bones.

Joint receptors Receptors that sense the position of limbs.

Juvenile onset diabetes (Type I diabetes or insulin-dependent diabetes) Diabetes characterized by a severe deficit in the ability of the pancreas to secrete normal amounts of insulin.

Killer T cell A class of T cell that can destroy a foreign cell.

Kilocalorie (kcal or calorie) The amount of heat needed to raise the temperature of 1 liter of water 1°C.

Kinocilium (ki,no-sil'e-um) A specialized extension of a hair cell in the semicircular canals, closely associated with the hairs.

Kupffer cells (koop'ferz) Phagocytic cells in the liver.

Labia majora (la'be'a mag-ora) The larger of two skin folds surrounding the vaginal orifice.

Labia minora (la'be'a mi'nora) The smaller fold surrounding the vaginal orifice.

Lactase (lak-t'ās) An enzyme present in the small intestine that catalyzes the conversion of lactose to glucose and galactose.

Lactation (lak-t'ashun) Secretion of milk by the breasts.

Lactose intolerant (lak'tōs) A condition in human adults in which the enzyme lactase is missing or deficient, leading to an inability to break down lactose.

Laryngopharynx The lower part of the throat, just above the esophagus.

Larynx (lar'ingks) A conducting airway segment that contains the vocal cords.

Lens The transparent, biconvex structure of the eye that focuses light on the retina.

Leukemia (loo-ke'me-ah) A disease characterized by a large increase in the number of circulating leukocytes.

Leukocytes (loo'ko-sits) White blood cells.

Leydig cells (li'digz) Testicular cells that secrete testosterone.

Ligament Connective tissue that joins bones together.

Limbic system (lim'bik) A set of interconnecting pathways in the brain that play an important role in the emotional, behavioral, and motivational aspects of a person.

Lipase (li'pās) An enzyme in saliva and pancreatic juice that contributes to the digestion of triglycerides.

Lipid (lih'pid) An organic molecule composed of a backbone of carbon and hydrogen; fats and oils.

Lipid bilayer The basic structure of a cell membrane, consisting of a double layer of phospholipids.

Lipogenesis (lip,o-jen'ĕ-sis) The manufacture of fatty acids from a variety of smaller molecules such as glucose and amino acids.

Lipolysis (lĭ-pol'ĭ-sis) A process, occurring mostly in adipocytes, that degrades triglycerides, releasing fatty acids into the blood.

Long-term memory Ability to recall events in long past.

Loop of Henle (hen'lēz) The portion of the renal tubule between the proximal and distal tubules.

Low-density lipoprotein (LDL) (lip'o-pro'tēn) Sometimes called "bad" cholesterol; LDL carries cholesterol to cells, where it may be deposited in the arteries.

Lower esophageal sphincter (ĕ-sof,ah-je'al) The sphincter at the lower end of the esophagus, where the esophagus joins the stomach.

Lower motor neuron The final motor neuron that synapses with the pyramidal neuron.

Lumbar (lum'bar) An area of the body pertaining to the level of the back.

Lumen (loo'min) The space within a tube or tubular organ.

Luteinizing hormone (LH) (lu'te-in-īz,ing) A hormone that stimulates the leydig cells of the testes.

Lymph nodes (limf') Tissue located in many parts of the body in which phagocytes are fixed and aid in host defense.

Lymph vessels Channels that take up the excess tissue fluid that is filtered out of the capillaries and carry it back to the circulatory system.

Lymphatics (lim-fat'iks) Intricate networks of lymph vessels.

Lymphocyte A white blood cell that produces antibodies.

Lysosome (li'so-sōmz) A membrane-bound vesicle in a cell that contains digestive enzymes.

Lysozyme (li'so-zim) An enzyme present in saliva that functions as an antibacterial agent.

Macrophages (mak'ro-fāj,) Large white blood cells that are able to engulf foreign particles.

Macula (mak'u-lah) A region of the retina that contains primarily cones; at the center of the macula is the fovea.

Major histocompatibility complex (MHC) A protein fragment brought to the surface of an infected cell aiding the binding of a T cell to the infected cell.

Malignant tumor (muh-lig'nent) A metastasizing and therefore dangerous mass of tissue that grows independently of its surrounding tissues and has no physiologic use; a cancerous tissue.

Malleus (mal,e-us) The first of the three bones in the middle ear that connect the eardrum to the oval window.

Mast cells Derivatives of basophils located throughout the body that synthesize and release histamine.

Mean blood pressure The average arterial blood pressure, which is approximately equal to diastolic pressure plus one-third of the difference between systolic and diastolic pressures.

Mechanoreceptor (meh,kĕ-no-rē-sep-ter) A receptor that responds to mechanical deformation of the cell membrane.

Medulla (mĕ-dul'ah) Also known as medulla oblongata; the brain stem located just above the spinal cord.

Medullary cavity (me-dul'ah-re) The hollow space at the center of some bones.

Meiosis (mi-o'sis) The process of cell division in which the chromosome number of the new cells is half that of the parent cell.

Membrane potential The difference in electrical charge between the inside and outside of a cell.

Memory Ability to recall events.

Memory cells Beta cells that remain in the body for a long time following an antigen encounter.

Menarche (mĕ-nar'ke) The first menstrual cycle.

Meninges (mĕ-nin'-jēz) A composite of four tough membranes that surrounds and protects the brain.

Menopause The cessation of menstruation.

Menstrual cycle (men'stroo-al) The periodic maturation of an oocyte in the ovary coupled with physiological changes that occur in the uterus.

Menstruation (men,stroo-a'shun) The periodic shedding of blood and the thickened lining of the uterus.

Mesencephalon (mes,en-sef'ah-lon) The midbrain.

Metabolic rate The total energy expended per unit time, usually calculated by measuring oxygen consumption during that time.

Metastasis (mĕ-tas'tah-sis) The spread of cells from one organ to another, usually cancerous cells.

Metastasize (mĕ-tas'tah-sīz) To form new sites at distant locations.

Metencephalon (met-en-sef'ah-lon) The hindbrain.

Micelle (mi-sel') A small droplet of bile salts and cholesterol that acts to emulsify fats.

Microfilaments (mi,kro-fil'ah-ment) Protein fibers that form part of the structural support for a cell.

Microtubules Small tubes composed of protein that form part of the structural support for a cell.

Microvillus (*pl.* **microvilli**) (mi,kro-vil'us) The general term for a small projection from the surface of a cell membrane.

Milk let-down The ejection of milk from the breasts.

Mineralocorticoids (min,er-al,ō-kor'tih-koyds) Steroid hormones affecting mineral metabolism.

Mitochondria (mi,to-kon'dre-ah) The organelles responsible for energy production in a cell.

Mitosis (mi-to'sis) The process of division of a cell into two cells, each with an identical set of DNA and the same number of chromosomes.

Mole 6.02×10^{23} particles of any substance; a mole of any substance has a mass in grams numerically equal to the substance's atomic or molecular mass.

Molecule The smallest unit of a substance, composed of two or more atoms.

Monocyte (mon'o-sīt) A phagocytic white blood cell.

Monosaccharide (mon,o-sak'ah-rid) The simplest form of carbohydrates, generally consisting of a backbone of five or six carbons arranged in a ring.

Monosynaptic (mon,o-sĭ-nap'tik) A neural pathway of a reflex in which only a single synapse is involved.

Mons pubis (mon'tes pu'bis) A small prominence of adipose tissue that lies over the external genitalia of the female.

Morphine The most active, addicting ingredient of opium.

Morula (mor'u-lah) The compact multicellular body of about sixteen cells formed following fertilization of the ovum.

Motility Movement.

Motor neuron A neuron that activates a muscle cell or cells, and hence causes movement.

Motor unit A single motor neuron and the muscle cells it controls.

Mucosa (mu-ko'sah) The innermost layer of the GI tract, so named because it generally secretes mucus.

Mucous cells The layer of mucus-secreting cells that line the inner surface of the stomach and the upper part of the gastric glands.

Multiple sclerosis (MS) (skle-ro'sis) A disease in which the myelin sheath of nerves becomes progressively damaged at multiple sites in the body.

Murmur An abnormal heart sound that may signal some cardiac pathology.

Muscle fiber A single muscle cell.

Muscle spindle A specialized bundle of modified muscle cells that sense involuntary changes in muscle length.

Muscularis (mus,ku-la'ris) The muscular layer of the GI tract, consisting primarily of smooth muscle tissue.

Mutation A change in the DNA sequence.

Myelencephalon (mi,el-en-sef'ah-lon) The posterior brain.

Myelin (mi'ĕ-lin) An insulating lipid found around many neurons, produced by a separate, myelin-forming cell.

Myelinated axon (mi'ĕ-lĭ-nāt'ed) An axon covered at regular intervals with myelin.

Myocardial infarction (mi,o-kar'de-al) The formation of an area of cardiac tissue damage, often brought about by compromise of the blood supply to the heart; a heart attack.

Myofibril (mi,o-fi'bril) A long, slender element in a muscle cell, composed of thousands of sarcomeres arranged end to end.

Myoglobin (mi,o-glo'bin) An iron-containing protein molecule similar to hemoglobin that stores oxygen in cardiac and skeletal muscle cells.

Myopia Nearsightedness; the inability to see distant objects clearly.

Myosin (mi'o-sin) The protein of the thick filament in a sarcomere.

Nasopharynx (na,zo-far'ingks) The upper part of the throat; the part of the throat that lies above the soft palate.

Natriuresis (na,tre-u-re'sis) An increase in salt excretion by the kidney.

Negative feedback A control system that serves to oppose the initial input to the system. It stabilizes the system under stress.

Nephron (nef'ron) The tubular, structural, and functional units of the kidney.

Nerve growth factor A protein that promotes the growth and development of the nervous system.

Nerve impulse The common name for an action potential in a neuron.

Neuroendocrine cell (nu,ro-en'do-krin) A class of modified neurons that release their chemical messengers directly into the bloodstream instead of into the synaptic cleft near another neuron.

Neurohormone (nu'ro-hor,mōn) The chemical messenger substance of a neuroendocrine cell.

Neuromuscular junction (nu,ro-mus'ku-lar) The synapse between a motor neuron and a muscle cell.

Neuromuscular transmission The process whereby an action potential in a neuron produces an action potential in a muscle cell.

Neuron (nu-ron') A nerve cell; neurons are specialized to transmit electrical signals.

Neurotransmitter The chemical messenger released by a neuron to influence another nearby nerve cell or other effector cell, such as muscle cell.

Neutrophils (nu'trah-fils) White cells that are most actively engaged in phagocytosis.

Nicotine An addicting chemical in tobacco, liberated when tobacco is smoked or chewed.

Nitric oxide (NO) A very reactive, highly toxic compound known as a free radical. It is also a neurotransmitter and a vasodilator.

Nociceptor (no,se-sep'tor) A receptor for pain.

Node of Ranvier (ron-vē-ā) The regularly spaced region of a myelinated axon devoid of myelin.

Non-insulin-dependent diabetes A diabetes which occurs in older people and is frequently the result of insensitivity to insulin.

Nonspecific immune response The body's ability to ward off invasion by a new pathogen, such as when many bacteria enter the body through cut skin.

Norepinephrine (nor,ep-ĭ-nef'rin) A common neurotransmitter in the brain and the sympathetic branch of the peripheral nervous system.

Nucleotide (nu'kle-o-tīd) The basic building block of DNA and RNA; consists of a sugar group, a phosphate group, and a nitrogenous base.

Nucleus (nu'kle-us) The membrane-bound, roughly spherical structure within the cell that contains the cell's DNA.

Obesity The condition of being considerably overweight.

Occipital lobe (ok-sip'ĭ-tal) The rearmost lobe of the brain.

Olfactory glands (ol-fak'to-re) Mucus-producing glands in the nasal passages.

Olfactory hairs Projections from the surface of olfactory receptor cells that contain receptors for specific odor molecules.

Olfactory nerve The nerve that carries action potentials from olfactory receptor cells to the brain.

Olfactory receptor cells Modified neurons located in the nasal passages that respond to odor molecules in the air.

Oligodendrocyte (ol,ĭ-go-den'dro-sīt) A glial cell that produces the myelin sheath.

Oocytes (o'ĭ-sits) Potential ova.

Oogonia (o'ĭ-goni'a) Primary germ cells that give rise to ova.

Opioid (o'pe-oid) Describing neurotransmitters that bind to the same receptor sites as the opium derivative morphine and the more potent chemical heroin.

Opioid neurotransmitter A CNS neurotransmitter that is associated with the same sites that bind opium derivatives.

Optic disc The point in the retina where the axons of the ganglion cells converge and pass through the retina.

Optic nerve The nerve that carries action potentials from an eye to the brain.

Organ A structure formed when two or more tissues are joined together to perform a specific function or functions.

Organ system A group of organs that serve a similar broad function.

Organelle (or,gah-nel) The general term for a membrane-bound structure in a cell that performs a specialized function.

Orgasm (or'gazm) The act of emission and its associated sensations.

Origin The end of a muscle that is attached to the more stationary part of the skeleton.

Osmoreceptor (oz,mo-re-sep'tor) A receptor that responds to the osmolality of body fluids.

Osmosis (oz-mo'sis) The net diffusion of water across a selectively permeable membrane.

Osmotic diuresis (oz-mo'tik) An increase in the flow rate of urine due to an increase in the number of osmotic particles appearing in the urine.

Osteoblast (os'te-o-blast) A bone-forming cell, capable of secreting new cartilaginous material and depositing calcium in bone.

Osteoclast (os'te-o-klast,) A cell arising from a type of white blood cell that dissolves the mineral deposits and digests the cartilage matrix of bone.

Osteocyte (os'te-o-sīt) A mature bone cell that has lost the ability to secrete new cartilaginous material; osteocytes can still deposit calcium in bone.

Osteoporosis (os,te-o-por-o'sis) A disease caused by excessive bone loss over time, characterized by a tendency toward fractures and a hunched posture.

Otoliths (o'to-liths) Crystals of calcium carbonate embedded in the gelatinous material of the utricle and saccule of the inner ear.

Ovaries (ov,er-e's) The female sex organs that produce the ova.

Ovulation (ov,u-la-shun) The release of a mature ovum from the ovary.

Ovum (o'vum) An unfertilized female germ cell or egg.

Oxidative phosphorylation (ok'sĭ-da'tiv fos,for ĭ-la-shun) A major mitochondrial biochemical process involved in aerobic respiration.

Oxygen debt The amount of oxygen that must be supplied to make up for the oxygen not supplied during a period of anaerobic metabolism.

Oxyhemoglobin (ok,sĭ-he,mo-glo'bin) Hemoglobin containing oxygen.

Oxytocin (ok,sĭ-to'sin) A neurohormone from the pituitary gland that stimulates uterine contraction and milk ejection.

P-wave The electrical activity of the atrial contraction as seen on an EKG.

Pacemaker cell The part of the heart muscle that initiates each contraction; the sinoatrial node as it normally sets the heart rate.

Pacinian corpuscle (pa-sin-e-an) An encapsulated sensory receptor located in deep layers of the skin that responds to pressure and vibration.

Pancreatic juice (pan,kre-at'ik) A watery fluid containing bicarbonate and digestive enzymes.

Papilla (pah-pil'ah) A general term for a small projection or elevation.

Papillary muscle Muscular tissue in the heart connected to the chordae tendinae.

Paracrine factor (par'ah-krin) A chemical messenger released by a cell that acts on a nearby cell without first circulating in the blood.

Parafollicular cells (par'ah-fol-ikler) Small clusters of cells near the follicular cells of the thyroid that secrete calcitonin.

Parasympathetic (par,ah-sim,pah-thet'ik) A subdivision of the autonomic nervous system.

Parathyroid gland (par,ah-thi'roid) An endocrine gland imbedded in the thyroid that secretes a hormone that regulates Ca^{++} concentrations.

Parathyroid hormone (PTH) Also known as parathormone; its rate of secretion is controlled by the concentration of Ca^{++} in plasma and regulates the calcium balance of blood.

Parietal cells (pah-ri'ĕ-tal) Cells of the gastric glands of the stomach that produce a watery fluid containing hydrochloric acid and intrinsic factor.

Parietal lobe The cortical lobe just posterior to the central sulcus.

Parkinson's disease A chronic condition marked by muscular rigidity and paralysis.

Partial pressure The pressure exerted by a single gas.

Parturition (par,tu-rish'un) The process of birth.

Pathogen (path'o-jen) A harmful organism.

Pathophysiology (path,o-fiz,e-ol'o-je) The abnormal functioning of an organ.

Pectoral girdle The encircling bony structure supporting the upper limbs.

Pelvic girdle The encircling bony structure supporting the lower limbs.

Pepsin A protein-degrading enzyme present in the stomach.

Pepsinogen (pep-sin'o-jen) A protein produced by chief cells of the gastric glands of the stomach; the inactive precursor of the enzyme pepsin.

Peptic ulcer A damaged region of the gastrointestinal tract, caused by gastric juice.

Peptide (pep'tid) A chain of two or more amino acids.

Peptide hormones Hormones composed of amino acids.

Periosteum (per,e-os'te-um) A thin, fibrous layer covering bone that contains blood vessels, nerves, and osteoblasts.

Peripheral chemoreceptors Chemical-sensing cells located outside the central nervous system.

Peripheral nervous system (PNS) The part of the nervous system that includes all of the neural elements not contained in the CNS.

Peristalsis (per,ĭ-stal'sis) A wave-like pattern of muscular contraction that moves in a sequential fashion along a tubular or hollow organ.

Peritubular capillaries (per,i-tub-ular) Capillaries that surround the tubular portions of the nephron.

Permissive hormone A hormone that allows another hormone to exert its action.

Peroxisome (pĕ-roks'ĭ-sōm) A membrane-bound vesicle in a cell that contains oxidative enzymes.

pH The logarithm of the reciprocal of the hydrogen ion concentration.

Phagocyte (fag'o-sīt) A cell capable of ingesting foreign matter.

Phagocytosis (fag,o-si-to'sis) The engulfment of large particles by specialized cells.

Pharynx (fayr'inks) The throat.

Phenylketonuria (PKU) (fen,il-ke,to-nu're-ah) An inheritable disease in which the person is unable to metabolize the amino acid phenylalanine properly.

Pheromones (fer'o-mōn) Chemicals released by animals that influence specific patterns of behavior in other members of the same species.

Phosphocreatine (fos,fo-kre'ah-tin) A high-energy molecule that can be used to replenish ATP.

Phospholipid (fos,fo-lip'id) A lipid composed of a glycerol backbone, two triglyceride tails, and a phosphate group; the primary structure of the cell membrane.

Photopigments Special molecules in the rods and cones of the eye that change shape when they absorb energy in the form of light.

Photoreceptor (fo,to-re-sep'tor) A receptor cell that responds to light.

Physiology (fiz,e-ol'o-je) The science devoted to the study of the function of the body and its parts.

Pituitary gland (pĭ-tu'ih-tayr,e) A pea-sized endocrine gland lying in a small depression of the skull located at the base of the brain.

Placenta (plah-sen'tah) The structure that connects the developing embryo to the mother and nourishes it from the mother's blood until birth.

Plaque (plak) The deposition of cholesterol in an artery.

Plasma The straw-colored liquid part of blood.

Plasma cell A B cell that enlarges and differentiates into and is carried by the lymph into the circulating blood and releases antibodies.

Platelet (plāt'let) The small cellular element of blood involved in clotting mechanisms.

Pleurae (ple'u-ra,e) Extremely thin tissues that line both the outside of the lung and the inside walls of the thorax.

Pneumothorax (nu,mo-he,mo-tho'raks) A dangerous condition where air enters the intrapleural space, often resulting in the collapse of a lung.

Podocytes (pod'o-sīt) Octopus-like cells that project foot-like processes enclosing the capillary basement membrane of the glomerulus.

Polar Describing a molecule having oppositely charged regions, or poles.

Polar body The smaller, nonfunctional female germ cell formed after a meiotic division.

Polymorphonuclear granulocyte (pol,e-mor,fo-nu'kle-ar) A type of white blood cell.

Polypeptide (pol,e-pep'tīd) A chain of up to approximately 100 amino acids.

Polysaccharide (pol,e-sak'ah-rīd) A chain of monosaccharides.

Pons Meaning *bridge;* a structure located immediately above the medulla and containing nerve tracts connecting the spinal cord with other areas of the brain.

Positive feedback A regulatory system in which the output enhances the input, causing an increasingly rapid move away from the steady state.

Posterior lobe The rear lobe of the pituitary that secretes ADH and oxytocin.

Postganglionic (post,gan-gle-ah'nik) Posterior to a ganglion.

Postganglionic fibers The axons running from the ganglia to the organs of innervation.

Postsynaptic neuron (post,sĭ-nap'tik) The neuron that is activated by the neurotransmitter at a synapse.

Precapillary sphincters The small rings of smooth muscle located at the junction of the arterioles and capillaries.

Preganglionic (prē'-gang-lē-on-ik) Anterior to a ganglion.

Preganglionic fibers The autonomic nerves that exit the spinal cord, enter ganglia, and synapse with cell bodies of postganglionic fibers in the ganglia.

Premotor cortex The brain area immediately in front of the primary motor area.

Presbyopia (pres,be-o'pe-ah) The inability to focus on very near objects.

Presynaptic neuron (prē-sin-ap-tik) The neuron that releases a neurotransmitter at a synapse in response to an action potential in its axon.

Primary follicles Oocytes embedded in a flattened layer of cells in the ovary.

Primary motor cortex The area of the frontal cortex that controls muscle movement.

Primary sexual characteristics The reproductive organs of the male and female.

Progesterone (pro-jes'ter-ōn) A female hormone that aids in stimulating growth, integrity, and glandular function of the endometrium.

Prolactin (pro-lak'tin) A hormone that stimulates milk production by the breasts.

Proprioceptors (pro,pre-o-sep'tors) A class of receptors that sense body position and movement.

Prostaglandins (pros,tah-glan'din) A family of fatty acid compounds present in many tissues that function as paracrine factors.

Prostate A gland that produces a fluid added to the sperm.

Protein (pro'ten) A specific chain or chains of amino acids, each with a particular function.

Prothrombin (pro-throm'bin) A proenzyme of the clotting system.

Proximal Closest to a point of reference.

Proximal convoluted tubule The renal tubule that is nearer to the glomerulus.

Pseudopods (soo-dō-pods) The foot-like processes that some white cells use to engulf foreign particles.

Puberty The onset of sexual maturity.

Pulmonary Pertaining to the lungs.

Pulmonary artery Blood vessel that leads from the heart to the lungs.

Pulmonary edema (e-de'mah) Excess fluid in the air spaces of the lung that increases the distance between alveolar air and the capillary blood, slowing down diffusion of O_2 and CO_2.

Pulmonary vein The vein that returns blood from the lungs to the left ventricle.

Pulmonary perfusion Blood flow to the lungs.

Pulmonary ventilation The volume of air that flows into and out of the lungs each minute.

Pulse pressure The difference in pressure between the systolic and diastolic pressures.

Pupil A circular hole at the center of the iris, through which light passes as it enters the eye.

Purkinje system (pur-kin'je) The specialized conducting system of the heart branching off the bundle of His that transmits electrical signals throughout the ventricles.

Pus A collection of dead white blood cells and tissue fluids.

Pyloric sphincter (pi-lor'ik sfink'ter) The sphincter between the stomach and the small intestine.

Pylorus (pi-lor'us) The third and smallest region of the stomach, where the stomach joins the small intestine.

Pyramidal system The descending tract of motor fibers running from the brain to a lower level in the spinal cord.

QRS wave Electrical activity of the EKG representing ventricular depolarization.

Receptor adaptation The decline in receptor potential that occurs during the maintenance of a constant stimulus.

Receptor potential The local graded change in membrane potential that occurs in a receptor cell in response to stimulation.

Receptor protein In the cell membrane, a protein that functions to transmit information across the cell membrane.

Red blood cell (RBC) The most abundant blood cell, red in color, which carries oxygen.

Reflex An involuntary reaction to a stimulus.

Reflex arc The series of neural signals involved in a reflex.

Relative refractory period The time period when it is more difficult than usual, though not impossible, to activate a recently activated section of cell membrane.

REM sleep A period of sleep characterized by dreaming, increased respiration and brain metabolism, and rapid eye movements.

Renal autoregulation (re'nal) The phenomenon in which renal blood flow and glomerular filtration rate are held relatively constant despite daily fluctuations of blood pressure.

Renal blood flow (RBF) The volume of blood that flows through the kidney per unit of time.

Renal cortex The outer region of the kidney.

Renal glucose threshold The plasma level of glucose at which glucose begins to escape complete reabsorption and enter the final urine.

Renal medulla The inner portion of the kidney.

Renin (re'nin) A protein secreted by the kidney that causes the formation of a substance called angiotensin I.

Replacement therapy The treatment of people with inadequate endocrine function to supply a missing hormone and restore function.

Repolarization The return of the membrane potential to the normal resting level.

Residual volume The volume of air remaining in the lung after we exhale as much as possible.

Respiration The process supplying oxygen to our cells.

Respiratory distress syndrome A lung disorder characterized by a very low lung compliance.

Respiratory system The physiological system that transports O_2 from the atmosphere to the cells.

Respiratory tract The structures that transport gases between the environment and the body and provide a gas exchange area between the air and blood.

Reticular formation (rĕ-tik'u-lar) A network of fibers composed of widely spread neurons that signal the cortex and act almost as a wake-up call, promoting arousal, alertness, and attention.

Retina (ret'ĭ-nah) The structure at the back of the eye, composed of several layers of cells, that receives incoming light and transduces it into action potentials.

Rh factor An antigen–antibody system first isolated in the rhesus monkey.

Rhodopsin (ro-dop'sin) The photopigment in rods.

Ribosomes (ri'bo-sōmz) Numerous small structures in a cell, consisting of RNA and protein, that are the sites of protein synthesis.

Ribs Paired bones attached to the thoracic vertebrae that surround and protect the heart and lungs.

Rigor mortis (rig'or mor'tis) The stiffening of muscles that occurs about four hours after death.

RNA (ribonucleic acid) (ri,bo-nu-kle'ik) A single-stranded molecule composed of nucleotides that functions in protein synthesis.

RNA translation The process of reading the RNA code and using it to produce a protein.

Rods Rod-shaped photoreceptor cells in the retina that are particularly sensitive to light; rods contain the photopigment rhodopsin.

Rough endoplasmic reticulum (ER) The region of the endoplasmic reticulum that has a granular or rough appearance by virtue of the many small ribosomes attached to it.

RU486 A drug that blocks the action of progesterone in preparing the uterus for implantation.

Rugae (ru'ge) Ridges or folds in the mucosal lining of the stomach, most prominent when the stomach is empty.

Saccule (sak'ul) One of two chambers of the inner ear that detect static position and changes in linear velocity of the head.

Sacral The part of the body pertaining to the level of the lower back.

Saliva A watery fluid containing bicarbonate, enzymes, and a small amount of mucus.

Salivary glands General term for all of the glands in the mouth that produce saliva.

Saltatory conduction (sal-ta-tō-rē) The form of nerve impulse conduction in a myelinated nerve in which a new action potential is generated at each node of Ranvier; it appears as if a single action potential "jumps" from node to node. Saltatory conduction is faster than conduction of an action potential in an unmyelinated nerve.

Sarcomere (sar'ko-mēr) The smallest contractile element in a skeletal or cardiac muscle fiber; the area between two Z-lines.

Sarcoplasmic reticulum (sar,ko-plaz'mik rĕ-tik'u-lum) The complex system of membrane-bound storage sacs for calcium in striated muscle cells.

Saturated fatty acid A fatty acid in which every carbon atom is bound to two hydrogen atoms.

Scientific method A method of testing hypotheses rigorously.

Second messenger The chemical that is released in response to the first messenger and causes a response in the target cell.

Secondary sexual characteristics Physical characteristics appearing at puberty but not implicated in gamete production.

Secretin A hormone secreted by cells of the small intestine into the bloodstream that inhibits gastric motility and stimulates the production of pancreatic juice.

Secretion (se-kre'shun) A general term for the process of elaborating a specific product, usually as the result of the activity of a gland; also the general term for that product.

Segmentation Alternating contraction and relaxation of short regions of the circular layer of smooth muscle in the intestine, producing both mixing and forward movement of the intestinal contents.

Semen (se'men) The whitish fluid secreted by the male reproductive tract.

Semicircular canals The three bony, fluid-filled canals of the inner ear, each oriented on a different plane, that detect movement of the head.

Semilunar valves (sĕ,me-loo'ner) The aortic and pulmonary valves of the heart.

Seminiferous tubule (sem,ĭ-nif'er-us) Testicular tubules that produce spermatozoa.

Serosa (se-ro'sah) The outermost layer of the gastrointestinal tract, consisting primarily of connective tissue, that attaches the muscularis to the body wall.

Serotonin (se,ro-to'nin) A class of neurotransmitters that is a modified amino acid.

Sertoli cells (ser-to'le) Testicular cells that support sperm development.

Set point The normal value about which a negative feedback system is set to operate.

Shock A condition of circulatory failure marked by pallor and clamminess of the skin, increased heart rate, and often unconsciousness.

Sickle cell anemia A genetic disorder in which the red blood cells may assume odd shapes, interfering with normal red blood cell function.

Signal reception–transduction The process of receiving and converting information into a form that can be used by the central nervous system.

Sinoatrial (SA) node (si,no-a'tre-al) A very small, well-delineated region in the wall of the right atrium near the entrance of the vena cava responsible for initiating the electrical activity of the cardiac cycle.

Skeletal muscle A striated type of muscle that is usually attached to bones.

Skull The bony framework of the head.

Slow-wave potential A type of slow, rhythmic electrical activity of smooth muscle caused by cyclical changes in the rate of sodium transport.

Smooth endoplasmic reticulum (ER) The region of the endoplasmic reticulum that is devoid of ribosomes.

Smooth muscle A nonstriated type of muscle commonly found around blood vessels and hollow organs.

Sodium-potassium ATP-ase The membrane protein that transports, or pumps, sodium and potassium across the cell membrane, using energy liberated from ATP.

Soft palate (pal'at) The upper part of the back of the mouth.

Somatic nervous system (so-ma'tik) The system that carries information from the CNS to the skeletal (voluntary) muscles, and information from muscle joints, skin, and bones back to the CNS.

Somatostatin (so,mah-to-stat'in) A hormone secreted by the delta cells of the pancreas.

Specific immune response The immunity conferred against a single pathogen.

Sperm The mature germ cell of the male.

Spermatocyte (sper-mat'o-sist) The precursor of sperm.

Spermatogonia The primary germ cells that give rise to sperm.

Sphincter (sfink'ter) The general term for a ring-like band of muscle around an orifice or passage that regulates flow or movement through the orifice or passage.

Sphincter of Oddi The sphincter around the common bile duct at its juncture with the small intestine, responsible for regulating the rate at which bile is delivered into the small intestine.

Sphygmomanometer (sfig,mo-mah-nom'ĕ-ter) An instrument used for measuring blood pressure.

Spindle A cytoplasmic structure made up of spindle fibers that connects each chromosome to a centriole during cell division.

Spindle fibers microtubular elements produced in a cell in preparation for cell division; spindle fibers form the spindle.

Spongy bone A form of bone in which the hard elements are interspersed with hollow spaces, giving it a spongy appearance.

Stapes (sta'pēz) The third of the three bones in the middle ear that connect the eardrum to the oval window.

Stem cells Cells that are located in the bone marrow and make all the cellular elements of blood.

Stenosis (stĕ-no'sis) An abnormal narrowing of a vessel or duct.

Sterilization Any method by which a person is made incapable of reproducing.

Steroid (stĕ'roid) A type of lipid with four interlocking rings of carbon, derived from cholesterol.

Steroid hormones The hormones derived from the cholesterol molecule.

Stress Exposure to any noxious stimuli.

Stretch receptor A receptor that responds to stretching and signals change in length.

Stroke A sudden onset of paralysis due to a clot or rupture of a cerebral blood vessel.

Stroke volume The volume of blood pumped with each beat of the heart.

Submucosa (sub,mu-ko'sah) A layer of connective tissue between the mucosa and the muscle layer of the gastrointestinal tract, containing numerous blood and lymph vessels.

Substance P A neurotransmitter involved in the perception of pain.

Sulci (su'l-se) The depressions between the gyri of the brain.

Summation Additivity in either time or space; applies to both graded potentials and muscle twitches.

Superior vena cava (ve'nah ka'vah) The vein that drains the upper part of the body, carrying blood to the right atrium.

Supporting cells In any organ, the general term for cells that do not carry out the unique function of the specialized cells.

Suppressor T cell A class of T cell that suppresses both killer and helper T cells.

Surface tension The force liquids exert when they are in contact with air.

Surfactant (ser-fak'tant) A substance in the liquid lining of the alveoli that reduces surface tension.

Sympathetic A subdivision of the autonomic nervous system that is largely concerned with excitation of the responses called fight or flight.

Sympathetic chain The lateral row of ganglia or cell bodies of the sympathetic nervous system that lie alongside the vertebral column.

Synapse (sin'aps) A specialized junction between a neuron and another excitable cell, across which an action potential in the presynaptic neuron influences the electrical activity in the postsynaptic cell, generally by the release of a chemical transmitter but occasionally by direct electrical connection.

Synaptic bulb (si-nap'tik) The rounded structure at the axon terminal of the presynaptic neuron; the synaptic bulb contains packets of chemical neurotransmitter, stored and ready for release.

Synaptic cleft The space between the presynaptic and postsynaptic cells in a synapse.

Synaptic transmission The process whereby information provided by an action potential is transmitted from a neuron to another cell.

Syncytium (sin-sish'e-um) A single large mass of cytoplasm containing many nuclei.

Synovial cavity (sǐ-no've-al) The fluid-filled space between bones in a synovial joint.

Synovial joint A special form of joint permitting free movement because of the presence of a lubricating fluid between the points of bone contact.

Systole (sis'to-les) The period of contraction of heart muscle, particularly of the ventricle.

Systolic pressure (s-tah'lik) The maximum pressure in the aorta during the cardiac cycle.

T cell A class of lymphocytes.

T-wave Electrical activity of the EKG representing repolarization of the ventricles.

Target cell The cell on which a hormone or drug acts.

Taste bud A cluster of cells located primarily on the tongue that contain the receptor cells for taste.

Taste receptor cells The cells in a taste bud that are specialized for sensing the presence of certain specific chemicals.

Tectorial membrane (tek-to're-al) The upper membrane of the inner ear against which the hairs of the hair cells bend in response to sound waves.

Telencephalon (tel,en-seh'fuh-lon) The endbrain.

Temporal lobe The cortical lobe in front of the central sulcus.

Tendon Connective tissue that attaches muscle to bone.

Teratogens (ter'ah-to-jens) Substances that produce birth defects.

Testes (tes'tēz) The male sex organ that produces sperm.

Testicle (tes'tǐ-k'l) The singular of testes.

Testosterone A major masculinizing hormone.

Tetanus (tet'ah-nus) A sustained muscular contraction of maximal strength.

Tetraiodothryonine (T4) (tě,trah-i,o-do-thi'ro-nēn) Thyroxine, the thyroid hormone, containing four iodine atoms.

Thalamus (tha'luh-mis) A large, oval mass of neural tissue lying alongside the third ventricle.

Theory A generally accepted explanation of facts.

Thermogenesis (ther,mo-jen'ě-sis) The heat produced in response to cold (usually by shivering) and for the digestion and absorption of a meal.

Thermoreceptor (ther,mo-re-sep'ter) A receptor that responds to heat or cold.

Thick filament One of two structural elements in a sarcomere, consisting of a protein called myosin, that interacts with thin filaments to produce shortening of a sarcomere during muscle contraction.

Thin filament One of two structural elements in a sarcomere, consisting of three proteins (actin, tropomyosin, and troponin), that interacts with myosin to produce shortening of a sarcomere during muscle contraction.

Thoracic (tho-ras'ik) Pertaining to the level of the chest.

Thorax (tho'raks) The chest.

Threshold The membrane potential at which an action potential is generated in an excitable cell.

Thrombin (throm'bin) An enzyme that converts fibrinogen to fibrin.

Thrombocyte (throm'bo-sīt) A platelet.

Thyroglobulin (TG) (thi,ro-glob'u-lin) A large glycoprotein molecule containing the amino acid tyrosine; it becomes the major thyroid hormone after combining with three or four atoms of iodine.

Thyroid glands (thi'roid) Endocrine glands located in the neck controlling general metabolism.

Thyroid stimulating hormone (TSH) A hormone that stimulates the thyroid gland to synthesize, store, and release thyroxine into the general circulation.

Thyrotropin-releasing hormone (TRH) (thi,ro-trōp'in) A tropic hormone that stimulates the anterior pituitary to secrete thyroid-stimulating hormone.

Thyroxine The major hormone of the thyroid gland.

Tidal volume The volume of air that is moved with each breath.

Tissue A group of cells with similar functions.

Tolerance An increasing resistance to the usual effect of a drug, leading to the need for more of the drug to achieve the same effect.

Total lung volume The most air with which we can fill our lungs.

Total peripheral resistance The total resistance to blood flow from the left ventricle to the right atrium.

Trachea (tra'ke-ah) A short conducting airway of the respiratory tract about 2–2.5 cm (~ 1 inch) in diameter that extends from the larynx and splits into two slightly smaller tubes called main bronchi.

Transducer Any device that converts energy input in one form into energy output in another form; in living systems, transducers convert various sensory modalities into action potentials.

Transport protein A membrane protein that transports molecules across the cell membrane.

Triglyceride (tri-glis'er-īd) The primary storage form of lipids, composed of one glycerol molecule and three fatty acids.

Triiodothyronine (T3) (tri,i-odo-thi'ro-nēn) A thyroid hormone containing three iodine atoms.

Trimester A three-month period of pregnancy.

Tropic hormone (trōp'ik) A hormone that stimulates the secretion of a second hormone.

Tropomyosin (tro,po-mi'o-sin) One of the three proteins that make up a thin filament; along with troponin, it is involved in the regulation of the binding of actin of the thin filament to myosin of the thick filament.

Troponin (tro'po-nin) One of the three proteins that make up a thin filament; along with tropomyosin, it is involved with the regulation of the binding of actin of the thin filament to myosin of the thick filament.

Tubal ligation Sterilization of the female by cutting the fallopian tubes.

Tubular fluid Fluid contained in the renal tubules.

Tubular reabsorption Transport of molecules from the tubular fluid into the renal interstitium.

Tubular secretion The transport of molecules in the kidney out of the peritubular capillaries into the renal tubular fluid.

Twitch The force developed by a muscle cell in response to a single action potential.

Tympanic membrane (tim-pan'ik) The eardrum.

Type I diabetes Juvenile onset diabetes due to inadequate insulin.

Type II diabetes Adult onset diabetes.

Unsaturated fatty acid A fatty acid in which there are fewer than two hydrogen atoms for every carbon atom.

Upper esophageal sphincter (e-sof,ah-je'al) The sphincter located at the entrance to the esophagus.

Uterine tubes (u'ter-in) Structures that conduct the ova from the ovary to the uterus.

Uterus (u'ter-us) The female organ in which the fetus develops.

Utricle (u'trĭ-k'l) One of two chambers of the inner ear that detect static position and changes in linear velocity of the head.

Vacuole (vak'u-ol) A membrane-bound space in a cell.

Vagina (vah-ji'nah) The canal leading from the uterus to the outside.

Vagus nerve (va'gus) The tenth cranial nerve that is largely parasympathetic.

Vascular endothelium Inner lining of blood vessels.

Vasectomy (vah-sek'to-me) Sterilization of the male by cutting the vas deferens.

Vasopressin (vas,o-pres'in) Another name for antidiuretic hormone.

Veins (vānz') Thin-walled vessels that return the blood to the right side of the heart.

Ventilation–perfusion matching Regulatory mechanisms that match pulmonary blood flow to pulmonary air flow.

Ventilatory rate (ven'tĭ-la,tore) The number of breaths taken each minute.

Ventral root Nerves that exit the spinal cord from the ventral side.

Ventricle (ven'trĭ-kul) A cavity; *in the heart,* the lower, thick muscular chamber.

Ventricle A cavity; *in the brain,* the hollow, liquid-filled chambers of the brain.

Ventricular fibrillation The process by which different segments of the ventricle contract at different times so that the ventricle of the heart appears almost as a mass of quivering jelly.

Ventricular systole (ven-trik'u-lar sis'to-le) The period of contraction of the ventricle of the heart.

Venules (ven'ūlz) The smallest veins.

Vertebrae (ver'te-brae) The individual bones of the vertebral column.

Vertebral column (ver'tĕ-bral) The backbone of vertebrates.

Vesicle (vĕ-sĭ-kul) A membrane-bound sphere containing the products of the Golgi apparatus or the endoplasmic reticulum.

Vestibular apparatus (ves-ti-byool) The structure consisting of three fluid-filled semicircular canals and two chambers, responsible for sensing static position, changes of position, and linear velocity of the head.

Villus (*pl.* villi) (vil'us) A general term for a protrusion or projection from the surface of a layer of tissue; in the small intestine of the GI tract, villi are composed of a layer of epithelial cells surrounding a central region containing blood and lymph vessels.

Vital capacity The volume of the biggest breath we can take. This is equal to total lung volume minus residual volume, which is usually about 5 liters.

Vulva (vul'vuh) The collective term for the female external genitals.

Wheal (hwēl) A whitened area that is slightly raised from the surrounding skin, that is usually due to direct mechanical stimulation of the underlying blood vessels.

White blood cells Cells involved in host defense system.

White matter A thick mass of brain tissue composed of myelinated fibers.

Withdrawal The symptoms that occur upon discontinuation of a drug.

X chromosome The larger of the two sex chromosomes.

Y chromosome The smaller of the two sex chromosomes.

Z-line A dark line in a myofibril that marks the junction between two sarcomeres.

Zygote (zi'gōt) The fertilized ovum.

Credits

LINE ART
All illustrations courtesy of Nadine B. Sokol.

PHOTOGRAPHS

Chapter 1
p. 4, Milestone 1.1: Stuart, Gilbert. Portrait of George Washington. Private Collection. Art Resource, NY. *p.5, figure 1.01*: Fildes, Sir Luke, The Doctor, exh. 1891. Tate Gallery, London/Art Resource NY. *figure 1.02*, © 1991 SIU/Photo Researchers.

Chapter 2
p. 20, figure 2.12a, John Durham/SPL/Photo Researchers. *figure 2.12b*, VU/© David M. Phillips. *figure 2.12c*, © Dr. David Scott/CNRI/Phototake NYC. *figure 2.12d*, © Eric Grave/Phototake NYC. *p. 24, figure 2.14b*, VU/© Don W. Fawcett. *p. 25, figure 2.15*, Scott, Foresman. *p. 33, figure 2.24a*, Ed Reschke. *figure 2.24b*, Ed Reschke. *p. 36, figure 2.28a*, VU/© H. Bernstel, D. Fawcett. *p. 37, figure 2.29a*, VU/© Don W. Fawcett. *p. 46, Milestone 2.1*, Nancy Kedersha/Immunogen/SPL/Photo Researchers.

Chapter 3
p. 63, figure 3.12a, Amer. J. Physiology 110:711 Bronk, D.W. and Stella, G. The Response to steady pressures of single end organs in the isolated carotid sinus. *figure 3.12b*, Amer. J. Physiology 110:711 Bronk, D.W. and Stella, G. The Response to steady pressures of single end organs in the isolated carotid sinus. *figure 3.12c*, Amer. J. Physiology 110:711 Bronk, D.W. and Stella, G. The Response to steady pressures of single end organs in the isolated carotid sinus. *figure 3.12d*, Amer. J. Physiology 110:711 Bronk, D.W. and Stella, G. The Response to steady pressures of single end organs in the isolated carotid sinus. *p. 64, Milestone 3.1*, © David Mechlin/Phototake NYC.

Chapter 5
p. 117, figure 5.05, Orci, L. (1984) Patterns of celluar and subcellular orgainization in the endocrine pancreas. J. Endocrin. 102:3-11. *p. 121, Milestone 5.1b*, VU/© A. Like, D.W. Fawcett. *Milestone 5.2*, Bettmann Archive. *Milestone 5.3a*, Courtesy, Richard Malvin. *Milestone 5.3b*, Courtesy, Richard Malvin. *p. 125, figure 5.09*, Mark Nielsen, University of Utah. *5.09B*, Ed Reschke. *p. 129, figure 5.11a*, Beverly M.K. Biller, MD, Massachusetts General Hospital. *figure 5.11b*, Beverly M.K. Biller, MD, Massachusetts General Hospital. *p. 131, figure 5.13a*, © 1981 Bettina Cirone/Photo Researchers. *figure 5.13b*, © Jerry Cooke/Photo Researchers. *p. 132, figure 5.14a*, Reprinted with permission from American Journal of Medicine. *figure 5.14b*, Reprinted with permission from American Journal of Medicine. *figure 5.14c*, Reprinted with permission from American Journal of Medicine. *figure 5.14d*, Reprinted with permission from American Journal of Medicine. *p. 133, figure 5.15B*, © Lester Bergman. *p. 135, Milestone 5.3*, © 1991 National Medical Slide/Custom Medical Stock. *p. 139, figure 5.18*, © Leonard Morse/Medical Images Inc. *figure 5.19*, Lester Bergman & Associates. *figure 5.20*, Lester Bergman & Associates.

Chapter 6
p. 148, figure 6.05, Lester Bergman & Associates. *p. 150, figure 6.08*, Lester Bergman & Associates. *p. 153, Highlight 6.3*, Lester Bergman & Associates. *p. 155, figure 6.11a*, Ed Reschke. *figure 6.11b*, Ed Reschke. *figure 6.11c*, Andrew J. Kuntzman. *p. 158, figure 6.13a*, © J. Venable/D. Fawcett/VU. *figure 6.13b*, D. Fawcett/VU. *figure 6.13c*, © J. Auber/D. Fawcett/VU. *p. 173, figure 6.25*, Brenda Russell.

Chapter 7
p. 191, figure 7.08, © John D. Cunningham/VU. *p. 196, figure 7.13a*, Robert Preston and Joseph E. Hawkins/Kresge Hearing Research Institute/Univ. of Michigan. *figure 7.13b*, Robert Preston and Joseph E. Hawkins/Kresge Hearing Research Institute/Univ. of Michigan. *p. 199, figure 7.16*, Dean H. Hillman. *p. 200, figure 7.17C*, Dean H. Hillman. *p. 208, figure 7.24a*, National Cancer Institute/Photo Researchers. *p. 209, figure 7.25b*, © Omikron/Photo Researchers. *p.212, Highlight 7.4*, Science VU/VU.

Chapter 8
p. 218, figure 8.02, The Granger Collection.

Chapter 9
p. 238, figure 9.03, © Dennis Kunkel/Phototake NYC. *p. 241, figure 9.07*, Lennart Nilsson Our Body Victorious © Boehringer Ingelheim International GmbH. *p. 242, Highlight 9.1*, Lewin/Royal Free Hospital /SPL/Photo Researchers. *p. 244, figure 9.10*, Lennart Nilsson Our Body Victorious © Boehringer Ingelheim International GmbH. *p. 246, figure 9.12a-b*, Lester Bergman & Associates. *p. 253, figure 9.17*, R. Umesh Chandran, TDR, WHO/SPL/Photo Researchers. *p. 259, figure 9.20a*, NASA. *figure 9.20b*, NASA.

Chapter 10
p. 267, figure 10.01c, Biology Media/Photo Researchers. *p. 268, figure 10.03a*, Ed Reschke. *figure 10.03b*, Biophoto/Photo Researchers. *figure 10.03c*, Biophoto/Photo Researchers. *figure 10.03d*, Biophoto/Photo Researchers. *figure 10.03e*,

Lester Bergman & Associates. *p. 269, figure 10.04b*, P. Motta/Dept. of Anatomy/University "La Sapienza", Rome/SPL/Photo Researchers.

Chapter 11

p. 282, figure 11.03a, © Fred Hossler/VU. *figure 11.03b*, CNRI/SPL/Photo Researchers. *p. 301, Milestone 11.2a*, © 1993/SPL/Custom Medical Stock. *Milestone 11.2b*, James Stevenson/Science Photo Library/Custom Medical Stock.

Chapter 12

p. 308, figure 12.03a, © F. Spinelli/D. Fawcett/VU. *figure 12.03b*, © D. Friend/D. Fawcett/VU. *p. 328, figure 12.16*, © Hank Morgan/Photo Researchers.

Chapter 13

p. 339, figure 13.06b, Medical Images Inc. *p. 341, figure 13.08C*, © Fred Hossler/VU. *p. 349, figure 13.14*, © Biophoto/Photo Researchers.

Chapter 14

p. 365, Milestone 14.1, Science Photo Library/Photo Researchers. *p. 374, Highlight 14.2a*, © Daemmrich/Stock Boston.

Highlight 14.2b, Hannon/Picture Cube. *p. 376, Milestone 14.2*, © Biophoto Associates/Photo Researchers.

Chapter 15

p. 386, Milestone 15.1.1, Jeffreys, A.J., Wilson, V. and Thien, S.L. Individual-specific "fingerprints" of human DNA. Nature 316:76-79, 1985. *p. 387, figure 15.02a*, Fawcett/Phillips/Photo Researchers. *p. 389, figure 15.04a-e*, Andrew Bajer. *figure 15.04f*, Christine J. Harrison, PhD., Patterson Institute for Cancer Research, Christie Hospital and Holt Radium Institute, Manchester, England. *p. 393, figure 15.08a*, Ed Reschke. *p. 398, figure 15.13*, Landis Keyes. *p. 399, figure 15.14*, Edward E. Wallach. *p. 403, figure 15.17a*, Petit Format/Nestlé/Photo Researchers. *figure 15.17b*, John Giannichi/Photo Researchers. *figure 15.17c*, Petit Format/Nestlé/Photo Researchers. *figure 15.17d*, Petit Format/Nestlé/Photo Researchers. *p. 405, figure 15.19*, Reprinted with permission of Ann Streissguth. From Streissguth, A.P., Aase, J.M., Clarren, S.K., Randels, S.P., LaDue, R.A., Smith, D.F., (1991). Fetal Alcohol Syndrome in Adolescents and Adults. Journal of the American Medical Association 265 (15): 1961-1967.

Index